CONCISE | PRACTICAL | INTEGRATED

General, Organic, and Biological

CHEMISTRY

Second Edition

Laura Frost
Florida Gulf Coast University

Todd Deal
Georgia Southern University

With Contributions By

Karen C. Timberlake

PEARSON

Boston Columbus Indianapolis New York San Francisco Upper Saddle River
Amsterdam Cape Town Dubai London Madrid Milan Munich Paris Montréal Toronto
Delhi Mexico City São Paulo Sydney Hong Kong Seoul Singapore Taipei Tokyo

Library of Congress Cataloging-in-Publication Data

Frost, Laura D.
 General, organic, and biological chemistry / Laura Frost, Florida Gulf Coast
University, Todd Deal, Georgia Southern University. -- 2e edition.
    pages cm
 Includes index.
 ISBN-13: 978-0-321-80303-0
 ISBN-10: 0-321-80303-5
 1. Chemistry--Textbooks. I. Deal, Todd S. II. Title.
 QD251.3.F76 2013
 540--dc23
                                2012033563

Editor in Chief: Adam Jaworski
Executive Editor: Jeanne Zalesky
Senior Marketing Manager: Jonathan Cottrell
Project Editor: Jessica Moro
Assistant Editor: Lisa Pierce
Editorial Assistant: Lisa Tarabokjia
Marketing Assistant: Nicola Houston
Associate Media Producer: Erin Fleming
Media Project Manager: Shannon Kong
Director of Development: Jennifer Hart
Development Editor: Donald Gecewicz
Managing Editor, Chemistry and Geosciences: Gina M. Cheselka

Production Project Manager: Connie M. Long
Full Service/Composition: GEX Publishing Services
Illustrator: Precision Graphics
Image Lead: Maya Melenchuk
Photo Researcher: Eric Schrader
Text Permissions Manager: Alison Bruckner
Text Permissions Researcher: Kati Benzer, S4Carlisle
Design Manager: Derek Bacchus
Interior and Cover Designer: Jeanne Calabrese
Operations Specialist: Jeffrey Sargent
Cover Image Credit: Mark Conklin/V & W/imagequestmarine.com

ISBN-10: 0-321-80303-5
ISBN-13: 978-0-321-80303-0

2 3 4 5 6 7 8 9 10—V011—16 15 14 13

Dedication

This book is dedicated to the students who inspire
me in the classroom and to my ever-supportive
family, Baxter, Iris, and Baxter.

Laura Frost

I dedicate this book to my loving wife, Karen, and to
our daughters, Abbie and Anna. Thank you for
believing in me. And to my students who inspired
me to help them learn; this book is written for you.

Todd Deal

 www.pearsonhighered.com

About the Authors

LAURA D. FROST is Professor of Chemistry at Florida Gulf Coast University and Director of the Whitaker Center for Science, Technology, Engineering, and Mathematics (STEM) Education. She has taught chemistry to allied health students since 2000. She received her bachelor's degree in chemistry from Kutztown University and a Ph.D. in chemistry with a biophysical focus from the University of Pennsylvania.

Professor Frost is actively engaged in the teaching and learning of chemistry and uses a guided inquiry approach in her classes. She is very involved in the scholarship of teaching and learning and has demonstrated that the use of inquiry-based activities increases student learning in her one-semester allied health chemistry course.

Dr. Frost is a member of the American Chemical Society and its Chemical Education division. In 2007, she was honored with the Regents' Award for the Scholarship of Teaching and Learning by the University System of Georgia and was inducted into the Regents' Hall of Fame for Teaching Excellence. In 2011, she received the Allen E. Paulson College of Science and Technology's Award for Excellence in Teaching at Georgia Southern.

In her current position she continues to advocate for STEM education reform through inquiry-based teaching and learning at all levels and through community outreach. She has spoken at numerous conferences and workshops on this topic.

When not teaching or writing, Dr. Frost enjoys running, disc golf, hiking, camping, and most importantly her family—her husband Baxter and their two elementary school-age children, Iris and Little Baxter.

TODD DEAL received his B.S. degree in chemistry in 1986 from Georgia Southern College (now University) in Statesboro, Georgia, and his Ph.D. in chemistry in 1990 from The Ohio State University. He joined the faculty of his undergraduate alma mater in 1992 where he currently serves as Director of the Office of Student Leadership and Civic Engagement. During his tenure at Georgia Southern, Professor Deal has also served as Associate Dean of the Allen E. Paulson College of Science and Technology.

Professor Deal has taught chemistry to allied health and preprofessional students for over 20 years. In 1994, he was selected Professor of the Year by the students at Georgia Southern University. Professor Deal is also the recipient of the Allen E. Paulson College of Science and Technology's Award for Excellence in Teaching (2003), the Georgia Southern University Award for Excellence in Contributions to Instruction (2003), and the Allen E. Paulson College of Science and Technology's Award for Excellence in Service (2006).

Professor Deal is a member of American Chemical Society and Chemical Education division. In 1996, he was named to the Project Kaleidoscope Faculty for the 21st Century in recognition of his innovative teaching in the sciences. He is also a member of the Omicron Delta Kappa National Leadership Honor Society.

When he is not teaching, Professor Deal enjoys spending as much time as possible with his wife Karen and their daughters, Abbie and Anna. He is an avid cyclist, a decent kayaker, a huge fan of Starbucks coffee, and is the "Voice of the Eagles" as public address announcer for Georgia Southern's football and baseball teams. Professor Deal also enjoys interacting with his students outside of class in his role as mentor and advisor for Georgia Southern's Baptist Collegiate Ministries.

Brief Contents

Contents

To the Student

Why do we take an antacid when we get heartburn? What are calories and how are they measured for food? How does kidney dialysis remove waste from the body? The key to understanding the answers to these questions starts with chemistry.

General, Organic, and Biological Chemistry was written especially for students interested in pursuing a health-science career like nursing, nutrition, dental hygiene, or respiratory therapy. Yet this textbook has applications for all students interested in discovering the concepts of chemistry in everyday situations. Throughout the text, you will find that we have integrated the concepts of general, organic, and biological chemistry to create a seamless framework to help you relate chemistry to your life.

One of our goals in writing this book is to help you become better problem solvers so that you can critically assess situations at your workplace, in the popular press, and in your world in general. The problem-solving approach has been expanded in this edition to help you organize and work out problems.

As you explore the pages of this book, you will encounter materials that

- apply chemistry to your life
- apply chemistry to health careers that interest you
- encourage you to develop problem-solving skills
- help you to work with and learn from your fellow students
- demonstrate how to be successful in this chemistry course and other courses.

As you read this book, you will notice the language is less formal. Wherever possible, we relate the chemical concepts to objects in everyday life to help you understand chemistry. Cognitive research in learning tells us that new ideas stick with us better if they are related to things that we already know.

New to This Edition

The Second Edition continues to build on our strategy of integrating concepts from general, organic, and biological chemistry to give students a focused introduction to the fundamental and relevant connections between chemistry and life. Emphasizing the development of problem-solving skills with distinct Inquiry Questions and Activities and a clear exposition of strategies, this text empowers students to solve problems in different and applied contexts relating to health and biochemistry.

- **Enhanced Chapter Guides** at the beginning of each chapter and visual **End-of-Chapter Summaries and Study Guides** work together to highlight the main concepts of each chapter and help students study more effectively. These eye-catching features give students a practical way to pinpoint and focus on the most important concepts in each chapter and then assess their own proficiency.

- **Inquiry Questions (IQs)** start each section, functioning as Learning Objectives while prompting student interactivity. Each IQ is revisited in the Chapter Summary to help students evaluate how well they understand the content.

- Revised inquiry activities precede sections as **Discovering the Concepts** features. These activities are intended to be used during class as an alternative way to introduce material to students as they work in groups. More information on the use of these activities can be found in the instructor's manual.

- **Integrating Chemistry** features throughout the chapters apply chemistry to common life experiences and biochemical contexts.

- **Consolidated Mathematics coverage and remediation in Chapter 1** provide students with a foundation for success throughout the course.

- **A new Chapter on Chemical Reactions (Chapter 5)** discusses common chemical reactions. The fundamentals of what causes reactions to start and what keeps reactions going are further applied and integrated in later chapters.

- **Solving a Problem** features at strategic points within each chapter provide students with a step-by-step approach to solving the more challenging topics in chemistry.

- **Health Icons** ✚ highlight specific end-of-chapter problems with application to health careers.

Chapter Organization and Revision

Throughout the text, the general, organic, and biological chemistry topics are integrated using relevant examples and applications to solidify concepts. This text intentionally contains only 12 chapters, allowing all chapters to be covered in a single semester. Each chapter builds upon conceptual understanding and skills learned from previous chapters, providing students with an efficient path through the content and a clear context for how all of the topics connect to one another.

This edition has undergone extensive chapter revision as discussed below.

1 Chemistry Basics—Matter and Measurement

In addition to being an introductory chapter highlighting the states, properties, types of matter, and the periodic table, Chapter 1 now includes more mathematics coverage. It begins with an introduction to the metric system and unit conversion, significant figures, scientific notation, and percent. This treatment is followed by application of mathematics to volume, mass, density, temperature, energy, and specific heat. The health-professions student sees the relevance of unit conversion when it is applied to dosing problems presented in Section 1.5. This application is extended into a new Integrating Chemistry feature on reading lab reports. Chapter 1 now includes a more extensive introduction to balancing chemical equations. Gas laws previously introduced in Chapter 1 now appear in Chapter 7.

2 Atoms and Radioactivity

Chapter 2 is an introduction to the atom. Introduction to basic atomic structure refocuses the student's attention on the periodic table through the introduction of atomic numbers and atomic mass. From this foundation, the chapter builds on its discussion of isotopes to lead into radioisotopes and the elementary concepts of nuclear chemistry. More complex concepts of nuclear chemistry are simplified with new problem-solving features on nuclear decay equations and half-life. Dosing is re-emphasized in a discussion of medical radioisotopes.

3 Compounds—Putting Particles Together

Building on the introduction to atoms and subatomic particles in Chapter 2, Chapter 3 starts with an introduction to the octet rule, leading to a discussion of bonding and compound formation. The initial focus is on ion formation and ionic compounds. Students get their first taste of chemical nomenclature as they learn to name ionic compounds. Ionic compounds are followed by covalent compounds, where a new flow chart is introduced to help students keep the naming rules straight. Students then learn to draw Lewis structures with a focus on bonding to C, O, and N. At this point, the focus shifts to counting and measuring compounds using the mole unit. This information is new to Chapter 3. Three new problem-solving features appear in this section to assist students with using unit conversion, molar mass, and Avogadro's number. The chapter then applies Lewis structures to determining molecular shape. The VSEPR concept develops with a continuing focus on carbon-containing compounds. The chapter concludes with an introduction to the concepts of electronegativity and bond polarity and the application of these concepts to the determination of molecular polarity.

4 Introduction to Organic Compounds

Chapter 4 utilizes the structural concepts developed in Chapter 3 to introduce students to organic compounds. The foundation of this chapter is structural and begins with a new step-by-step approach to drawing skeletal structures. The chapter then briefly introduces organic functional groups, focusing on unsaturated hydrocarbons. Further discussion of functional groups is integrated into later chapters as they appear in biomolecules. Students are also introduced to their first biomolecules—fatty acids—in the context of the structural and polarity concepts with which they are now familiar. Nomenclature of alkanes, cycloalkanes, and haloalkanes is now discussed through new problem-solving features and sample problems. Treatment of isomerism has been organized into one section in the second edition, but still includes structural isomers, conformational isomers, cis–trans isomers, and enantiomers.

5 Chemical Reactions

Chapter 5 is a new chapter on chemical reactivity. This chapter begins with thermodynamics and applies the concepts to calories in food. Thermodynamics is followed by a discussion of kinetics that includes reaction energy diagrams and introduces enzymes as biological catalysts. The chapter then turns to types of chemical reactions and identifies them in terms students may have seen in other courses: synthesis, decomposition, and exchange. Combustion then re-acquaints students with balancing chemical equations. Combustion is followed by the more specific reactions of oxidation, reduction, condensation, hydrolysis, and addition to alkenes, which includes hydrogenation and hydration. Chapter 5 also includes two new inquiry activities on the topics of chemical reaction types and oxidation–reduction.

6 Carbohydrates—Life's Sweet Molecules

Chapter 6 integrates the functional groups alcohol, aldehyde, and ketone into the context of carbohydrates to illustrate relevant structure, bonding, and chirality. Students are introduced to Fischer projections and diastereomers using carbohydrates as examples. Several of the reactions introduced in Chapter 5 are applied here to carbohydrates including oxidation–reduction, and condensation/hydrolysis. The ring formation of carbohydrates serves as the focus of hemiacetal formation reactions and leads to a consideration of the Haworth projections. Glycoside formation as a condensation reaction and the naming of glycosidic bonds introduce important disaccharides. The chapter continues with some important polysaccharides and concludes with treatment of relevant blood carbohydrates, the ABO blood groups, and heparin.

7 What's the Attraction? State Changes, Solubility, and Lipids

The term *intermolecular force* has been replaced in Chapter 7 of the second edition with the more general term *attractive force* to include ionic and ion–dipole forces in a more complete discussion. After an introduction to attractive forces, the chapter applies the forces to the states of matter, emphasizing that the attractive forces are present in the condensed states. The identification of attractive forces then allows students to predict relative boiling points and solubility. Three new problem-solving features have been added to the first two sections to assist students. Attractive forces are mainly absent in the gaseous state, but a brief treatment of gas laws is included here. Boyle's law and Charles's law are discussed, and an integrating chemistry feature shows how Boyle's law relates to breathing. The chapter then applies the understanding of attractive forces to dietary lipids, partial hydrogenation, and cell membranes, completing an integration of lipids.

8 Solution Chemistry—How Sweet Is Your Tea?

This chapter expands on the solubility concepts from Chapter 7 to introduce properties of solutions and the solubility of ionic compounds. Section 8.3 reintroduces students to balancing chemical equations by balancing solvation equations. The focus then shifts to concentration and the determination of solute concentration in a solution. Because equivalents are often used in medical applications, this concept is introduced and developed for ionic solutions. Percent concentrations are emphasized in the solution calculations because of their widespread use in health sciences. Two new problem-solving features have been added to this chapter to assist students with calculating molarity and using the dilution equation. The chapter concludes with a discussion of osmosis and diffusion and their application to cells and membrane transport.

9 Acids, Bases, and Buffers in the Body

Chapter 9 naturally follows solution chemistry as an introduction to acid–base chemistry. The reactivity of organic and biochemical compounds that act as acids and bases is included in this discussion. After an introduction to strong acids and strong bases, neutralization reactions are highlighted with a new problem-solving feature that emphasizes equation balancing once more. Equilibrium and Le Châtelier's principle are introduced as a prelude to weak acids and bases. A new feature applies Le Châtelier's principle to hemoglobin and oxygen transport. This is followed by a thorough treatment of pH and pK_a. Amino acids are introduced as examples of biochemical acids and bases. The chapter concludes with a relevant discussion of biological buffers, highlighting the bicarbonate buffer, acidosis, and alkalosis.

10 Proteins—Workers of the Cell

Chapter 10 has been reorganized from the first edition. Several concepts from previous chapters are integrated and applied in Chapter 10, including further consideration of amino acids (Chapter 9), chirality of amino acids (Chapter 4), and polarity of side chains (Chapters 3 and 7). Remaining functional groups (Chapter 4) not discussed in previous chapters appear here in amino-acid side chains. A new review of condensation and hydrolysis reminds students of the previous biochemical reactions through problem solving and introduces peptide bond formation in context. The ensuing material centers on the theme of protein structure with a focus on attractive forces at all levels of structure. Oxidation and reduction are integrated into disulfide bond formation. Protein denaturation is discussed. The structural treatment of proteins is now followed by a functional treatment, concluding with two sections discussing enzyme terminology, enzyme activity, and inhibition.

11 Nucleic Acids—Big Molecules with a Big Role

Chapter 11 provides the student with basic information regarding the structure and important functions nucleic acids play in living systems. The chapter builds a nucleotide from the simple biological molecules already introduced in previous chapters. Emphasis is placed on the chemical reactions that govern the building of nucleotides and nucleic acids (condensation reactions). A new problem-solving feature helps students master writing these condensation products. The discovery and structure of DNA are discussed in the familiar terms of primary, secondary, and tertiary structures. A new inquiry activity using the genetic code introduces students to protein synthesis. Mutations and viruses are included as applications of the concepts in the chapter. The chapter ends by discussing recombinant DNA technology, providing further relevance and understanding of information presented earlier in the chapter and the distinction between gene cloning and organism cloning.

12 Food as Fuel—A Metabolic Overview

Chapter 12 provides the student with an overview of the metabolism of carbohydrates, proteins, and lipids. This includes their digestion, catabolism, and use in energy production. The focus is the catabolic oxidation of glucose and ATP production. Two new inquiry activities have been added to this chapter to introduce students to catabolic ATP production and beta oxidation. A new problem-solving feature helps students count ATP production from beta oxidation. The chapter concludes with the alternative fuel sources of fatty acids and amino acids and discusses how these feed into catabolic oxidation and metabolism in general.

Resources in Print and Online

Supplement	Available in Print?	Available Online?	Instructor or Student Supplement	Description
Mastering Chemistry		X	Instructor and Student Supplement	MasteringChemistry from Pearson has been designed and refined with a single purpose in mind: to help educators create that moment of understanding with their students. The Mastering platform delivers engaging, dynamic learning opportunities—focused on your course objectives and responsive to each student's progress—that are proven to help students absorb course material and understand difficult concepts. By complementing your teaching with our engaging technology and content, you can be confident your students will arrive at that moment—the moment of true understanding.
Instructor Solutions Manual		X	Instructor Supplement	This Solutions Manual provides detailed solutions to all in-chapter as well as the end-of-chapter exercises in the text.
Test Bank		X	Instructor Supplement	This test bank contains over 600 multiple-choice, true/false, and matching questions. It is available in the TestGen program, in Word format, and included in the item library of MasteringChemistry.
Instructor Resources		X	Instructor Supplement	This provides an integrated collection of online resources to help instructors make efficient and effective use of their time. Includes all artwork from the text, including figures and tables in PDF format for high-resolution printing, as well as four pre-built PowerPoint™ presentations. The first presentation contains the images embedded within Power-Point slides. The second includes a complete lecture outline that is modifiable by the user. The final two presentations contain worked "in chapter" sample exercises and questions to be used with classroom iClicker systems. Also includes electronic files of the Instructor's Resource Manual, as well as the Test Bank. Can access resources through http://www.pearsonhighered.com/.
Study Guide	X		Student Supplement	This manual for students contains complete solutions to the selected odd-numbered end-of-chapter problems in the book.
Laboratory Manual	X		Student Supplement	Written by one of the text's authors (Deal), the lab manual continues the strategy of integration of concepts to help students understand chemistry. Several of the experiments included in the lab manual are original works developed by Professor Deal and his students in support of the integrated strategy. Most experiments are designed around a question, which is intended to engage students and to demonstrate the applicability of chemistry concepts to real-world problems. Many of the experiments highlight concepts from multiple chapters of the text, once again building on the strategy of integration.
Guided Inquiry Activities	X		Instructor and Student Supplement	Guided Inquiry Activities, written by Laura Frost, are available on the Pearson Custom Library (pearsoncustomlibrary.com). These activities are designed for in-class use by groups of students with facilitation by instructor. Students are asked to explore information, develop chemical concepts, and apply the concepts to further examples.

Acknowledgments

We have learned much since the first edition was published. Faculty and reviewer feedback has allowed us to enhance the second edition with some much-needed coverage while keeping the book length reasonable for a one-semester course. The expertise and professionalism of Karen Timberlake in editing and proofing the textbook cannot be understated. She has been an inspiration and mentor and we are deeply indebted.

The editorial staff at Pearson has been exceptional. We are grateful for the assistance of Don Gecewicz, freelance developmental editor, whose fresh look at the content allowed for better streamlining. His years of textbook development were apparent. Thanks goes to Carol Trueheart for assigning him to this project. We also want to acknowledge the enthusiasm of Jeanne Zalesky, executive editor in chemistry, who has convinced Pearson that the second edition of this textbook will be even better than the first. We also greatly appreciate the project organizers, first associate editor Jessica Moro and later assistant editor Lisa Pierce, who have gone through much of the material with a fine-tooth comb, making sure that author comments were interpreted correctly by production. Once again, Eric Schrader, photo researcher, sought out the perfect photos to help students visualize chemistry. Thank you. We want to thank the production team including Connie Long, project manager, and Marisa Taylor. They have been very patient with us as we embarked on the production process. We also appreciate the contributions of Lisa Tarabokjia, who organized and managed the review process and assisted Jeanne with many other tasks.

Laura Frost would like to also thank Adam Jaworski, editor-in-chief, for sharing the vision that an inquiry-based classroom can enhance student understanding of chemistry and has supported the inclusion of the inquiry activities.

Todd Deal would like, once again, to express a special appreciation to Jim Smith, our original editor, whose enthusiasm for the integrated strategy used in this project and belief in us as authors provided the foundation upon which this text is built.

This text reflects the contributions of many professors who took the time to review and edit the manuscript and provided outstanding comments, help, and suggestions. We are grateful for your contributions.

In addition, this project could not have been completed without the support of several exceptional colleagues in the Department of Chemistry at Georgia Southern University, who have taught using the first edition and offered many comments and corrections. Many of our GSU colleagues were always willing to discuss seemingly random chemistry topics, and they will always be held in high regard.

If you would like to share your experience using this textbook, either as a student or faculty member, or if you have questions regarding its content, we would love to hear from you.

Laura Frost
lfrost@fgcu.edu

Todd Deal
stdeal@georgiasouthern.edu

Reviewers

SECOND EDITION

Marsha Adrian, *University of the Incarnate Word*

Cynthia Ault, *Jamestown College*

Jay Baltisberger, *Berea College*

Julie Bezzerides, *Lewis-Clark State College*

Laura Boyd, *University of Texas, Tyler*

Martin Brock, *Eastern Kentucky University*

Yie-Hwa Chang, *Saint Louis University*

Robert Cook, *Louisiana State University*

Ron Duchovic, *Indiana University-Purdue University Fort Wayne*

Lisa Sharpe Elles, *Washburn University*

Mark Andrew Even, *West Liberty University*

George Craig Flowers, *Darton College*

Amy Graff, *Amarillo College*

Mary Graff, *Amarillo College*

Miriam Gulotta, *William Paterson University*

Melanie Harvey, *University of Saint Mary*

Michelle L. Hatley, *Sandhills Community College*

Meg Hausman, *University of Southern Maine*

Steven Heninger, *Allegany College of Maryland*

Richard Kashmar, *Wesley College*

Mary Ann Lee-Stanislav, *Rockhurst University*

Sara Lorenz, *Walla Walla Community College*

Laura J. Medhurst, *Marymount University*

Ed O'Connell, *Fairfield University*

Thomas Olmstead, *Grossmont-Cuyamaca Community College*

Julie Peyton, *Portland State University*

Ria Ramoutar, *Georgia Southern University*

Tanea Reed, *Eastern Kentucky University*

Eileen Reilly-Widow, *Fairfield University*

Abbey Rosen, *Marian University*

Scott C. Russell, *California State University, Stanislaus*

Ruth N. Russo, *Walla Walla Community College*

Elaine L. Schalck, *Alvernia University*

Upasna Sharma, *Saddleback College*

Emery Shier, *Amarillo College*

John Singer, *Jackson Community College*

Chad Snyder, *Western Kentucky University*

James D. Stickler, *Allegany College of Maryland*

Corey Stilts, *Elmira College*

Amy G. Taketomo, *Hartnell College*

Nathan Tice, *Eastern Kentucky University*

Jacqueline Wittke-Thompson, *University of St. Francis*

Accuracy Reviewers

Gordon Sproul, *University of South Carolina Beaufort*

Manuela Trani, *Georgia Perimeter College*

Focus Group Reviewers

Abbey Rosen, *Marian University*

Elizabeth Pulliam, *Tallahassee Community College*

Jim Zubricky, *University of Toledo*

Ed O'Connell, *Fairfield University*

Christine Hermann, *Radford University*

Shirley Hino, *Santa Rosa Junior College*

Rita Rhodes, *The University of Tulsa*

Dan Vandenberg, *Springfield High School*

FIRST EDITION

Eric Arnoys, *Calvin College*

George Bandik, *University of Pittsburgh*

Alessandra L. Barrera, *Georgia Gwinnett College*

Shay Bean, *Chattanooga State Technical Community College*

Joe C. Burnell, *The University of Indianapolis*

Kathy Carrigan, *Portland Community College*

Kent Chambers, *Hardin Simmons University*

Paul Chamberlain, *George Fox University*

Ron Paul Choppi, *Chaffey College*

Jeannie T. B. Collins, *University of Southern Indiana*

Christa L. Colyer, *Wake Forest University*

William Davis, *University of Texas at Brownville*

Brahmadeo Dewprashad, *Borough of Manhattan Community College*

P. K. Duggal, *Maple Woods Community College*

Stephen U. Dunham, *Moravian College*

George C. Flowers, *Darton College*

Hao Fong, *South Dakota School of Mines and Technology*

Don Fujito, *LaRoche College*

Emily Halvorson, *Pima Community College*

Kirk Hunter, *Texas State Technical College*

Louis Giacinti, *Milwaukee Area Technical College*

Christina Goode, *CSU Fullerton*

Matthew A. Johnston, *Lewis-Clark State University*

John R. Kiser, *Isothermal Community College*

Andrea D. Leonard, *University of Louisiana, Lafayette*

Charles F. Marth, *Western Carolina University*

Janice J. O'Donnell, *Henderson State University*

Rebecca O'Malley, *University of South Florida*

Julie Peyton, *Portland State University*

Jennifer Powers, *Kennesaw State University*

Deboleena Roy, *American River College*

Gillian E. A. Rudd, *Northwestern State University*

Michael Russell, *Mt. Hood Community College*

Karen Sanchez, *Florida State College, Jacksonville*

John Singer, *Jackson Community College*

Dan Stasko, *University of Southern Maine, Lewiston-Auburn College*

James D. Stickler, *Allegheny College of Maryland*

Koni Stone, *CSU Stanislaus*

Everett Shane Talbott, *Somerset Community College*

Kwok-tuen Tse, *Harold Washington College*

Maria Vogt, *Bloomfield College*

James Zubricky, *University of Toledo*

Concise. Practical. Integrated.

The **Second Edition** of *General, Organic, and Biological Chemistry* strengthens the proven strategy of integrating general, organic, and biological chemistry for a focused introduction to the fundamental connections between chemistry and life. Frost and Deal's streamlined approach offers students a clear path through the content over a single semester.

"This brief, yet comprehensive text gives the students just enough general and organic chemistry to get to the good stuff—biochemistry—and it doesn't wait until the end."

—Associate Professor, Abbey Rosen Ph. D.
Marian University, Fond du Lac, WI

"This is the most student-friendly, yet rigorous integrated GOB text out there!"

—Lecturer, Jim Zubricky
University of Toledo, OH

A concise approach to important concepts and skills

A Visual Framework for Studying includes both chapter-opening spreads and end-of-chapter study guides that introduce students to the main concepts of each chapter.

Table of Contents

Integration of Biochemistry into the general and organic chapters of the text creates a more efficient path through topics needed in the course. This text contains only 12 chapters, allowing all chapters to be covered in a single semester. Each chapter builds upon conceptual understanding and skills learned from the previous chapter, providing context for how all topics connect to one another.

Brief Contents

Visual Study Guides

The end of each chapter includes a visual Summary and Study Guide that work together to highlight the main concepts and help students study more effectively. These eye-catching features give students a practical way to pinpoint and focus on the most important concepts in each chapter and then assess their own proficiency.

---

68   **SUMMARY**

2.1 Atoms and Their Components

2.1 Inquiry Question: What characteristics of the subatomic particles make up an atom?

An atom consists of three subatomic particles: protons, neutrons, and electrons. Protons have a positive charge, neutrons have no charge, and electrons have a negative charge. Most of the mass of an atom comes from the protons and neutrons located in the center, or nucleus, of an atom. The unit for the mass of an atom is the atomic mass unit (amu); each proton and neutron present in an atom weighs approximately 1 amu.

2.2 Atomic Number and Mass Number

2.2 Inquiry Question: What do the atomic number and mass number indicate?

The atomic number of an atom indicates the number of protons present in an atom. All atoms of a given element have the same number of protons. The mass number of an atom is the total number of protons and neutrons present in a given atom of an element.

2.3 Isotopes and Atomic Mass

— Number of protons
— Symbol for carbon
— Average atomic mass of all carbon atoms

2.3 Inquiry Question: What is the difference between the mass number for an isotope and the atomic mass of an element?

The mass number is the number of protons and neutrons for a given isotope. For example, nitrogen-14 has 7 protons and 7 neutrons. The atomic mass is the average atomic mass for all the isotopes of an element found in nature. This number is found on the periodic table, often below the element symbol.

2.4 Radioactivity and Radioisotopes

2.4 Inquiry Question: How does a radioisotope emit radiation?

Some atomic isotopes emit radiation (a form of energy) spontaneously from their nucleus in a process called radioactive decay. Isotopes that undergo radioactive decay are called radioisotopes, and the high-energy particles given off in this process are referred to as ionizing radiation, or radioactivity. Three common forms of radioactivity are alpha (α) and beta (β) particles and gamma (γ)

rays. An X-ray is also a form of ionizing radiation, although it is not caused by a radioactive decay event. Different forms of ionizing radiation penetrate the body differently, producing different biological effects.

2.5 Nuclear Equations and Radioactive Decay

2.5 Inquiry Question: How is a nuclear decay equation written?

The radioactive decay of a radioisotope can be represented symbolically in the form of a nuclear decay equation. The number of protons and the mass number in the reactant (the decaying radioisotope) is equal to the sum of the number of protons and the mass numbers found in the products (the stable isotope and the radioactive particle).

2.6 Radiation Units and Half-Lives

2.6 Inquiry Question: How can the amount of radiation be determined from radiation units and half-life?

Radioactive decay is measured as the number of decay events, or disintegrations, that occur in 1 second. The common unit for measuring radioactive decay is the curie (Ci), and the SI unit is the becquerel (Bq). For the smaller quantities in medical applications, the microcurie (μCi) is often used. A becquerel is equal to a disintegration per second. A curie is equivalent to 3.7×10^{10} becquerels. The half-life of a radioisotope is the amount of time it takes for one-half of the radiation in a given sample to decay. Most radioisotopes used in medicine have short half-lives, allowing the radioactivity to be more quickly eliminated from the body.

Cold spot

2.7 Medical Applications for Radioisotopes

2.7 Inquiry Question: How does the medical field use radioisotopes?

Certain elements concentrate in particular organs of the body. If a radioisotope of this element can be made, this area of the body can be imaged using that radioisotope. A patient can be injected with a trace amount of a radioisotope to diagnose a diseased state. Radioisotopes can also be used to treat diseases. Radioisotopes can be applied externally (external beam radiation therapy) or internally (brachytherapy) by applying radiation directly at the tumor site in high doses destroying cancerous cells. Positron emission tomography (PET) uses a radioisotope to image tissues that are not functioning normally.

Presenting chemistry in a practical, real-world context

Practical connections and applications show students how to use their understanding of chemistry in their everyday lives and future health professions.

3.71 Magnesium salicylate is sold as an over-the-counter drug to treat pain and inflammation. If there are two salicylate ions for each magnesium ion in the compound, what is the charge on the salicylate ion?

3.72 Two common additives to bottled water are the ionic compounds potassium chloride and magnesium sulfate. Write the formulas for these ionic compounds.

3.73 Sodium metabisulfite is used in processed foods as a preservative and antioxidant. The metabisulfate anion has the formula $S_2O_5^{2-}$. Provide the formula for this compound.

3.74 Stannous or tin(II) fluoride is a compound found in many toothpastes. Write the formula for this compound.

NEW!
Health Application Problems Denoted by Health Icons

Health Application Problems at the end of the chapter promote critical thinking and help students apply their problem-solving skills to real world career contexts.

NEW!
A Chapter on Chemical Reactions

A new chapter on Chemical Reactions (Chapter 5) deepens student understanding of chemical reactivity. Thermodynamics is followed by a discussion of kinetics that includes reaction energy diagrams and introduces enzymes as biological catalysts. Fundamentals of reactions are applied and integrated throughout the chapter.

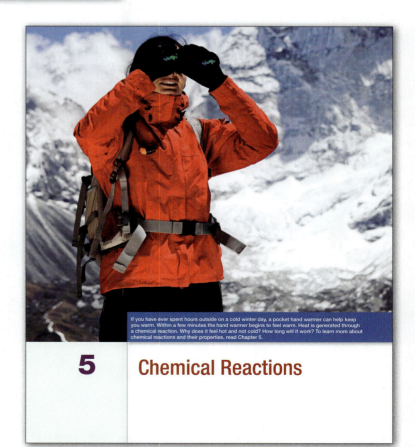

If you have ever spent hours outside on a cold winter day, a pocket hand warmer can help keep you warm. Within a few minutes the hand warmer begins to feel warm. Heat is generated through a chemical reaction. Why does it feel hot and not cold? How long will it work? To learn more about chemical reactions and their properties, read Chapter 5.

5 **Chemical Reactions**

Integrating chemical concepts promotes students' interest

General, Organic, and Biological Chemistry, **Second Edition** gives students a focused introduction to the connections between general chemistry concepts and reactivity and organic and biological chemistry.

integrating chemistry

Important Ions in the Body

A number of ions found in the fluids and cells of the human body perform important functions. The main cations found in the body include Na^+, K^+, Ca^{2+}, and Mg^{2+}, which are important in maintaining solution concentrations inside and outside cells. The main anions are Cl^-, HCO_3^-, and HPO_4^{2-}. These ions help maintain the charge neutrality between the blood and fluids inside cells. The phosphate ion, also called inorganic phosphate and abbreviated P_p is involved in cellular energy transfer during chemical reactions as well. In contrast to ions in fluids, ions can also form hard minerals. Tooth enamel contains the mineral hydroxyapatite, consisting of calcium (Ca^{2+} hydroxide (OH^-) ions. **Table 3.3** shows a listing of the r their functions.

General, Organic, and Biological Chemistry Topics

Throughout the text, the general, organic, and biological chemistry topics are integrated using relevant examples and applications to solidify concepts. This text uses biochemistry to enliven topics of general and organic chemistry, as well as health-related applications to show how all branches of chemistry work together.

Integrating chemistry

Boyle's Law and Breathing

Do you know that you use Boyle's law every day? Breathing is a very practical application of Boyle's law. When you breathe in, you are actually not forcibly drawing air into your lungs. Instead, the muscles of your rib cage and your diaphragm contract to cause the volume of your chest cavity to increase (see **Figure 7.19**). When this happens, the air pressure inside your lungs decreases (volume increase results in pressure decrease), and the pressure of the atmosphere causes air to rush into your lungs to equalize the internal and external pressures. When the muscles relax, the volume of your chest cavity decreases, increasing the pressure in your lungs above that of the outside, and air flows out to the lower pressure (atmosphere).

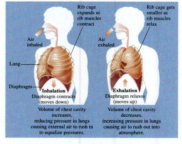

FIGURE 7.19 Boyle's law and breathing. Breathing is controlled by contraction and relaxation of the diaphragm muscles.

sample problem 7.6 **Boyle's Law Calculations**

The lungs of a normal adult can hold 5.0 L of air under typical atmospheric pressure (760 mmHg). If a diver dives to a depth where his lungs compress to a volume of 4.0 L, what is the pressure surrounding her?

Solution

Changes in the volume and pressure of a gas are related using Boyle's law. First, check the problem to determine what is given and what information is requested.

STEP 1: Determine the given information. We are given P_i = 760 mmHg, V_i = 5.0 L, and V_f = 4.0 L.

STEP 2: Solve for the missing variable using the Boyle's law relationship. In this case, P_f.

$$\frac{P_i \times V_i}{V_f} = P_f$$

STEP 3: Substitute the given information into the equation and solve.

$$\frac{760 \text{ mmHg} \times 5.0 \text{ L}}{4.0 \text{ L}} = P_f$$

$$P_f = 950 \text{ mmHg}$$

Discovering the Concepts

Discovering the Concepts guides students through inquiry and exploration to develop chemical concepts. The activities can be used during class as an alternate way to introduce material to students as they work in groups. A full complement of Guided-Inquiry Activities by coauthor Laura Frost are available through the Pearson Custom Library for the instructor who wishes to explore more concepts through cooperative learning in the classroom.

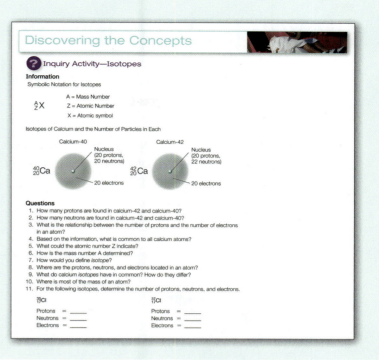

Inquiry Questions (IQ)

Inquiry Questions accompany each section, piquing interest and functioning as learning targets while prompting student interactivity. Each IQ is revisited in the Chapter Summary to help students evaluate their understanding of the section's content.

2.1 Atoms and Their Components

2.1 Inquiry Question: What are the characteristics of the subatomic particles that make up an atom?

Imagine the number of fine grains of sand it would take to cover the entire east and west coastlines of the mainland United States with beaches as wide as a football field and 1 meter deep. This is about the same number of atoms of carbon that are found in a half-carat diamond (Figure 2.1a). That is a lot of grains of sand and a lot of carbon atoms.

As you can guess, individual atoms of any element are incredibly small, so a lot of them fit into a small space. Carbon atoms (element number 6 on the periodic table) are found in combination with other elements in virtually all living substances. The element carbon can also be found in pure form in the nonliving substances diamond and graphite (see Figure 2.1a and b). If we were able to crush either of these substances into a very fine powder, we could view the outline of individual atoms through a powerful microscope called a scanning tunneling microscope (see Figure 2.1c).

An Integrated GOB Lab Manual

An Integrated GOB Lab Manual by coauthor Todd Deal specifically correlates with *General, Organic, and Biological Chemistry,* **Second Edition.** The lab manual also includes two labs that focus on math, providing additional math practice for the course.

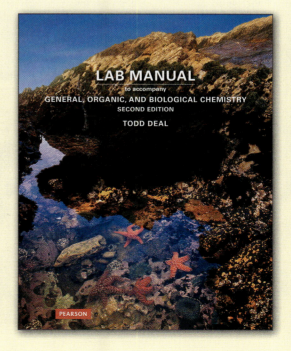

Develop and reinforce math and problem-solving skills

An extensively revised chapter in the textbook coupled with tutorials in MasteringChemistry® gives students practice with math skills, establishing a solid foundation for the course.

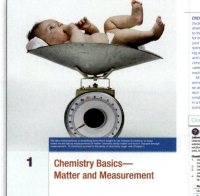

NEW!
Mathematics Coverage and Remediation

Mathematics coverage and remediation in Chapter 1 helps students establish needed mathematical skills early in the course and begins with an introduction to:

- Metric system and unit conversion
- Significant figures
- Scientific notation percentage

These topics are followed by application of mathematics to volume, mass, density, temperature, energy, and specific heat. Chapter 1 now also includes a more extensive introduction to balancing chemical equations and dosing problems tied to a new Integrating Chemistry feature on reading lab reports.

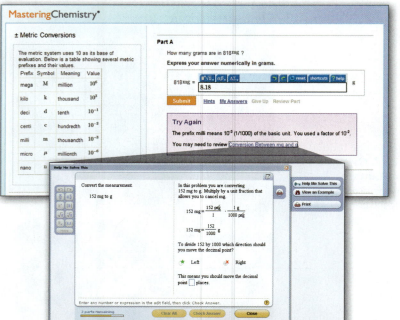

Math Remediation Links

In MasteringChemistry, Math Remediation links found in selected tutorials launch algorithmically generated practice exercises from MathXL to help students acquire the quantitative skills to succeed in the course. Exercises include guided solutions, sample problems, and additional learning aids, and they also offer helpful feedback when students enter incorrect answers.

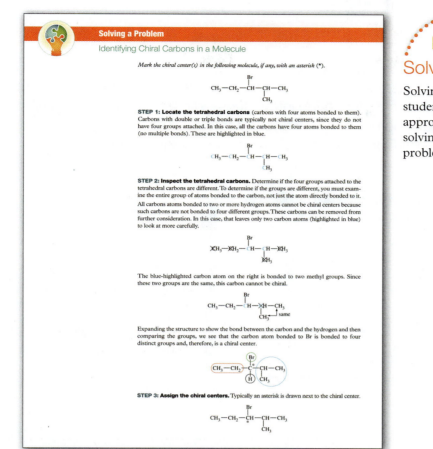

Mastering Chemistry®

www.masteringchemistry.com

MasteringChemistry motivates students to learn outside of class and arrive prepared for lecture. The textbook works with tutorial activities and assessment in MasteringChemistry to reinforce the integrated approach and tie chemistry to students' real-world experiences. New tutorial topics include Dosing Calculations Using Unit Conversions and Balancing Neutralization Reactions.

NEW!
Integrated Tutorials

Integrated Tutorials in MasteringChemistry reinforce connections between general chemistry, organic chemistry, and biochemistry principles while helping students develop their problem-solving skills.

NEW!
Pause and Predict Video Quizzes

Pause and Predict Video Quizzes in MasteringChemistry bring chemistry to life with lab demonstrations illustrating key topics in general, organic, and biological chemistry. Students are asked to predict the outcome of experiments as they watch the videos; a set of multiple-choice questions challenges students to apply the concepts from the video to related scenarios. Videos are also provided in the study area for browser and mobile viewing.

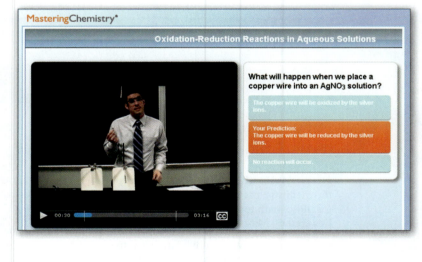

Reading Questions

Reading Questions cover key concepts from each chapter, ensuring that students stay on track with assigned readings and come prepared to lecture.

General, Organic, and Biological

CHEMISTRY

Second Edition

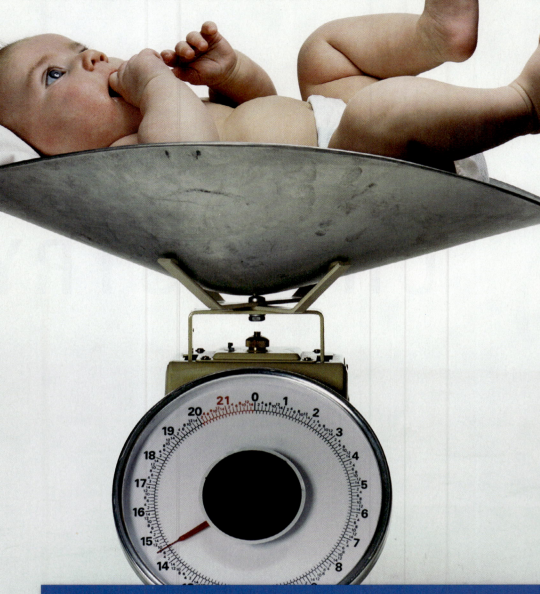

We take measurements of everything from infant weight to car mileage to cooking. In these cases we are taking measurements of matter. Chemists study matter and how it changes through measurement. To introduce yourself to the study of chemistry, begin with Chapter 1.

1

Chemistry Basics— Matter and Measurement

DID YOU KNOW that everything you do every day involves chemistry? Yes, everything. From the water and shampoo in your shower, to the food you ate for breakfast, to the gasoline that powers your car, to the therapeutic drugs for treating diseases, to the sunscreen lotion that protects your skin, and even the clothes that you wear—all of these somehow involve chemistry. Learning chemistry is really learning about everyday life and how chemistry impacts our lives and even provides life itself. Chemistry helps us understand concepts as diverse as how our bodies function and the wide variety of conveniences that make our lives easier. So, come explore with us. It will be challenging, but fun. We promise!

All of the "stuff" that we just mentioned is composed of something that chemists call matter. **Matter** can be defined as anything that takes up space (scientists call this volume) and weighs something (scientists call this mass). From the smallest tablet dispensed by a pharmacist to the shampoo in a bottle to the air in a balloon, each of these takes up some amount of space and is a form of matter.

Discovering the Concepts

? Inquiry Activity—Classifying Matter

Information

Chemistry is the study of matter and its changes. There are two main types of matter: *pure substances* and *mixtures* of substances. A mixture that is mostly water is called an *aqueous solution.* Matter can exist in several different *states,* the three most common being solid, liquid, or gas.

Questions

1. Consider the examples on the flow chart. How does the formula of an element differ from that of a compound?
2. Can an element be a pure substance? Can a compound be a pure substance?
3. Using the information given, how might you define a pure substance? How does a pure substance differ from a mixture?
4. Based on the information in the data set, would you classify the following substances as (a) an element or compound or (b) atom or molecule?

Matter	Element or Compound	Atom or Molecule
He		
N_2		
CH_2O (formaldehyde)		
CH_3COOH (vinegar)		

Types of Matter Flow Chart with Examples, Their Formulas, and States

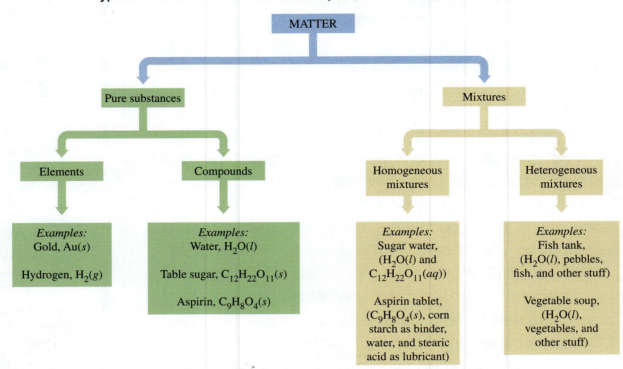

Elements can exist as individual *atoms* (neon, Ne), in combination with other atoms, or as pairs (hydrogen, H_2) forming *molecules*.

5.   As a group, devise a definition for a compound.
6.   In your own words, describe the difference between a homogeneous and heterogeneous mixture.
7.   Would you classify the following matter as element, compound, or mixture? If you classify it as a mixture, classify it as homogeneous or heterogeneous.
     a.   table salt, (NaCl)
     b.   nickel (Ni)
     c.   chocolate chip cookie dough
     d.   air
8.   What do you think the labels (*s*), (*l*), (*g*), and (*aq*) on the formulas in the data set mean?

1.1 Classifying Matter: Mixture or Pure Substance

What's an Inquiry Question?
Inquiry Questions are designed to focus your reading on the main concepts by section. An Inquiry Question appears at the beginning of each section.

1.1 Inquiry Question:
How is matter classified?

Sometime during your education, you may have had a course in biology where you classified an organism taxonomically—according to its family, genus, and species as well as several other larger groupings. For example, the genus and species name for humans is *Homo sapiens.* In chemistry, we classify matter, but not quite to the extent that biologists classify organisms. The system used in chemistry provides information about the characteristics of a sample of matter.

While the states of matter—solid, liquid, and gas—may be thought of as a classification of matter, we usually use much broader terms when classifying it. The flow chart shown in **Figure 1.1** gives you an idea of how the classification of matter works. Use this flow chart as a guide as we work to develop the scheme that chemists use to classify matter.

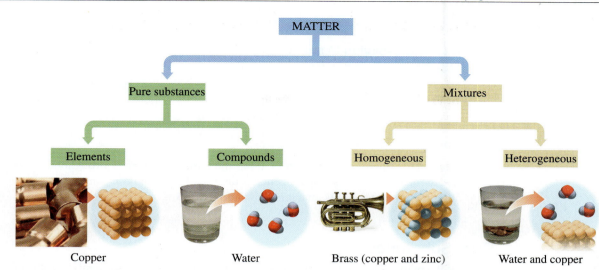

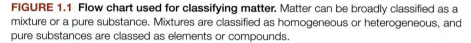

| Copper | Water | Brass (copper and zinc) | Water and copper |

FIGURE 1.1 Flow chart used for classifying matter. Matter can be broadly classified as a mixture or a pure substance. Mixtures are classified as homogeneous or heterogeneous, and pure substances are classed as elements or compounds.

As two examples of matter, consider blood and a diamond. If you are a blood donor, you may have heard blood discussed in terms of whole blood, plasma, white cells, and red cells. These terms give you a hint that blood is a mixture containing many components. In contrast, if you have ever bought a diamond, you know that one of the measures used to grade these gemstones is clarity, which is a measure of purity. The clearer (more pure) a diamond is, the more it costs. So, how do the complexity of a sample of blood and the purity of a diamond contribute to how they are classified?

Take a look at the flow chart. The first thing you notice is that all matter is divided into two large categories or classifications—mixtures and pure substances. Let's consider these one at a time.

Mixtures

A **mixture** is a combination of two or more substances. Blood contains red blood cells, white blood cells, plasma (which is mostly water), and several other things. One of the defining characteristics of a mixture is that it can be separated into its different components. Therefore, blood is a mixture. A specialist in blood banking (SBB) separates blood into red cells and plasma for storage and use in different medical procedures at a blood bank.

Mixtures can be further classified as homogeneous or heterogeneous. A **homogeneous mixture** is one whose composition is the same throughout. The air that you breathe is a homogeneous mixture. Air contains nitrogen, oxygen, carbon dioxide, argon, and other gases. Every time you take a breath, you get the same substances in the same amount as another person breathing the same air. In contrast, grandma's chunky garden-style spaghetti sauce is a **heterogeneous mixture** because its composition is not uniform, but varies throughout. One spoonful of spaghetti sauce may contain onions, mushrooms, and large chunks of tomato, while another spoonful contains celery, onions, and large chunks of green pepper. No two samples contain the same substances in the same amount.

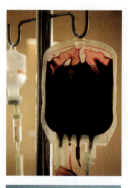

Blood is a mixture of many components whereas a diamond is a single component of the element carbon. A specialist in blood banking stores blood.

sample
problem
1.1   **Classifying Mixtures**

Classify each of the following mixtures as homogeneous or heterogeneous. Briefly justify your answer.

a. olive oil   b. rocky road ice cream

Solution

a. Olive oil is a homogeneous mixture. It has the same composition throughout.
b. Rocky road ice cream is a heterogeneous mixture. Each spoonful contains different amounts of nuts, marshmallow, and ice cream.

Pure Substances

Much of the matter that we encounter in our everyday lives is some sort of mixture. From the air we breathe to the food we eat to the concrete we walk on, all are mixtures. However, because of their varied composition, mixtures are difficult to study. While chemists routinely work with and study mixtures, pure substances are easier to describe.

A **pure substance** is matter that is made up of only one substance. Diamonds are composed only of carbon. Because a diamond is composed of a single substance, it is not a mixture but a pure substance.

Note from the flow chart that pure substances can be one of two types—elements or compounds. An **element** is the simplest type of matter because it is made up of only one type of atom. An **atom** is the smallest unit of matter that can exist and keep its chemically unique characteristics. A **compound** is a pure substance that is made of two or more elements that are chemically joined together.

sample
problem
1.2   **Mixture Versus Pure Substance**

Classify each of the following substances as a mixture or a pure substance:

a. cake batter   b. helium gas inside a balloon

Solution

a. Cake batter is a mixture of flour, butter, sugar, and other ingredients.
b. The helium gas is a pure substance.

practice
problems

1.1   Classify each of the following mixtures as homogeneous or heterogeneous. Briefly justify your answer.
a. a bowl of vegetable soup   b. mouthwash
c. an unopened can of cola   d. a dinner salad

1.2   Classify each of the following mixtures as homogeneous or heterogeneous. Briefly justify your answer.
a. a bottle of sports drink   b. a blueberry
                                  pancake
c. gasoline   d. a box of raisin
                                  bran

1.3   Classify each of the following substances as a pure substance or a mixture:
a. copper   b. ice cream
c. salt

1.4   Classify each of the following substances as a pure substance or a mixture:
a. salt water   b. purified water
c. concrete

Discovering the Concepts

? Inquiry Activity—The Periodic Table

All *elements* are listed individually on the *periodic table of the elements.* There is a periodic table on the inside front cover of your textbook and an alphabetical listing of all the elements on the facing page. Refer to this as you answer the questions in this activity.

The rows on the periodic table are referred to as *periods*, and the columns are referred to as *groups.* The groups are numbered across the top of the periodic table and the periods are numbered down the left side of the table. The table is organized with metals on the left and nonmetals on the right, with a staircase dividing line between the two. Find this staircase on the periodic table in your textbook.

Chemical formulas show the type and number of each element present in a compound. For example, water's chemical formula is H_2O, and it contains two hydrogen atoms and one oxygen atom.

Questions

1. In which group on the periodic table are the following elements found?
   a. sodium   b. oxygen   c. calcium   d. carbon
2. In which period are the following elements found?
   a. hydrogen   b. nitrogen   c. sulfur   d. phosphorus
3. Provide names for the following elements and identify them as metals or nonmetals:
   a. Cu   b. Na   c. Cl   d. C   e. K   f. P
4. Look at the periodic table of the elements. About how many elements are there?
5. Identify the number and name of each element in the formulas given:
   a. $C_6H_{12}O_6$ (dextrose)   b. NaOH (lye, found in drain cleaners)
   c. $NaHCO_3$ (baking soda)   d. $C_{15}H_{21}NO_2$ (Demerol, a painkiller)
6. Are most of the elements on the periodic table metals or nonmetals? Considering that most of the Earth's crust is made up of the elements silicon and oxygen in the form of sand and that Earth's biomass is made up of mostly carbon, does this seem surprising?

1.2 Elements, Compounds, and the Periodic Table

If a pure substance can be an element or a compound, how do we distinguish which is which? We have a tremendous tool to guide us in this process: the **periodic table of the elements.** At its most basic level, the periodic table is a listing of all the elements found on Earth.

Somewhere along your educational journey, you may have encountered the periodic table, most likely hanging on the wall in a science classroom. But have you ever tried to decipher it? This strangely shaped chart contains an amazing collection of data, both in the way that it is organized and in the actual information listed. We will make extensive use of the periodic table and the information it contains. Take a look at the periodic table in Figure 1.2.

The periodic table consists of many small blocks in a basic layout. Each has a letter or two in its center and numbers above and below these letters. The letters are known as the **chemical symbol** and represent the name of each element. For many of the elements, the symbols are derived directly from the name of the element—for example, lithium = Li, carbon = C, hydrogen = H, oxygen = O, and so on. A few of the elements have symbols that do not match the first few letters of their name—for example, sodium = Na, gold = Au. Their elemental symbols are derived from the Latin names for the element, natrium and

? 1.2 Inquiry Question:
How is the periodic table organized?

Periodic Table of Elements

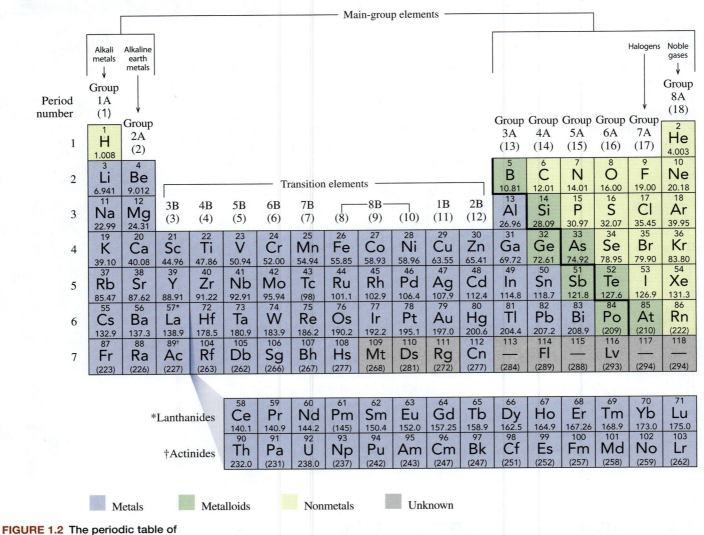

Metals **Metalloids** **Nonmetals** **Unknown**

FIGURE 1.2 The periodic table of the elements.

aurum respectively. Because the elements form the basic vocabulary of chemistry, it is important to know the names and symbols of the more common elements, especially those that are found in living things (carbon, hydrogen, oxygen, nitrogen, phosphorus, and sulfur).

The blocks on the periodic table are organized in specific arrangements. A vertical column of blocks is known as a **group** of elements. The elements in a group have similar chemical behaviors. Each group has a number and letter designation. The groups with A designations (1A–8A) are known as the main-group elements, and the groups with B designations are the transition elements. This system of numbering the groups is most common in North America and will be used throughout this textbook. The system using the numbers 1 through 18 for the columns has been recommended by the International Union of Pure and Applied Chemistry (IUPAC) and is also used. Several of the groups have special names to designate the members of the group as shown at the top of Figure 1.2.

A horizontal row of blocks is known as a **period.** The periods are numbered from 1 to 7 with sections of Periods 6 and 7 set apart at the bottom of the periodic table. As we continue our exploration, we will see that each of the groups and periods has special characteristics and helps us better understand the chemistry of the elements.

In the periodic table of the elements, a different element is in each block. The bold staircase-shaped line, which begins at the element boron and runs diagonally down and to the right, separates the metals from the nonmetals. Elements bordered by the line, with the exception of aluminum (Al), are metalloids.

Metals and nonmetals show different properties. Aluminum can be made into sheets or blocks and is a metal. Carbon is more brittle, and we meet carbon as a powder or as the "lead" in a pencil. Carbon is a nonmetal.

> ### sample problem
> ### 1.3 Using the Periodic Table
>
> Use the periodic table to answer the following questions:
>
> a. What is the elemental symbol for oxygen?
> b. To what group does oxygen belong?
> c. What period is oxygen in?
>
> #### Solution
>
> a. Oxygen's symbol is O.
> b. Oxygen is located in Group 6A.
> c. Oxygen is in Period 2.

Compounds

How do we classify substances like water (H_2O) or table salt (NaCl) that contain more than one element from the periodic table? Remember that we said a pure substance containing two or more elements chemically combined is classified as a compound. Therefore, both water and table salt are compounds. Compounds combine elements in specific ratios. How and why elements combine in these ratios is discussed in Chapter 3.

> ### sample problem
> ### 1.4 Classifying Pure Substances
>
> Classify each of the following as an element or compound and explain your classification:
>
> a. carbon dioxide (the gas we breathe out)
> b. helium (the gas sometimes used to inflate balloons)
>
> #### Solution
>
> a. Carbon dioxide is a compound. From its name, you can tell that this substance contains carbon and "something else" (even if you did not know that oxide is a name often used for oxygen when it is in compounds). Because carbon dioxide contains two different elements, it is a compound.
> b. Helium is an element. Look for helium on the periodic table. You will find it on the upper right. The gas used to inflate a balloon held by a string is pure helium—an element that is lighter than air.

Before we continue, let's take a quick look at the representations we use for compounds. Why does the representation for water contain the subscript 2, but the one for table salt doesn't? These representations, known as **chemical formulas,** show that water is composed of two particles of hydrogen and one particle of oxygen and table salt is composed of one particle of sodium and one particle of chlorine. The absence of a subscript on oxygen, sodium, and chlorine is understood to mean *one* particle. A compound's chemical formula identifies both the type and number of particles of each of the elements in a compound.

sample problem 1.5   **Particles in Compounds**

Identify and give the number of atoms of each element in the following compounds:

a. $C_2H_4O_2$, acetic acid, a component of vinegar   b. H_3PO_4, phosphoric acid

Solution

a.   Acetic acid contains 2 atoms of carbon, 4 atoms of hydrogen, and 2 atoms of oxygen.
b.   Phosphoric acid contains 3 atoms of hydrogen, 1 atom of phosphorus, and 4 atoms of oxygen.

practice problems

1.5   Classify each of the following as an element or compound and explain your classification:

    a.   aluminum, used in soft drink cans and foil

    b.   sodium hypochlorite, found in bleach

    c.   hydrogen, fuel of the sun

    d.   potassium chloride, a table salt substitute

1.6   Classify each of the following as an element or a compound:
    a. Fe   b. $CaCl_2$ c. Si   d. KI

1.7   Use the periodic table to supply the missing information in the following chart:

Name	Elemental Symbol	Group	Period	Metal or Nonmetal
Fluorine				
		1A	2	
	Cl			
		4A	2	Nonmetal

1.8   Use the periodic table to supply the missing information in the following chart:

Name	Elemental Symbol	Group	Period	Metal or Nonmetal
Sodium				
		2A	4	
	P			
		5A	2	Nonmetal

1.9   Identify and give the number of particles of each element in the following compounds:

    a. $NaNO_2$, sodium nitrite, a preservative

    b. $C_{12}H_{22}O_{11}$, table sugar

    c. $CaCO_3$, calcium carbonate, limestone

1.10   Identify and give the number of particles of each element in the following compounds:

    a. Na_2SO_4, sodium sulfate, a drying agent

    b.   CO, carbon monoxide

    c. NH_3, ammonia

1.3 Math Counts

The proper tools can make a job easier. Mathematics gives us some tools that can make understanding chemistry easier. This first section covers some basic principles from mathematics that are applicable to the field of chemistry.

Units, Prefixes, and Conversion Factors

In science, health care, and business, we often make a measurement or report a value. For measurements to be consistently and easily compared, a defined set of standards or a measurement system is required. The **metric system** used throughout the world by scientists and health care professionals alike is part of the ***Système International d'Unités*** (**SI**) or International System of Units. SI units and the metric system will be used throughout this text, so it is important that you become familiar with them.

The SI system is based on a series of standard units for each type of measurable quantity like mass (weight—see section 1.4 for mass vs. weight) and volume. The standard SI unit for mass is the **kilogram** (kg). The standard SI unit for volume is the **liter** (L), and the standard SI unit for length is the **meter** (m). Although these measures may seem unfamiliar now, a kilogram is about 2.2 pounds, a liter is a little more than a quart, and a meter is slightly more than 3 feet.

If a paper clip has a mass of about 1 gram, imagine how many paper clips it would take to equal your body mass. This would be a huge number and would be awkward to work with. Or think of trying to get the mass of one human hair by comparing it to a paper clip. It would take a very large number of hairs to equal the mass of a paper clip.

To deal efficiently with quantities that are much larger or much smaller than one another, the SI system employs a set of prefixes that can be applied to the base unit that change the meaning of the unit by a power of ten. The power-of-ten concept is similar to our currency in that 10 pennies equal 1 dime, and 10 dimes equal 1 dollar. Table 1.1 lists the common SI prefixes, as well as the abbreviation and value for each.

For example, because the average person has a mass much greater than a paper clip, we apply the prefix **kilo** to the base unit of mass, the gram. "Kilo" means 1000 times greater than, so a *kilo*gram is 1000 g. Thus, an average person's mass would be expressed as 70 kg, which is a much more manageable number than 70,000 g. Note that the kilogram is the standard SI unit of mass, but that it is 1000 times larger than the base unit, which has no prefix, the gram.

1.3 Inquiry Question:
What basic math concepts are important to chemistry?

TABLE 1.1 Metric Prefixes

Prefix	Abbreviation	Relationship to Base Unit
giga	G	$1,000,000,000 \times$
mega	M	$1,000,000 \times$
kilo	k	$1000 \times$
base unit (has no prefix)		$1 \times$ (gram, liter, meter)
deci	d	$\div 10$
centi	c	$\div 100$
milli	m	$\div 1000$
micro	μ	$\div 1,000,000$
nano	n	$\div 1,000,000,000$

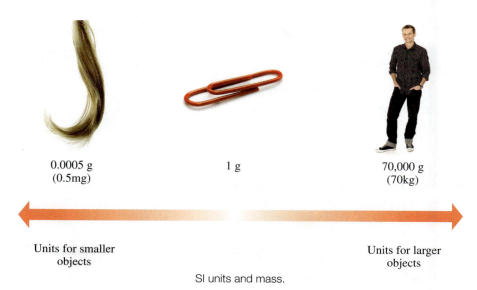

0.0005 g
(0.5mg)

1 g

70,000 g
(70kg)

Units for smaller objects

Units for larger objects

SI units and mass.

We can use Table 1.1 to obtain equivalency relationships between different units. The numbers in the right-hand column provide the relationship between any unit and the base unit. For example, when moving from decigrams to grams, the last column says we should divide by 10; therefore, the equivalency is 1 dg = 0.1 g. Quantities that can be related to each other by an equal sign are called **equivalent units.** By multiplying both sides by 10, we can avoid working with decimals, and say that 10 dg = 1 g.

Such equivalencies can be used as **conversion factors** to convert one unit to another using one or more of these factors. For example, when the equivalency stated previously, 10 dg = 1 g, is written as a conversion factor it can take the following forms:

$$\frac{10\ dg}{1\ g} \quad or \quad \frac{1\ g}{10\ dg}$$

These conversion factors are read as, "There are 10 decigrams in 1 gram" or "1 gram contains 10 decigrams," respectively. Conversion factors allow you to convert a quantity in one unit to the equivalent quantity in a larger or smaller unit. This use of converting units to an equivalent unit is also called **dimensional analysis.** Let's try an example.

Solving a Problem

Converting Units

How many grams of vitamin C are in a tablet containing 100 mg?

STEP 1: Determine the units on your final answer. In this case, the units will be grams.

STEP 2: Establish the given information. In this case, you have a value of 100 mg given in the problem. You also have an equivalency relationship (Table 1.1) showing that 1 mg = 0.001 g or 1000 mg = 1 g, which gives the following conversion factors:

$$\frac{1000\ mg}{1\ g} \quad or \quad \frac{1\ g}{1000\ mg}$$

STEP 3: Decide how to set up the problem. Which conversion factor should be used? The given units must cancel out (appearing in both a numerator and denominator), leaving the desired unit in the answer. This problem gives a number of milligrams and asks for the number of grams. We must *convert* from milligrams to grams. To do this, set up an equation using a conversion factor with the desired unit in the numerator. The form of this equation is as follows:

$$\text{Given unit} \times \left(\frac{\text{desired unit}}{\text{given unit}}\right) \longleftarrow \text{Units of answer in numerator}$$

This is the conversion factor.

For our example, the equation is as follows:

$$100\ mg \times \left(\frac{1\ g}{1000\ mg}\right) =$$

STEP 4: Solve the problem. When the calculation is carried out, the given unit in the denominator of the conversion factor will cancel the given unit from the problem.

$$100\ \cancel{mg} \times \left(\frac{1\ g}{1000\ \cancel{mg}}\right) = 0.1\ g$$

sample problem 1.6 — Unit Conversions

Using the appropriate conversion factor, convert each of the following quantities to the indicated unit:

a.   10,000 cg = _____ g

b.   0.005 L = _____ mL

Solution

a.   **STEP 1: Determine the units on your final answer.** In this case, the units will be grams.

STEP 2: Establish the given information. In this case, you have a value of 10,000 cg given. You also have an equivalency relationship (Table 1.1) showing that 1 g = 100 cg.

$$\frac{100\ cg}{1\ g} \quad or \quad \frac{1\ g}{100\ cg}$$

STEPS 3 & 4: Decide how to set up the problem and then solve the problem. The answer must be in grams, so grams must be in the numerator of our conversion factor. The equation is set up and solved like this:

$$10,000\ \cancel{cg} \times \left(\frac{1\ g}{100\ \cancel{cg}}\right) = 100\ g$$

b.   **STEP 1: Determine the units on your final answer.** In this case, the units will be mL.

STEP 2: Establish the given information. In this case, you have a value of 0.005 L given. You also have an equivalency relationship (Table 1.1) showing that 1 L = 1000 mL.

STEPS 3 & 4: Decide how to set up the problem and then solve the problem. The answer must be in mL, so it must appear in the numerator of the conversion factor. The equation is set up and solved like this:

$$0.005\ \cancel{L} \times \left(\frac{1000\ mL}{1\ \cancel{L}}\right) = 5\ mL$$

Significant Figures

Before proceeding further with any calculations, it is important that we understand that measuring matter relies on the precision of the instruments that we use to measure it. For example, a bathroom scale only measures to the closest half-pound, so we wouldn't consider recording a person's weight as 150.2356 pounds! It is important that we report calculated answers reasonably, or as scientists refer, to the correct number of significant figures.

Look at the man in the picture. How old would you say he is?

Would you agree that he looks to be about 50 years old? Does that mean that you know he is *exactly* 50? Unless you know him personally and also know that this picture was taken on his fiftieth birthday, you cannot know that. So, you've guessed that he is about 50 years old.

In estimating this person's age, you encounter the concept of uncertainty in a measurement. All measurements have some level of uncertainty. We can understand the level of uncertainty in any measurement by the number of significant figures used in the number. In any measurement, the **significant figures** are the digits known with certainty plus the estimated digit. When we estimate that the man in the picture is 50 years old, we feel confident that his age is somewhere between 40 and 60 years. This means that our determination has one significant figure. In other words, we are confident that he is neither younger than 40 nor older than 60, but we cannot say for sure that he is, for example, 48 or 54.

We can only estimate the first digit of his age. The zero serves to hold the "ones" place in the number to give the correct magnitude. We do not know the actual value of either number with absolute certainty.

How certain are you of this man's age?

If you were told that the man in the picture was born in 1952 and that this photo was taken in 2004, you could say with some certainty that he is 52 years old. Because you do not know the month of his birth or the month that the picture was taken, you know that he is older than 51, but not yet 53 years old. You know with absolute certainty the first digit of his age, and the second digit is an estimate. Thus, you now know his age to two significant figures.

In any type of measurement (age, mass, volume, length, and so forth), there will be some uncertainty. Working with significant figures allows us to represent and convey to others the uncertainty in a given measurement. In any measurement, all nonzero numbers are considered significant. Zeros tend to be the "troublemakers" because their significance depends on their position in a number. For example, when an object's weight is reported as 20 g, the number has only one significant figure, so its weight is between 10 g and 30 g. However, if an object's weight is reported as 21.0 g, the last zero is significant and tells us that the object's weight is between 20.9 g and 21.1 g.

If a zero in a number without a decimal point is significant, its significance can still be shown. To indicate that the zero in a measurement of 10 inches is significant, simply put in a decimal point: 10. inches means that both digits are significant. Otherwise, to indicate significant zeros in large numbers, draw a line above the significant zero: $10,0\overline{0}0$ has 4 significant figures.

It should be noted that numbers like those used in conversion factors and when counting items are exact numbers with an infinite number of significant figures. Significant figures should not be considered for these numbers in calculations. Only measured numbers determine the number of significant figures given in a final answer.

The rules for counting significant figures in measurements are summarized in **Table 1.2**.

TABLE 1.2 Counting Significant Figures in Measurements

Rule	Measurement	Number of Significant Figures
1. A digit is significant if it is		
a. not a zero	41 g	2
	15.3 m	3
b. a zero between nonzero digits	101 L	3
	6.071 kg	4
c. a zero at the end of a number with a decimal point	20. g	2
	9.800 °C	4
2. A zero is not significant if it is		
a. at the beginning of a number with a decimal point	0.03 L	1
	0.00024 g	2
b. in a large number without a decimal point	12,000 km	2
	3,450,000 m	3

sample problem 1.7   **Determining Significant Figures**

Give the number of significant figures in each of the following measurements:

a. 380 min   b. 10,800 m   c. 9.317 kg   d. 0.0580 L

Solution

a.   Two significant figures—see rule 2b in Table 1.2.
b.   Three significant figures—the zero between the 1 and 8 is significant, but the other two aren't. See rules 1b and 2b in Table 1.2.
c.   Four significant figures—see rule 1a.
d.   Three significant figures—the last zero is significant, but the first two are not. See rules 1c and 2b in Table 1.2.

Calculating Numbers and Rounding

When measurements are used in calculations, each measurement has some uncertainty. Adding, subtracting, multiplying, or dividing numbers that are uncertain can result in numbers that seem more certain than they actually are. Look again at the man in the photo. Before knowing his birth year, we estimated that he was 50 years old. To calculate how many days he has been alive, we would multiply his age in years by 365.24 days/year. Multiplying these numbers in your calculator gives an answer of 18,262 days.

Consider the answer from the calculator. We know his age within ±10 years, but now by multiplying we now know it within one-half of a day! Somehow that doesn't seem logical. Manipulating measurements (which have some built-in uncertainty) with arithmetic cannot *increase* their certainty. A calculator does not know the certainty in your measurement and will often make an attempt at more certainty than makes sense, so the person doing the calculation must be prepared to round the answer correctly.

How many of the resulting numbers from our previous calculation of the man's age are certain and, therefore, meaningful? Keeping all of the numbers results in an answer with six significant figures, yet we started with a number (50) that has only one.

In general, your answer can be no more certain than the least certain number that you started with. The rules for determining the number of significant figures in a calculated answer depend on the arithmetic operation being used.

Rules for Significant Figures in Calculations
- **Addition and Subtraction.** Answers should be given to the least number of decimal places in the measured numbers.

$$
\begin{array}{ll}
1.002 & \longleftarrow \text{ three decimal places} \\
2.5 & \longleftarrow \text{ one decimal place} \\
\underline{10.16} & \longleftarrow \text{ two decimal places} \\
13.662 & \text{report as } 13.7
\end{array}
$$

Because one of the numbers contains only one decimal place, the final answer must be given to one decimal place.

- **Multiplication and Division.** Answers should be given to the least number of significant digits in the measured numbers. For our age-in-days calculation, the answer can have only one significant digit because we estimated the man's age at 50.

$$50 \text{ years old } \times \frac{365.24 \text{ days}}{1 \text{ year}} = 18{,}262 \text{ days old report as } 20{,}000 \text{ days old}$$

Rules for Rounding Numbers
- If the leftmost digit to be dropped (the one following your last retained significant digit) is 4 or less, simply remove it and the remaining digits.
- If the leftmost digit to be dropped is 5 or greater, increase the last retained digit by 1 and remove all other digits.
- If rounding a large number with no decimal point, zeros are substituted for numbers that are not significant.
- When conducting multiple-step calculations, do not round answers until the end of the calculation. *Rounding at each step introduces rounding errors and produces incorrect answers.*

sample problem 1.8 — Rounding Calculated Numbers

Complete each of the following calculations and give the answer to the correct number of significant figures:

a. $75 - 21.3 =$

b. $8.24 \times 3.0 =$

c. $0.005 + 1.23 =$

d. $12.4 \div 0.06 =$

Solution

a. Answer is 54. The calculator result of 53.7 is rounded so that the answer has no decimal places because 75 has no numbers after the decimal point.

b. Answer is 25. The calculator result of 24.72 is rounded to two significant figures because 3.0 has only two significant figures.

c. Answer is 1.24. The calculator result of 1.235 is rounded to the least number of decimal places, which is two.

d. Answer is 200. The calculator result of 206.66666 is rounded to one significant digit as 200, which is the number of significant digits in 0.06.

Scientific Notation

In chemistry, we sometimes use extremely large and extremely small numbers in the same calculation. Suppose that we wanted to figure out how many sheets of paper we could stack in a cabinet that is 3 m high. Each sheet of paper has a thickness of one-tenth of a millimeter or 0.0001 m. The number of sheets of paper would be

$$\frac{1 \text{ sheet of paper}}{0.0001 \text{ m}} \times \frac{3 \text{ m}}{1 \text{ cabinet}} = \frac{30,000 \text{ sheets of paper}}{\text{cabinet}}$$

1 piece of paper = 0.0001m

This is a lot of pieces of paper! We can represent this with this large number. Yet it is more easily written using scientific notation. To explore scientific notation, consider that 30,000 pieces of paper is $3 \times 10 \times 10 \times 10 \times 10$ (try it on your calculator). In scientific notation, we can express the four times we multiplied by 10 as an exponent, showing it as 10^4. The scientific notation for 30,000 would be written as 3×10^4.

A number smaller than 1, like the 0.0001 meter for each sheet of paper can be determined by dividing 1 by $10 \times 10 \times 10 \times 10$ or $1/10^4$. Dividing by a number is equivalent to multiplying by its reciprocal. So, in exponential form, $1/10^4$ can be written as 1×10^{-4}. The general form for scientific notation is

$$C \times 10^n$$

where C is called the **coefficient** and is a number between 1 and 9 and n is the exponent telling us the number of tens places that apply.

A positive exponent tells us that the actual number is greater than 1, and a negative exponent tells us the number is between 0 and 1. In scientific notation, only significant figures are shown in the coefficient (any place-holding zeros are not part of the notation). For example, the scientific notation of the number 0.00650 would be 6.50×10^{-3} because the zeros between the decimal point and the 6 are placeholders and the zero after the 5 is significant.

We could rewrite an equivalent calculation in scientific notation as

$$\frac{1 \text{ sheet of paper}}{1 \times 10^{-4} \text{ m}} \times \frac{3 \text{ m}}{1 \text{ cabinet}} = \frac{3 \times 10^4 \text{ sheets of paper}}{\text{cabinet}}$$

Table 1.3 shows how numbers and scientific notation are related for a wide range of numbers.

Scientific notation can be displayed on your calculator as long as you know which buttons to press. Most scientific calculators have a button labeled either EE , EXP , or SCI , which displays the coefficient (C) and the power of ten exponent (n). It is worth exploring your calculator to be sure that you can input scientific notation correctly. To input 3×10^4, try typing 3 EE or EXP 4. To input 1×10^{-4}, try typing 1 EE or EXP +/- 4. With a scientific calculator (which requires a reverse input), type 3 EE or EXP 4 +/- .

TABLE 1.3 Relating Numbers to Scientific Notation

Number	Meaning	Scientific Notation
1,000,000	$10 \times 10 \times 10 \times 10 \times 10 \times 10$	1×10^6
100,000	$10 \times 10 \times 10 \times 10 \times 10$	1×10^5
10,000	$10 \times 10 \times 10 \times 10$	1×10^4
1000	$10 \times 10 \times 10$	1×10^3
100	10×10	1×10^2
10	10	1×10^1
1	1	1×10^0
0.1	$1/10$	1×10^{-1}
0.01	$1/(10 \times 10)$	1×10^{-2}
0.001	$1/(10 \times 10 \times 10)$	1×10^{-3}
0.0001	$1/(10 \times 10 \times 10 \times 10)$	1×10^{-4}
0.00001	$1/(10 \times 10 \times 10 \times 10 \times 10)$	1×10^{-5}
0.000001	$1/(10 \times 10 \times 10 \times 10 \times 10 \times 10)$	1×10^{-6}

sample problem 1.9 **Expressing Numbers in Scientific Notation**

Express the following numbers in scientific notation:

a. 820,000   b. 0.00096

Solution

a. To determine the correct scientific notation for a number greater than 1, count the number of places to the right of the first number, 8. In this case, there are 5 places to the right of 8. This becomes the exponent on the number 10 in scientific notation. Any other nonzero numbers go after a decimal point. The answer is 8.2×10^5.

b. To determine the correct scientific notation for a number less than 1, count the number of places to the left of the first number, 9, and add 1. In this case, there are 3 places to the left of the 9 and adding 1 would give a total of 4. The negative of this number becomes the exponent on the number 10 in scientific notation. Any other nonzero numbers go after the decimal point. The answer is 9.6×10^{-4}.

sample problem 1.10 **Changing Numbers from Scientific Notation to Decimal Format**

Express the following numbers in decimal format:

a. 6.5×10^4 b. 8.34×10^{-6}

Solution

a. The number in the exponent tells you how many places are beyond the decimal place in the scientific notation when converting from scientific notation to decimal format. Notice that the number of significant figures does not change when the format of the number changes. In this problem, the number 5 will occupy one of the places, so 3 more zeros after the 5 are necessary. The answer is 65,000.

b. Negative exponents indicate that this number is smaller than 1, but greater than zero. The number in the exponent tells you how many places should appear to the left of the decimal place when converting from scientific to decimal format. The number 8 will occupy one of these places, so 5 more zeros should appear in the final answer to the right of the decimal place. All digits to the right of the decimal place in the scientific notation should appear in the final answer. The answer is 0.00000834.

Percent

If your chemistry class has 40 students enrolled and 20 students get an A on the first test, you might comment that one-half, or 50%, of the class got an A. Percent, represented by the symbol %, means the part out of 100 total, or hundredths. So, 50% means 50 out of a total of 100 or 50 hundredths (0.50), which reduces to ½, the same fraction as our original example using grades. Percent allows us to directly compare two sets of numbers that have different total sizes.

$$\text{Percent (\%)} = \frac{\text{part}}{\text{whole}} \times 100$$

A fraction can be converted to a percent by dividing the numerator by the denominator, multiplying by 100, and adding a percent sign. A decimal number can be converted to a percent by multiplying by 100 and adding a percent sign.

sample problem 1.11   Expressing Fractions and Decimals as a Percent

a. Express 1/5 as a percent.

b. Express 0.35 as a percent.

Solution

a. For a fraction, divide the numbers of the fraction, multiply by 100, and add a percent sign.

$$1/5 = 0.20 \times 100 = 20\%$$

b. For a decimal, multiply by 100 and add the percent sign.

$$0.35 \times 100 = 35\%$$

sample problem 1.12   Expressing Percents as Fractions and Decimals

a. Express 25% as a decimal.

b. Express 25% as a fraction.

Solution

a. To express 25% as a decimal, divide the value by 100.

$$\frac{25}{100} = 0.25$$

b. Percent means parts out of a total of 100 parts, so the fraction is

$$\frac{25}{100} \text{ which reduces to } \frac{1}{4}$$

Solving a Problem
Calculating Percent

Suppose that a couple is trying to lose weight. The husband is told by his doctor to lose 40 pounds. The wife is told by her doctor to reduce her weight by 12%. The husband weighs 315 pounds and the wife weighs 285 pounds. Which one actually has to lose more weight in pounds to meet the goal?

STEP 1: Determine the units on your final answer. In this case, the final answer is in pounds.

STEP 2: Establish the given information. The husband's weight is given in pounds, so we know how much he has to lose. The wife's total weight (285 lb) and percent she is to lose is given (12%).

STEP 3: Decide how to set up the problem. To determine the number of pounds the wife must lose from 12%, the percent is expressed as a decimal and multiplied by her total weight.

$$12\% = 0.12$$
$$0.12 \times 285 \text{ pounds total weight} = 34 \text{ pounds to lose}$$

Lose 40 lb   Lose 12%

STEP 4: Solve the problem. The husband has to lose more weight (40 pounds) than the wife does (34 pounds).

---

sample problem 1.13   Determining the Percent of a Number

a.   What is 40.0% of 275?   b.   What is 80.0% of 65?

Solution

To determine the percent of a number, change the percent to a decimal and multiply by the number.

a. $40.0\% = 0.400$
$0.400 \times 275 = 110.$

b. $80.0\% = 0.800$
$0.800 \times 65 = 52$

---

sample problem 1.14   Determining the Percent That One Number Is of Another Number

a.   What percent of 1500 is 30?   b.   Thirty-five is what percent of 140?

Solution

To determine what percent one number is of a second number, a fraction is created to represent parts out of the whole. This gives a decimal answer that can then be converted to a percent by multiplying by 100.

a. $\dfrac{30}{1500} = 0.02$
$0.02 \times 100 = 2\%$

b. $\dfrac{35}{140} = 0.25$
$0.25 \times 100 = 25\%$

1.11 Give the number of significant figures in each of the following measurements:

   a. 25 minutes   b. 44.30 °C

   c. 0.037010 m   d. 800. L

1.12 Give the number of significant figures in each of the following measurements:

   a. 0.000068 g   b. 100 °F

   c. 9,237,200 years   d. 25.00 m

1.13 Complete each of the following calculations and give your answer to the correct number of significant figures:

   a. $100 \times 23 =$

   b. $97.5 - 43.02 =$

   c. $8064/0.0360 =$

   d. $54.00 + 78 =$

1.14 Complete each of the following calculations and give your answer to the correct number of significant figures:

   a. $340/96 =$

   b. $0.305 + 43.0 =$

   c. $0.0065 \times 11 =$

   d. $19.029 - 0.00801 =$

1.15 A typical aspirin tablet contains 325 mg of the active ingredient acetylsalicylic acid. Convert to grams.

1.16 On average, an adult's lung volume is 5 L. Convert to mL.

1.17 Write scientific notation for the following numbers:

   a. 203,000,000   b. 12.4   c. 0.0000000278

1.18 Write scientific notation for the following numbers:

   a. 1,400,000   b. 0.079   c. 0.00000354

1.19 Express the following numbers in a decimal format:

   a. 1.56×10^{-5} b. 2.8×10^{5} c. 9.0×10^{-2}

1.20 Express the following numbers in a decimal format:

   a. 7.4×10^{-2} b. 3.75×10^{3} c. 1.19×10^{-8}

1.21 Express the following numbers as percents:

   a. ¼   b. 3/8   c. 66/100

1.22 Express the following numbers as percents:

   a. 0.58   b. 0.36   c. 0.125

1.23 Express the following numbers in decimal form:

   a. 4.5%   b. 13.0%   c. 66%   d. 78%

1.24 Express the following numbers as a fraction:

   a. 20%   b. 75%   c. 40%   d. 12%

1.25 Determine numbers from the percent given:

   a. What is 25% of 80?

   b. What is 0.9% of 1000?

   c. What is 5.0% of 750?

   d. What is 66.6% of 200.?

1.26 Determine the percent from two numbers given here.

   a. Fifty is what percent of 125?

   b. Six is what percent of 600?

   c. What percent of 300 is 15?

   d. What percent of 400 is 30?

1.4 Matter: The "Stuff" of Chemistry

? 1.4 Inquiry Question:
What properties of matter can be measured?

Chemists study matter that we cannot see, even with a high-powered microscope. Even though we cannot see it, we can see and measure its properties in bulk. Here we discuss some of those properties and characteristics of matter.

Mass

Measuring mass at home

Consider that anything that takes up space can also be placed on a scale and weighed—that is, it has mass. So, matter can be more completely defined as anything that takes up space and has mass. **Mass is a measure of the amount of material in an object.** As we saw in Section 1.3, a common unit used to measure the mass of a substance is the gram (g).

A raisin or a paperclip has a mass of about 1 gram. The term *weight* may be more familiar to you than the term *mass*. These terms are related, but they do not mean the same thing. The weight of an object is determined by the pull of gravity on the object, and that force changes depending on location. For instance, astronauts weigh less on the moon than on Earth. This is because the gravitational pull of the moon is less than that of Earth. The astronauts' mass did not change, but their weight did.

So, how are mass and weight related to each other? When you step on the bathroom scale, are you determining your mass or your weight?

A scale or balance is used to measure mass. When balances are manufactured, they are set at the factory so that a 1-kilogram (1000-g) object has the same mass on all of the balances. This is known as "calibrating" the balance and accounts for the gravitational pull of Earth. After calibrating, the mass and the weight of the 1-kg object is the same. If the object and balance were transported to the moon, the balance would have to be adjusted to account for the lesser gravitational pull of the moon.

As long as an object is weighed in roughly the same location on Earth's surface, its mass and weight will have the same measured value. To avoid confusion, this book will always refer to weighable quantities using the term *mass*.

Measuring mass in the lab

Volume

Another way of saying that matter "takes up space" is to say that it has **volume.** Large soft drink bottles are often called "2-liter" bottles. The volume of the soft drink in the bottle is 2 liters (1 liter = 1.057 quart). Along the same lines, as you blow up a balloon, it gets bigger. The volume of air in the balloon increases. So, volume is a three-dimensional measure of the amount of space occupied by matter.

When measuring volumes at home, you would use a measuring cup or a measuring spoon. In the lab, volumes are routinely measured with a graduated cylinder or a pipet. The unit typically used in the lab is the milliliter (mL). In a clinical setting, volumes are often measured with calibrated syringes. The typical unit in the clinical setting is the cubic centimeter (cc or cm³). One milliliter equals one cubic centimeter (see **Figure 1.3**). A teaspoon of cough syrup contains 5 mL or 5 cc.

Measuring
spoons

Measuring
cup

Measuring volume
at home

Graduated
cylinder   Syringe

Measuring volume
in the lab

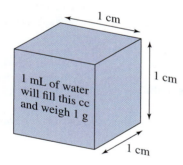

1 cm

1 cm

1 cm

1 mL of water
will fill this cc
and weigh 1 g

FIGURE 1.3 **The relationship between volume and length.**
One cubic centimeter (abbreviated cc or cm³) equals 1 milliliter of volume. For water, this volume weighs 1 gram.

Density

Regardless of its mass or volume, most types of wood will float on water while a piece of metal will sink. This observation implies that the mass of wood that fits into a certain space is less than that for metal. This property of matter is called density. **Density** (*d*) is a comparison (also called a ratio) of a substance's mass (*m*) to its volume (*V*). This can be displayed mathematically as

$$d = \frac{m}{V}$$

One gram of water has a volume of one milliliter, so the density of water is 1.00 g/mL. A piece of wood will float on water because it is less dense than water, while a piece of metal will sink because it is more dense than water (see **Figure 1.4**).

FIGURE 1.4 **Visualizing density.**
Objects that are less dense than water will float while objects more dense than water will sink.

We can think of density as a measure of the packing of matter. For the most part, the size of our bones doesn't change, but the density can. As we age, minerals are lost from our bones faster than they are replenished. The bones' strength and integrity can become compromised. In this case, the mass of the bone lessens as the volume stays the same, making bones less dense. This can lead to the condition known as *osteoporosis*.

---

sample problem 1.15 — Calculating Density

A brick stamped "14-karat gold" has a mass of 1100 g and a volume of 85 mL. Calculate its density. Will the brick float or sink when placed in water?

Solution

You are given the mass and volume and asked to calculate the density in g/mL. Insert the given values into the density equation.

$$d = \frac{m}{V}$$

$$d = \frac{1100\ \text{g}}{85\ \text{mL}} = 13\ \text{g/mL}$$

The final answer is reported to two significant figures because the minimum number of significant figures measured in the problem is two. The gold brick is more dense than water, so it will sink.

Because the density of a substance does not change, we can use density values as conversion factors to determine either the mass or the volume of a substance.

---

sample problem 1.16 — Problem Solving with Density

Table sugar has a density of 1.29 g/mL. Calculate the mass of one teaspoon of sugar. Recall that 1 teaspoon has a volume of 5 mL.

Solution

These problems are worked similarly to the unit conversion problems solved in the previous section.

STEP 1: Determine the units on your final answer. In this case, the units will be the mass in grams.

STEP 2: Establish the given information. In this case, the density is given and you will solve the density equation for mass. You are also reminded of the volume equivalency that 1 tsp = 5 mL.

STEP 3: Decide how to set up the problem. You are given the density and the volume. To solve for mass, the density equation must be rearranged.

$$d = \frac{m}{V}$$

Multiply both sides by V, as shown:

$$d \times V = \frac{m}{V} \times V$$

$$m = d \times V$$

STEP 4: Solve the problem. The values are plugged in and mL units cancel out.

$$m = \frac{1.29\ \text{g}}{1\ \cancel{\text{mL}}} \times 5\ \cancel{\text{mL}} = 6.45\ \text{g}$$

Because the density contains three significant figures, the answer must be given to three significant figures. Note that 5 mL is part of a conversion factor and is not considered when determining significant figures (refer to Table 1.2).

Specific Gravity

Which is more dense, water or honey? Both contain water, but if you pour honey into water (or tea, which is mostly water), it drops to the bottom. From this experiment, we can observe that honey must be more dense than liquid water. Honey, although a liquid, possesses the property of density because it has mass and takes up space.

For liquids, density often is measured with respect to water. The density of water is 1.00 g/mL at 4 °C and close to that value at body temperature. The ratio of the density of a sample to the density of water is called **specific gravity** (sp gr).

$$\text{Specific gravity} = \frac{\text{density of sample}}{\text{density of water}}$$

To calculate the specific gravity of a liquid, the density of the liquid and the density of water must have the same units. Specific gravity is a unitless quantity because it is a ratio of two densities.

The specific gravity of a liquid can be measured with a simple instrument called a *hydrometer* (see **Figure 1.5**). This instrument is commonly used in the wine- and beer-making industry. Yeast produces alcohol from sugar during the fermentation process. As beer or wine ferments, samples are tested with a hydrometer to determine when the fermentation process is completed.

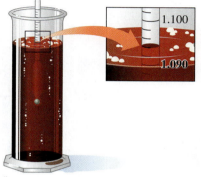

FIGURE 1.5 Red wine that measures a specific gravity of 1.090 has completed its fermentation process.

---

sample problem 1.17   Calculating Density and Mass from Specific Gravity

A 25 mL urine sample has a specific gravity of 1.025. What is its mass?

Solution

STEP 1: Determine the units on your final answer. In this case, the units will be the mass in grams.

STEP 2: Establish the given information. In this case, the specific gravity and the volume of urine are given. You must (a) solve for the density of the urine, and then (b) use the density equation to solve for the mass.

STEP 3 & 4: Decide how to set up the problem and solve.

(a) Density of urine:

$$\text{Specific gravity} = \frac{\text{density of urine}}{\text{density of water}}$$

$$\text{Specific gravity} \times \text{density of water} = \text{density of urine}$$

$$1.025 \times 1.00\ \text{g/mL} = 1.025\ \text{g/mL}$$

(b)   Solve for mass:

$$d = \frac{m}{V}$$

$$d \times V = m$$

$$1.025 \text{ g/mL} \times 25 \text{ mL} = 26 \text{ g}$$

The fewest number of significant figures measured in the problem is two, so the final answer is reported to two significant figures.

Temperature

We measure the **temperature** of a substance to determine its hotness or coldness. This is often done using a thermometer or an electronic temperature probe. In different parts of the world, temperature is measured using different scales. In the United States, we continue to use the Fahrenheit scale, while the rest of the world uses the Celsius scale. However, scientists use still another scale called the absolute, or Kelvin, scale where the temperature unit is the kelvin. Kelvin is the SI unit for temperature.

The most straightforward way to compare the temperature scales is to compare temperatures that we are most familiar with and observe their values on each scale. Consider the thermometers in **Figure 1.6**.

Pay particular attention to the freezing point and boiling point of water on the three scales. Notice that the Celsius and Kelvin scales have degrees of the same size, but the two scales are offset by 273 degrees. Mathematically, this relationship is expressed as follows:

$$\text{Kelvin (K)} = \text{Celsius (°C)} + 273 \text{ °C}$$

The relationship between the Fahrenheit scale and the others is not so straightforward. A unit of 1 degree (1°) on the Celsius scale is equal to 1.8 degrees (1.8°) on the Fahrenheit

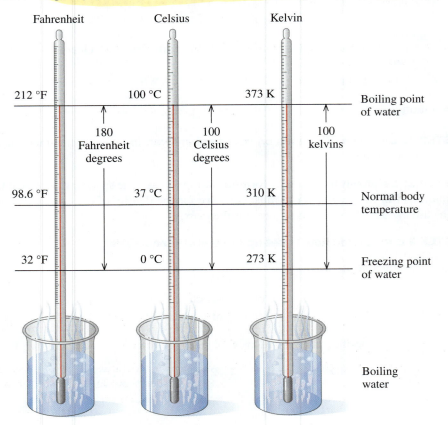

FIGURE 1.6 Comparison of temperature scales. The zero point (freezing of water) is different for each of the scales. Comparing the boiling point to the freezing point of water on each scale, we see that the "size" of a degree is the same on the Celsius and Kelvin scales, but smaller on the Fahrenheit scale.

scale, or 5 °C = 9 °F. This relation provides a conversion factor. Also, in Figure 1.6, the "zero points" are offset by 32 degrees. The relationship between °F and °C is shown:

$$°C = (°F - 32 °F) \times \frac{1 °C}{1.8 °F}$$

$$°F = \left(\frac{1.8 °F}{1 °C} \times °C \right) + 32 °F$$

After converting a Fahrenheit temperature to degrees Celsius, you can convert the Celsius temperature to kelvins using the Celsius-to-Kelvin scale equation. A separate conversion equation to from Fahrenheit to kelvins is not needed.

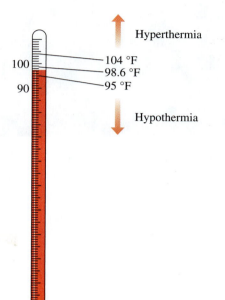

Body Temperature

Normal body temperature is considered to be 98.6 °F or 37.0 °C. However, body temperature varies from person to person and changes throughout the day. Body temperature also changes based on level of exertion. Our bodies expend a lot of effort maintaining constant body temperature within a narrow range. If human body temperature rises over 40.0 °C (104 °F), a condition known as *hyperthermia* exists. This condition can be caused by heatstroke where sweat production ceases. At body temperatures this high, a person can go into convulsions or coma and experience permanent brain damage. One immediate remedy for this situation is to immerse the person in ice-cold water.

If body temperature drops below 35 °C (95 °F), hypothermia is present. A person in this condition feels cold, has an irregular heartbeat, and a slow breathing rate. Because these bodily functions are operating slower than normal, the person may lapse into unconsciousness. Treatments may include increasing blood volume, administering oxygen, or, in extreme cases, injection of warm (37 °C) fluid.

---

sample problem 1.18   Temperature Conversion

On a hot summer day in Georgia, the thermometer often rises as high as 102 °F. What is this temperature in °C and kelvins?

Solution

Use the Fahrenheit-to-Celsius equation and substitute the appropriate number:

$$°C = (102 °F - 32 °F) \times \frac{1 °C}{1.8 °F}$$

$$°C = (70 °F) \times \frac{1 °C}{1.8 °F}$$

$$°C = 39 °C$$

Now that we have the temperature in °C, we can convert it to the Kelvin scale using Equation 1.2:

$$K = 39 °C + 273$$
$$K = 312$$

Energy

After you eat, your stomach breaks up food and absorbs the nutrients. We often say that food is converted to energy by the body. We use the energy produced by food to move muscles or do other work in the body. What does this actually mean?

FIGURE 1.7 Potential and kinetic energy. The runner contains potential energy in her muscles waiting for the start of the race. Her body movements during the race transform the potential energy into kinetic energy as she runs.

Energy is defined as the ability to do work. Food contains chemical substances that can transform chemical energy into work. This stored chemical energy is called **potential energy.** When you flex a muscle you are using the chemical energy to produce motion. The energy of motion is called **kinetic energy** (see **Figure 1.7**). Energy takes various forms, but it is never created or destroyed. This concept is called the law of **conservation of energy.**

The SI unit for energy is the **joule** (J) and is used to measure energy in chemistry. You may be more familiar with the energy unit called the **calorie** (cal). The calorie was originally defined as the amount of energy required to raise the temperature of one gram of water one degree Celsius. A calorie can be defined in terms of a joule.

$$1 \text{ cal} = 4.184 \text{ J}$$

We can convert between the two units by applying a conversion factor, as follows:

$$\frac{1 \text{ calorie}}{4.184 \text{ joules}} \quad \text{or} \quad \frac{4.184 \text{ joules}}{1 \text{ calorie}}$$

A nutritional Calorie as found on nutrition labels is termed a **Calorie** (Cal) and is 1000 times larger than a calorie. The nutritional Calorie uses a capital C in its unit, and the calorie uses a lowercase c.

$$1 \text{ Calorie} = 1000 \text{ calories}$$

The two units can be converted using a conversion factor, as follows:

$$\frac{1000 \text{ cal}}{1 \text{ Cal}} \quad \text{or} \quad \frac{1 \text{ Cal}}{1000 \text{ cal}}$$

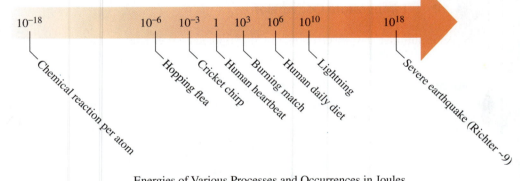

Energies of Various Processes and Occurrences in Joules

sample problem 1.19   Converting Energy Units

The nutritional label on a granola bar says it contains 220 Calories.

a.   Calculate the number of calories (lowercase) in the granola bar.
b.   Calculate the number of joules in the granola bar.

Solution

a.   There are 1000 calories in 1 nutritional Calorie.

$$220 \text{ Calories} \times \frac{1000 \text{ calories}}{1 \text{ Calorie}} = 220,000 \text{ calories}$$

b. One calorie is equivalent to 4.184 joules. Converting from Calories to joules requires two conversion factors, as follows:

$$220 \text{ Calories} \times \frac{1000 \text{ calories}}{1 \text{ Calorie}} \times \frac{4.184 \text{ joules}}{1 \text{ calorie}} = 920{,}000 \text{ joules}$$

The final answer is reported to two significant figures because the measured value in the problem, 220 Calories, contains two significant figures.

Heat and Specific Heat

When you get into a parked car on a sunny day, the car feels warm. If you get into the same car on a cold winter's day, the car is cold. In a warm car, because you are not as hot as the car, the excess warmth of the car is being transferred to your body. In a cold car, your body is warmer than the car and the warmth from your body is transferred to the car, so you perceive the car as cold. This example describes the phenomenon we know as heat. **Heat** is kinetic energy flowing from a warmer body to a colder one. Note that heat is a form of energy.

When you cook food or store the leftovers in the refrigerator, heat is being transferred into the food during cooking and out of the food in the cool air of the fridge. Every substance has this ability to absorb or lose heat as the temperature changes. Each substance, though, is different and these values can be measured. The **specific heat capacity** or specific heat of a substance is the amount of heat needed to raise the temperature of exactly 1 gram of a substance by exactly 1 °C. To represent a change in temperature, we include the Greek capital delta, Δ, with the symbol for temperature, T, as ΔT. The specific heat of a substance is represented as

$$\text{Specific Heat } (SH) = \frac{\text{heat}}{\text{grams} \times \Delta T}$$

Table 1.4 shows the specific heat values for several different substances. Notice that metals have low specific heat values and water has a very high specific heat. We enjoy this property of water on a hot summer day when we cool off in a swimming pool. Because it takes a lot of heat to raise the temperature of water, the water does not warm up as fast as the surrounding air and refreshes us when we are hot.

In warm weather, water stays cooler than air due to its high specific heat.

TABLE 1.4 Specific Heats of Various Substances

Substance	Specific Heat (cal/g °C)
Water (liquid)	1.00
Human body	0.83
Paraffin wax	0.60
Wood, soft	0.34
Wood, hard	0.29
Air	0.24
Aluminum	0.21
Table salt	0.21
Brick	0.20
Stainless steel	0.12
Iron	0.11
Copper	0.092
Silver	0.056
Gold	0.031

sample problem

1.20 Specific Heat

Based on Table 1.4, would a silver or stainless steel spoon stay cooler if you wanted to stir sugar into a hot cup of coffee? Support your answer.

Solution

The stainless steel spoon stays cooler. Stainless steel has a higher heat capacity than silver, so it takes more heat energy to raise its temperature. It will warm more slowly, giving you time to stir your coffee.

States of Matter

Now that we have seen some of the properties of matter, we can begin to understand its behavior. Next, we examine its composition. At the most basic level, matter is a collection or assembly of particles too small to be seen. In Chapters 2 and 3, we will explore these particles in more depth. What we can see or touch in our everyday lives is the result of many of these individual particles gathered together. The look, feel, and behavior of matter that we can see and touch depend on the composition and behavior of these particles.

Solid

The particles in a solid are tightly packed and barely moving.

Liquid

The particles in a liquid are less orderly than those in a solid and are freely moving.

Gas

The particles in a gas are disordered and rapidly moving.

FIGURE 1.8 Particles of a solid, liquid, and gas.

You can see and hold a brick or a glass of water, and you are aware that each has mass and volume. Based on our balloon example earlier in the chapter, air and other gases have volume (they have mass as well) and are therefore matter, too.

The brick, water, and air represent different forms of matter that are physically different. The brick has a definite, unchanging shape and volume, while water has a definite volume, even as its shape changes, depending on its container. Because you cannot "see" air, you can only witness the changes that it causes. For example, it fills up and takes the shape of a container, like a balloon, into which it is placed. Each of these examples represents a different state of matter. A **state of matter** is the physical form in which the matter exists. The three most common states of matter are solid (the brick), liquid (water), and gas (air).

Let's take a look at the submicroscopic particles that make up substances like the brick, the water, and the air, as shown in **Figure 1.8**.

The particles in a solid—for example, a brick—are in an orderly arrangement and are tightly packed together. They are moving, but only very slightly, and not enough to affect the shape of the solid. Therefore, **solids** have a definite shape and a definite volume.

The particles in a liquid—for example, water—are somewhat less orderly and are only loosely associated with each other. They are moving freely, colliding with and sliding over each other. These behaviors of its particles mean that a **liquid** has a definite volume, but it takes the shape of its container.

The particles in a gas—for example, air—have no orderly arrangement and are far apart from each other. They move at high rates of speed and often collide with each other and with the walls of their container. Therefore, a **gas** has no definite shape or volume. Instead, it expands to fill the container in which it is placed.

An object's state of matter can be determined by the motion and association of the particles in a substance. A substance's properties, such as shape, volume, and kinetic energy, are greatly influenced by particle motion. Some of the basic properties of matter are summarized in **Table 1.5**.

TABLE 1.5 Properties of Solids, Liquids, and Gases

Property of a Substance	Solid	Liquid	Gas
Shape	Definite shape	Adopts shape of container	Adopts shape of container
Volume	Definite volume	Definite volume	Fills volume of container
Kinetic energy	Lowest of the three states	More than solid, less than gas	Highest of the three states
Positioning of particles	Closely packed and fixed	Loosely packed, but random	Far apart and random
Attraction between particles	Very strong	Strong	Practically none

sample problem
1.21   **States of Matter**

Based on the properties of matter in Table 1.5, in which state of matter are the particles moving the most?

Solution

Energy of motion is kinetic energy. Because the kinetic energy present in gases is the highest of the three states of matter, the particles in a gas are moving the most. In fact, they are in constant motion.

1.27   Based on your experience, are each of the following more or less dense than liquid water?

   a. vegetable oil

   b. piece of gold jewelry

   c. cork

1.28   Based on your experience, are each of the following more or less dense than liquid water?

   a. paper clip

   b. table sugar

   c. ice

1.29   Calculate the grams of sugar present in a 355 mL can of soda if the water in the can weighs 355 g and the density of the soda is 1.11 g/mL.

1.30   Calculate the density in g/mL of 2.0 L of gasoline that weighs 1.32 kg.

1.31   The specific gravity of honey is 1.40. If a jar of honey contains 624 grams of honey, calculate the volume of honey in the jar.

1.32   A liquid has a mass of 54.5 g and a volume of 50.0 mL. Calculate its specific gravity.

1.33   A family visiting Europe goes to the hospital because their child is sick. The child has a temperature of 40.3 °C. Calculate the temperature in °F.

1.34   Fetal cord blood is stored at −112 °F. Calculate this temperature in kelvin.

1.35   Discuss the differences in the type of energy present in a snack before, during, and after you eat it.

1.36   Discuss the differences in the type of energy present as a child climbs to the top of a slide, stands at the top, and slides down.

1.37   A defibrillator delivers about 360 joules per shock. Calculate the number of calories.

1.38   A moderate walk burns 225 Calories per hour. Calculate the number of joules burned by this activity.

1.39   Two warehouses have space available for storage. One is made of brick, the other of pine, which is a soft wood. Which one will keep its contents cooler on a 90 °F day? Support your answer.

1.40   Which has a higher specific heat, water or salt water? Support your answer.

1.41   Indicate if each of the following describes a solid, liquid, or gas:

   a. This state of matter has no definite volume or shape.

   b. The particles in this state of matter are rigidly fixed in place.

1.42   Indicate if each of the following describes a solid, liquid, or gas:

   a. This state of matter has a definite volume, but takes the shape of its container.

   b. The particles in this state of matter are so far apart they do not interact with each other.

1.5 Measuring Matter

In this section, we apply the math concepts in Section 1.3 with the properties of matter from Section 1.4 to work some practical problems in health and wellness.

? 1.5 Inquiry Question: How are the properties and measurement of matter applied to problems in health?

Accuracy and Precision

Accuracy and precision are often equated in everyday language, but from a scientist's perspective, they have different meanings. Have you ever known someone who constantly runs late even after setting watches and clocks ahead 20 minutes? All the timepieces are off from the true time by 20 minutes, but they are all *exactly* 20 minutes ahead. This example illustrates the difference between accuracy and precision.

In a set of experimental measurements, accurate measurements are those that are close to the actual or true value. Precise measurements are measurements that are similar in value—that is, results easy to reproduce—but they may not even be close to the actual value. In taking measurements, it is best to measure with both **accuracy** and **precision.** This can be accomplished by taking measurements several times and averaging their values. This requires good technique during measurement and making sure the instrument being used is calibrated correctly (set to a standard value).

In the case of the late person, the clocks are precise with respect to each other, but none of them are reading the true time, so none of the clocks are accurate. The difference between accuracy and precision is also illustrated in **Figure 1.9**.

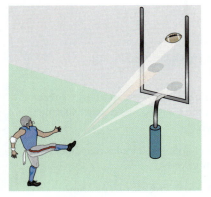

The kicker is accurate

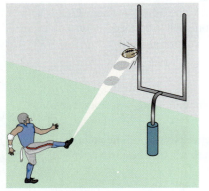

The kicker is precise

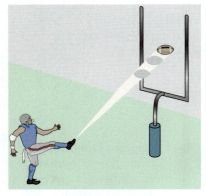

The kicker is both
accurate and
precise

**FIGURE 1.9 Accuracy and precision
have different meanings in science.**
A kicker is shown kicking a football three
times to illustrate the difference between
accuracy and precision.

---

sample problem 1.22 **Distinguishing Accuracy and Precision**

Identify the following as examples of good accuracy, good precision, or both.

a. Adjusting a balance 5 pounds lighter and reporting weights of 125, 125, and 124.5 when stepped on three times in a row.
b. Throwing four baseballs outside the upper right corner of the strike zone.
c. Using medical syringes that deliver 5.000 cc (mL) of medicine to patients requiring 5.00 cc of medicine.

Solution

a. Good precision, but not good accuracy. If the balance is offset by 5 pounds, it is not measuring the true value.
b. Good precision, but not good accuracy. The pitcher is aiming for the strike zone (the true value), but throws all the balls to the same spot off center (precise).
c. Both good precision and good accuracy. By the number of significant figures in the measurement, the device measures to the nearest 1/1,000 of a cc. This measurement is both precise and accurate because each patient will get that amount of medicine from a syringe.

Measurement in Health: Units and Dosing

Imagine that a thermometer for measuring body temperature is offset by 2 °F. It would not be a very accurate instrument. Measurements in health care require great accuracy and a level of precision that leads to correct diagnoses. Next, we take a look at common units outside the SI system that are used in measurement.

Comparing Metric and U.S. Customary Systems

Health care professionals use SI or metric units like the cc and the mg in many applications, but must also be familiar with the U.S. customary system of measurement. Similar to the metric system, the U.S. customary system (formerly called the English system) has a set of basic units including the pound (lb) and ounce (oz) for weight, and the foot (ft) and inch (in.) for length.

Unlike the metric system, the U.S. system is not a "power of 10" system allowing us easily to move from smaller to larger or larger to smaller units simply by moving the decimal. The U.S. system has a complex set of equivalencies. Perhaps the most complex of these are the equivalencies for volume. U.S. units of volume include teaspoons, tablespoons, cups, fluid ounces, pints, quarts, and gallons. The equivalencies for conversions between units of volume are not intuitive and may require memorization.

Table 1.6 shows some U.S. customary units for mass, volume, and length and equivalent metric units. Recall from Section 1.1 that the metric units can be adjusted to larger or smaller units by using prefixes such as *kilo* (1000 × larger) or *milli* (1000 × smaller).

TABLE 1.6 Equivalent Units in SI and U.S. Customary Systems

Property	U.S Customary Unit	Metric Equivalent	U.S. Customary Equivalent Unit
Mass	Pound (lb)	2.205 lb = 1 kg	1 lb = 16 oz
Volume	Quart (qt)	1.06 qt = 1 L	1 qt = 4 cups
	Fluid ounce (fl oz)	1 oz = 29.6 mL	1 cup = 8 oz
	Teaspoon (tsp)	1 tsp = 5 mL	1 fl oz = 6 tsp
Length	Mile (mi)	1 mi = 0.62 km	1 mi = 5,280 ft
	Inch (in.)	1 in. = 2.54 cm	1 ft = 12 in.

The metric prefixes allow us to work with smaller, more manageable numbers, instead of forcing us to use numbers with lots of zeros at the beginning or end.

Reading Lab Reports

Blood can be tested for any number of substances including chemical substances, cells, and cholesterols. Often this type of analysis is called "blood work." Depending upon the test, the report can be long and complex to the uneducated reader. The results for the individual are given along with normal limits for the patient. If the patient is out of the range, the value is usually highlighted. As an introduction, Table 1.7 shows a portion of a blood chemistry sample report. It is called blood chemistry because these substances are simple chemicals (versus cells or cholesterols) found in the liquid portion of the blood.

It is easier to read numbers that are greater than one to a patient, so all the values are listed this way, but this makes the units on the numbers vary widely. Nurse practitioners and other health care professionals *must* be well acquainted with units of measure to read and interpret the report. Notice that most of the units are metric. Often the deciliter (dL) is used in clinical lab reports as the volume measure. One deciliter is equal to a tenth of a liter, or 100 mL. One of the units, mmole (millimole) is equivalent to 1000 moles. A mole is a unit used to count the particles in matter. A similar unit to the millimole is the milliequivalent (mEq). For the electrolytes using mmole/L in Table 1.7, mEq/L is an equivalent unit used for these substances. These units will be described in detail later in Chapter 8.

TABLE 1.7 Sample Blood Chemistry Lab Results

Patient Name: Jane Patient		Age: 40	
Test	**Result**	**Nrml-Range**	**Units**
Blood Chemistry			
Sodium	137	135–145	mmole/L or mEq/L
Potassium	3.9	3.5–5.2	mmole/L or mEq/L
Chloride	100	97–108	mmole/L or mEq/L
Glucose	88	65–99	mg/dL
Urea Nitrogen	14	5–26	mg/dL
Calcium	9.1	8.5–10.6	mg/dL
Phosphorus	3.5	2.5–4.5	mg/dL
Protein, Total	7.1	6.0–8.5	g/dL
Iron	113	35–155	μg/dL

In the United States, body weight is usually measured in units of pounds. Because pharmaceuticals are often dispensed by body weight in kilograms, it is important for health-care professionals to be able to convert between the two systems. Let's try a problem calculating dosage.

sample problem 1.23   Calculating a Dosage

A doctor's prescription for acetaminophen is 7.5 mg per kg of body weight per dose. For a child who weighs 42 lb, calculate the number of milligrams to be given in a dose.

Solution

You are given a mass in lb and a mass in kg. This requires the conversion factor for pounds to kilograms. Because the final answer is in mg, place the mg unit in the numerator and set up the rest of the units so that they will cancel.

$$\frac{7.5 \text{ mg}}{1 \text{ kg body weight}} \times \frac{1 \text{ kg}}{2.205 \text{ lb}} \times 42 \text{ lb body weight} = 140 \text{ mg}$$

The final answer is reported to two significant figures because the minimum number of significant figures measured in the problem is two.

Percents in Health

Percent values are prevalent in health and nutrition. Here are some applications.

Percent Active Ingredient

Because of the high potency of many medicines, these are administered in very small doses (in the range of mg). However, such a small amount of material would be extremely difficult to handle, so inert binders such as dextrose, starch, or microcrystalline cellulose are often added to increase the size of a pill.

---

sample problem 1.24   Percent Active Ingredient

An aspirin tablet weighs 0.670 g. It contains 325 mg of the active ingredient aspirin. Calculate the percent aspirin in the tablet.

Solution

Recall from Section 1.1 that percent is a fraction of the part to the whole times 100. In this case, the total mass is the tablet mass, which is 0.670 g. The aspirin is a part of that mass, 325 mg. Units must be the same to calculate percent, so first convert 0.670 g to mg:

$$0.670 \text{ g} \times \frac{1000 \text{ mg}}{1 \text{ g}} = 670. \text{ mg}$$

Then calculate the percent aspirin, as follows:

$$\frac{325 \text{ mg}}{670. \text{ mg}} \times 100 = 48.5$$

The final answer is reported to three significant figures because both measured values contain three significant figures.

Percent of an Adult Dose

Because children weigh less than adults, they are often administered a percent of the dose necessary for an adult.

---

sample problem 1.25   Percent of an Adult Dose

The adult dose for sodium thiopental, an anesthetic, is 280 mg. A 19-pound child should only receive 17% of the adult dose. Calculate this dose.

Solution

Remember that percent means part divided by the whole. Here, 17% means for each 100 mg of the adult dose, the child should receive only 17 mg. Mathematically, this is shown as

$$\frac{17 \text{ mg}}{100 \text{ mg}}$$

We can solve for the missing dose by using a ratio

$$\frac{17 \text{ mg}}{100 \text{ mg}} = \frac{? \text{ mg}}{280 \text{ mg}}$$

$$\frac{(17 \text{ mg})(280 \text{ mg})}{100 \text{ mg}} = ? \text{ mg}$$

$$= 48 \text{ mg}$$

The final answer is reported to two significant figures because both given values contain two significant figures.

Percent in Nutrition Labeling

Nutrition labels give us a set of nutrition facts about foods that we eat. They list the amount of carbohydrate, protein, and fat found in the food per serving. The labels also indicate percent of the recommended daily allowance (RDA) for vitamins present in a serving.

Nutrition Facts

Serving Size: 1 cup (193g)

Amount Per Serving

Calories 670	Calories from Fat 21

	% Daily Value*
Total Fat 2.37 g	4%
Saturated Fat 0.45 g	2%
Trans Fat	
Cholesterol 0 mg	0%
Sodium 23.16 mg	1%
Potassium 2688.49 mg	77%
Total Carbohydrate 120.72 g	40%
Dietary Fiber 29.92 g	120%
Sugars 4.07 g	
Sugar Alcohols	
Protein 41.34 g	
Vitamin A 0 IU	0%
Vitamin C 12.16 mg	20 %
Calcium 218.09 mg	22%
Iron 9.79 mg	53%

Nutrient Information

Percent RDA

Nutrition facts from a bag of raw pinto beans

sample problem 1.26   Percent in Nutrition

A package of six cheese crackers weighs 39 g. The crackers contain 11 g of fat, 3 g protein, 22 g carbohydrate, and 3 g other ingredients. Calculate the percent fat, protein, and carbohydrate present in the crackers.

Solution

Percent is part over the whole times 100. So for each nutrient, we can calculate the percent in the package.

$$\text{Fat:} \qquad \frac{11 \cancel{g}}{39 \cancel{g}} \times 100 = 28$$

$$\text{Protein:} \qquad \frac{3 \cancel{g}}{39 \cancel{g}} \times 100 = 8$$

$$\text{Carbohydrate:} \qquad \frac{22 \cancel{g}}{39 \cancel{g}} \times 100 = 56$$

The final answer for % protein can only be reported to one significant figure.

practice problems

1.43 Identify the following as either good accuracy, good precision, both, or neither:

   a. Tossing three horseshoes, none of which land near the center pole. The horseshoes are scattered on either side.

   b. Tossing three horseshoes and all three are ringers on the center pole.

   c. Tossing three horseshoes and all three land 3 feet in front of the center pole on top of each other.

1.44 Consider the following measurements determined for a known volume of 10.0 mL:

Volumes Measured with a Graduated Cylinder	Volumes Measured with a Beaker
10 mL	12 mL
9.5 mL	8 mL
10.5 mL	9 mL

   a. Which set of measurements is more precise?

   b. Which set of measurements is more accurate?

   c. Which measuring device for volume is more accurate?

1.45 A low dose of aspirin is often recommended for patients who are at risk for coronary artery disease. Low-dose aspirin contains 81 mg of aspirin per tablet. How many grams of aspirin will a patient have taken over the course of 5.0 years if taking one aspirin daily?

1.46 A mother is to give her child 10 cc's of medicine every 6 hours. Unfortunately, she lost the syringe provided with the liquid medicine. She calls your clinic for help. How much medicine in teaspoons can you instruct her to give to the child per dose? Per day?

1.47 Give "Drug X" 5.0 mg/kg per day in two divided doses. The patient weighs 44 lb. How many mg should be given per dose? Per day?

1.48 A 38-lb child is prescribed acyclovir for chicken pox in an amount of 80 mg/kg body weight per day to be divided in 4 doses. Each tablet contains 700 mg of the medicine. How many tablets should be given per day? Per dose?

1.49 A tablet of cold medicine weighs 0.700 g. It contains 0.500 g active ingredient. What percent of the tablet is the active ingredient?

1.50 A medium-sized carrot weighs 61 g and contains 6 g of carbohydrate. What percent of the carrot is carbohydrate?

1.6 How Matter Changes

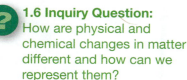

1.6 Inquiry Question: How are physical and chemical changes in matter different and how can we represent them?

On a hot summer day, you add several ice cubes to your glass of tea and come back later to find that the ice has disappeared. Now your glass seems to have two liquids—the tea at the bottom and a new, clear liquid on top. You may have seen someone at a science fair pour a clear, vinegar-smelling liquid into a papier-mâché volcano and then watched a foamy "lava" flow out. Each of these instances represents a change in matter, but each is a different type of change: One is a physical change, and the other is a chemical reaction.

Physical Change

The ice in a glass of tea is water in the solid state. As it melts, the ice undergoes a physical change and becomes a liquid (the clear liquid on top of your tea). In changing from solid to liquid, the identity of the water did not change—both ice and liquid are still H_2O (see **Figure 1.10**). In fact, if you add enough heat to the liquid water, it will eventually become steam—a gas—but it is still water (H_2O). A change in the state of matter—from a solid to a liquid or a liquid to a gas, for example—represents a physical change of matter. In a **physical change,** the form of the matter is changed, but its identity remains the same.

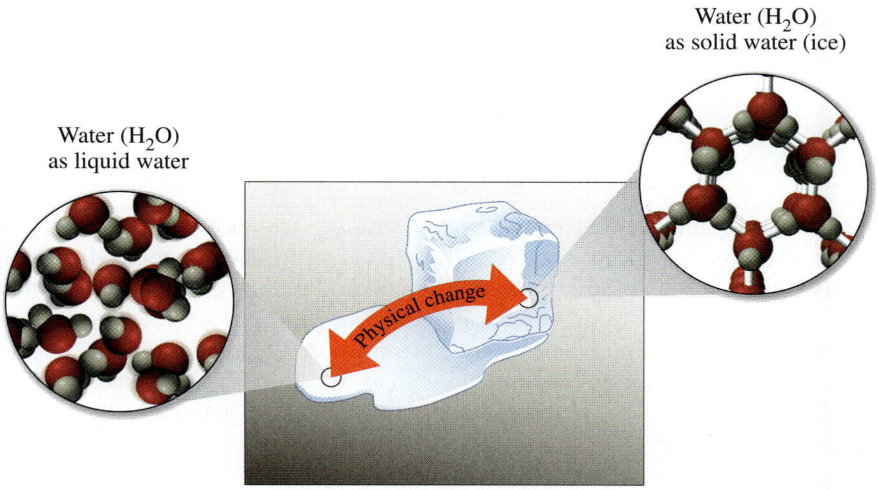

Water (H_2O)
as solid water (ice)

Water (H_2O)
as liquid water

Physical change

FIGURE 1.10 Physical change of water from solid to liquid to gas. As water changes from solid to liquid to gas, the arrangement of the particles becomes less orderly and their motion increases.

Chemical Reaction

The volcano in our example was most likely filled with baking soda before the vinegar was poured into it. When the vinegar and baking soda combine, water and carbon dioxide bubbles form. Chemically, neither vinegar nor baking soda is the same as water or carbon dioxide. So when these two substances combined, new substances were formed. Unlike a physical change, a chemical change results in a change in the chemical identity of the substance (or substances) involved. When a substance (or substances) undergoes such a change, we more commonly refer to it as a **chemical reaction. Table 1.8** gives some examples of physical changes and chemical reactions.

TABLE 1.8 Comparing Physical Changes and Chemical Reactions

Physical Change

Sawing a log in half

An ice cube melting

Molten iron solidifying to form a nail

Mixing oil, water, and eggs to make mayonnaise

Mixing cooked spaghetti and sauce to make a meal

Chemical Reaction

Burning a log in your fireplace

Hydrogen and oxygen combining to form water

A nail rusting

Butter becoming rancid

A spaghetti supper being metabolized to provide energy for a morning run

sample problem 1.27   Physical Change or Chemical Reaction

Determine whether each of the following is a physical change or a chemical reaction:

a. a plant using carbon dioxide and water to make sugar

b. water vapor condensing on a cool evening to form dew on your grass

Solution

a. This is a chemical reaction. Carbon dioxide and water are chemically different substances from sugar.

b. This is a physical change. Water vapor is a gas that condenses to form dew, which is simply liquid water.

Chemical Equations

Do you enjoy a thick, juicy hamburger that has been grilled over a charcoal fire? Have you ever wondered how the big pile of charcoal that you start with is reduced to a small pile of gray dust when the fire finally dies out? When charcoal burns with oxygen, or combusts, a chemical reaction is happening. To show what is happening to the element carbon (the main component of charcoal) as it burns, we use a **chemical equation** like the one shown here.

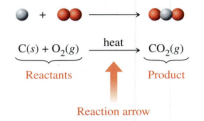

$$\underbrace{C(s) + O_2(g)}_{\text{Reactants}} \xrightarrow{\text{heat}} \underbrace{CO_2(g)}_{\text{Product}}$$

Reaction arrow

The number of atoms in reactants must equal number in products.

A chemical equation is the chemist's way of writing a sentence about what happens in a chemical reaction. It is meant to be read from left to right. The accompanying chemical equation explains that solid carbon (from charcoal) and oxygen gas when heated react to form carbon dioxide gas. In this equation, carbon and oxygen are the **reactants,** and carbon dioxide is the **product.** The reaction arrow means "react to form."

Special reaction conditions are often written as words or symbols above the reaction arrow. (Typical examples include heat, light, or the name of a catalyst or other substance required for the reaction to occur.)

The labels in parentheses after each substance (state labels) indicate its physical state—(s)olid, (l)iquid, or (g)as. A fourth label, (aq), meaning aqueous or dissolved in water, will be encountered later. So, when solid charcoal is burned, it combines with oxygen in the air and forms carbon dioxide, a gas.

Balancing Chemical Equations

The reaction of carbon with oxygen clearly illustrates the *parts* of a chemical reaction. Notice something else. There is one atom of carbon and there are two atoms of oxygen on the reactant side and the same number on the product side even though a chemical change took place. The fact that the number of each element in the reactants equals the number of elements in the products when represented in a chemical reaction illustrates the **law of conservation of mass.** Matter can neither be created nor destroyed. Matter only changes form, so the amount of matter on the reactant side and the product side must be equal.

The number of atoms in reactants must equal number in products

For any chemical equation that we write, the number of each element must be the same on both sides of the equation, or balanced. We can balance chemical equations when necessary by adding a number, called a **coefficient,** in front of the chemical formula for a substance in the chemical equation. This is easiest to understand by working some examples.

Balancing a Chemical Equation

Balance the chemical equation for the combustion of natural gas, also called methane which occurs when oxygen (O_2) reacts with methane (CH_4) producing carbon dioxide and water. The unbalanced equation is written as follows:

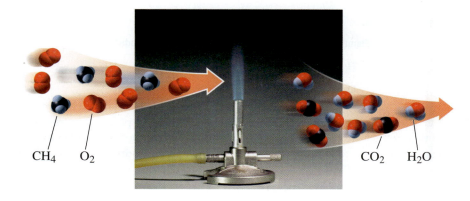

$$CH_4(g) + O_2(g) \rightarrow CO_2(g) + H_2O(g)$$

Once the equation is written, proceed through steps 1 to 3.

STEP 1: Examine the original equation. Is it balanced? If not, proceed to step 2.

Element	Number in Reactants	Number in Products
H	4	2
C	1	1
O	2	3

There are different numbers of elements in the reactants and products, so no, it is not balanced.

STEP 2: Balance the equation one element at a time by adding coefficients. Balance elements that appear only once on a side first. For our example, we save oxygen for last because it appears in both CO_2 and H_2O in the products. The number of carbon is balanced, but the number of H is not. Adding the coefficient 2 in front of H_2O will give us 4 Hs (2 × H_2O) in the products. This changes the total number of oxygen on the product side to four. Adding a 2 in front of the O_2 in the reactants gives 4 oxygen atoms (2 × O_2) on the reactant side.

$$CH_4(g) + 2O_2(g) \rightarrow CO_2(g) + 2H_2O(g)$$

STEP 3: Check to see if the equation is balanced. The coefficients should represent the least common denominator (no common factor).

Element	Number in Reactants	Number in Products
H	4	4
C	1	1
O	4	4

Now the number of elements in the reactants equals that in the products, so the equation is balanced.

1.28 Balancing Chemical Equations

Add coefficients to balance the chemical equations shown.

a. $H_2(g) + O_2(g) \rightarrow H_2O(g)$

b. $C(s) + SO_2(g) \rightarrow CS_2(l) + CO(g)$

Solution

a.

STEP 1: Examine.

Element	Number in Reactants	Number in Products
H	2	2
O	2	1

STEP 2: Balance. The element not balanced is oxygen. If we place a 2 in front of the H_2O in the product, we balance the oxygen, but now the hydrogen is doubled to 4. The hydrogen can then be balanced by putting a 2 in front of the hydrogen in the reactants.

$$2H_2(g) + O_2(g) \rightarrow 2H_2O(g)$$

STEP 3: Check.

Element	Number in Reactants	Number in Products
H	4	4
O	2	2

The equation written in STEP 2 is balanced.

b.

STEP 1: Examine.

Element	Number in Reactants	Number in Products
C	1	2
S	1	2
O	2	1

STEP 2: Balance. None of the elements are balanced. If we place a 2 in front of the SO_2 in the reactant to balance the sulfur, we get a total of 4 oxygens on the reactant side. This can be balanced by placing a 4 in front of the CO in the products. Now we have 5 carbons in the product, which can be balanced by placing a 5 on the reactant side.

$$5C(s) + 2SO_2(g) \rightarrow CS_2(l) + 4CO(g)$$

STEP 3: Check.

Element	Number in Reactants	Number in Products
C	5	5
S	2	2
O	4	4

The equation written in STEP 2 is balanced.

1.51 Determine whether each of the following is a physical change or a chemical reaction:
   a. A copper penny turns green.
   b. Sugar is melted to make candy.
   c. An antacid tablet is dropped into water and bubbles of carbon dioxide form.

1.52 Determine whether each of the following is a physical change or a chemical reaction:
   a. Cream, sugar, and flavorings are mixed and frozen to make ice cream.
   b. Milk spoils and becomes sour.
   c. A puddle of spilled nail polish remover "disappears."

1.53 Balance each of the following reactions by adding coefficients:

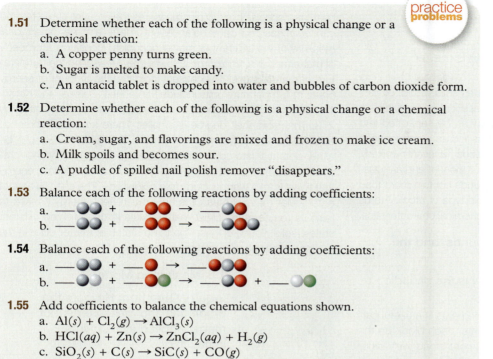

1.54 Balance each of the following reactions by adding coefficients:

1.55 Add coefficients to balance the chemical equations shown.
   a. $Al(s) + Cl_2(g) \rightarrow AlCl_3(s)$
   b. $HCl(aq) + Zn(s) \rightarrow ZnCl_2(aq) + H_2(g)$
   c. $SiO_2(s) + C(s) \rightarrow SiC(s) + CO(g)$

1.56 Add coefficients to balance the chemical equations shown.
   a. $NO_2(g) + H_2O(l) \rightarrow HNO_3(aq) + NO(g)$
   b. $NH_3(g) + O_2(g) \rightarrow N_2(g) + H_2O(l)$
   c. $C_7H_{16}(g) + O_2(g) \rightarrow CO_2(g) + H_2O(g)$

SUMMARY

1.1 Classifying Matter: Mixture or Pure Substance

Inquiry Question: How is matter classified?

Besides classifying matter as solid, liquid, or gas, chemists use a broader classification of mixture or pure substance. Mixtures can be separated into their component parts and are further classified as homogeneous (evenly mixed throughout) or heterogeneous (unevenly mixed). Pure substances are made up of a single component and are classified as elements or compounds. Compounds contain more than one element while elements contain a single type of atom. Atoms are the smallest unit of matter with unique chemical characteristics.

1.2 Elements, Compounds, and the Periodic Table

Inquiry Question: How is the periodic table organized?

The periodic table of the elements is a useful catalogue of all of the elements. Each block on the table contains the symbol of a single element along with other useful information about that element. The blocks are arranged in columns known as groups and in rows known as periods. The elements in Groups 1A through 8A are known as the main-group elements and those in the B groups are known as the transition elements. A staircase line on the right side of the periodic table separates the elements that are metals from nonmetals. Compounds are chemical combinations of elements represented by chemical formulas, which provide the identity and number of each element in the compound.

1.3 Math Counts

Inquiry Question: What basic math concepts are important to chemistry?

Some basic mathematical concepts apply to chemistry. The SI system of units is based on powers of 10 and includes the metric system. In both, the prefixes reflect the powers of ten. Conversion factors are used to convert SI units. Significant figures allow us to designate the certainty of measurements in chemistry. Rounding answers to no more certainty than measured ensures meaningful answers. Scientific notation is used to write very large and very small numbers economically. This notation also allows a direct comparison of extremely large and very small numbers. Direct comparison of different sample sizes can also be examined using percent.

$$\text{Percent (\%)} = \frac{\text{part}}{\text{whole}} \times 100$$

1.4 Matter: The "Stuff" of Chemistry

Inquiry Question: What properties of matter can be measured?

Matter is anything that takes up space and has mass. Chemists measure properties such as mass, volume, density, temperature, energy,

heat, and specific heat. Mass measures the amount of matter and can be measured on a balance. Volume is a three-dimensional measure of the space that matter occupies. Density is a property of matter and is a ratio of mass to volume. Measuring the temperature of matter is useful because it indicates the amount of energy present. Temperature units include Fahrenheit, Celsius, and Kelvin. The Kelvin and Celsius degree are the same size unit, but offset by 273. The Fahrenheit degree is smaller. There are nine Fahrenheit degrees for every five degrees Celsius. Energy in matter can be either potential (stored) or kinetic (moving). Energy is neither created nor destroyed but simply changes form. Heat is kinetic energy that flows from a warmer body to a colder one. The specific heat of a particular material measures how much heat energy it takes to raise its temperature. Most matter exists in one of three different states: solid, liquid, or gas. The three states of matter differ in the motion, kinetic energy, and the positioning of their particles.

1.5 Measuring Matter

Inquiry Question: How are the properties and measurement of matter applied to problems in health?

This section applies conversion factors and the units for measuring matter from Sections 1.1 and 1.2 to solve problems in health. The U.S. system of units is introduced and compared to the SI system. Practical dosing calculations show how conversions, units, and percent can be applied in health care.

1.6 How Matter Changes

Inquiry Question: How are physical and chemical changes in matter different and how can we represent them?

Matter can undergo two types of changes: a physical change and a chemical change called a reaction. In a physical change, the substance changes states, but its identity remains the same. In a chemical reaction, the identity of the reacting substance (or substances) is changed. A chemical reaction is represented using a chemical equation, which identifies the reacting substance(s) and the product(s), the physical state of all substances in the reaction, and any conditions necessary for the reaction to occur. The law of conservation of mass dictates that a balanced chemical equation must have equal numbers of each atom in both the reactants and products. An outline for balancing chemical equations is given as three steps: (1) Examine. (2) Balance. (3) Check.

STUDY GUIDE CHAPTER 1

The study guide will help you check your understanding of the main concepts in Chapter 1. You should be able to perform all the items described this list.

1.1 Classifying Matter: Mixture or Pure Substance

- Classify matter as a pure substance or a mixture.
- Classify mixtures as homogeneous or heterogeneous.
- Classify pure substances as elements or compounds.

1.2 Elements, Compounds, and the Periodic Table

- Distinguish between groups and periods.
- Locate metals and nonmetals on the periodic table.
- Identify the number of elements in a chemical formula.

1.3 Math Counts

- Convert between metric units.
- Apply the appropriate number of significant figures to a calculation.
- Convert numbers to scientific notation.
- Convert numbers and fractions to percent.

1.4 Matter: The "Stuff" of Chemistry

- Define mass and its measurement.
- Define volume and its measurement.
- Calculate and solve problems using density.
- Convert temperatures among the three temperature scales.
- Distinguish between kinetic and potential energy.
- Convert between energy units.
- Compare specific heat values of various materials.
- Contrast the properties of solids, liquids, and gases.

1.5 Measuring Matter

- Distinguish between accuracy and precision.
- Convert between SI and U.S. units.
- Apply conversion factors, units, and percent to measurements in health.

1.6 How Matter Changes

- Distinguish between physical changes and chemical reactions.
- Balance a given chemical equation.

Key Terms

accuracy—The closeness of an experimental measurement to the true value.

atom—The smallest unit of matter that can exist and keep its chemically unique characteristics.

calorie—The amount of energy required to raise the temperature of 1 gram of water 1 degree Celsius.

chemical equation—A chemist's shorthand for representing a chemical reaction. It shows reactants and products, and often chemical conditions.

chemical formula—A representation showing both the identity and number of elements in a compound.

chemical reaction—A change involving the reaction of one or more substances to form one or more new substances; identity of substances is changed.

chemical symbol—The one- or two-letter abbreviation for an element found on the periodic table.

coefficient—(a) A number used as a multiplier in scientific notation or (b) a number preceding a chemical formula in a chemical equation.

compound—A pure substance made up of two or more elements that are chemically joined together.

conservation of energy—Energy is neither created nor destroyed. It merely changes forms.

conservation of mass—Matter is neither created nor destroyed. It merely changes forms.

conversion factor—An equivalency written in the form of a fraction showing the relationship between two units.

density—The comparison of a substance's mass to its volume.

dimensional analysis—A tool to convert between equivalent units of the same quantity.

energy—The ability to do work.

element—The simplest type of matter; a pure substance composed of only one type of atom.

equivalent units—Two quantities that can be related to each other with an equal sign.

gas—A state of matter composed of particles that are not associated with each other and are rapidly moving, giving the substance no definite volume or shape.

gram—The base metric unit for measuring mass; roughly equivalent to the mass of a paper clip or raisin.

group—One of the vertical columns of elements on the periodic table.

heat—Kinetic energy flowing from a warmer body to a colder one.

heterogeneous mixture—A mixture whose composition varies throughout.

homogeneous mixture—A mixture whose composition is the same throughout.

kinetic energy—Energy of motion.

liquid—A state of matter composed of particles that are loosely associated and freely moving, giving the substance a definite volume, but no definite shape.

liter—The base metric unit for measuring volume; a liter is about 6% larger than a quart.

mass—The measure of the amount of material in an object.

matter—Anything that takes up space and has mass.

meter—The base metric unit for measuring length; a meter is approximately 10% larger than a yard.

metric system—Part of the International System (SI) of measurement.

mixture—A combination of two or more substances.

periodic table of the elements—A table of the elements organized by their properties into periods and groups.

period—One of the horizontal rows of elements on the periodic table.

physical change—A change in the state of matter, but not its identity.

potential energy—Stored energy.

precision—The closeness of a set of measurements to each other.

product—The new substance(s) produced by a chemical reaction; shown on the right side of a chemical equation.

pure substance—Contains only one type of substance.

reactant—The substance(s) that react to form different substance(s) in a chemical reaction; shown of the left side of a chemical equation.

significant figures—The digits in any measurement that are known with certainty plus the first estimated digit.

solid—A state of matter composed of particles that are arranged orderly and have very little motion, giving the substance a definite shape and a definite volume.

specific heat capacity—The amount of heat needed to raise the temperature of exactly 1 g of a substance by exactly 1 °C.

state of matter—The physical form in which a substance exists, either solid, liquid, or gas.

***Système International* (SI) units**—The preferred system of units for measurement in science, which includes the metric system.

temperature—A measure of the hotness or coldness of a substance. The three scales of temperature in use are Celsius, Fahrenheit, and Kelvin.

volume—The amount of space occupied by a substance.

Important Equations

Percent

$$\text{Percent (\%)} = \frac{\text{part}}{\text{whole}} \times 100$$

Temperature Conversions

$$\text{Kelvin (K)} = \text{Celsius (°C)} + 273 \text{ °C}$$

$$°C = (°F - 32 \text{ °F}) \times \frac{1 \text{ °C}}{1.8 \text{ °F}}$$

$$°F = \left(\frac{1.8 \text{ °F}}{1 \text{ °C}} \times °C \right) + 32 \text{ °F}$$

Additional Problems

1.57 Classify each of the following as a mixture or a pure substance:

a. ink in pen   **b.** wood   **c.** blood

d. ammonia (NH_3)   **e.** silicon

1.58 For each of the substances that you classified as a mixture in problem 1.57, determine whether it is a homogeneous or heterogeneous mixture.

1.59 For each of the substances that you classified as a pure substance in problem 1.57, determine whether it is an element or a compound.

1.60 Classify each of the following as a mixture or a pure substance:

a. paint   **b.** $CaCO_3$ (calcium carbonate)

c. tin   **d.** milk

e. table sugar ($C_{12}H_{22}O_{11}$)

1.61 For each of the substances that you classified as a mixture in problem 1.60, determine whether it is a homogeneous or heterogeneous mixture.

1.62 For each of the substances that you classified as a pure substance in problem 1.60, determine whether it is an element or a compound.

1.63 Use the periodic table to supply the missing information in the following chart:

Name	Elemental Symbol	Group	Period	Metal or Nonmetal
Nitrogen				
		3A	3	
	S			
	P	5A		Nonmetal

1.64 Use the periodic table to supply the missing information in the following chart:

Name	Elemental Symbol	Group	Period	Metal or Nonmetal
Oxygen				
		2A	3	
	Si			
		1A	4	Metal

1.65 Give the name and number of particles of each element in the following compounds:

a. TiO_2—titanium dioxide, white pigment in paint

b. NH_3—ammonia, a household cleaner

c. Na_2CO_3—sodium carbonate, pool additive

d. N_2O—nitrous oxide, laughing gas

e. KOH—potassium hydroxide, used in some drain cleaners

1.66 Give the name and number of particles of each element in the following compounds:

a. H_2O_2—hydrogen peroxide, a disinfectant

b. N_2H_4—hydrazine, a rocket fuel

c. $MgSO_4$—magnesium sulfate, Epsom salts

d. $C_6H_{12}O_6$—glucose

e. $C_{14}H_{18}N_2O_5$—aspartame, an artificial sweetener

1.67 When completing each of the following conversions, would you expect the number to get larger or smaller?

a. 50 mg to g   **b.** 0.23 g to μg

c. 5.2 m to km   **d.** 800 kL to dL

1.68 When completing each of the following conversions, would you expect the number to get larger or smaller?

a. 789 cm to m   **b.** 0.45 kg to mg

c. 694 mL to μL   **d.** 800 dm to m

1.69 Supply the missing information in each of the following conversion factors:

a. $\dfrac{? \text{ mL}}{1.0 \text{ L}}$ b. $\dfrac{? \text{ kg}}{1000 \text{ g}}$ c. $\dfrac{100 \text{ cm}}{? \text{ m}}$

d. $\dfrac{1.0 \text{ L}}{? \text{ dL}}$ e. $\dfrac{? \text{ } \mu g}{1.0 \text{ g}}$

1.70 Supply the missing information in each of the following conversion factors:

a. $\dfrac{? \text{ g}}{1000 \text{ mg}}$ b. $\dfrac{10 \text{ dg}}{? \text{ g}}$ c. $\dfrac{? \text{ g}}{1{,}000{,}000 \text{ } \mu g}$

d. $\dfrac{? \text{ m}}{1.0 \text{ km}}$ e. $\dfrac{1.0 \text{ L}}{? \text{ cL}}$

1.71 The typical runner's stride is about 1 meter. If a runner completes a 5K (5-kilometer) race, how many strides did he take (assume all of his strides were the same length)?

1.72 If a prescription calls for 2000 mg of aspirin per dose, how many grams of aspirin will a patient take if she takes one dose?

1.73 Soft drinks are sold in plastic bottles that will hold 1.0 L or 2.0 L. How many mL of soft drink are contained in each size bottle?

1.74 Over-the-counter cough syrups often have doses listed as a number of teaspoons. A teaspoon contains 5.0 mL of cough syrup. Write a conversion factor relating number of teaspoons and number of mL.

1.75 A raisin has a mass of approximately 1 g.

a. Write a conversion factor relating number of raisins and number of grams.

b. If a box contains 4500 cg of raisins, how many 1-g raisins are present in the box?

1.76 If a drop of blood is 0.05 mL, how many drops of blood are in a blood collection tube that holds 2 mL?

1.77 Round the following numbers to three significant figures:

a. 651,457   b. 0.004500

c. 6.665   d. 2000.

1.78 Round the following numbers to two significant figures:

a. 0.33333   b. 300,000

c. 155,555   d. 0.02105

1.79 Consider the following data set for three bags of jelly beans:

Bag Number	Number of Black Jelly Beans	Total Number of Jelly Beans
1	25	320
2	50	532
3	60	680

Which bag of jelly beans contains the highest percent of black jelly beans?

1.80 It is flu season. Professor F has a class with 50 students enrolled, and on Monday, 12 students are absent. Professor D has a class with 30 students enrolled, and on Monday, 9 students are absent. Calculate the percent of students absent from each class. Which class has the higher percent of students attending on Monday?

1.81 Calculate the volume in liters that 5.00×10^2 g of air will occupy if the density of air is 1.30 g/L.

1.82 An adult femur weighs about 225 g and has a volume of 118 cm^3. Calculate the density of this bone.

1.83 The specific gravity of olive oil is 0.703. How many grams of olive oil will fit into a container that holds 1L?

1.84 A children's liquid cold medicine has a specific gravity of 1.23. If a child is to take 1.5 tsp in a dose, what is the mass (in grams) of the cold medicine? Remember, 1 tsp = 5 mL.

1.85 To keep a room comfortable, the air is heated or cooled to remain at approximately 75 °F. This is often referred to as ambient or room temperature. What is room temperature in °C?

1.86 On the Kelvin scale, the lowest possible temperature is zero—a temperature so cold, it has never been achieved. Zero Kelvin is known as absolute zero. What is this temperature in °C? In °F?

1.87 Indicate the main type of energy found in each of the following examples:

a. firewood   b. running

c. standing on a diving board preparing to dive

d. a battery on the shelf

1.88 Indicate the main type of energy found in each of the following examples:

a. riding a bicycle   b. preparing dinner

c. dinner   d. propane fuel

1.89 A cup of yogurt contains 130 Calories. Calculate the number of joules present.

1.90 A person can burn 550 Calories by running at a moderate pace for one hour. Calculate the number of Calories that would be burned if the person only ran for 22 minutes. Calculate the number of joules burned in 22 minutes.

1.91 A single heartbeat requires about 0.5 joules of energy. Calculate the number of calories.

1.92 A banana contains 420,000 joules of energy. Calculate the number of nutritional Calories.

1.93 If you are shopping for a new set of cookware and one set is made of aluminum, one is made of stainless steel, and one is made of copper, which one would use the least heat energy from your stove during cooking? Support your answer.

1.94 Softwood trees have a higher specific heat than hardwood trees. Provide a reasonable explanation for this.

1.95 Contrast the kinetic energy of particles in a liquid with those in a gas.

1.96 Contrast the kinetic energy of particles in a liquid with those in a solid.

1.97 A student is weighing a standard 5.00 gm weight four times on two different balances to check them for accuracy and precision. The following shows the data:

Trial	Balance A	Balance B
1	4.2 gm	5.1 gm
2	4.1 gm	4.9 gm
3	4.3 gm	4.7 gm
4	4.2 gm	5.2 gm

Which of the balances is more accurate? Which is more precise?

1.98 Your patient gets a prescription for 62.5 mcg (micrograms, μ) of digoxin in liquid form. The label reads 0.0250 mg/mL. How many milliliters of digoxin should you give?

1.99 A mother calls you to ask about a proper dosage of cough medicine for her 2-year-old. The bottle says to administer 5.0 mg/kg three times a day for children under 3 years of age. The cough syrup is supplied in a liquid that contains 50.0 mg/mL. How many teaspoons of cough syrup should she give the child, who weighs 28 lb, in each dose?

1.100 A prescription for amoxicillin comes in an oral suspension that is 250 mg/5.0 mL. A patient is prescribed 500 mg every six hours. How many milliliters of amoxicillin should the patient take per dose? How many teaspoons? How many teaspoons per day?

1.101 How does the arrangement of particles in a liquid differ from that of a solid?

1.102 Which contains more kinetic energy: a solid, liquid, or gas? Support your answer.

1.103 Identify each of the following as a physical change or a chemical reaction:
  a. bleaching your hair
  b. cutting your finger
  c. a cut healing
  d. nitric acid dissolving copper metal to form copper nitrate
  e. striking a match

1.104 Identify each of the following as a physical change or a chemical reaction:
  a. water in a cloud forming into snow
  b. gasoline burning in a car's engine
  c. chewing a stick of gum
  d. a hand grenade exploding
  e. breaking a china plate

1.105 Add coefficients to balance the chemical equations shown.
  a. $LiOH(s) + CO_2(g) \rightarrow LiHCO_3(s)$
  b. $HCl(aq) + Mg(s) \rightarrow MgCl_2(aq) + H_2(g)$
  c. $Fe(s) + S(s) \rightarrow Fe_2S_3(s)$

1.106 Add coefficients to balance the chemical equations shown.
  a. $NO_2(g) + H_2O(l) \rightarrow HNO_3(aq) + NO(g)$
  b. $P(s) + O_2(g) \rightarrow P_2O_5(s)$
  c. $C_2H_2(g) + O_2(g) \rightarrow CO_2(g) + H_2O(g)$

1.107 If red spheres represent oxygen atoms and blue spheres represent nitrogen atoms:

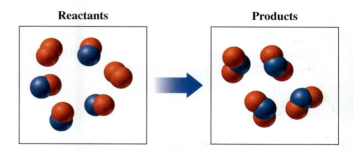

Reactants   Products

Write a balanced chemical equation for the reaction.

1.108 If red spheres represent oxygen atoms and green spheres represent chlorine atoms:

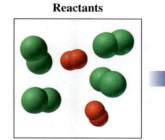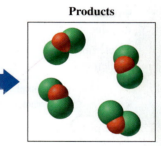

Reactants   Products

Write a balanced chemical equation for the reaction.

Challenge Problems

1.109 To donate blood, your blood must have a density greater than 1.053 g/mL, indicating that reasonable levels of hemoglobin and red blood cells are present. Before donating blood, a simple test is done to determine eligibility by placing a drop of the donor's blood into a tube containing a copper sulfate solution having a density of 1.053 g/mL. How does the technician decide whether or not the person can donate blood?

1.110 A child who weighs 66 lb is receiving a dose of 125 mg of a drug three times a day. The safe range is 10–20 mg/kg/day. Is the child receiving a safe dose?

1.111 The brakes in your car work on a hydraulic system. When you depress the brake pedal, fluid in your brake line is compressed. In turn, this fluid presses against the brake pads, which then press against the wheels, causing them to stop. If air gets into your brake lines, the brakes will work, but will require you to step on the brake pedal harder to make your car stop. Using your knowledge of the difference in the molecular arrangement in liquids and gases and your understanding of gas behavior, explain why air in your brake lines causes your brakes to work less efficiently.

1.112 The following equation shows the reaction of magnesium metal and chlorine to form magnesium chloride, which is used as a de-icer for roads and airport runways:

$$Mg(s) + Cl_2(g) \rightarrow MgCl_2(s)$$

a. Is magnesium metal an element or a compound? How do you know?

b. Is chlorine an element or a compound? How do you know?

c. What do the labels (s) and (g) represent?

d. What do you notice about the number of atoms on either side of the chemical equation?

e. Is this a physical change or a chemical reaction?

1.113 Table sugar ($C_{12}H_{22}O_{12}$) is completely dissolved in water to make sugar water.

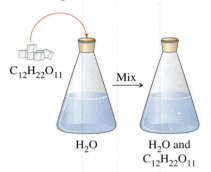

a. Is water an element or a compound? How do you know?

b. Is sugar water a homogeneous or heterogeneous mixture? How do you know?

c. When sugar is dissolved in water, is it a physical change or a chemical reaction?

Answers to Odd-Numbered Problems

Practice Problems

1.1 a. heterogeneous   b. homogeneous
     c. homogeneous   d. heterogeneous

1.3 a. pure substance   b. mixture
     c. pure substance

1.5 a. element   b. compound
     c. element   d. compound

1.7

Name	Elemental Symbol	Group	Period	Metal or Nonmetal
Fluorine	F	7A	2	Nonmetal
Lithium	Li	1A	2	Metal
Chlorine	Cl	7A	3	Nonmetal
Carbon	C	4A	2	Nonmetal

1.9 a. 1 sodium, 1 nitrogen, 2 oxygens
     b. 12 carbons, 22 hydrogens, 11 oxygens
     c. 1 calcium, 1 carbon, 3 oxygens

1.11 a. 2   b. 4   c. 5   d. 3

1.13 a. 2000   b. 54.5   c. 224,000   d. 132

1.15 a. 0.325 g

1.17 a. 2.03×10^8 b. 1.24×10^1 c. 2.78×10^{-8}

1.19 a. 0.0000156   b. 280,000   c. 0.090

1.21 a. 25%   b. 37.5%   c. 66%

1.23 a. 0.045   b. 0.13   c. 0.66   d. 0.78

1.25 a. 20   b. 9   c. 38   d. 133

1.27 a. less   b. more   c. less

1.29 39 g

1.31 446 mL

1.33 104.5 °F

1.35 Before eating, the food is mostly potential energy, during eating it is a mixture of potential and kinetic energy as it is broken down, and after eating the food gets transformed into mostly kinetic energy to carry out the functions of the body.

1.37 86 cal

1.39 The warehouse made of pine. The pine has a higher specific heat, so it will take more heat energy to raise the temperature of the wood versus the brick. This will result in a slower transfer of the heat to the contents of the warehouse.

1.41 **a.** gas **b.** solid

1.43 **a.** neither **b.** both **c.** good precision

1.45 150 g

1.47 50 mg/dose; 100 mg/day

1.49 71.4%

1.51 **a.** chemical reaction
b. physical change
c. chemical reaction

1.53 **a.**
b.

1.55 **a.** $2 \, Al(s) + 3 \, Cl_2(g) \rightarrow 2 \, AlCl_3(s)$
b. $2 \, HCl(aq) + Zn(s) \rightarrow ZnCl_2(aq) + H_2(g)$
c. $SiO_2(s) + 3 \, C(s) \rightarrow SiC(s) + 2 \, CO(g)$

Additional Problems

1.57 **a.** mixture **b.** mixture **c.** mixture
d. pure substance **e.** pure substance

1.59 **d.** compound **e.** element

1.61 **a.** homogeneous **d.** homogeneous

1.63

Name	Elemental Symbol	Group	Period	Metal or Nonmetal
Nitrogen	N	5A	2	Nonmetal
Aluminum	Al	3A	3	Metal
Sulfur	S	6A	3	Nonmetal
Phosphorus	P	5A	3	Nonmetal

1.65 **a.** 1 titanium, 2 oxygens
b. 1 nitrogen, 3 hydrogens
c. 2 sodiums, 1 carbon, 3 oxygens
d. 2 nitrogens, 1 oxygen
e. 1 potassium, 1 oxygen, 1 hydrogen

1.67 **a.** smaller **b.** larger **c.** smaller **d.** larger

1.69 **a.** 1000 **b.** 1 **c.** 1 **d.** 10
e. 1,000,000

1.71 5000 strides

1.73 1 L = 1000 mL and 2 L = 2000 mL

1.75 **a.** $\dfrac{1 \text{ raisin}}{1 \text{ g}}$ **b.** 45 raisins

1.77 **a.** 651,000 **b.** 0.00450 **c.** 6.67 **d.** $2,0\overline{0}0$

1.79 Bag 2

1.81 385 L

1.83 700 g

1.85 24 °C

1.87 **a.** potential **b.** kinetic **c.** potential **d.** potential

1.89 5.4×10^5 J

1.91 0.1 cal

1.93 The copper cookware. Copper has the lowest specific heat of the three metals. It will take the copper cookware the shortest time to heat up, which will result in a shorter overall cooking time, reducing the amount of time the stove must be on to warm up the pan.

1.95 The gas particles contain more kinetic energy than the liquid particles. In a gas, the particles are not touching each other and can move more freely and quickly.

1.97 Balance B is more accurate. Balance A is more precise.

1.99 Approximately 1.25 mL equals ¼ teaspoon in each dose.

1.101 In a solid, the particles are packed together more rigidly than in a liquid; their movement is more limited in a solid.

1.103 **a.** chemical reaction **b.** physical change
c. chemical reaction **d.** chemical reaction
e. chemical reaction

1.105 **a.** $LiOH(s) + CO_2(g) \rightarrow LiHCO_3(s)$ (balanced)
b. $2HCl(aq) + Mg(s) \rightarrow MgCl_2(aq) + H_2(g)$
c. $2Fe(s) + 3S(s) \rightarrow Fe_2S_3(s)$

1.107 $O_2(g) + 2NO(g) \rightarrow 2NO_2(g)$

Challenge Problems

1.109 Because the density of healthy blood is greater than the density of the copper sulfate solution, healthy blood will drop to the bottom of the tube. This happens within a few seconds.

1.111 Stepping on the brake pedal when no air is present causes the particles of the liquid to be compressed (move closer together). Because the particles in a liquid are already together, very little pressure on the brake pedal is required to make the car stop. With air in the brake lines, depressing the pedal first compresses the gas particles, which have much more space between them. The air must be compressed with maximum pressure before the liquid is compressed to stop the car.

1.113 **a.** It is a compound. It contains more than one element in its formula (H_2O).
b. It is homogeneous. The sugar and water particles are evenly distributed when mixed.
c. It is a physical change.

Radiation from radioisotopes is used in medicine to identify and treat diseases. To find out more about atoms, isotopes, and the properties of radioisotopes, begin reading Chapter 2.

2

Atoms and Radioactivity

TODAY, MANY FORMS of cancer are treatable by implanting radioactive "seeds" into a tumor in a procedure called brachytherapy. What are radioactive seeds? How does the procedure work? To understand the answers to these questions, first we have to take a closer look at elements in their simplest form: the atom. In the first part of this chapter, the basic parts and properties of atoms are introduced so that we can then begin to explore radioactive atoms. Radioactivity and radioactive elements used in medicine are the focus of the remainder of the chapter. Radioactive elements emit energy as radiation. In medicine, such elements can be implanted (the seeds mentioned earlier) into cancerous tissues, selectively killing the cancerous cells. Our discussion begins with the structure of the atom.

2.1 Atoms and Their Components

2.1 Inquiry Question: What are the characteristics of the subatomic particles that make up an atom?

Imagine the number of fine grains of sand it would take to cover the entire east and west coastlines of the mainland United States with beaches as wide as a football field and 1 meter deep. This is about the same number of atoms of carbon that are found in a half-carat diamond (**Figure 2.1a**). That is a lot of grains of sand and a lot of carbon atoms.

As you can guess, individual atoms of any element are incredibly small, so a lot of them fit into a small space. Carbon atoms (element number 6 on the periodic table) are found in combination with other elements in virtually all living substances. The element carbon can also be found in pure form in the nonliving substances diamond and graphite (see **Figure 2.1a** and **b**). If we were able to crush either of these substances into a very fine powder, we could view the outline of individual atoms through a powerful microscope called a scanning tunneling microscope (see **Figure 2.1c**).

Subatomic Particles

If we could crush our half-carat diamond into particles even smaller than the atoms themselves, the atoms would no longer be recognizable as carbon. Instead, we would have a collection of small parts called **subatomic particles** that organize to form all atoms. The three basic subatomic particles are the proton, neutron, and electron. **Protons** and **electrons** are charged particles; that is, they have electrical properties. **Neutrons,** as their name implies, are neutral or uncharged. Protons have a positive (+) charge, and electrons have an opposite negative (−) charge. Overall, atoms have *no charge* because the number of protons is *equal* to the number of electrons (**Table 2.1**).

Structure of an Atom

If an atom were the size of a passenger hot-air balloon, the protons and neutrons would be clustered together in an area the size of a grain of salt in the center of the balloon in a location known as the **nucleus** of an atom. The electrons would be dispersed throughout the balloon away from the nucleus. The space occupied by the electrons is called the **electron cloud** since the electrons are constantly moving in this space and it is difficult to pinpoint

What's an Inquiry Question?
Inquiry Questions are designed to focus your reading on the main concepts by section. An Inquiry Question appears at the beginning of each section.

(a)

(b)

(c)

FIGURE 2.1 Carbon atoms.
Photographs of (a) a half-carat diamond ring and (b) carbon in the form of graphite. These are both forms of pure carbon. (c) Atoms in graphite, a form of pure carbon, taken with a scanning tunneling microscope.

TABLE 2.1 Properties of Particles in an Atom

Subatomic Particle	Symbol	Electrical Charge	Relative Mass	Location in Atom
Electron	e^-	1−	0.0005 (1/2000)	Outside nucleus
Proton	p or p^-	1+	1	Nucleus
Neutron	n or n^0	0	1	Nucleus

their exact location (see **Figure 2.2**). The rest of the balloon's interior would represent empty space. In fact, most of an atom consists of empty space.

Even though atoms are small and their subatomic particles are even smaller, both still qualify as matter (they have mass and take up space). Most of the mass of an atom lies in the nucleus. The electrons contribute very little to the mass of an atom since their mass is about 2000 times less than a proton or neutron. For this reason, we will ignore the electron's contribution to the mass of an atom.

Chemists use a unit called the **atomic mass unit,** or **amu,** when discussing the mass of atoms. An amu is defined as one-twelfth of the mass of a carbon atom containing six protons and six neutrons. Therefore, the mass of a carbon atom would be 12 amu. Because a proton and a neutron have approximately the same mass, each is defined as weighing about 1 amu.

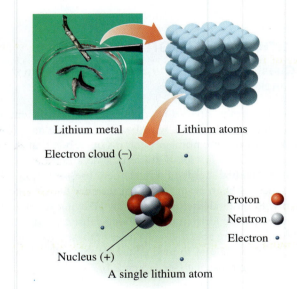

FIGURE 2.2 The location of subatomic particles in an atom. In an atom, the protons and neutrons are packed into the center (nucleus), and the electrons are in motion outside of this nucleus in an area called the electron cloud that surrounds the nucleus. An atom is mainly empty space between the nucleus and the electron cloud.

---

sample problem 2.1

Characterizing Subatomic Particles

Name the subatomic particles that make up all atoms and the charge associated with each.

Solution

The subatomic particles found in all atoms are the proton, electron, and neutron. Protons have a positive charge, electrons have an opposite negative charge, and neutrons have no charge (neutral).

practice problems

2.1 Where are the subatomic particles located in an atom?

2.2 How does the mass of a proton compare to the mass of a neutron?

2.3 How does the mass of an electron compare to the mass of a proton?

2.4 What units are used for describing the mass of a proton or neutron?

2.2 Atomic Number and Mass Number

All atoms of the same element always have the same number of protons. This feature distinguishes atoms of one element from atoms of all other elements.

2.2 Inquiry Question:
What do the atomic number and mass number indicate?

Atomic Number

The number of protons in an atom of any given element can be determined from the periodic table. (There is a periodic table on the inside front cover of this text.) The number that appears above each element within its block is its **atomic number.** The atomic number indicates the *number of protons* present in an atom of each element. The number of protons gives an atom its unique properties. For example, a carbon atom, atomic number 6, contains six protons. *All* atoms of carbon have six protons. Lithium, atomic number 3, contains three protons. *All* atoms of lithium have three protons. Because atoms are neutral (no charge), the number of electrons in an atom is equal to the number of protons. Carbon must contain six electrons, and lithium must contain three electrons to be neutral.

Mass Number

How many neutrons are present in an atom of carbon? Remember that almost all of an atom's mass is in the protons and neutrons and both particles weigh about the same. Suppose that the mass of a carbon atom is given as 12. The number of neutrons present must be six because the number of protons in a carbon atom is six. The number of neutrons in an atom can be found from an atom's **mass number,** which is simply the number of protons plus the number of neutrons. This information is summarized in Table 2.2.

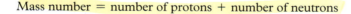

Mass number = number of protons + number of neutrons

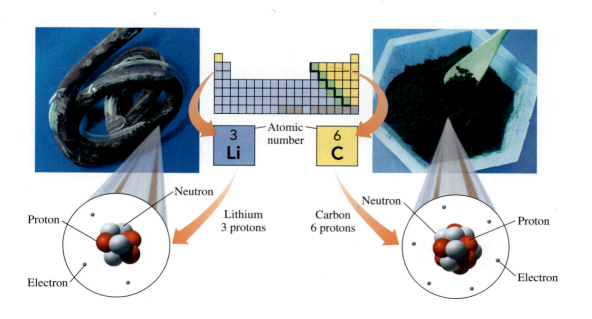

Lithium
3 protons

Carbon
6 protons

TABLE 2.2 Determining the Number of Subatomic Particles in an Atom

Subatomic Particle	Number of Subatomic Particles in Atom
Protons	Atomic number (*found above element in periodic table*)
Electrons	Equal to number of protons in an atom
Neutrons	Mass number minus number of protons (*mass number designated for a given atom*)

Once the atomic number and the mass number for a given atom are known, you can determine the number of subatomic particles present. We can represent an atom with its mass number and atomic number in *symbolic notation*:

$$\text{Mass number} \longrightarrow {}^{12}_{6}\text{C} \longleftarrow \text{Atomic symbol}$$
$$\text{Atomic number} \longrightarrow$$

sample problem 2.2 **Determining the Number of Subatomic Particles**

Determine the number of protons, electrons, and neutrons present in a fluorine atom that has a mass number of 19.

Solution

Find fluorine on the periodic table (its atomic symbol is F).

The atomic number = the number of protons, so there are nine protons in this fluorine atom.
The number of electrons = the number of protons, so there are nine electrons.
The number of neutrons = mass number − number of protons.
The mass number is given as 19; therefore, the number of neutrons = $19 - 9$, which is 10.

sample problem 2.3 **Symbolic Notation**

How would you represent the fluorine atom and its subatomic particles described in Sample Problem 2.2 in symbolic notation?

Solution

The atomic symbol for fluorine is F. The mass number appears as a superscript to the left of the symbol, and the atomic number appears as a subscript to the left of the symbol:

$$^{19}_{9}\text{F}$$

practice problems

2.5 How can you determine the following?

a. the number of protons present in an atom

b. the number of neutrons present in an atom

c. the number of electrons present in an atom

2.6 What can be determined from the following?

a. the mass number

b. the atomic number

c. the mass number minus the atomic number

2.7 Provide the name and atomic symbol of the element that has the following atomic number:

a. 8   b. 12   c. 10   d. 29   e. 47

2.8 Provide the name and atomic symbol of the element that has the following atomic number:

a. 79   b. 19   c. 82   d. 33   e. 11

2.9 What is the mass number for a nitrogen atom that contains seven neutrons?

2.10 What is the mass number for a sulfur atom that contains 17 neutrons?

2.11 Determine the number of protons, neutrons, and electrons in the following atoms:

a. bromine atom that has a mass number of 80

b. a sodium atom that has a mass number of 23

2.12 Determine the number of protons, neutrons, and electrons in the following atoms:

a. a hydrogen atom that has mass number of 1

b. a sulfur atom that has a mass number of 32

2.13 Write the symbolic notation for atoms with

a. 2 protons and 2 neutrons.

b. 17 protons and 18 neutrons.

c. 16 protons and 16 neutrons.

d. 55 protons and 78 neutrons.

2.14 Write the symbolic notation for atoms with

a. 3 protons and 4 neutrons.

b. 7 protons and 10 neutrons.

c. 15 protons and 16 neutrons.

d. 46 protons and 60 neutrons.

2.15 Determine the number of protons, neutrons, and electrons for each of the following atoms:

a. $^{18}_{8}O$ b. $^{40}_{20}Ca$ c. $^{108}_{47}Ag$ d. $^{207}_{82}Pb$

2.16 Determine the number of protons, neutrons, and electrons for each of the following atoms:

a. $^{2}_{1}H$ b. $^{27}_{13}Al$ c. $^{23}_{11}Na$ d. $^{79}_{35}Br$

Discovering the Concepts

 Inquiry Activity—Isotopes

Information

Symbolic Notation for Isotopes

$$^{A}_{Z}X$$

A = Mass Number

Z = Atomic Number

X = Atomic symbol

Isotopes of Calcium and the Number of Particles in Each

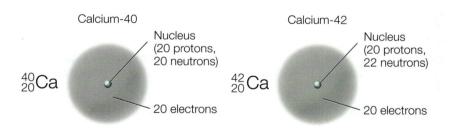

Calcium-40

Nucleus (20 protons, 20 neutrons)

$^{40}_{20}Ca$

20 electrons

Calcium-42

Nucleus (20 protons, 22 neutrons)

$^{42}_{20}Ca$

20 electrons

Questions

1. How many protons are found in calcium-42 and calcium-40?
2. How many neutrons are found in calcium-42 and calcium-40?
3. What is the relationship between the number of protons and the number of electrons in an atom?
4. Based on the information, what is common to all calcium atoms?
5. What could the atomic number Z indicate?
6. How is the mass number A determined?

7. How would you define *isotope*?
8. Where are the protons, neutrons, and electrons located in an atom?
9. What do calcium *isotopes* have in common? How do they differ?
10. Where is most of the mass of an atom?
11. For the following isotopes, determine the number of protons, neutrons, and electrons.

$^{35}_{17}Cl$

Protons   = _____
Neutrons = _____
Electrons = _____

$^{37}_{17}Cl$

Protons   = _____
Neutrons = _____
Electrons = _____

2.3 Isotopes and Atomic Mass

2.3 Inquiry Question:
What is the difference between the mass number for an isotope and the atomic mass of an element?

Among the carbon atoms found in nature, most have a mass number of 12. Yet a few have a mass number of 13. Fewer still have a mass number of 14. These atoms must have the same number of protons to be considered carbon atoms, so what is different about them? It must be the number of neutrons. Atoms of the same element can have different numbers of neutrons, so not all atoms of the same element have the same mass number. Atoms of the same element with different mass numbers are called **isotopes.** Isotopes have the same number of protons (therefore, the same atomic number), but different numbers of neutrons (different mass numbers). We can indicate isotopes in two ways. The isotopes of carbon with mass numbers 12, 13, and 14 can be written in symbolic notation as

$$^{12}_{6}C, \ ^{13}_{6}C, \text{ and } ^{14}_{6}C$$

Or, we can simply state the mass numbers after the name of the element—carbon-12, carbon-13, or carbon-14—since carbon atoms always have six protons.

sample problem 2.4   **Subatomic Particles in Isotopes**

Determine the number of protons, electrons, and neutrons present in the following isotopes of nitrogen:

a. $^{15}_{7}N$

b. $^{14}_{7}N$

c. $^{16}_{7}N$

Solution

Nitrogen's atomic number is 7, so all have seven protons. A nitrogen atom has no overall charge, so they all also have seven electrons. The number of neutrons is different for each of the isotopes.

$$\text{Number of neutrons} = \text{mass number} - \text{number of protons}$$

a.   seven protons, seven electrons, eight neutrons
b.   seven protons, seven electrons, seven neutrons
c.   seven protons, seven electrons, nine neutrons

Atomic Mass

To explore isotopes further, consider the large number of carbon atoms present in the half-carat diamond shown earlier. All the atoms will have six protons and six electrons, but the number of neutrons will differ for different isotopes. Carbon has three natural isotopes, carbon-12, carbon-13, and carbon-14, which have six, seven, and eight neutrons, respectively.

How do we know how common each isotope is? The periodic table gives us a clue. The number below each element on the periodic table shows the *average* atomic mass for that element. Carbon's average atomic mass is 12.01 amu. If the most common isotopes were the heavier isotopes, we would expect the average mass of carbon given in the periodic table to be closer to 13 or 14 amu. The atomic mass of carbon depends on the proportion of each isotope in a sample of carbon. Therefore, carbon's atomic mass of 12.01 amu indicates that the most abundant isotope is carbon-12.

In our half-carat diamond, there are mostly carbon-12 atoms. However, there are a few carbon-13 atoms (1 in every 100) and even fewer carbon-14 atoms (1 in every 1,000,000,000,000). Because the carbon-13 and carbon-14 isotopes have more neutrons, the *average* atomic mass of the carbon atoms in a sample of pure carbon such as diamond is 12.01, slightly more than 12.00. The **atomic mass** shown below each of the elements on the periodic table is the average atomic mass weighted for all the isotopes of that element found naturally (see **Figure 2.3**).

(a)

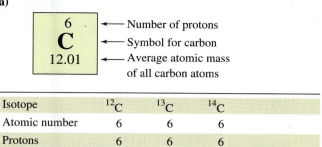

Isotope	^{12}C	^{13}C	^{14}C
Atomic number	6	6	6
Protons	6	6	6
Neutrons	6	7	8
Abundance	Most	1/100	1/1,000,000,000,000

(b)

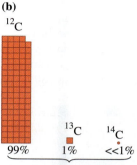

Atomic mass = 12.01 amu

FIGURE 2.3 The periodic table and atomic mass. (a) Each symbol on the periodic table gives information regarding the number of protons and the average atomic mass of the element. (b) Because the atomic mass of carbon depends on the proportion of each isotope in a sample of carbon, the atomic mass of 12.01 amu is closest to that of the most abundant isotope ^{12}C.

---

sample problem 2.5   Isotopes and Atomic Mass

Copper has two naturally occurring isotopes: copper-63 and copper-65. Based on the atomic mass from the periodic table, which of these is more prevalent?

Solution

The atomic mass of copper (element number 29) is 63.55 amu. Because the average atomic mass is closer to 63 than 65, there must be more copper-63 atoms present in the natural world than copper-65 atoms.

---

practice problems

2.17 The mass of the isotope carbon-12 is exactly 12 amu. The atomic mass for carbon (as seen on the periodic table) is 12.01 amu. Explain this difference.

2.18 How are atomic mass and mass number similar? How are they different?

2.19 There are three naturally occurring isotopes of magnesium: magnesium-24, magnesium-25, and magnesium-26.
   a. How many neutrons are present in each isotope?

   b. Write complete symbolic notation for each isotope.
   c. Based on the average atomic mass given in the periodic table, which isotope of magnesium is the most abundant?

2.20 Some of the naturally occurring isotopes of tin are tin-118, tin-119, tin-120, and tin-124.
   a. How many neutrons are present in each isotope?
   b. Write complete symbolic notation for each isotope.

2.4 Radioactivity and Radioisotopes

Do you recognize the symbol shown?

This symbol is called the *trefoil* and is the international symbol for radiation. What is radiation? Energy given off spontaneously from the nucleus of an atom is called **nuclear radiation.** Nuclear radiation is something that we are encouraged to avoid in everyday life because it can sometimes be life-threatening. Elements that emit radiation are said to be radioactive. What is radioactivity? What elements possess this property and why? If radiation is so dangerous, why do we see these stickers in different areas in a hospital? These questions are answered in the next few sections.

Radiation is a form of energy that we get from both natural and human-made sources. Light and heat from the Sun are both natural forms of radiation. Microwaves used in ovens and radio waves used for communication are examples of human-made radiation. Similarly, the nuclei of some atoms can also "radiate." Radiating nuclei were first detected in the late 1800s with materials that are still used today in professional film photography.

In 1896, Henri Becquerel got an exposure on a photographic plate (in those days they had plates, not film or digital images), not from interaction with visible light, but by exposing the plate to a rock in the dark. Elements that spontaneously emit radiation from their nucleus are said to be **radioactive** (see **Figure 2.4**). The rock that Becquerel used in his experiment contained the element uranium, which we know today to be a radioactive element.

Are all atomic nuclei radioactive? No. Most naturally occurring isotopes actually have a stable nucleus, and these isotopes are therefore *not* radioactive. Isotopes that are not stable become stable by spontaneously emitting radiation from their nuclei. This process is called **radioactive decay.** These radioactive isotopes that emit radiation are also called **radioisotopes.** Several naturally occurring radioisotopes can be mined from the earth. Many more can be prepared in scientific laboratories. All the isotopes of elements with atomic number 83 and higher are radioactive. Some of the lower elements also have radioisotopes. The same symbolic notation used to distinguish isotopes (Section 2.2) is also used to distinguish radioisotopes. For example, the most abundant isotope of gallium is $^{69}_{31}$Ga, which is not radioactive. The radioisotope $^{67}_{31}$Ga is used in medicine to image tumor cells. Some radioisotopes used in medicine are shown in **Table 2.3**.

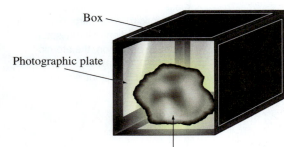

Box

Photographic plate

Radioactive rock containing uranium

FIGURE 2.4 Radioactive decay. When a radioactive element like the uranium in a rock is placed in a light-tight container, it emits radiation, chemically altering photographic film.

TABLE 2.3 Some Radioisotopes and Their Use in Diagnosing and Treating Disease

Radioisotope	Clinical Use
Chromium-51	Red blood cell labeling for determining blood volume
Cobalt-60	Cancer therapy
Gallium-67	Soft tissue tumor (lymphoma) detection
Iodine-123, iodine-131	Thyroid imaging and uptake
Technetium-99m	Bone imaging, kidney imaging, breast tumor imaging, heart perfusion imaging
Xenon-133	Lung function imaging

Source: nuclearpharmacy.uams.edu/RPLIST.html

Forms of Radiation

The first three forms of nuclear radiation that were discovered are the positively charged alpha (α) particle, the negatively charged beta (β) particle, and the neutral gamma (γ) ray. There are two other less common forms; the positron (+ charge) and the subatomic particle the neutron (no charge). The properties of these five types are summarized in Table 2.4.

An **alpha particle** is represented by the Greek letter α or as $^4_2$He, a helium nucleus containing two protons and two neutrons. Because it is a helium atom without electrons, it carries a 2+ charge from its two protons.

In some atoms with an unstable nucleus, one of the neutrons may eject a high-energy electron. This electron is called a **beta particle.** The neutron that emits the beta particle becomes a proton as a result. A beta particle is a high-energy electron and carries a charge of 1−. It is represented by the Greek letter β, or symbolically as $_{-1}^{0}$e since e$^−$ is the symbol for an electron.

Gamma rays are high-energy radiation emitted during radioactive decay, but they contain no particle. Gamma rays are released when an unstable nucleus rearranges to a more stable state. Neither the charge nor the mass of the isotope will change, so the gamma ray is simply represented by the Greek letter γ. Gamma rays are even higher-energy radiation than X-rays. X-rays are generated by high-energy electron processes and are not a form of nuclear radiation; that is, they are not emitted from a radioactive nucleus. They are included in this section because they are commonly used in the clinical setting and share many properties with gamma rays.

A **positron** has the same mass as a beta particle but is positively charged. It is represented as $_1^0$e. Some radioisotopes with an unstable nucleus will change a proton to a neutron and emit a positron. During the decay process, the positron collides with an electron (equal and opposite in charge and mass) ultimately emitting energy in the form of gamma rays. Positron decay events, commonly called *emissions*, are used in the medical imaging technique Positron Emission Tomography, or PET (see Section 2.7).

TABLE 2.4 Forms and Properties of Nuclear Radiation

Emission	Symbol	Charge
Alpha	α or $^4_2$He	2+
Beta	β or $_{-1}^{0}$e	1−
Gamma	$_0^0\gamma$	0
Positron	$_1^0$e	1+
Neutron	$_0^1$n	0

Biological Effects of Radiation

Why is radiation dangerous to living organisms? Radioactive emissions like the ones just described contain a lot of energy. When emitted, they will interact with any atoms they come into contact with, whether in a living organism or not. Alpha and beta particles, neutrons, and gamma rays and X-rays are also known as **ionizing radiation.** When they interact with another atom, they have the effect of ejecting one of that atom's electrons. (Remember, electrons are on the outside of the atom, far from the nucleus.) This makes the atom more reactive and less stable.

The loss of too many electrons from too many atoms over long periods of time in living cells can affect a cell's chemistry and genetic material. In humans, this can cause a number of health problems, the most common of which is cancer.

Not all ionizing radiation has the same amount of energy. Radiation of higher energy can penetrate farther into a tissue, affecting cells located deeper in the body. The penetrating power for different forms of ionizing radiation is shown in Figure 2.5. Persons who work with radioactive materials—for example, radiologists—must take special precautions to protect themselves from radiation exposure

FIGURE 2.5 Penetration of radiation. Different forms of ionizing radiation have different energies and penetrate the body to different extents. Alpha particles are stopped at the skin surface, while beta particles penetrate slightly into the skin, and gamma rays penetrate deeply into tissue. X-rays penetrate skin but not bone. Ionizing radiation can be stopped by shielding with different materials.

integrating chemistry

by wearing a heavy lab coat, lab glasses, and gloves. Depending on the type of radiation with which they are working, they may have to stand behind a plastic or lead shield (see Table 2.5). People who routinely work with radioactive materials or X-rays usually wear a film badge to monitor their total exposure to radiation over a given time period (see Figure 2.6).

FIGURE 2.6 Film badge. A film badge can be used to monitor exposure to nuclear radiation or X-rays over a given time period.

TABLE 2.5 Properties of Common Ionizing Radiation

	Travel Distance through Air	Tissue Penetration	Protective Shielding
Alpha (α)	A few centimeters	Stops at the skin surface; only dangerous if inhaled or eaten	Paper, clothing
Beta (β)	A few meters	Will not penetrate past skin layer	Heavy clothing, plastic, aluminum foil, gloves
X-ray	Several meters	Penetrates tissues, but not bone	Lead apron, concrete barrier
Gamma (γ)	Several hundred meters	Fully penetrates body	Thick lead, concrete, layer of water

2.6   **Properties of Radiation**

Based on its penetrating power into tissue, which form of ionizing radiation is the most dangerous?

Solution

An examination of Table 2.5 indicates that gamma radiation is the most penetrating. It will pass through a person's whole body.

practice problems

2.21 If the symbol for an alpha particle is $^4_2$He, explain how an alpha particle differs from a helium atom.

2.22 What is the charge on a beta particle? An alpha particle? A positron?

2.23 Which form of nuclear radiation has the greatest penetrating power?

2.24 Which form of nuclear radiation is most similar to X-ray radiation? How is it different?

Discovering the Concepts

 Inquiry Activity—Radioactivity

Information

Types of Ionizing Radiation Produced in Nuclear Reactions

Emission	Symbol	Charge	Mass Number
Alpha	α or $^{4}_{2}He$	2+	4
Beta	β or $^{0}_{-1}e$	1−	0
Gamma	$^{0}_{0}\gamma$	0	0
Positron	$^{0}_{1}e$	1+	0
Neutron	$^{1}_{0}n$	0	1

Nuclear Equation

$$^{238}_{92}U \longrightarrow {}^{234}_{90}Th + {}^{4}_{2}He$$
$$\text{Reactant} \qquad \text{Products}$$

Questions

1. Using the table provided, what type of ionizing radiation (radioactivity) is produced in the nuclear equation shown?
2. Using the nuclear equation shown, compare the following in the reactants versus the products. Answer as either the same, more in products, or more in reactants.
   a. Total number of protons
   b. Total number of neutrons
   c. Total mass number
3. Balance the mass numbers and the atomic numbers, and write the correct atomic symbol to complete the following equation by filling in the blanks.

$$^{131}_{53}I \rightarrow \underline{\hspace{1cm}} + {}^{0}_{-1}e$$

4. Write a nuclear equation for the following statement. Iron-59 emits a beta particle to form cobalt-59.

2.5 Nuclear Equations and Radioactive Decay

Uranium-238 is a radioactive isotope that emits an alpha particle when it undergoes radioactive decay. We can represent this process with a special type of chemical equation, called a nuclear decay equation, as follows:

$$^{238}_{92}U \longrightarrow {}^{234}_{90}Th + {}^{4}_{2}He$$
$$\text{Reactant} \qquad \text{Products}$$

Notice that the mass number of the reactant (238 for uranium) is equal to the total mass number of the products (234 + 4: thorium and an alpha particle) and the atomic number of the reactant (92) is equal to the sum of the atomic numbers of the products (90 + 2). The symbol changed from uranium to thorium because the number of protons present changed (92 to 90). Atoms with different numbers of protons are different elements.

The general form for a **nuclear decay equation** is

Radioactive nucleus undergoing decay → new nucleus formed + radiation emitted.

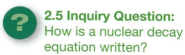 **2.5 Inquiry Question:**
How is a nuclear decay equation written?

Remember that the number of protons plus neutrons on both sides of the nuclear decay equation will be the same whether they are all in the same atom or divided up due to the decay. Let's practice writing nuclear decay equations for ionizing radiation.

Alpha Decay

In alpha decay, an alpha particle, consisting of two protons and two neutrons, is emitted. The mass number of the isotope produced decreases by 4 and the atomic number decreases by 2.

Solving a Problem

Writing a Nuclear Decay Equation for Alpha Decay

Write a nuclear decay equation for radium-224 undergoing alpha decay.

STEP 1: Write the symbolic notation for the radioisotope undergoing decay. Radium-224 has an atomic number of 88, so the symbolic notation is

$$^{224}_{88}\text{Ra} \longrightarrow$$

STEP 2: Place the ionizing radiation on the product side of the equation. An alpha particle is ^4_2He.

$$^{224}_{88}\text{Ra} \longrightarrow \square + ^4_2\text{He}$$

STEP 3: Determine the missing product radioisotope. Remember that the mass and atomic numbers must be equal on both sides of the equation. If radium (Ra) has a mass number of 224 and it loses 4 particles in an alpha decay, the resulting atom will have a mass number of 220 (224 minus 4). During alpha decay 2 protons are lost, so the resulting atom will have two fewer protons than Ra, or 86 (88 minus 2). The new symbol for the element will be the element that has an atomic number of 86, radon. The symbol of the resulting atom is $^{220}_{86}\text{Rn}$.

$$^{224}_{88}\text{Ra} \longrightarrow ^{220}_{86}\text{Rn} + ^4_2\text{He}$$

sample problem
2.7   **Writing Nuclear Decay Equations**

Write a nuclear decay equation for americium-241 undergoing alpha decay.

Solution

STEP 1: Write the symbolic notation for the radioisotope undergoing decay. Americium is element number 95, so the reactant is written as

$$^{241}_{95}\text{Am} \longrightarrow$$

STEP 2: Place the ionizing radiation on the product side of the equation. An alpha particle is ^4_2He.

$$^{241}_{95}\text{Am} \longrightarrow \square + ^4_2\text{He}$$

STEP 3: Determine the missing product radioisotope. If americium (Am) has a mass number of 241 and it loses 4 particles in an alpha decay, the resulting atom will have a mass number of 237 and an atomic number of 93. This element is neptunium, $^{237}_{93}\text{Np}$. The nuclear decay equation is

$$^{241}_{95}\text{Am} \longrightarrow ^{237}_{93}\text{Np} + ^4_2\text{He}$$

Beta Decay and Positron Emission

In beta decay, a high-energy electron is emitted from the nucleus and, in the process, a neutron becomes a proton. A radioisotope that produces a positron is also considered a beta emitter because, in both, the mass number of the isotope does not change. Although the two forms of ionizing radiation are identical except for their charge, typically beta decay refers to the emission of an electron.

Electron   Positron

$_{-1}^{0}e$ $_{1}^{0}e$

Electrons and positrons are both beta emitters.

Solving a Problem

Writing a Nuclear Decay Equation for Beta Decay

Write a nuclear equation for manganese-56 undergoing beta decay.

STEP 1: Write the symbolic notation for the radioisotope undergoing decay. Manganese is element 25, so the reactant is written as

$$_{25}^{56}\text{Mn} \longrightarrow$$

STEP 2: Place the ionizing radiation on the product side of the equation. A beta particle is $_{-1}^{0}e$.

$$_{25}^{56}\text{Mn} \longrightarrow \square + _{-1}^{0}e$$

STEP 3: Determine the missing product radioisotope. Remember: The total mass (56) does not change, but one more proton is present. The number of protons increases on the right side by one so the resulting isotope has a total of 26 protons. The symbol for the element that has 26 protons is Fe, iron. The symbol of the resulting atom is $_{26}^{56}\text{Fe}$. The nuclear decay equation is

$$_{25}^{56}\text{Mn} \longrightarrow _{26}^{56}\text{Fe} + _{-1}^{0}e$$

---

sample problem
2.8   **Writing Nuclear Decay Equations**

Write a nuclear decay equation for the positron emission of fluorine-18.

Solution

STEP 1: Write the symbolic notation for the radioisotope undergoing decay. Fluorine is element 9, so the reactant is written as

$$_{9}^{18}\text{F} \longrightarrow$$

STEP 2: Place the ionizing radiation on the product side of the equation. A positron is $_{1}^{0}e$.

$$_{9}^{18}\text{F} \longrightarrow \square + _{1}^{0}e$$

STEP 3: Determine the missing product radioisotope. Remember: The total mass (18) does not change, and one less proton is present in the resulting isotope if a positron is emitted, meaning 8 protons. The symbol for the element that has 8 protons is O, oxygen. The symbol of the resulting atom is $_{8}^{18}\text{O}$. The nuclear decay equation is

$$_{9}^{18}\text{F} \longrightarrow _{8}^{18}\text{O} + _{1}^{0}e$$

Gamma Decay

Because gamma rays are energy only, an isotope that is a pure gamma emitter will not change its atomic number or mass number upon decay. Radioactive technetium (atomic symbol Tc) is one of the few pure gamma emitters. It is used extensively in medical imaging, since most (~96%) of the radioactivity decays from the metastable (semistable) state containing the gamma ray within 24 hours. Short decay times mean that there is less radioactivity in the patient for a shorter time period, which is a desirable feature for medical radioisotopes. This radioactive isotope is designated in its metastable state with the symbol m, technetium-99m ($^{99m}_{43}$Tc). By emitting a gamma ray, the technetium becomes more stable. The equation for such a gamma decay is

$$^{99m}_{43}\text{Tc} \longrightarrow ^{99}_{43}\text{Tc} + ^{0}_{0}\gamma$$

Producing Radioactive Isotopes

Although some radioisotopes occur in nature, many more are prepared in chemical laboratories. Radioisotopes can be prepared by bombarding stable isotopes with fast-moving alpha particles, protons, or neutrons. The source for technetium-99m is the radioisotope molybdenum-99, which is produced in a nuclear reactor by bombarding the stable isotope molybdenum-98 with neutrons.

$$^{98}_{42}\text{Mo} + ^{1}_{0}\text{n} \longrightarrow ^{99}_{42}\text{Mo}$$

Here, the particle appears on the reactants side of the equation since a radioisotope is being produced. The nuclear equation is still balanced. (The sums of the mass numbers and the atomic numbers on both sides of the equation are equal.)

Because technetium-99m is used so much in nuclear medicine, many labs keep a supply of molybdenum-99 on hand, which decays by beta emission to give the technetium-99m.

$$^{99}_{42}\text{Mo} \longrightarrow ^{99m}_{43}\text{Tc} + ^{0}_{-1}\text{e}$$

sample problem
2.9 Producing Radioisotopes

Cobalt-60 is produced by bombarding naturally occurring cobalt-59 with neutrons. Write the nuclear equation for this reaction.

Solution

The symbol for a neutron is $^{1}_{0}$n (Table 2.4). A neutron is added to cobalt-59 and cobalt-60 is produced:

$$^{59}_{27}\text{Co} + ^{1}_{0}\text{n} \longrightarrow ^{60}_{27}\text{Co}$$

practice problems

2.25 Write a balanced nuclear equation for the decay of each of the following:
a. carbon-14 undergoing beta decay
b. polonium-212 undergoing alpha decay
c. copper-66 undergoing beta decay
d. carbon-11 undergoing positron emission

2.26 Write a balanced nuclear equation for the decay of each of the following:
a. thorium-232 undergoing alpha decay
b. strontium-92 undergoing beta decay
c. nitrogen-13 undergoing positron emission
d. californium-251 undergoing alpha decay

2.6 Radiation Units and Half-Lives

Radioactivity Units

The activity of a radioactive sample is measured as the number of radioactive emissions or decay events, often called disintegrations, that occur in a second. For example, the loss of one alpha particle from a nucleus is one disintegration. The unit for measuring disintegrations is called the **curie (Ci).** The curie was named for the Polish scientist Marie Curie, who studied radioactivity in France at the turn of the twentieth century. The SI unit for measuring disintegrations is called the **becquerel (Bq)** after Henri Becquerel.

The *activity* of a radioactive isotope defines how quickly (or slowly) it emits radiation. A curie is a unit of activity equal to 3.7×10^{10} becquerel or disintegrations per second. One curie is a dangerously high level of radiation, so a fraction of a curie, like a millicurie (mCi—one-thousandth of a curie) or a microcurie (μCi—one-millionth of a curie), is often used in medical applications. See Table 2.6.

2.6 Inquiry Question:
How can the amount of radiation be determined from radiation units and half-life?

TABLE 2.6 Units for Radiation Activity

Common Unit	Relationship to Other Units
becquerel (Bq)	1 Bq = 1 disintegration per second
curie (Ci)	1 Ci = 3.7×10^{10} disintegrations per second
millicurie (mCi)	1 Ci = 1000 mCi
microcurie (μCi)	1 Ci = 1,000,000 μCi

sample problem 2.10 — Determining Doses of Radioisotopes

A 10-mL sample of gallium-67 for tumor treatment contains 15 mCi.

a. Calculate the number of becquerels present in the sample.
b. If a patient is to receive a 3 mCi dose, how many milliliters should be injected?

Solution

a. Use the relationships defined in Table 2.6 to convert millicuries to becquerels.

$$\frac{15 \text{ mCi}}{\text{sample}} \times \frac{1 \text{ Ci}}{1000 \text{ mCi}} \times \frac{3.7 \times 10^{10} \text{ Bq}}{1 \text{ Ci}} = 5.6 \times 10^8 \text{ Bq per sample}$$

b. In this type of dosage conversion, the units on the final answer are mL. This unit should appear in the numerator of the problem.

$$\frac{3 \text{ mCi}}{\text{injection}} \times \frac{10 \text{ mL}}{15 \text{ mCi}} = 2 \text{ mL/injection}$$

Half-Life

Every radioactive isotope emits its radiation at a different rate. In other words, some isotopes are more unstable than others and emit radiation more rapidly. This rate of decay can be measured as the **half-life,** the time it takes for one-half (50%) of the atoms in a radioactive sample to decay (emit radiation). This amount is measured on an instrument called a Geiger counter, which measures the number of decay events over a period of time in curies or becquerels (see Figure 2.7). Those radioisotopes that are more unstable than others emit radiation more rapidly and have shorter half-lives.

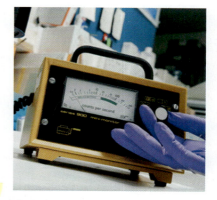

FIGURE 2.7 A Geiger counter. The Geiger counter measures radioactive disintegrations per second.

Naturally occurring radioisotopes tend to have long half-lives, disintegrating slowly over a number of years. Radioisotopes used in medicine tend to have much shorter half-lives, decaying rapidly (in hours or days). Their shorter half-lives allow radioactivity to be eliminated quickly from the body. The half-lives of several radioactive elements are shown in Table 2.7.

Knowing the half-life allows us to determine how much radioactivity is left after a given amount of time has passed. Let's look at an example.

TABLE 2.7 Half-Lives of Some Radioisotopes

Radioisotope	Symbol	Half-Life	Radioisotope	Symbol	Half-Life
Naturally occurring radioisotopes			**Radioisotopes used in medicine**		
Hydrogen-3 (tritium)	^{3}H	12.3 years	Chromium-51	^{51}Cr	28 days
Carbon-14	^{14}C	5730 years	Fluorine-18	^{18}F	110 minutes
Radium-226	^{226}Ra	1600 years	Iron-59	^{59}Fe	45 days
Uranium-238	^{238}U	4.5 billion years	Phosphorus-32	^{32}P	14.3 days
			Technetium-99m	^{99m}Tc	6.0 hours
			Iodine-123	^{123}I	13.2 hours
			Iodine-131	^{131}I	8 days

Solving a Problem

Determining Half-Life

Radioactive iodine-131 has a half-life of eight days. If a dose with an activity of 200 μCi is given to a patient today, about how much of the radioactivity will still remain active after 32 days?

STEP 1: Determine the total number of half-lives. How many half-lives are in 32 days? Since 1 half-life is 8 days, there are 4 half-lives in 32 days. Mathematically, you could arrive at this answer in the following way:

$$32 \text{ days} \times \frac{1 \text{ half-life}}{8 \text{ days}} = 4 \text{ half-lives}$$

STEP 2: Determine the amount of isotope remaining. Divide the original dose (200 μCi) by 2 four successive times since four half-lives were determined in step 1. This can be diagrammed as

$$200 \text{ μCi} \xrightarrow[\text{1 half-life}]{\text{8 days}} 100 \text{ μCi} \xrightarrow[\text{2 half-lives}]{\text{16 days}} 50 \text{ μCi} \xrightarrow[\text{3 half-lives}]{\text{24 days}} 25 \text{ μCi} \xrightarrow[\text{4 half-lives}]{\text{32 days}} 12.5 \text{ μCi}$$

Another way this can be solved is by using the equation

$$\text{isotope remaining} = \left(\frac{1}{2}\right)^n \times \text{ starting amount}$$

Where $n =$ the number of half-lives determined

$$\text{amount remaining} = \left(\frac{1}{2}\right)^4 \times 200 \text{ μCi}$$

$$\text{amount remaining} = 12.5 \text{ μCi}$$

sample problem 2.11   Half-Life

Iron-59 has a half-life of 45 days. If 96 g of a radioactive iron (^{59}Fe) is received in the lab today, how many grams of the iron will still be radioactive after 135 days?

Solution

STEP 1: Determine how many half-lives will be in 135 days.

$$135 \text{ days} \times \frac{1 \text{ half-life}}{45 \text{ days}} = 3 \text{ half-lives}$$

STEP 2: Determine the amount of isotope remaining. Diagram the solution using the number of half-lives you determined in the first step:

$$96 \text{ g} \xrightarrow[\text{1 half-life}]{45 \text{ days}} 48 \text{ g} \xrightarrow[\text{2 half-lives}]{90 \text{ days}} 24 \text{ g} \xrightarrow[\text{3 half-lives}]{135 \text{ days}} 12 \text{ g}$$

practice problems

2.27 How do the radioisotopes used in medical imaging differ from naturally occurring radioisotopes?

2.28 Define half-life in your own words.

2.29 A brain scan uses the radioisotope oxygen-15. The recommended dosage is 50 mCi. A supply of 250 mCi in 20 mL arrives at the lab. How many mL will be injected into a patient?

2.30 Xenon-133 is used for imaging the lung. The recommended dose is 15 mCi. If xenon-133 arrives in the laboratory in a package containing 120 mCi, how many patients can be treated with the contents of this package?

2.31 Radioactive iodine-131 has a half-life of 8 days. If a dose with an activity of exactly 400 μCi of ^{131}I is given to a patient on October 1, how much of the ^{131}I will still be active on November 1 (32 days later)?

2.32 Technetium-99m is very useful in diagnostic imaging since it has a short half-life of 6 hours. If a patient receives a dose with an activity of 25 mCi of technetium-99m for cardiac imaging, how much radioactivity will be left in the patient's body 48 hours after injection?

2.7 Medical Applications for Radioisotopes

Nuclear radiation has many forms. It can be high-energy particles or high-energy rays. Like the uranium rock that left a pattern on Becquerel's photographic plate, some radioisotopes of elements are useful in medical imaging. Certain radioisotopes concentrate in particular tissues. The emitted radiation from those locations can create an image on a photographic plate or be detected by scanning sections of the body. An actual image is shown in **Figure 2.8**.

Because it is important to expose patients to the smallest possible dose of radiation for the shortest time period, radioisotopes with short half-lives are selected for use in nuclear medicine. Specific radioisotopes can provide images of specific body tissues. For example, iodine is used in the body only by the thyroid gland, so any iodine—radioactive or nonradioactive—put into the body will accumulate in the thyroid (see **Figure 2.9**). Radioisotopes used in medicine are prepared in a laboratory (unlike naturally occurring isotopes). Several radioisotopes and their medical uses were shown earlier in Table 2.3.

The two main uses of medical radioisotopes are (1) in diagnosing diseased states and (2) in therapeutically treating diseased tissues.

? 2.7 Inquiry Question: How does the medical field use radioisotopes?

FIGURE 2.8 Chest image taken using a radioactive element. Radioactive elements like xenon-133 can be used to image lung tissue. Notice the large dark spot on the right side indicating a tumor in this patient who was diagnosed with lung cancer.

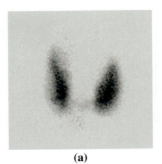

(a)

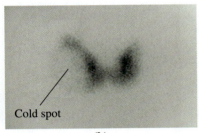

Cold spot

(b)

FIGURE 2.9 The thyroid gland has a butterfly shape. Nuclear radiation scans of (a) a normal thyroid and (b) a thyroid gland with a cold spot on the left lobe. Can you see a difference?

When diagnosing a diseased state, a minimum amount of radioisotope is administered because the isotope is used for detection only and thus should have minimal effect on body tissue. A radioisotope used in this way is called a **tracer.** As an example, iodine-123 can be used to diagnose proper thyroid function. It is a gamma emitter and has a half-life of 13.2 hours. Gamma emitters are useful for diagnosis because gamma radiation is highly penetrating and thus can more easily exit the body. When injected, the iodine-123 will concentrate in the thyroid. The ionizing radiation (gamma rays) emitted from the iodine-123 is detected by a scanner that produces an image of the thyroid (see Figure 2.9a). If the thyroid is functioning normally, the radioisotope will be evenly distributed throughout the organ. If there is a nonfunctioning area in the thyroid, the iodine will not be distributed in this area, and it shows up as a "cold" spot (absence of tracer) on the scan (see Figure 2.9b). Unusually high areas of activity, like an area with rapidly dividing cancer cells, would show up as a "hot" spot (more tracer than normal).

Radioisotopes can be used in therapy to destroy diseased or cancerous tissues. In the case of thyroid cancer, iodine-131, a beta emitter, is administered in a dose approximately 1000 times higher than that of the iodine-123 tracer used to detect the disease. The radioactive iodine will be absorbed only by the thyroid and the beta emissions will destroy cells in that specific location. Cells that are rapidly dividing (cancer cells) are more susceptible to ionizing radiation damage because their genetic material is exposed more often during cell division. These cells are destroyed at a higher rate.

sample problem 2.12   Radioisotopes in Medicine

A patient suspected of having thyroid cancer can be diagnosed and, if necessary, treated with radioactive iodine. Compare the dose and radioisotope that would be given for (a) diagnosis and (b) treatment.

Solution

a. For diagnosis, a patient would be receiving only a low dose or trace amount (typically in the μCi range) of radioactive iodine to capture a scan of the thyroid gland. A gamma emitter is preferred because gamma radiation travels farther. Iodine-123, a gamma emitter, works well for this purpose.
b. For treating cancer, a higher dose (typically in the mCi range) and a less penetrating radioisotope is more suitable. Iodine-131, a beta emitter, works well for treatment.

integrating chemistry

Radioisotopes and Cancer Treatment

In the therapeutic treatment of cancer, radioisotopes can also be applied both externally and internally without delivering the radioisotope through the bloodstream. In *external beam radiation therapy,* gamma radiation generated from the radioisotope cobalt-60 can be aimed directly at a tumor area, destroying the tissue. In *brachytherapy,* small titanium capsules or "seeds" containing radioisotopes are implanted at a tumor site. In this way, a high dose of radiation can be localized to a cancer tumor while minimizing damage to surrounding tissue. These therapies are used in the treatment of some cancers such as lung, prostate, and breast.

In brachytherapy, small titanium seeds containing a radioisotope like Pd-103 or I-125 are implanted at the tumor site.

Positron Emission Tomography

Another scan that uses radioisotopes is positron emission tomography, or PET. PET scans are used to identify functional abnormalities in organs and tissues. One of the tracers used in PET is fluorine-18, which has a short half-life of 110 minutes. The fluorine isotope emits a positron as it decays to form oxygen-18. During the emission, the positron comes into contact with an electron (its opposite) and gamma radiation is produced and detected by the scanner. One of the areas where this type of scan is commonly used is the brain. It has been used successfully in the diagnosis of Alzheimer's patients where areas of the brain do not scan as normal brain tissue (see **Figure 2.10**).

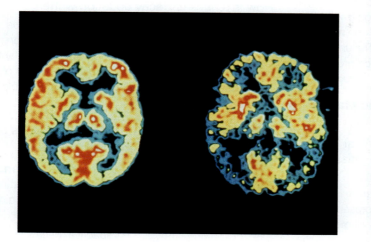

FIGURE 2.10 PET scans can reveal abnormalities in tissue function.
A cross-section of a brain with normal activity (left) is compared with the brain of an Alzheimer's patient (right). Can you see a difference?

sample problem 2.13   Medical Isotopes and Half-Life

Gold-198 is a beta emitter used in the treatment of leukemia. It has a half-life of 2.7 days. How long would it take for a dose with an activity of 96 mCi to decay to an activity of 3.0 mCi?

Solution

This problem is most easily done by diagramming the problem to determine the number of half-lives that will have passed. After the passing of each half-life, one-half of the previous amount is still present.

$$96 \text{ mCi} \xrightarrow[\text{1 half-life}]{2.7 \text{ days}} 48 \text{ mCi} \xrightarrow[\text{2 half-lives}]{5.4 \text{ days}} 24 \text{ mCi} \xrightarrow[\text{3 half-lives}]{8.1 \text{ days}} 12 \text{ mCi} \xrightarrow[\text{4 half-lives}]{10.8 \text{ days}} 6.0 \text{ mCi} \xrightarrow[\text{5 half-lives}]{13.5 \text{ days}} 3.0 \text{ mCi}$$

Therefore, it would take about 14 days.

practice problems

2.33   Because calcium is an element in bone, why do you think it might be useful to use the radioisotope calcium-47 in the diagnosis and treatment of bone diseases?

2.34   What is a diagnostic tracer? What are the preferable characteristics?

2.35   Chromium-31 is used in imaging red blood cells and has a half-life of 28 days. If a dose with an activity of 40 μCi is given to a patient, how long will it take for the patient to have less than 5 μCi present?

2.36   Iodine-123 is used for thyroid imaging and has a half-life of 13.2 hours. How much would be left after 40 hours (3 half-lives) if a dose with an activity of exactly 300 μCi was initially administered?

2.1 Atoms and Their Components

2.1 Inquiry Question: What characteristics of the subatomic particles make up an atom?

An atom consists of three subatomic particles: protons, neutrons, and electrons. Protons have a positive charge, neutrons have no charge, and electrons have a negative charge. Most of the mass of an atom comes from the protons and neutrons located in the center, or nucleus, of an atom. The unit for the mass of an atom is the atomic mass unit (amu); each proton and neutron present in an atom weighs approximately 1 amu.

2.2 Atomic Number and Mass Number

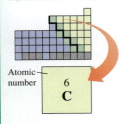

Atomic number — 6 C

2.2 Inquiry Question: What do the atomic number and mass number indicate?

The atomic number of an atom indicates the number of protons present in an atom. All atoms of a given element have the same number of protons. The mass number of an atom is the total number of protons and neutrons present in a given atom of an element.

2.3 Isotopes and Atomic Mass

6 C 12.01 ← Number of protons
← Symbol for carbon
← Average atomic mass of all carbon atoms

2.3 Inquiry Question: What is the difference between the mass number for an isotope and the atomic mass of an element?

The mass number is the number of protons and neutrons for a *given* isotope. For example, nitrogen-14 has 7 protons and 7 neutrons. The atomic mass is the average atomic mass for *all* the isotopes of an element found in nature. This number is found on the periodic table, often below the element symbol.

2.4 Radioactivity and Radioisotopes

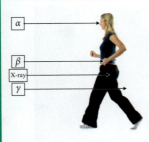

 α
β
X-ray
γ

2.4 Inquiry Question: How does a radioisotope emit radiation?

Some atomic isotopes emit radiation (a form of energy) spontaneously from their nucleus in a process called radioactive decay. Isotopes that undergo radioactive decay are called radioisotopes, and the high-energy particles given off in this process are referred to as ionizing radiation, or radioactivity. Three common forms of radioactivity are alpha (α) and beta (β) particles and gamma (γ)

rays. An X-ray is also a form of ionizing radiation, although it is not caused by a radioactive decay event. Different forms of ionizing radiation penetrate the body differently, producing different biological effects.

Ionizing radiation
$$^{238}_{92}U \longrightarrow {}^{234}_{90}Th + {}^{4}_{2}He$$
Decaying isotope (Reactant)
Decay product

2.5 Nuclear Equations and Radioactive Decay

2.5 Inquiry Question: How is a nuclear decay equation written?

The radioactive decay of a radioisotope can be represented symbolically in the form of a nuclear decay equation. The number of protons and the mass number found in the reactant (the decaying radioisotope) is equal to the sum of the number of protons and the mass numbers found in the products (the stable isotope and the radioactive particle).

2.6 Radiation Units and Half-Lives

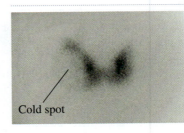

2.6 Inquiry Question: How can the amount of radiation be determined from radiation units and half-life?

Radioactive decay is measured as the number of decay events, or disintegrations, that occur in 1 second. The common unit for measuring radioactive decay is the curie (Ci), and the SI unit is the becquerel (Bq). For the smaller quantities in medical applications, the microcurie (μCi) is often used. A becquerel is equal to a disintegration per second. A curie is equivalent to 3.7×10^{10} becquerels. The half-life of a radioisotope is the amount of time it takes for one-half of the radiation in a given sample to decay. Most radioisotopes used in medicine have short half-lives, allowing the radioactivity to be more quickly eliminated from the body.

Cold spot

2.7 Medical Applications for Radioisotopes

2.7 Inquiry Question: How does the medical field use radioisotopes?

Certain elements concentrate in particular organs of the body. If a radioisotope of this element can be made, this area of the body can be imaged using that radioisotope. A patient can be injected with a trace amount of a radioisotope to diagnose a diseased state. Radioisotopes can also be used to treat diseases. Radioisotopes can be applied externally (external beam radiation therapy) or internally (brachytherapy) by applying radiation directly at the tumor site in high doses destroying cancerous cells. Positron emission tomography (PET) uses a radioisotope to image tissues that are not functioning normally.

The study guide will help you check your understanding of the main concepts in Chapter 2. You should be able to

2.1 Atoms and Their Components

- Name the kind of subatomic particles that make up an atom.
- Locate the subatomic particles in an atom.
- Predict the mass of an atom from the number of subatomic particles.

2.2 Atomic Number and Mass Number

- Define atomic number.
- Determine the mass number for a given atom.

2.3 Isotopes and Atomic Mass

- Define isotope.
- Distinguish between mass number and atomic mass.

2.4 Radioactivity and Radioisotopes

- Define radioactivity.
- Distinguish the forms of ionizing radiation.
- Differentiate the penetrating power of the forms of ionizing radiation.

2.5 Nuclear Equations and Radioactive Decay

- Write a balanced nuclear decay equation for alpha, beta, gamma, and positron emissions.

2.6 Radiation Units and Half-Lives

- Perform dosing calculations using radiation activity units.
- Determine the remaining dose of a radioactive isotope given the half-life.

2.7 Medical Applications for Radioisotopes

- Contrast the use of radioisotopes for the diagnosis and treatment of disease.

Key Terms

alpha (α) particle—A form of nuclear radiation consisting of two protons and two neutrons (a helium nucleus).

atomic mass—The weighted average mass of all naturally occurring isotopes of an element.

atomic mass unit (amu)—The small unit of mass used to quantify the mass of very small particles; 1 amu is equal to one-twelfth of a carbon atom containing six protons and six neutrons.

atomic number—The number above the symbol of an element on the periodic table. It is equal to the number of protons in the element.

becquerel (Bq)—A unit of radiation activity equal to 1 nuclear decay event (disintegration) per second.

beta (β) particle—A form of nuclear radiation consisting of a high-energy electron emitted from an unstable neutron.

curie (Ci)—A unit of radiation activity equal to 3.7×10^{10} disintegrations per second.

electron—A negatively charged subatomic particle, symbol e^-.

electron cloud—The area of an atom outside of the nucleus where the electrons can be found.

gamma (γ) ray—High-energy nuclear radiation emitted to stabilize a radioactive nucleus.

half-life—The time it takes for one-half of the atoms in a radioactive sample to decay.

ionizing radiation—A general name for high-energy radiation of any kind (for example, nuclear, X-ray).

isotope—Two or more atoms of an element that have the same number of protons but different numbers of neutrons.

mass number—The total number of protons and neutrons in an atom.

nuclear decay equation—An equation where the reactant is a radioactive isotope and the products include a new isotope and an ionizing emission particle.

nuclear radiation—Energy emitted spontaneously from the nucleus of an atom.

neutron—A subatomic particle with no charge, symbol n.

nucleus—The central part of an atom containing protons and neutrons.

positron—A form of nuclear radiation having the same mass as a beta particle, but with opposite charge.

proton—A positively charged subatomic particle, symbol p.

radioactive—Elements that spontaneously emit nuclear radiation.

radioactive decay—The process of a nucleus spontaneously emitting radiation.

radioisotope—A radioactive isotope of an element.

subatomic particles—The small parts that organize to form atoms: protons, neutrons, and electrons.

tracer—A minimal amount of a radioactive substance used for detection purposes only.

Additional Problems

2.37 Complete the following statements:

a. A _____ is a subatomic particle that has a neutral charge.

b. The atomic number on the periodic table equals the number of _____ in the atom.

c. Two atoms that have the same number of protons and electrons but a different number of neutrons are called _____.

2.38 Complete the following statements:

a. The mass number is the sum of the number of _____ and _____.

b. Most of the mass of an atom is found in the _____.

c. In an atom, the number of _____ equals the number of _____.

2.39 Provide the number of protons and electrons in the following atoms:

a. gold   b. Zn

c. element number 29

2.40 Provide the number of protons and electrons in the following atoms:

a. chlorine   b. element number 11   c. S

2.41 Determine the number of protons, neutrons, and electrons in the following atoms:

a. $^{55}_{26}\text{Fe}$ b. $^{15}_{7}\text{N}$ c. $^{52}_{24}\text{Cr}$ d. $^{137}_{56}\text{Ba}$

2.42 Determine the number of protons, neutrons, and electrons in the following atoms:

a. $^{39}_{19}\text{K}$ b. $^{16}_{8}\text{O}$ c. $^{81}_{35}\text{Br}$ d. $^{4}_{2}\text{He}$

2.43 Complete the following table:

Symbol	Number of Protons	Number of Neutrons	Number of Electrons	Mass Number	Name
$^{1}_{1}\text{H}$					
	12		12	24	
$^{9}_{4}\text{Be}$		5			

2.44 Complete the following table:

Symbol	Number of Protons	Number of Neutrons	Number of Electrons	Mass Number	Name
$^{14}_{6}\text{C}$				14	
	35	40	35		
$^{59}_{27}\text{Co}$			27		

2.45 Identify the radiation associated with the following:

    a. particle that has the same number of protons and neutrons as a helium nucleus

    b. an electron that is emitted from an atom's nucleus

    c. has no mass

2.46 Identify the radiation associated with the following:

    a. can be stopped by paper

    b. the highest energy (most penetrating) emission

    c. can be stopped by plastic or aluminum foil

2.47 Write the symbolic notation for the following radioactive isotopes:

    a. phosphorus-32

    b. cobalt with 33 neutrons

    c. element number 24 containing 27 neutrons

2.48 Write the symbolic notation for the following radioactive isotopes:

    a. uranium-238

    b. xenon with 89 neutrons

    c. element number 53 containing 78 neutrons

2.49 Complete the following table:

Isotope Name	Symbolic Notation	Number of Protons	Number of Neutrons	Mass Number	Medical Use
Thallium-201					Tumor imaging
	$^{123}_{53}I$				Thyroid imaging
		54		133	Lung function imaging
Fluorine-18					Positron emission tomography (PET) imaging

2.50 Complete the following table:

Isotope Name	Symbolic Notation	Number of Protons	Number of Neutrons	Mass Number	Medical Use
Carbon-14					Used in breast tumor detection
	$^{103}_{46}Pd$				Prostate cancer treatment
Ytterbium-169		70			Gastrointestinal tract diagnoses
Xenon-127	$^{127}_{54}Xe$				Brain imaging for mental disorders

2.51 Complete the following nuclear equations:

    a. $^{15}_{8}O \rightarrow {}^{15}_{7}N + ?$ **b.** $^{46}_{23}V \rightarrow ? + {}^{0}_{-1}e$

    c. $^{234}_{92}U \rightarrow ? + {}^{4}_{2}He$ **d.** $^{8}_{4}Be \rightarrow ? + {}^{4}_{2}He$

2.52 Complete the following nuclear equations:

    a. $^{214}_{82}Pb \rightarrow ? + {}^{0}_{-1}e$ **b.** $^{18}_{8}O + {}^{0}_{-1}e \rightarrow ? + {}^{0}_{1}n$

    c. $^{218}_{84}Po \rightarrow {}^{214}_{82}Pb + ?$ **d.** $^{60}_{27}Co \rightarrow ? + {}^{0}_{0}\gamma$

2.53 Write balanced nuclear decay equations for each of the following emitters:

    a. copper-66 (β)

    b. platinum-192 (α)

    c. tin-126 (β)

    d. gallium-72 produces germanium-72

2.54 Write balanced nuclear decay equations for each of the following emitters:

    a. thorium-225 (α)   **b.** bismuth-210 (α)

    c. cesium-137 (β)   **d.** sulfur-35 (β)

2.55 When boron-10 is bombarded with alpha particles, nitrogen-12 and neutrons are produced. Write a nuclear equation for this reaction.

2.56 Iron-52 is a positron emitter. Write a nuclear equation for this reaction.

2.57 A 25-mL sample of chromium-51 contains 1.00 mCi. If a patient is to receive a 50 μCi dose to undergo a red blood cell count, how many mL should be injected?

2.58 Thallium-201 is a radioisotope used in brain scans. If the recommended dose is 3.0 mCi and a vial contains 60 mCi in 50 mL, how many milliliters should be injected?

2.59 Fluorine-18, which has a half-life of 110 minutes, is used in PET scans. If exactly 100 mg of fluorine-18 is shipped at 8:00 A.M., how many milligrams of the radioisotope are still active if the sample arrives at the nuclear medicine laboratory at 1:30 P.M. the same day?

2.60 A 120-mg sample of technetium-99m is used for a diagnostic test. If technetium-99m has a half-life of 6.0 hours, how much of the technetium-99m remains 24 hours after the test?

2.61 If the amount of radioactive iodine-123 in a sample decreases from 0.400 g to 0.100 g in 26.4 hours, what is the half-life of iodine-123?

2.62 If the amount of radioactive phosphorus-32 in a sample decreases from 1.2 g to 0.30 g in 28.6 days, what is the half-life of phosphorus-32?

2.63 What is the importance of a cold spot when using a radioactive tracer? A hot spot?

2.64 How are external beam radiation therapy and brachytherapy the same? How are they different?

Challenge Problems

2.65 Iron-59 has a half-life of 45 days. If 168 g of a radio-active iron (^{59}Fe) is received in the lab today, what percentage of the original is left after 270 days?

2.66 On an archaeological dig, a wooden canoe is unearthed and analyzed for carbon-14. About 25% of the carbon-14 that was initially present remains. What is the approximate age of the canoe? The half-life of carbon-14 is 5730 years.

2.67 PET scans are useful for imaging areas of high activity in the brain. One of the compounds commonly used is an F-18 labeled isotope of the sugar glucose called fludeoxyglucose, or FDG. Glucose is used in the brain for energy. How do you think a PET scan of the brain of a patient with higher than normal activity would compare to that of a normal person? Support your answer.

Answers to Odd-Numbered Problems

Practice Problems

2.1 Protons and neutrons are located in the nucleus (center) of an atom, and the electrons are found in a cloud outside of the nucleus called the electron cloud.

2.3 The mass of an electron is about 2000 times smaller than that of a proton.

2.5 **a.** The number of protons is the atomic number.
  b. The number of neutrons is the mass number minus the atomic number.
  c. The number of electrons is the same as the number of protons in an atom.

2.7 **a.** oxygen, O   **b.** magnesium, Mg
  c. neon, Ne   **d.** copper, Cu
  e. silver, Ag

2.9 14

2.11 **a.** 35 protons, 45 neutrons, 35 electrons
  b. 11 protons, 12 neutrons, 11 electrons

2.13 **a.** $^4_2$He   **b.** $^{35}_{17}$Cl   **c.** $^{32}_{16}$S   **d.** $^{133}_{55}$Cs

2.15 **a.** 8 protons, 10 neutrons, 8 electrons
  b. 20 protons, 20 neutrons, 20 electrons
  c. 47 protons, 61 neutrons, 47 electrons
  d. 82 protons, 125 neutrons, 82 electrons

2.17 The isotope carbon-12 contains exactly 6 protons and 6 neutrons. The atomic mass as seen on the periodic table is an average mass taking into consideration the abundance of all the carbon isotopes. Because about 1% of the carbon isotopes are carbon-13, the atomic mass is slightly higher than 12 amu.

2.19 **a.** Magnesium-24 has 12 neutrons, magnesium-25 has 13 neutrons, and magnesium-26 has 14 neutrons.
  b. $^{24}_{12}$Mg, $^{25}_{12}$Mg, $^{26}_{12}$Mg
  c. magnesium-24

2.21 An alpha particle is a helium nucleus. There are no electrons in an alpha particle.

2.23 gamma

2.25 **a.** $^{14}_6$C $\longrightarrow$ $^{14}_7$N + $^0_{-1}$e
  b. $^{212}_{84}$Po $\longrightarrow$ $^{208}_{82}$Pb + $^4_2$He
  c. $^{66}_{29}$Cu $\longrightarrow$ $^{66}_{30}$Zn + $^0_{-1}$e
  d. $^{11}_6$C $\longrightarrow$ $^{11}_5$B + $^0_1$e

2.27 The half-lives are shorter, and they are prepared in the lab.

2.29 4 mL

2.31 25 μCi

2.33 Ca-47 will concentrate in bone.

2.35 After three half-lives (84 days), the patient will have less than 5 μCi present.

Additional Problems

2.37 **a.** neutron   **b.** protons   **c.** isotopes

2.39 **a.** 79 protons, 79 electrons
  b. 30 protons, 30 electrons
  c. 29 protons, 29 electrons

2.41 **a.** 26 protons, 29 neutrons, 26 electrons
  b. 7 protons, 8 neutrons, 7 electrons
  c. 24 protons, 28 neutrons, 24 electrons
  d. 56 protons, 81 neutrons, 56 electrons

2.43

Symbol	Number of Protons	Number of Neutrons	Number of Electrons	Mass Number	Name
$^1_1$H	1	0	1	1	Hydrogen-1
$^{24}_{12}$Mg	12	12	12	24	Magnesium-12
$^9_4$Be	4	5	4	9	Beryllium-9

2.45 **a.** alpha particle   **b.** beta particle   **c.** gamma ray

2.47 **a.** $^{32}_{15}$P   **b.** $^{60}_{27}$Co   **c.** $^{51}_{24}$Cr

2.49

Isotope Name	Symbolic Notation	Number of Protons	Number of Neutrons	Mass Number	Medical Use
Thallium-201	$^{201}_{81}\text{Tl}$	81	120	201	Tumor imaging
Iodine-123	$^{123}_{53}\text{I}$	53	70	123	Thyroid imaging
Xenon-133	$^{133}_{54}\text{Xe}$	54	79	133	Lung function imaging
Fluorine-18	$^{18}_{9}\text{F}$	9	9	18	Positron emission tomography (PET) imaging

2.51 **a.** $^{15}_{8}\text{O} \longrightarrow ^{15}_{7}\text{N} + ^{0}_{1}\text{e}$

b. $^{46}_{23}\text{V} \longrightarrow ^{46}_{24}\text{Cr} + ^{0}_{-1}\text{e}$

c. $^{234}_{92}\text{U} \longrightarrow ^{230}_{90}\text{Th} + ^{4}_{2}\text{He}$

d. $^{8}_{4}\text{Be} \longrightarrow ^{4}_{2}\text{He} + ^{4}_{2}\text{He}$

2.53 **a.** $^{66}_{29}\text{Cu} \longrightarrow ^{66}_{30}\text{Zn} + ^{0}_{-1}\text{e}$

b. $^{192}_{78}\text{Pt} \longrightarrow ^{188}_{76}\text{Os} + ^{4}_{2}\text{He}$

c. $^{126}_{50}\text{Sn} \longrightarrow ^{126}_{51}\text{Sb} + ^{0}_{-1}\text{e}$

d. $^{72}_{31}\text{Ga} \longrightarrow ^{72}_{32}\text{Ge} + ^{0}_{-1}\text{e}$

2.55 $^{10}_{5}\text{B} + ^{4}_{2}\text{He} \longrightarrow ^{12}_{7}\text{N} + 2^{1}_{0}\text{n}$

2.57 1.3 mL

2.59 12.5 mg

2.61 13.2 hours

2.63 A cold spot indicates a diseased area of an organ, and a hot spot indicates an area of the organ where the cells are rapidly dividing or a cancerous growth.

2.65 1.56%

2.67 Because a person with a more active brain uses more glucose, more FDG will be present in the brain. The PET image would be brighter in more areas than for the normal person.

Why is the formula for water H_2O and not H_3O or H_4O? Why do most elements exist in combination with others as compounds? What is so special about the arrangement of the periodic table? Read on to explore these ideas in Chapter 3.

3

Compounds—Putting Particles Together

IN THE PREVIOUS two chapters, you were introduced to matter and its component parts—atoms. Interestingly, most elements do not occur in uncombined states. In other words, most elements do not exist as individual atoms. Instead atoms tend to combine with other atoms to form compounds. For example, the formula for water is H_2O, which means that water is the unique combination of two hydrogen atoms with one oxygen atom.

What advantage does an atom gain through combining with other atoms? What is special or unique about the Group 8A elements that do exist as single atoms?

As we explore why most elements exist in combination with other elements, we will see that this behavior is related to the element's position on the periodic table. We will also see that metals and nonmetals combine differently from the way nonmetals combine with other nonmetals.

3.1 Electron Arrangements and the Octet Rule

? 3.1 Inquiry Question: How do electrons arrange in an atom?

In Chapter 2, we saw that nuclei can be reactive (radioactive), but in this chapter, we turn our focus to the electron cloud. The arrangement of electrons around the nucleus of an atom is one of the reasons that elements are "driven" to form compounds. In Chapter 2, we saw that the electrons occupy a large space outside the nucleus of an atom. The exact location of any given electron outside the nucleus is difficult to determine, but most of them lie within a space outlined by an electron cloud. Because electrons are charged and are in constant motion, they possess energy. The electrons in an atom are found in distinct energy levels based on the amount of energy the electrons possess. Think of the energy levels as an uneven staircase (see **Figure 3.1**) where the height of the steps in the staircase decreases as you ascend.

In atoms, the electrons can exist only at distinct energy levels (steps on the staircase), not in-between. In general, electrons will occupy the lowest energy step first. This lowest energy level is found closest to the nucleus. Unlike a real staircase, the energy levels get closer together the farther you get from the nucleus. The maximum number of electrons that can be found in any given energy level can be calculated by the formula $2n^2$ where n is the number of the energy level. In the first energy level ($n = 1$), the maximum number of electrons present is two; in the second energy level ($n = 2$), the maximum number of electrons is eight; in the third energy level ($n = 3$), the maximum number of electrons is 18; and so on.

Table 3.1 shows the electron arrangements for the first 20 elements. In the table, look at element 19, potassium (K), and element 20, calcium (Ca). They have electrons in the fourth energy level even though they do not have the maximum 18 electrons filling up the third level. Once there are eight electrons in $n = 3$, the next two electrons are placed in the higher energy shell, $n = 4$. This trend continues as the energy levels increase down the periodic table.

? What's an Inquiry Question?
Inquiry Questions are designed to focus your reading on the main concepts by section. An Inquiry Question appears at the beginning of each section.

75

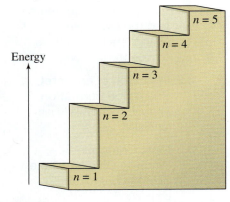

FIGURE 3.1 An energy staircase.
Electrons in an atom occupy the lower energy levels first. Electrons are only found at energy levels represented by each step, not between the steps. The number of the energy level is represented by n.

TABLE 3.1 Energy-Level Arrangements of Electrons for the First 20 Elements (Valence Electrons in Blue)

Element	Group Number	Total Number of Electrons	$n=1$	$n=2$	$n=3$	$n=4$
				Number of Electrons in Energy Level		
H	1A	1	1			
He	8A	2	2			
Li	1A	3	2	1		
Be	2A	4	2	2		
B	3A	5	2	3		
C	4A	6	2	4		
N	5A	7	2	5		
O	6A	8	2	6		
F	7A	9	2	7		
Ne	8A	10	2	8		
Na	1A	11	2	8	1	
Mg	2A	12	2	8	2	
Al	3A	13	2	8	3	
Si	4A	14	2	8	4	
P	5A	15	2	8	5	
S	6A	16	2	8	6	
Cl	7A	17	2	8	7	
Ar	8A	18	2	8	8	
K	1A	19	2	8	8	1
Ca	2A	20	2	8	8	2

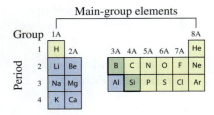

The number of electrons in the highest energy level (the valence electrons) is the same as the group number for the main-group elements.

sample problem 3.1 Electrons and Energy Levels

How many electrons are in each energy level of the following elements?

a. O   b. Ca

Solution

a. Oxygen has an atomic number of 8, so it has eight electrons total. Two are located in the first energy level, while six are in the second energy level.

n	Number of Electrons
1	2
2	6

b. Calcium has an atomic number of 20, so it has 20 electrons total. Two are located in the first energy level, eight in the second, eight in the third, and two in the fourth.

n	Number of Electrons
1	2
2	8
3	8
4	2

Examining the energy levels and number of electrons present in those energy levels provides insight into how the elements in a group or period are related to each other. Look again at Table 3.1 and notice that the elements with the same number of electrons in their highest energy level are in the same group on the periodic table. For example, boron (B) and aluminum (Al) have three electrons in their highest energy level. Both are in Group 3A.

The highest energy level that contains electrons is called the **valence shell,** or valence level, and the electrons residing in that energy level are called **valence electrons.** The valence electrons are farthest from the nucleus and are the ones responsible for combining elements and making compounds.

The group (column) number represents the number of valence electrons for the main-group elements. Group 1A elements have one valence electron, Group 2A elements have two, and so on across the periodic table. The period (row) represents the outermost energy level or shell containing electrons. The electrons in the elements H and He (Period 1) are located in energy level 1, which can hold a maximum of only two electrons. The electrons in the elements in Period 2, lithium (Li) through neon (Ne), have their valence electrons in energy level 2 which can hold a maximum of eight electrons. Valence electrons for the main-group elements in Periods 3 and higher are similarly found in energy level 3 and higher. Our focus will be on the electron arrangements in the main-group elements.

As we will see throughout this chapter and the remainder of this book, the reactivity of an atom is determined by the arrangement of the electrons, specifically the valence electrons, in its electron cloud.

Groups indicate the number of valence electrons in atoms for the main-group elements

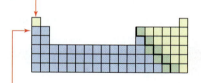

Periods indicate the energy level for atoms for the main-group elements

sample problem 3.2   Valence Electrons

How many valence electrons are in the following elements?

a. N   b. Si

Solution

The number of valence electrons can be determined by the group number for the main-group elements.

a. Nitrogen is in group 5A, so the number of valence electrons is five.

b. Silicon is in group 4A, so the number of valence electrons is four.

The Octet Rule

Chemists have long understood that if an element is more stable, it is less likely to react with another element, and if an element is less stable, it is more likely to react. One group of elements, Group 8A, does not readily form compounds. The Group 8A elements are also called the **noble** or inert **gases** and are chemically unreactive under all but extreme conditions. Elements that are unreactive are stable as atoms. In other words, the behaviors of stability and reactivity are *inversely related.* What makes the noble gas elements so much more stable and less reactive than other atoms? The answer is found in the valence electrons of the noble gases. The valence electrons are the electrons in atoms that combine when forming compounds.

Recall that the group number of the main-group elements indicates the number of valence electrons. The noble gases (8A) have eight valence electrons (except for helium, which has only two valence electrons). Experimentally it has been observed that possessing eight valence electrons is a highly stable state for an atom. Most of the atoms we will see in this book will react with other atoms to achieve a total of eight electrons in their valence shell. This is known as the **octet rule.**

To achieve a valence octet, some atoms give away electrons, others will accept or take electrons, and still others share electrons with other atoms. These different modes of achieving a valence octet lead to different kinds of compounds.

practice problems

3.1 How many electrons constitute a full energy level $n = 1$? Energy level $n = 2$?

3.2 Which electrons have higher energy, those in energy level $n = 1$ or those in energy level $n = 2$?

3.3 How many electrons are in each energy level of the following elements?

   a. He   b. C   c. Na   d. Ne

3.4 How many electrons are in each energy level of the following elements?

   a. H   b. F   c. Ar   d. K

3.5 How many valence electrons are present in the following atoms?

   a. O   b. C   c. P   d. Na

3.6 How many valence electrons are present in the following atoms?

   a. S   b. Mg   c. Be   d. Cl

3.2 In Search of an Octet, Part 1: Ion Formation

3.2 Inquiry Question:
How do ions form?

First let's take a look at atoms that gain or lose electrons to achieve a valence octet. In Section 2.1, we saw that an atom of a given element has an equal number of protons and electrons, which means that an atom is electrically neutral—it has no charge. If an atom gains or loses electrons to achieve an octet, this newly formed particle would have an unequal number of protons and electrons and, therefore, would have a net charge. We refer to these charged atoms as **ions.**

As our first example, consider chlorine, element number 17 on the periodic table. As an element in Group 7A, a chlorine atom has seven valence electrons. To achieve an octet in its valence shell, the chlorine atom must gain one electron. By gaining that one electron, the chlorine atom would then have an extra electron in its electron cloud, which means that it has more negative charges than positive charges and, therefore, a net negative charge. Ions formed when atoms gain electrons and become negatively charged are called **anions.** When chlorine gains the one electron necessary to complete its octet, the newly formed ion has a net charge of 1−. As **Figure 3.2** shows, adding one electron to the chlorine atom means that it now has 18 electrons but still only 17 protons, resulting in a net charge of 1−.

Notice that when chlorine gains an electron to complete the octet in its valence shell, it achieves not only the same number of valence electrons as argon but also has the same number of total electrons as argon. The ion formed by chlorine is **isoelectronic** (*iso*, a prefix meaning "same") with the argon atom. The chlorine ion is more stable than the chlorine atom because it has satisfied the octet rule.

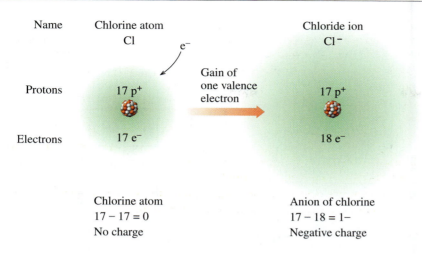

Chlorine atom
17 − 17 = 0
No charge

Anion of chlorine
17 − 18 = 1−
Negative charge

FIGURE 3.2 Chlorine forming an anion. A chlorine atom gains an electron when forming an anion.

As a second example, consider sodium, element number 11 on the periodic table. The sodium atom is in Group 1A and has one valence electron. To complete its octet, the sodium atom would have to gain seven more electrons. Such a process takes more energy than the atom has available for attracting electrons and, therefore, is not favorable. However, look at the possibility shown in **Figure 3.3**. If the sodium atom loses one electron, it becomes isoelectronic with neon. This means that the ion formed by the loss of that one electron would have a complete octet in its valence shell—exactly like neon. After losing the electron, the ion would have only 10 electrons but still 11 protons. Thus, it would have a net charge of 1+ (11 protons − 10 electrons = 1+). Ions formed when atoms lose electrons and become positively charged are called **cations.**

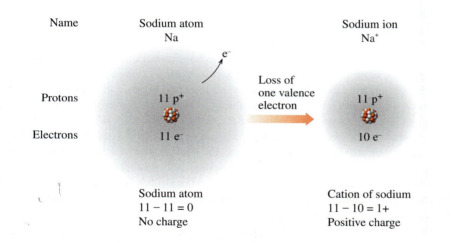

Sodium atom
11 − 11 = 0
No charge

Cation of sodium
11 − 10 = 1+
Positive charge

FIGURE 3.3 Sodium forming a cation. A sodium atom gives up an electron when forming a cation.

Trends in Ion Formation

Which atoms gain electrons to form anions and which atoms lose electrons to form cations? The periodic table is a useful tool to help answer this question.

Recall from Chapter 1 that we can use the periodic table to distinguish the elements that are nonmetals from those that are metals (see **Figure 3.4**). As a rule, when elements form ions, those that are metals lose electrons to form cations, while elements that are nonmetals gain electrons to form anions. Among the main-group elements, those in Groups 1A, 2A, and 3A form cations. On the other hand, the nonmetallic elements in Groups 5A, 6A, and 7A form anions. (The elements of Group 4A tend not to form ions but gain their stability in ways we will discover later in this chapter.)

How is the charge on an ion determined? The periodic table is our tool for determining the charges on ions.

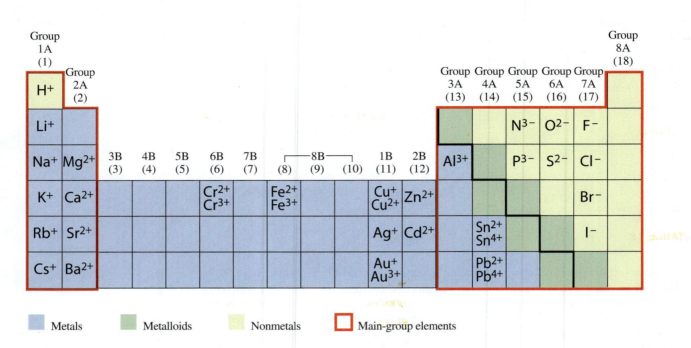

Group 1A (1)	Group 2A (2)	3B (3)	4B (4)	5B (5)	6B (6)	7B (7)	8B (8)	8B (9)	8B (10)	1B (11)	2B (12)	Group 3A (13)	Group 4A (14)	Group 5A (15)	Group 6A (16)	Group 7A (17)	Group 8A (18)
H^+																	
Li^+														N^{3-}	O^{2-}	F^-	
Na^+	Mg^{2+}											Al^{3+}		P^{3-}	S^{2-}	Cl^-	
K^+	Ca^{2+}				Cr^{2+} Cr^{3+}		Fe^{2+} Fe^{3+}			Cu^+ Cu^{2+}	Zn^{2+}					Br^-	
Rb^+	Sr^{2+}									Ag^+	Cd^{2+}		Sn^{2+} Sn^{4+}			I^-	
Cs^+	Ba^{2+}									Au^+ Au^{3+}			Pb^{2+} Pb^{4+}				

☐ Metals   ☐ Metalloids   ☐ Nonmetals   ☐ Main-group elements

FIGURE 3.4 Trends in ion formation. Positive ions are produced from metals and negative ions are formed from nonmetals. Common ions formed are shown.

Previously, we saw that a sodium atom loses one electron to form a cation with a 1+ charge. In fact, the same is true of all the elements of Group 1A. Because all members of that group have one valence electron, they all behave the same way, giving up that electron and forming a cation with a charge of 1+. The elements of Group 2A all have two valence electrons. During ion formation, these elements lose those two electrons to form cations with a 2+ charge. Similarly, the metallic elements of Group 3A lose three electrons to form cations with a 3+ charge. Notice that there is a pattern here. The metallic elements in Groups 1A to 3A form cations with a charge that is the same as the element's group number. In each case the cation formed is isoelectronic with one of the noble gases.

Recall that nonmetals gain electrons to form anions. Previously, we used the example of chlorine gaining one electron to complete its octet and form an anion with a 1− charge. The other members of Group 7A behave the same way, each gaining one electron to form an anion with a 1− charge. Next consider the elements in Group 6A. Recall that each of these elements has six valence electrons. For an element of Group 6A to complete its octet, it must gain *two* electrons. In so doing, the element forms an anion with two more electrons than the atom from which it originated and, therefore, has a charge of 2−. Following this trend, the nonmetallic elements of Group 5A (nitrogen and phosphorus) form anions with a charge of 3−. The charge on the anions formed by nonmetallic main-group elements is equal to the group number of the element minus eight.

The rules for predicting charges on ions formed by the main-group elements can be summarized as follows:

Predicting Charge on an Ion Formed by a Main-Group Element

Cations: Charge = Group Number;

for example, K = 1+, Mg = 2+, Al = 3+

Anions: Charge = Group Number − 8;

for example, Br (7 − 8) = 1−, O (6 − 8) = 2−, N (5 − 8) = 3−

For elements other than the main-group elements, the charge on the ions they form is not as simple to predict. These elements do not follow the octet rule. In fact, some of the transition metals form more than one ion. For example, both copper and iron can each form more than one cation. Copper forms the ions Cu^+ and Cu^{2+}, while iron forms Fe^{2+} and Fe^{3+}. Others like silver (Ag) and zinc (Zn) do not vary their charge and form only Ag^+ and Zn^{2+} ions respectively. Unlike the Group 1A-3A metals, it is not possible to determine the charge of the ions formed by the transition metals from their group number. Later, we will show how it is possible to predict the charges on these metal ions. For now, common charges on some of the transition metal ions and main-group metals in Periods 5 and 6 are shown in Figure 3.4.

Another group of common ions whose charges cannot be determined from the periodic table are the **polyatomic ions.** These are a special group of ions consisting of a group of nonmetals that interact with each other to form an ion. Examples include HCO_3^- (bicarbonate) and NH_4^+ (ammonium). Notice that each ion contains more than one type of nonmetal, but they interact to form an ion with a single charge. The charge shown is for the entire group of atoms together. See Table 3.2 for a listing of some common polyatomic ions.

TABLE 3.2 Common Polyatomic Ion Names and Formulas

Nonmetal	Formula of Ion[a]	Name of Ion
Hydrogen	OH^-	Hydroxide
Nitrogen	NH_4^+	Ammonium
	$\boxed{NO_3^-}$	**Nitrate**
	NO_2^-	Nitrite
Chlorine	$\boxed{ClO_3^-}$	**Chlorate**
	ClO_2^-	Chlorite
Carbon	$\boxed{CO_3^{2-}}$	**Carbonate**
	HCO_3^-	Hydrogen carbonate (or bicarbonate)
	CN^-	Cyanide
	$C_2H_3O_2^-$	Acetate
Sulfur	$\boxed{SO_4^{2-}}$	**Sulfate**
	HSO_4^-	Hydrogen sulfate (or bisulfate)
	SO_3^{2-}	Sulfite
	HSO_3^-	Hydrogen sulfite (or bisulfite)
Phosphorus	$\boxed{PO_4^{3-}}$	**Phosphate**
	HPO_4^{2-}	Hydrogen phosphate
	$H_2PO_4^-$	Dihydrogen phosphate
	PO_3^{3-}	Phosphite

[a]Boxed formulas are the most common polyatomic ion for that element.

sample problem 3.3 — Determining Protons and Electrons in Ions

How many protons and electrons are in the following ions?

a. Mg^{2+} b. O^{2-}

Solution

Recall that the number of protons is represented by the atomic number of the element as shown on the periodic table.

a. Protons = 12. An ion with a 2+ charge has two fewer electrons than protons or, in this example, 10 electrons.
b. Protons = 8. An ion with a 2− charge has two more electrons than protons or, in this example, 10 electrons.

Naming Ions

To distinguish an ion and the atom from which it was formed, we give the ion a different name. For metal ions, this simply involves adding the word *ion* to the name of the metal, so Na^+ is called sodium *ion*. Transition metals that form more than one ion use a Roman numeral in parentheses following the name of the metal to indicate the charge on the ion. So, Fe^{2+} is the iron(II) ion, and the Fe^{3+} ion is the iron(III) ion.

For the nonmetals, the suffix *ide* replaces the last few letters of the name of the element. For example, the anion formed from fluorine is fluor*ide* and that of oxygen is ox*ide*. In most cases, the first syllable of the element is kept and the *ide* suffix is applied for the dropped letters.

Most of the common polyatomic ions end in *ate* (see Table 3.2). The *ite* ending is used for the names of related ions that have one fewer oxygen atom. By recognizing these endings you can identify when a polyatomic ion is present in a compound. The hydroxide (OH^-), hydronium (H_3O^+), and cyanide ions (CN^-) are exceptions to this naming pattern. Just like vocabulary words, you may have to memorize the number of oxygens and charges associated with some of these ions. By memorizing the formulas in the boxes, the other related ions can be derived. For example, the sulfate ion is SO_4^{2-}. The formula for the sulfite ion with one fewer oxygen atom is, therefore, SO_3^{2-}.

---

sample problem 3.4 **Naming Ions**

Name the following ions:

a. Ca^{2+} b. O^{2-} c. Cr^{3+} d. NO_3^-

Solution

a. Calcium is a main-group metal, so the name will be the metal name plus the word *ion*. The name is *calcium ion*.
b. Oxygen is a nonmetal, so the ending *ide* will replace the current ending. The name is *oxide*.
c. Chromium is a transition metal, so it is necessary to identify the charge using a Roman numeral after the metal name. The name is *chromium(III) ion*.
d. This is a polyatomic ion. It has the highest number of oxygens of any polyatomic nitrogen ion. The name is *nitrate*.

Important Ions in the Body

integrating chemistry

A number of ions found in the fluids and cells of the human body perform important functions. The main cations found in the body include Na^+, K^+, Ca^{2+}, and Mg^{2+}, which are important in maintaining solution concentrations inside and outside cells. The main anions are Cl^-, HCO_3^-, and HPO_4^{2-}. These ions help maintain the charge neutrality between the blood and fluids inside cells. The phosphate ion, also called inorganic phosphate
and abbreviated P_i, is involved in cellular energy transfer during chemical reactions as well. In contrast to ions in fluids, ions can also form hard minerals. Tooth enamel contains the mineral hydroxyapatite, consisting of calcium (Ca^{2+}), phosphate (PO_4^{3-}), and hydroxide (OH^-) ions. Table 3.3 shows a listing of the main ions in bodily fluids and their functions.

TABLE 3.3 Biologically Important Ions

Ion	Function	Sources
Cations		
Na^+	Regulates fluids outside cells	Table salt, seafood
K^+	Maintains ion concentration in cells; induces heartbeat	Dairy, bananas, meat
Ca^{2+}	Found outside cells; involved in muscle contraction, formation of bones and teeth; regulates heartbeat	Dairy, whole grains, leafy vegetables
Mg^{2+}	Found inside cells; involved in transmission of nerve impulses	Nuts, seafood, leafy vegetables
Fe^{2+}	Found in the protein hemoglobin, which is responsible for oxygen transport from lungs to tissue	Liver, red meat, leafy vegetables
Anions		
Cl^-	Found in gastric juice and outside cells; involved in fluid balance in cells	Table salt, seafood
HCO_3^-	Controls acid–base balance in blood	Body produces own supply through breathing and breakdown of foods
HPO_4^{2-}	Controls acid–base balance in cells	Fish, poultry, dairy

practice problems

3.7 What is the difference between a metal atom and a cation of the same element?

3.8 What is the difference between a nonmetal atom and an anion of the same element?

3.9 How are the names of a nonmetal and its anion different?

3.10 How are the names of a transition metal and its cation different?

3.11 How many protons and electrons are present in the following ions?
a. Ca^{2+} b. I^- c. S^{2-} d. Zn^{2+}

3.12 How many protons and electrons are present in the following ions?
a. Al^{3+} b. Br^- c. Hg^+ d. N^{3-}

3.13 Name the ions in Problem 3.11.

3.14 Name the ions in Problem 3.12.

3.15 Give the name and symbol of the ion with the following number of protons and electrons:
a. 9 protons, 10 electrons
b. 24 protons, 21 electrons

3.16 Give the name and symbol of the ion with the following number of protons and electrons:
a. 11 protons, 10 electrons
b. 33 protons, 36 electrons

3.17 Name the following ions:
a. NH_4^+ b. $C_2H_3O_2^-$ c. CN^-

3.18 Name the following ions:
a. Cu^{2+} b. SO_4^{2-} c. HPO_4^{2-}

Discovering the Concepts

? Inquiry Activity — Ionic Compounds

Part 1. Information

Ionic compounds form between elements that can give or take electrons (exist as cations or anions) to yield a full valence shell. In an ionic compound, the cation gives its electron(s) to the anion, and a strong attraction of opposites is formed between the two. This type of chemical bond is called an *ionic bond.*

Examine the following data set. Notice that none of the compounds have a net charge (+ and − charges add up to zero).

Data Set	
Formula	**Name**
NaI	Sodium iodide
K_2S	Potassium sulfide
CaF_2	Calcium fluoride
Mg_3N_2	Magnesium nitride
$CuBr_2$	Copper(II) bromide
Fe_2O_3	Iron(III) oxide
FeO	Iron(II) oxide
AlP	Aluminum phosphide

Questions

1. In naming ionic compounds, which goes first in the name, the metal or the nonmetal? The anion or the cation?
2. How are anions and cations the same? How are they different?
3. a. In potassium sulfide, the charge on the potassium ion is _____ and the charge on the sulfide is _____.
    b. Why does the formula for potassium sulfide have two potassium ions and one sulfide ion?
4. Some of the names of ionic compounds have Roman numerals after the metals and others do not. Consider their position on the periodic table. How do these two groups of metals differ?
5. a. The charge on one Fe in Fe_2O_3 is _____.
    b. The charge on the Fe in FeO is _____.
6. a. Based on your answer to question 5, what does the Roman numeral in the names in the data set represent?
    b. Based on your answer to question 4, when is the Roman numeral used in the name of an ionic compound?
7. How does the name of the nonmetal (anionic) part of the compound differ from that of its original element?

Part 2. Information

Polyatomic ions are groups of nonmetal atoms that together act like an anion or cation. When combined with a cation or anion respectively, they form ionic compounds.

Formula	Name	Formula	Name
NH_4^+	Ammonium	OH^-	Hydroxide
$C_2H_3O_2^-$	Acetate	HPO_4^{2-}	Hydrogen phosphate
HCO_3^-	Bicarbonate	NO_3^-	Nitrate
CO_3^{2-}	Carbonate	SO_4^{2-}	Sulfate

8. Complete the following table:

Formula	Name
KCl	
	Sodium sulfate
$Fe(OH)_3$	
$NaHCO_3$	
	Ammonium nitrate
	Silver acetate
$CaHPO_4$	
	Mercury(II) chloride

3.3 Ionic Compounds—Electron Give and Take

In the previous section, we saw that main-group metal atoms lose electrons, forming cations to achieve an octet, whereas nonmetal atoms gain electrons, forming anions for the same reason. Where do the electrons go, or where do they come from? Ion formation does not occur in isolation. In other words, cations do not form unless anions are also formed and vice versa. When a metal and a nonmetal combine, a compound is formed and electrons are transferred between the atoms forming oppositely charged ions as shown in **Figure 3.5**.

Because opposite charges attract, the newly formed cation and anion are strongly attracted to each other. This attraction of cation and anion is called an **ionic bond.** The resulting combination of cation and anion held together by this strong attraction is called an **ionic compound. Figure 3.6** demonstrates this process for the formation of common table salt, sodium chloride.

3.3 Inquiry Question:
How do ionic compounds form?

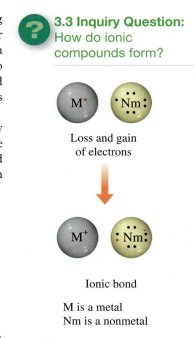

Loss and gain of electrons

Ionic bond

M is a metal
Nm is a nonmetal

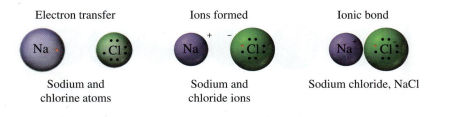

Electron transfer   Ions formed   Ionic bond

Sodium and chlorine atoms   Sodium and chloride ions   Sodium chloride, NaCl

FIGURE 3.6 Formation of the ionic compound sodium chloride. Both sodium and chlorine are able to complete their octet by giving or taking electrons, respectively.

FIGURE 3.5 Ion formation. When ions form, metals transfer electrons to nonmetals.

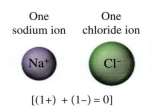

One   One
sodium ion   chloride ion

$[(1+) + (1-) = 0]$

Formulas of Ionic Compounds

In the example shown in Figure 3.6, a sodium atom gives up one electron to a chlorine atom to form a sodium ion (charge = 1+) and a chloride ion (charge = 1−). The oppositely charged ions now have completed octets of valence electrons through this give and take of electrons. The ions are attracted to each other, and an ionic bond is formed between them. The result is a compound with an overall charge that is neutral (zero) as shown at left.

Sodium chloride, NaCl, is an ionic compound because it contains a metal combined with a nonmetal, but the formula does not indicate the charges on the ions. The formula is written with no charges shown. For sodium chloride, only one sodium ion (Na^+) and one chloride ion (Cl^-) combine to form a neutral compound. When writing formulas for chemical compounds, we represent the number of each particle (ions in this case) in the compound with a subscript. Just as a charge of 1+ on an ion is represented as +, the "1" is understood in the formula for the compound and is written as NaCl instead of Na_1Cl_1.

How would we write a formula if one of the ions has a greater charge (although still opposite sign) than the other? The number of ions that combine to form an ionic compound is determined by the charge of the cation and the charge of the anion. Cations and anions combine so that the compound has a net charge of zero. No charges appear in the compound formula.

For example, we saw that calcium forms ions with a 2+ charge, and fluorine, like chlorine, forms ions with a 1− charge. When calcium and fluorine react to form an ionic compound, calcium must give away two electrons, but fluorine can accept only one. Therefore, a calcium atom reacts with *two* fluorine atoms, giving up one electron to each of the fluorine atoms to form a calcium ion (Ca^{2+}) ion and two fluoride (F^-) ions. The resulting mineral, called fluorite, forms a hard transparent crystal used in telescope and camera lenses. **Figure 3.7** demonstrates the formation of the ionic compound from calcium and fluorine.

Once again the charges on the ions that were formed and the resulting ionic bond formation result in a compound whose net (overall) charge is zero as shown:

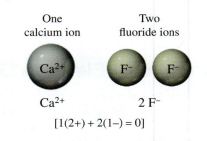

One   Two
calcium ion   fluoride ions

Ca^{2+} $2\ F^-$

$[1(2+) + 2(1-) = 0]$

Because the neutral ionic compound has one calcium ion and two fluoride ions, the correct formula is CaF_2.

If we examine the formulas more closely, we see that the subscript on each ion is actually the magnitude of the charge (number without the sign) on the other ion. Using CaF_2 as an example, the calcium ion has a charge magnitude of 2. Notice the subscript on fluorine in the formula. It is a 2. Similarly, fluoride has a charge magnitude of 1, and the subscript on calcium in the formula is also 1.

In many cases, this pattern can be used as a check that you have determined the correct formula for an ionic compound.

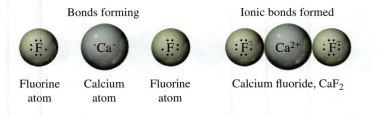

Bonds forming   Ionic bonds formed

Fluorine   Calcium   Fluorine   Calcium fluoride, CaF_2
atom   atom   atom

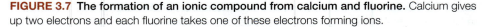

FIGURE 3.7 **The formation of an ionic compound from calcium and fluorine.** Calcium gives up two electrons and each fluorine takes one of these electrons forming ions.

Checking a Chemical Formula

Determine the correct formula for the ionic compound formed from aluminum ions and oxide ions.

STEP 1: Determine the charge of the ions from the periodic table if they are main-group elements. In this case, aluminum is in Group 3A and has a 3+ charge, and oxygen is in Group 6A and the oxide ion has a 2− charge.

STEP 2: Combine the ions in a formula so that the total charge present sums to zero. In this case, two aluminum ions would provide a total 6+ charge and three oxide ions would provide a total 6− charge.

$$(2 \times (3-)) + (3 \times (2-)) = 0.$$ The formula is written as Al_2O_3.

STEP 3: Check your formula by examining the charges on the ions and subscripts in the formula.

The charge on this ion Al^{3+} O^{2-} The charge on this ion

becomes the subscript on this ion Al_2O_3 becomes the subscript on this ion

When writing ionic formulas, in many cases, the charge on the anion becomes the subscript on the cation and vice versa.

To write the formula for an ionic compound correctly, you must be able to use the periodic table to determine the charges on the main-group elements and remember the charges on the polyatomic ions.

So far, we have written formulas for ionic compounds containing main-group elements. Next, we address transition metals where the charges often vary. If given the formula, can we predict the charge on a transition metal in a compound? To demonstrate this, let's work through an example within the Solving a Problem below.

Predicting Number and Charge in an Ionic Compound

Predict the number and charge on the ions present in the compound CuO.
We can determine the transition metal charge in an ionic compound by remembering a few rules.

RULE 1. The formula of an ionic compound has no net charge (+ and − charges add up to zero).

RULE 2. The charge of the anion is known.

Apply **RULE 1** and **RULE 2.** In the case of CuO, an oxygen anion (Group 6A) has a 2− charge. This compound contains one copper ion and one oxide. Since the + charge of the copper ion must equal the negative charge of the oxide for the compound to have no net charge, the charge of the copper is 2+.

sample problem 3.5 **Writing Formulas for Ionic Compounds of Main-Group Elements**

Determine the correct formula of the ionic compound formed from the following combination of ions:

a. sodium ions and sulfide ions
b. magnesium ions and oxide ions
c. calcium ions and hydroxide ions

Solution

a. **STEP 1: Determine the charge.** Sodium ions have a 1+ charge, Na^+, and sulfide ions have a 2− charge, S^{2-}.

STEP 2: Combine the ions in a formula. Two sodium ions (total charge 2+) combine with one sulfide (total charge 2−) to form the neutral ionic compound sodium sulfide. The two sodium ions are represented in the formula by the subscript 2 after the symbol for sodium, Na. The formula is written Na_2S.

STEP 3: Check your formula.

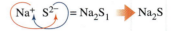

b. **STEP 1: Determine the charge.** Magnesium ions have a charge of 2+, Mg^{2+}, and oxide ions have a charge of 2−, O^{2-}.

STEP 2: Combine the ions in a formula. Because they have the same charge, one magnesium ion (total charge 2+) will combine with one oxide (total charge 2−) to form the neutral ionic compound magnesium oxide. The formula is written MgO.

STEP 3: Check your formula.

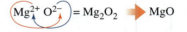

This formula reminds us of an important exception. When the charges on the ions are equal but opposite, the correct formula should contain the smallest common ratio (here 1:1). The correct formula for this compound combines one Mg^{2+} ion and one O^{2-} ion in a neutral compound with the formula MgO.

c. **STEP 1: Determine the charge.** Calcium ions have a 2+ charge, Ca^{2+}, and the polyatomic ion, hydroxide, has a 1− charge, OH^-.

STEP 2: Combine the ions in a formula. One calcium ion (total charge 2+) will combine with two hydroxides (total charge 2−) to form the neutral ionic compound calcium hydroxide. In this case, parentheses are used around the hydroxide to indicate multiple polyatomic ions. The formula is written $Ca(OH)_2$.

STEP 3: Check your formula.

$$\left(Ca^{2+} \quad OH^- \right) = Ca_1(OH)_2 \quad \Rightarrow \quad Ca(OH)_2$$

sample problem 3.6   Predicting Charges of Ions

Determine the number and charge of each ion present in the following formulas:

a.   KBr, used to treat epilepsy in dogs
b. $Al_2(SO_4)_3$, used in water purification
c. $CoCl_2$, used in the synthesis of vitamin B_{12}

Solution

a.   For the main-group elements, the charge can be determined from the periodic table. Potassium is in Group 1A so its charge as an ion is 1+. Bromine is in Group 7A so its charge as an ion is 1−. The ionic compound of potassium and bromide contains one potassium ion (K^+) and one bromide (Br^-).

b.   Aluminum is in Group 3A so its charge as an ion is 3+. This compound contains sulfate, a polyatomic ion (see Section 3.2). This group of atoms in sulfate functions together as a single ion and has a 2− charge as an anion. The formula for the ionic compound contains two aluminum ions and three sulfate ions.

c.   Cobalt is a transition metal, so we must also determine its charge by applying the rules previously discussed. Chlorine is an element in Group 7A, so its charge is 1−. Because the compound combines two chlorides (total charge 2−) with one cobalt ion, the cobalt must have a 2+ charge for the compound to have a net charge of zero. The compound contains one cobalt ion (Co^{2+}) and two chloride ions (Cl^-).

Naming Ionic Compounds

Now that we can write formulas for ionic compounds, the next step is to name them. We can already name cations and anions. To name an ionic compound, put the names of the two ions together, always remembering to list the cation first. So, NaCl is sodium chloride and MgO is magnesium oxide. Notice that you do not include the word *ion*—sodium ion—in the name. For transition metals, recall that a Roman numeral is used to designate the charge on the transition metal, since it can vary depending on the compound. The example compound CuO used earlier would be written as copper(II) oxide since the charge on the copper ion is 2+.

If a polyatomic ion is present, the name of the ion remains unchanged in the name of the compound. For example, $Ca_3(PO_4)_2$ is named calcium phosphate.

sample problem 3.7   Naming Ionic Compounds

Name the ionic compounds represented by the formulas given.

a.   LiF, used in production of ceramics
b. MnO_2, used in disposable batteries
c. $PbSO_4$, found in car batteries
d. NH_4Cl, used to treat electrolyte imbalances

Solution

a.   This compound is lithium fluoride.
b.   This compound contains the transition metal manganese. Before naming this compound we must determine the charge on the manganese ion. Because an oxygen ion (oxide) always has a 2− charge, and two oxides (total 4− charge) combine with one manganese ion, the manganese ion has a 4+ charge. (This gives the compound no net charge.) This compound is manganese(IV) oxide.

c. Lead is a transition metal. The polyatomic anion sulfate has a 2– charge. Since this combines one-to-one with the lead, the charge on the lead must be 2+ to give a neutral ionic compound. The compound is lead(II) sulfate.

d. Remembering the names of the polyatomic ions, this compound is ammonium chloride. Note that this is an example of an ionic compound that does not contain a metal combined with a nonmetal. Ammonium compounds are ionic compounds that do not contain a metal ion.

practice problems

3.19 Give the formula and name for the ionic compounds formed from the following pairs of ions:
a. gold(III) and chloride, an antimicrobial
b. calcium and carbonate, the mineral limestone
c. magnesium and hydroxide, the main compound in milk of magnesia

3.20 Give the formula and name for the ionic compounds formed from the following pairs of elements:
a. potassium and chloride, the main ingredients in salt substitutes used by people on low-sodium diets
b. zinc and oxide, used in some sunscreens *(Note: The charge on zinc does not vary; see Figure 3.4.)*
c. copper(I) and sulfide, a common copper mineral mined for copper metal

3.21 Give the number of each type of ion present in each of the ionic compounds in Problem 3.19.

3.22 Give the number of each type of ion present in each of the ionic compounds in Problem 3.20.

3.23 Give the formula and name for the ionic compound formed from the combination of the indicated metal and carbonate, CO_3^{2-}, a polyatomic ion.
a. Na   b. Fe(II)   c. Al

3.24 Repeat Problem 3.23 using the acetate, $C_2H_3O_2^-$, ion.

3.25 Give the number of each type of ion present in each of the ionic compounds in Problem 3.23.

3.26 Give the number of each type of ion present in each of the ionic compounds in Problem 3.24.

3.4 Inquiry Question: How do covalent compounds form?

Nonliving mineral
Mostly ionic and metallic compounds

Living frog and plant
Mostly nonmetal compounds

3.4 In Search of an Octet, Part 2: Covalent Bonding

Previously we saw that a stable octet of electrons can be achieved by an atom by forming an ion. Cations are usually formed from metals, and anions are formed from nonmetals. An ionic bond between cation and anion results from this give-and-take of electrons due to the strong attraction of the oppositely charged ions. While such ionic compounds are abundant in nature in nonliving things like rocks and minerals, compounds containing mostly nonmetal components are found in living things.

Consider carbon dioxide, CO_2. This compound is a byproduct of our cellular energy production process and is used by plants to manufacture their food source. Carbon dioxide is everywhere. Notice that CO_2 contains one atom of carbon (a nonmetal) and two atoms of oxygen (also a nonmetal). This combination means that it is *not* an ionic compound.

Earlier in this chapter, we noted that the Group 4A elements like carbon do not form ions. But, carbon *does* form compounds like CO_2 with other elements. How does it form compounds without forming ions?

Nonmetals combine by *sharing* valence electrons to achieve an octet. This sharing of electrons results in the formation of a **covalent bond.** Unlike an ionic bond, no electrons are transferred between the atoms in a covalent bond; instead, electrons are shared, making both atoms more stable. In other words, the valence electrons in the bond belong to both of the atoms.

When atoms share electrons to form covalent bonds, the resulting new compound is called a **covalent compound.** The smallest (or fundamental) unit of a covalent compound is a **molecule.** So, carbon dioxide, CO_2, is a molecule and contains atoms held together by covalent bonds.

sample problem
3.8 **Distinguishing Covalent and Ionic Compounds**

Determine whether each of the following is a covalent or ionic compound:

a. NaBr   b. CCl_4 c. NH_4CN

Solution

a. NaBr contains a metal (sodium) and a nonmetal (bromine). Compounds formed from the combination of a metal and a nonmetal are ionic.
b. CCl_4 is made up of carbon (a nonmetal) and chlorine (also a nonmetal). Compounds formed from the combination of nonmetals are covalent compounds.
c. Initial inspection of the formula of NH_4CN shows that all the atoms involved are nonmetals (N, H, and C). However, in Section 3.2, we saw that both NH_4^+ and CN^- are polyatomic ions. Despite the fact that this compound is composed entirely of nonmetals, it is an ionic compound because it contains polyatomic ions.

Covalent Bond Formation

Atoms share electrons with other atoms, forming covalent bonds to complete an octet, but how many covalent bonds does an atom form? A simple rule of thumb is this: *The number of covalent bonds that an atom will form equals the number of electrons necessary to complete its octet.*

We can see how this works with atoms by using dots to represent electrons. The use of electron-dot symbols was first developed by chemist G. N. Lewis to show how atoms share electrons in covalent bonding. The electron-dot symbol for any atom consists of the elemental symbol plus a dot for each valence electron. The electron-dot symbol for chlorine is as follows:

$$:\overset{\cdot\cdot}{\underset{\cdot\cdot}{Cl}}\cdot$$

We know that chlorine, a member of Group 7A, has seven valence electrons. We construct its electron-dot symbol by placing one electron on each side of the elemental symbol (top, bottom, left, and right) and then adding the remaining electrons to form pairs until all the valence electrons are represented. After the first four electrons are in place, each of the remaining electrons can be placed with any one of the existing electrons to form pairs. So, these two electron-dot symbols for chlorine are equivalent:

$$:\overset{\cdot\cdot}{\underset{\cdot\cdot}{Cl}}\cdot \quad \text{or} \quad \cdot\overset{\cdot\cdot}{\underset{\cdot\cdot}{Cl}}:$$

To complete its octet through a covalent bond, chlorine must share its unpaired electron with another nonmetal atom, here a chlorine atom. A shared pair of electrons, known as a **bonding pair** or simply a bond, is represented as a dash or line connecting the two electron-dot symbols (see **Figure 3.8**). Note that each of the atoms in the newly formed molecule now has an octet—three **lone pairs** of electrons, which are not shared, and the shared bonding pair.

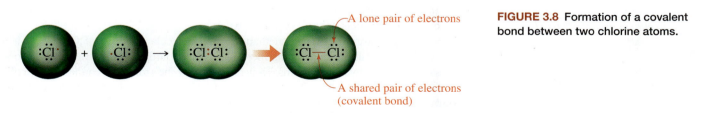

A lone pair of electrons

A shared pair of electrons (covalent bond)

FIGURE 3.8 Formation of a covalent bond between two chlorine atoms.

A glance at the properly constructed electron-dot symbols of the main-group nonmetals in the second period of the periodic table helps to determine the preferred covalent bonding patterns for these elements.

Let's look at the element on which all life on Earth is based, carbon. With its four unpaired valence electrons, carbon must share its four electrons and form four covalent bonds to complete its octet. To accomplish this, carbon can share one electron each with four other atoms to form four **single bonds.** Or, it can form **double bonds** by sharing two of its electrons with one atom. Carbon can also share three of its electrons with one atom to form a **triple bond.** The most common covalent bonding patterns observed for carbon are shown in **Figure 3.9**. Note that in each case carbon has four bonding pairs—a complete octet—surrounding it.

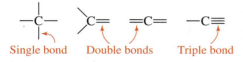

Single bond   Double bonds   Triple bond

FIGURE 3.9 Preferred bonding patterns used by carbon to complete its octet when forming bonds in a molecule.

Returning to the electron-dot symbols, we see that nitrogen forms three covalent bonds, oxygen forms two covalent bonds, and fluorine, like its group member chlorine, forms one covalent bond. Note that hydrogen is an exception to the octet rule. As a member of Period 1, hydrogen requires only two electrons to complete its valence shell; therefore, hydrogen forms just one covalent bond.

Table 3.4 shows the common bonding patterns for the main-group nonmetals that are encountered frequently in living systems.

TABLE 3.4 Preferred Covalent Bonding Patterns for Main-Group Elements

Group 1A	Group 4A	Group 5A	Group 6A	Group 7A
H—	—C̵—	—N̈—	—Ö—	:F̈—
	C̵=	—N̈=	Ö=	
	=C=	:N≡		
	—C≡			
			—S̈—	:C̈l—
				:B̈r—
				:Ï—

Formulas and Structures of Covalent Compounds

Like the chemical formulas for ionic compounds, the **molecular formula** for a covalent compound completely identifies *all* the components in a molecule. Glucose, the sugar we use for energy, is a covalent compound with the formula $C_6H_{12}O_6$. This tells us that a molecule of glucose has 6 carbon atoms, 12 hydrogen atoms, and 6 oxygen atoms. This is not the same formula as CH_2O. Molecular formulas do not reduce to the smallest whole number ratio like the chemical formulas for ionic compounds.

While the molecular formula does tell us the number of atoms in a molecule, it does not tell us *how* the atoms are joined together. The formula does not show the structure. The electron-dot symbols help us visualize structure.

Consider the natural gas called methane with a molecular formula of CH_4. This is a covalent compound (all nonmetals). Its molecular formula indicates that methane has one carbon atom and four hydrogen atoms. To determine how the atoms are connected, start by drawing the electron-dot symbols for all the atoms involved in the molecule. Because carbon has four unpaired valence electrons, it must make four bonds to complete its octet. Each of

the hydrogen atoms has one valence electron and must have two. Therefore, each hydrogen atom will make one bond. So, carbon forms four bonds and each of the *four* hydrogen atoms forms one bond. Connecting the atoms as shown fulfills these requirements. Recall that a line connecting two atoms represents a bonding pair of electrons,

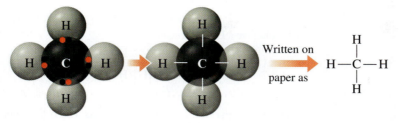

The drawing shows the number and type of each atom in the molecule and their connectivity. This representation is called a **Lewis structure.**

sample problem **3.9** **Drawing Lewis Structures for Covalent Compounds**

Draw the correct Lewis structure for each of the following covalent compounds:

a. NH_3
b. C_2H_4

Solution

a. Start by drawing the correct electron-dot symbol for each of the atoms involved in the formula.

Next, arrange the atoms so that unpaired electrons are next to each other, much as you line up the pieces of a puzzle to fit them together.

To complete the Lewis structure, replace the pairs of dots between atoms with lines representing bonds connecting the two atoms.

Finally, compare the structure with those in Table 3.4, which shows that a structure with three single bonds and one lone pair is one of nitrogen's preferred bonding patterns.

b. Once again, start by drawing the correct electron-dot symbol for each of the atoms involved in the formula.

In this case, two carbon atoms are present, and since each hydrogen atom can form only one bond, the two carbons must bond to each other. Arrange the atoms as presented in part a using just single bonds and you would get

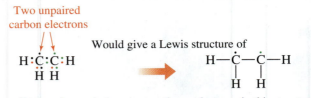

Note that each of the carbon atoms in this structure still has an unpaired electron *and* each has only seven valence electrons (three bonding pairs = 6 electrons + 1 unpaired electron = 7 electrons). In this book, carbon will always form four bonds to complete its octet. So, to complete their octets in this case, the carbons share their unpaired electrons to give

Finally, when we compare the structure with Table 3.4, we find that the arrangement with two single bonds and one double bond is one of carbon's preferred bonding patterns.

TABLE 3.5 Greek Prefixes Used When Naming Binary Covalent Compounds

Prefix	Meaning
mono	1
di	2
tri	3
tetra	4
penta	5
hexa	6
hepta	7

Naming Covalent Compounds

Naming ionic compounds is a relatively straightforward process—name the cation and then name the anion. Naming covalent compounds is a little different. Here we focus on naming the simplest covalent compounds and look at a few common exceptions.

Covalent compounds composed of only two elements are known as **binary compounds.** These compounds can be named by a three-step procedure. Let's try an example to illustrate the procedure.

Solving a Problem

Naming Covalent Compounds

Name the covalent compounds SO_2 and SO_3.

STEP 1: Name the first element in the formula. In both compounds, the first element is **sulfur**.

STEP 2: Name the second element in the formula and change the ending to *ide*. In both compounds, the second element is oxygen, so **oxide**.

STEP 3: Designate the number of each element present using one of the Greek prefixes shown in Table 3.5. In the first compound there are two oxides, in the second compound there are three. The two compounds are named as follows:

SO_2: Sulfur dioxide
SO_3: Sulfur trioxide

Indicating the number of each element present is important when naming covalent compounds because nonmetals can combine with each other in multiple ways. Notice that in both cases, the *mono* prefix for sulfur was understood and was not included in the name. This exception (for *mono*) holds true for only the first element in the name. The compound CO is named carbon *mono*xide.

Some binary covalent compounds are so prevalent in our world that their long-standing traditional names have never been replaced by the rules-derived names. One of the most notable exceptions to the rules for naming binary covalent compounds is H_2O. According to the naming steps, this compound is dihydrogen monoxide, but of course it is recognized as water. Ammonia—a component of smelling salts, many household cleaners, and agricultural fertilizers—has a molecular formula of NH_3 and is also an exception.

sample problem 3.10   Naming Binary Covalent Compounds

Give the correct name for each of the following binary covalent compounds:

a. N_2O_4, used as a rocket fuel
b. CO_2, produced in respiring cells
c. Cl_2O, a degradation product found in the ozone layer

Solution

a. dinitrogen tetroxide
b. carbon dioxide
c. dichlorine monoxide

sample problem 3.11   Writing Formulas of Binary Covalent Compounds

Give the correct formula for each of the following binary covalent compounds:

a. carbon monoxide, a poisonous gas
b. phosphorus trichloride, used in the manufacture of flame retardants
c. dinitrogen pentoxide, a greenhouse gas

Solution

a. CO b. PCl_3 c. N_2O_5

We have introduced the identification of naming binary ionic and covalent compounds as separate topics, but in practice, before naming any compound, we must first determine whether the compound is ionic (metal with nonmetal) or covalent (all nonmetals). The flow chart in **Figure 3.10** offers guidance for correct identification and naming of compounds.

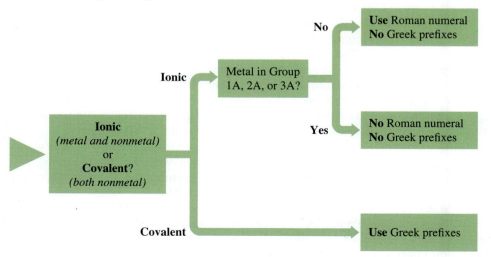

FIGURE 3.10 Flow chart for naming ionic and binary covalent compounds.

3.27 Draw the correct Lewis structure for each of the
following covalent compounds:
a. $CHCl_3$, chloroform
b. H_2S, smell of rotting eggs
c. CO_2, dry ice
d. C_2H_2, acetylene, a fuel used in welding

3.28 Draw the correct Lewis structure for each of the
following covalent compounds:
a. H_2O
b. C_3H_8, propane, a fuel used for heating
c. N_2, nitrogen gas
d. SiF_4, used in making integrated circuits

3.29 Using Table 3.4, decide whether molecules with
the following formulas are likely to exist (form Lewis
structures that obey the octet rule). Briefly explain
your answer.
a. CH_2 b. NH_2 c. $BrCl$ d. H_2S

3.30 Using Table 3.4, decide whether molecules with
the following formulas are likely to exist (form Lewis
structures that obey the octet rule). Briefly explain
your answer.
a. CCl_3 b. HF c. OF_4 d. C_2H_6

3.31 Determine whether each of the following is a covalent
or ionic compound:
a. NCl_3 b. Fe_2O_3 c. Cs_2CO_3 d. PBr_3
e. C_5H_{12} f. NH_4OH g. SeO_2 h. BaS
i. CBr_4 j. OF_2

3.32 Determine whether each of the following
is a covalent or ionic compound:
a. Li_2O b. CS_2 c. $AlCl_3$ d. XeF_2
e. CrO_3 f. $Ca_3(PO_3)_2$ g. ICl h. NO_2
i. NH_4NO_3 j. PH_3

3.33 Complete the table by supplying the missing name
or formula for each covalent compound.

Formula	Name
?	Sulfur trioxide
P_2O_5	?
?	Selenium tetrafluoride
CO	?
?	Dinitrogen trioxide

3.34 Complete the table by supplying the missing name
or formula for each covalent compound.

Formula	Name
CF_4	?
?	Nitrogen dioxide
PCl_5	?
?	Carbon disulfide
N_2O_5	?

3.5 The Mole: Counting Atoms and Compounds

3.5 Inquiry Question:
How can the molar mass
establish the number of
moles or particles present
in a known mass of
a compound?

So far in this chapter we have explored the formation of chemical bonds, ionic and
covalent. In the remainder of the chapter we examine some properties of these compounds.
We begin by asking the question, "Can we measure the number of atoms present in a
compound?" We can get the mass of a substance by weighing it on a balance, but how
many atoms are in that amount?

Chemists use a unit called the **mole** to relate the mass of an element in grams to the
number of atoms it contains. The mole is a unit for counting atoms just as a *dozen* is a unit for
counting things like eggs. The number of eggs in 1 dozen is 12. Suppose that each egg weighs
approximately 50 grams (g). We could determine how many eggs are present in a closed egg
carton by weighing it. For example, if the contents of a carton of eggs weigh 600 g, we could
calculate that there were 12 eggs or 1 dozen eggs inside even without being able to see the
eggs inside the carton. We have used the mass of the eggs to count the number of eggs.
Similarly, exactly 12 grams of the isotope carbon-12 contains a certain number of atoms;
this is called a mole. By measuring a mass of a substance, we can count the number of atoms.

We cannot actually see atoms, but now, by definition, we can relate the atomic mass
from the periodic table to the number of atoms using the mole unit. For example, on the
periodic table, carbon has an atomic mass of 12.01 amu. One mole of any element has a
molar mass in grams numerically equal to the atomic mass of that element. The molar
mass tells us the mass of one mole of a substance and corresponds to the mass of a single
atom in amu. So, similar to the fact that a dozen eggs in our example has a mass of 600 g, we
can say the molar mass of carbon is 12.01 grams per mole of carbon atoms (12.01 g/mole).

600 g = 12 eggs = 1 dozen

Just as we know that there are 12 objects in the unit a dozen, the number of atoms present in the unit called the mole is experimentally found to be about 602,000,000,000,000,000,000,000 atoms. Wow! That is a very large number. Huge numbers like this are awkward to handle. It is more convenient to express such a large number in scientific notation. The number of atoms in a mole can be represented in scientific notation as 6.02×10^{23} atoms.

Avogadro's Number

The large number of atoms defined as one mole is known as **Avogadro's number (N)** in honor of the Italian physicist, Amedeo Avogadro.

$$6.02 \times 10^{23} \text{ atoms} = 1 \text{ mole of atoms}$$

In general, for any substance, Avogadro's number is

$$N = 6.02 \times 10^{23} \text{ particles/mole}$$

We have used the mass in grams for the carbon atoms to count the number of carbon atoms.

$$12.01 \text{ g C} = 1 \text{ mole of C} = 6.02 \times 10^{23} \text{ atoms of C}$$

Just as a dozen eggs, a dozen paper clips, and a dozen bowling balls have different masses, a mole of silver atoms and a mole of carbon atoms will not have the same mass. Since elements have different atomic masses, the same numbers of atoms of two different elements have different masses.

Let's solve a problem to see how atoms and moles are related to each other.

Two quantities like atoms and moles that can be related to each other with an equal sign are referred to as **equivalent units.**

From our definition of Avogadro's number we know the equivalent unit relating atoms and moles is

$$6.02 \times 10^{23} \text{ atoms} = 1 \text{ mole of atoms}$$

This equivalent unit can be rearranged to

$$\frac{6.02 \times 10^{23} \text{ atoms}}{1 \text{ mole}} \quad \text{or} \quad \frac{1 \text{ mole}}{6.02 \times 10^{23} \text{ atoms}}$$

and can be used as a conversion factor when we solve a problem.

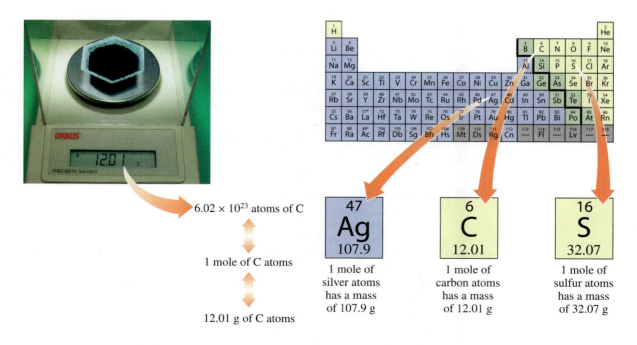

6.02×10^{23} atoms of C

1 mole of C atoms

12.01 g of C atoms

47	6	16
Ag	C	S
107.9	12.01	32.07

1 mole of silver atoms has a mass of 107.9 g

1 mole of carbon atoms has a mass of 12.01 g

1 mole of sulfur atoms has a mass of 32.07 g

Solving a Problem

Converting between Units

How many atoms are present in 2.00 moles of carbon?

To determine the answer to a problem like this, the problem can be set up using the following guidelines:

STEP 1: Examine the problem. Moles of carbon are given and we are to determine atoms.

STEP 2: Establish the desired units. Atoms must appear in the numerator.

STEP 3: Determine the conversion factors. Place the equivalent unit with with the desired unit (atoms) in the numerator so that the other units cancel each other out, leaving the desired unit.

STEP 4: Set up the equation.
The equation can be set up as:

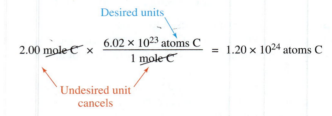

Desired units

$$2.00 \text{ mole C} \times \frac{6.02 \times 10^{23} \text{ atoms C}}{1 \text{ mole C}} = 1.20 \times 10^{24} \text{ atoms C}$$

Undesired unit
cancels

STEP 5: Check your answer. When solving a problem for the number of atoms, your answer should be a large number having a large positive exponent.

---

sample problem 3.12

Converting from Moles to Grams

What is the mass of 4.00 moles of carbon in grams?

Solution

STEP 1: Examine the problem. In this case, moles of carbon are given and grams of carbon are required.

STEP 2: Establish the desired units. The unit for the final answer (g) must appear in the numerator and other units (mole) must cancel out. In this problem, we are looking for the number of grams.

STEP 3: Determine the conversion factor. The equivalent unit that we have been working with that contains grams is the mass where from experiment

$$12.01 \text{ g C} = 1 \text{ mole C}$$

Two possible conversion factors are

$$\frac{12.01 \text{ g C}}{1 \text{ mol C}} \text{ or } \frac{1 \text{ mol C}}{12.01 \text{g C}}$$

STEP 4: Set up the equation. The unit for the final answer (g) must appear in the numerator and other units (mole) must cancel out. Using the conversion factor that cancels the given unit (mole), we can set up the problem as

$$4.00 \text{ mole C} \times \frac{12.01 \text{ g C}}{1 \text{ mole C}} = 48.0 \text{ g C}$$

STEP 5: Check your answer. If 1 mole of carbon weighs about 12 g, then 4 moles should weigh four times that amount.

sample
problem
3.13   **Using Avogadro's Number to Find the Number of Atoms**

a.   How many atoms of aluminum are present in 1.00 g of aluminum?
b.   How does this compare to the number of atoms in 1.00 g of lead?

Solution

STEP 1: Examine the problem. In this case, grams of aluminum and lead are given and we are to determine atoms of each.

STEP 2: Establish the desired units. Atoms must appear in the numerator.

STEP 3: Determine the conversion factors. To solve these two problems, we use two conversion factors, the molar mass for the element from the periodic table (located beneath the atomic symbol with units of g/mole) and the conversion factor for Avogadro's number to relate the mole to the number of atoms. The units of Avogadro's number will be atoms per mole.

STEP 4: Set up the equation.

a.   How many atoms of aluminum?

$$1.00 \ \text{g Al} \ \times \ \frac{1 \ \text{mole Al}}{26.98 \ \text{g Al}} \times \frac{6.02 \times 10^{23} \ \text{atoms of Al}}{1 \ \text{mole Al}} = 2.23 \times 10^{22} \ \text{atoms of Al}$$

Information   Molar mass   Avogadro's number
given   (conversion factor)

b.   How many atoms of lead?

$$1.00 \ \text{g Pb} \ \times \ \frac{1 \ \text{mole Pb}}{207.2 \ \text{g Pb}} \times \frac{6.02 \times 10^{23} \ \text{atoms of Pb}}{1 \ \text{mole Pb}} = 2.91 \times 10^{21} \ \text{atoms of Pb}$$

Information   Molar mass   Avogadro's number
given   (conversion factor)

STEP 5: Check your answer. The exponents in the two answers show that there are about 10 times more atoms in 1.00 g of aluminum than in 1.00 g of lead. Because aluminum atoms are lighter, it takes more aluminum atoms to obtain the same mass.

Molar Mass and Formula Weight

We can also use Avogadro's number to determine the number of molecules or particles present in a compound. To do this, we must first determine the molar mass of the compound. Just as the atomic mass for one atom of an element is expressed in atomic mass units, the formula weight for a compound is the sum of the atomic masses. The molar mass of a compound is numerically equal to the formal weight with units of grams/mole. Let's look at some examples.

The molar mass can be used to calculate the number of molecules or atoms in a given sample.

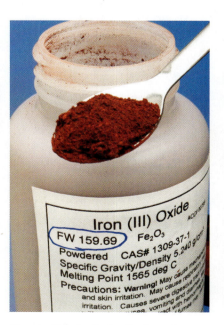

Chemical labels display the formula weight of the compound.

Solving a Problem

Calculating Molar Mass for a Compound

Calculate the molar mass for (a) NaCl and (b) H_2O.

(a) The molar mass for NaCl is the sum of the atomic masses of the elements.

22.99	Atomic mass for Na
+ 35.45	Atomic mass for Cl
58.44	g/mole Molar mass of NaCl

(b) The molar mass for H_2O is the sum of the atomic masses for the elements.

2 × 1.01	Atomic mass for two H
+ 16.00	Atomic mass for O
18.02 g/mole	Molar mass for H_2O

---

sample problem
3.14   Calculating Molar Mass for a Compound

Determine the molar mass for the compounds shown.

a.   KOH   b. NH_3 c. $NaHCO_3$

Solution

a.

39.10	Atomic mass for K
16.00	Atomic mass for O
+ 1.01	Atomic mass for H
56.11 g/mole	Molar mass of KOH

b.

14.01	Atomic mass for N
+ 3 × 1.01	Atomic mass for three H
17.04 g/mole	Molar mass for NH_3

c.

22.99	Atomic mass for Na
1.01	Atomic mass for H
12.01	Atomic mass for C
+ 3 × 16.00	Atomic mass for three O
84.01 g/mole	Molar mass of $NaHCO_3$

Note that because the numbers are added, the answers are reported to two decimal places. For a review, see significant figures in Section 1.3.

Solving a Problem

Finding the Number of Molecules in a Sample

Determine the number of molecules of H_2O present in an ice cube of pure water that weighs 4.00 grams.

STEP 1: Calculate the molar mass of the compound. As calculated in the previous example, the molar mass of H_2O is 18.02 g/mole.

STEP 2: Apply conversion factors to reach the desired unit. To determine the number of molecules in a sample, we would use the molar mass (g/mole) and Avogadro's number (molecules/mole). The unit on the final answer is molecules, so this must go in the numerator and all other units must cancel.

$$4.00 \text{ grams} \times \frac{1 \text{ mole}}{18.02 \text{ grams}} \times \frac{6.02 \times 10^{23} \text{ molecules}}{1 \text{ mole}} = 1.34 \times 10^{23} \text{ molecules}$$

A Unit from Biochemistry: The Dalton

We briefly bring up the dalton, because it is commonly used in biochemistry as a unit to measure mass of biological molecules. The **dalton** (Da) is the same as the amu (Chapter 2). It resembles molar mass in that it is a measure of the mass of a compound. It is mainly used in biochemistry to measure the weight of large molecules like proteins. It is defined as one-twelfth the mass of a carbon-12 atom, which, in large molar masses, gives approximately the same numerical value as molar mass.

practice problems

3.35 Compare (a) the number of atoms and (b) the number of grams present in 1.00 mole of silver (Ag) and 1.00 mole of gold (Au).

3.36 Compare (a) the number of atoms and (b) the number of grams present in 1.00 mole of helium and 1.00 mole of argon.

3.37 Calculate the following:
a. the number of Na atoms in 0.250 mole of Na
b. the number of moles in 4.0 g of Pb
c. the number of atoms in 12.0 g of Si

3.38 Calculate the following:
a. the number of S atoms in 0.660 mole of S
b. the number of moles in 15.0 g of Fe
c. the number of atoms in 66.0 g of Cu

3.39 Determine the molar mass for the following compounds:
a. magnesium sulfate, $MgSO_4$
b. glucose, $C_6H_{12}O_6$
c. carbon dioxide, CO_2

3.40 Determine the molar mass for the following compounds:
a. methane, CH_4
b. lactic acid, $C_3H_6O_3$
c. methylamine, CH_3NH_2

3.41 Determine the number of molecules in a 325-mg dose of aspirin, molecular formula $C_9H_8O_4$.

3.42 Determine the number of molecules in a 200-mg dose of ibuprofen, molecular formula $C_{13}H_{18}O_2$.

Discovering the Concepts

? Inquiry Activity — Molecular Shape

Information

Electron pairs joining atoms (whether in single, double, or triple bonds) will get as *far away from each other as possible* when forming bonds in a molecule. Electrons in bonds will also repel lone pair electrons. This is known as the *valence-shell electron-pair* repulsion theory, or *VSEPR*. VSEPR allows us to predict the three-dimensional shape (also called molecular geometry) that a molecule will adopt.

Lone pairs repel each other more than they repel electrons involved in bonds.

Activity

Using the toothpicks for bonds, a half toothpick for lone pairs on the central atom(s), large marshmallows for C, N, and O, and small marshmallows for H, build the five structures outlined in the following. Acquire 24 toothpicks, 8 large marshmallows, and 13 small marshmallows. Keep in mind the bonding patterns outlined in Section 3.4.

a. Make a model of CO_2, being sure to get all bonds (toothpicks) as far away from each other as possible. What is the bond angle between O–C–O?

b. Make a model of H_2CCH_2, being sure to get all bonds (toothpicks) as far away from each other as possible. What is the bond angle between H–C–H?

c. Make a model of CH_4, being sure to get all bonds (toothpicks) as far away from each other as possible. Do you think the bond angle between H–C–H less than, equal to, or greater than 90°?

d. Make a model of NH_3, being sure to get all bonds and lone pairs (toothpicks) as far away from each other as possible. Do you think the H–N–H bond angle is greater or less than the H–C–H bond angle in CH_4? Support your answer.

e. Make a model of H_2O, being sure to get all bonds and lone pairs (toothpicks) as far away from each other as possible. Do you think the H–O–H bond angle is greater or less than the H–C–H bond angle in CH_4? Support your answer.

Keep your structures for reference.

Questions

You have built the following molecular shapes with your toothpicks and marshmallows: tetrahedral, trigonal planar, linear, pyramidal, and bent.

1. Based on the names given and the molecules that you built, assign each of the following molecules a shape:

   a. CO_2 _____   b. H_2CCH_2 _____   c. CH_4 _____
   d. NH_3 _____   e. H_2O _____

2. Complete the following table:

Molecular Formula	Lewis Structure	Molecular Shape (geometry) [from question 1]	Number of Atoms Bonded to Carbon or Central Atom	Number of Lone Pairs on Carbon or Central Atom
CO_2				
H_2CCH_2				
CH_4				
NH_3				
H_2O				

3. Complete the following table:

Molecular Formula	Lewis Structure	Molecular Shape (geometry)	Number of Atoms Bonded to Carbon or Central Atom	Number of Lone Pairs on Carbon or Central Atom
CH_3F				
HCCH				
CH_2CHCl				
NH_4^+				
CH_3CH_3				

4. Based on the tables in questions 2 and 3, does the number of atoms attached to a central carbon atom affect its shape? List the shapes that carbon atoms can adopt and the number of atoms attached in each case.

3.6 Getting Covalent Compounds into Shape

Just as a road map allows us to see how towns and cities are connected by roads but gives very little information about the topography, the Lewis structures discussed in Section 3.4 tell us how atoms are connected in molecules but do not tell us anything about a molecule's three-dimensional shape. Our world is three-dimensional, so to understand molecules it is important to identify the three-dimensional shapes they can adopt.

3.6 Inquiry Question: How is the shape of a covalent compound determined?

For example, the Lewis structure for methane appears flat or two-dimensional on paper. If this were how methane was actually arranged in three dimensions, the angle between two adjacent pairs of electrons would be about 90°.

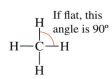

If flat, this angle is 90°

To understand how methane looks in three dimensions, first remember that electrons are negatively charged and that like charges repel each other. So the four bonding pairs of electrons around the carbon atom in CH_4 form four clouds of negative charge that want to get *as far apart as possible*.

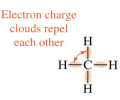

Electron charge clouds repel each other

If, however, the electrons are allowed to rearrange in three-dimensional space to get as far apart as possible, the result is an arrangement of the atoms of CH_4 that achieves an angle of 109.5° between the negatively charged clouds of electrons. The molecule is no longer two dimensional; instead, it has adopted a three-dimensional shape called **tetrahedral,** as shown in Figure 3.11.

This reasoning describes the **valence-shell electron-pair repulsion model** that is often referred to simply as **VSEPR.** This model is used to predict the shape of a molecule based on the number of electron **charge clouds** on a given atom. According to VSEPR, in the valence shell of an atom the charge clouds formed by groups of electrons will arrange themselves to be as far away from each other as possible in order to reduce repulsions.

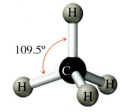

109.5°

FIGURE 3.11 The tetrahedral shape of methane, CH_4.

Determining the Shape of a Molecule

To determine the shape (also called geometry) of a molecule using VSEPR, a molecule is assigned a VSEPR form. The VSEPR form is a tool used to relate the number and type of charge clouds on an atom to the shape around that atom. To determine the VSEPR form and, from that, the shape of a molecule, first determine the atom around which you want to know the shape. For small molecules, this is the central atom. (We discuss larger molecules later.) The central atom is represented with a capital *A* in the VSEPR form.

Next, determine the number of charge clouds around the central atom. When counting charge clouds, keep in mind that a single bond, a double bond, and a triple bond are counted as one charge cloud. Charge clouds in bonds are called *bonding clouds* and are represented with a capital *B* in the VSEPR form.

A lone pair of electrons is also counted as one charge cloud, but as we will see, these charge clouds affect the shape of the molecule in a slightly different way. Charge clouds from lone pairs of electrons are called *nonbonding clouds* and are represented with a capital *N* in the VSEPR form.

Here are some examples to guide you as you learn to count charge clouds and determine a molecule's VSEPR form.

Formaldehyde has three charge clouds (all bonding clouds) around the central atom, carbon

Formaldehyde, a preservative

The VSEPR form for formaldehyde is AB_3—a central atom surrounded by three bonding charge clouds.

Ammonia

Ammonia has three bonding charge clouds and one nonbonding charge cloud (its lone pair) around the central atom, nitrogen

The VSEPR form for ammonia is AB_3N—a central atom surrounded by three bonding charge clouds and one nonbonding charge cloud.

The molecular shapes associated with the VSEPR forms are shown in **Table 3.6** for the common forms containing the elements C, N, and O found in living things.

TABLE 3.6 Predicting Molecular Shape Using VSEPR

VSEPR Form	Molecular Shape	Bond Angle	Example
AB_4	Tetrahedral	109.5°	
AB_3N	Pyramidal	< 109.5°	
AB_2N_2	Bent	< 109.5°	
AB_3	Trigonal planar	120°	
AB_2	Linear	180°	

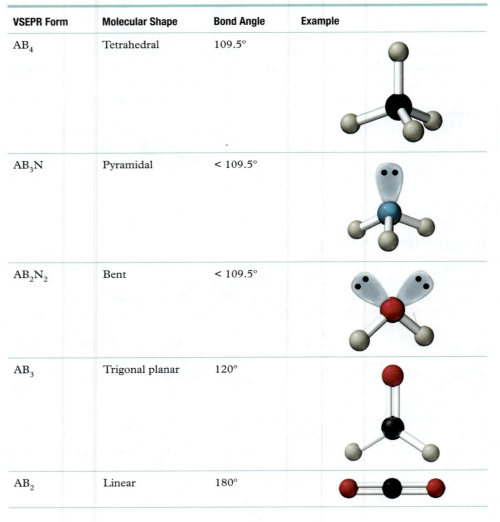

sample problem 3.15 Determining a Molecule's Shape

From the following Lewis structures, assign the VSEPR form for each molecule and, using Table 3.6, determine the shape of each molecule:

a. $:\ddot{O}=C=\ddot{O}:$ b. $H-\ddot{S}-H$ c. $:\ddot{Br}-\overset{\textstyle .}{P}-\ddot{Br}:$
                                                              $|$
                                                          $:\ddot{Br}:$

Solution

a. The central atom in the molecule is carbon. Carbon is participating in two double bonds, indicating it has two bonding charge clouds in its valence shell. So this molecule has the VSEPR form AB_2. This VSEPR form corresponds to the molecular shape of linear (Table 3.6). Note that we did not concern ourselves with the oxygen atoms in this molecule. Because oxygen is not the central atom, its arrangement of charge clouds is not included when we assign the VSEPR form. The molecular shape is linear.

b. Sulfur is the central atom in this molecule. Careful examination shows that sulfur is participating in two single bonds and has two nonbonding pairs (also called lone pairs) of electrons. Therefore, the sulfur is surrounded by four charge clouds with a VSEPR form of AB_2N_2. This VSEPR form corresponds to the molecular shape of bent (Table 3.6). Note that the Lewis structures of CO_2 and H_2S look fairly similar—a central atom connected to two other atoms. Both Lewis structures can be drawn with all the atoms in a straight line, but only CO_2 actually has a linear shape. It is important to remember that Lewis structures show only the *connectivity* of the atoms in a molecule, not the molecule's *shape*. The molecular shape is bent.

c. Phosphorus is the central atom in this molecule; it is participating in three single bonds and has one nonbonding pair of electrons. The VSEPR form of this molecule is AB_3N. This VSEPR form corresponds to the molecular shape of pyramidal (Table 3.6).

Carbon atoms adopt one of three molecular shapes determined by the number of atoms bonded directly to the carbon. If four atoms are bonded to a carbon, the shape is tetrahedral; if it is bonded to three atoms, the shape is trigonal planar; and if it is bonded to two atoms, the shape is linear (see **Table 3.7**).

Nonbonding Electrons and Their Effect on Molecular Shape

The first three entries in Table 3.6 have the VSEPR forms AB_4, AB_3N, and AB_2N_2. Each has a total of four charge clouds around a central atom, but each has a different shape. How can this be? Isn't there one optimal arrangement of four charge clouds around a central atom? To explore this, let's look at the molecules methane (CH_4), ammonia (NH_3), and water (H_2O) that represent these three VSEPR forms.

Nonbonding electrons do affect the shape of a molecule. The tetrahedral shape of methane, which has four atoms around a central atom, changes to pyramidal for ammonia because a nonbonding pair occupies the fourth position (see **Figure 3.12**). Because nonbonding pairs of electrons take up space but are invisible in the molecule's shape, the VSEPR forms are different. A molecule's shape is determined by the relative positions of the *atoms* in the molecule.

Nonbonded pairs have a subtle effect on the shape of a molecule. In the Bond Angle column of Table 3.6, notice that the presence of nonbonded pairs on the central atom changes the bond angle, making it smaller. A non-bonded pair of electrons forces the bonded pairs closer together. But why?

A bonding pair of electrons is confined to the space directly between the two atoms being held. A nonbonded pair of electrons has no such restriction, so the nonbonded pair can move around more in the valence shell.

TABLE 3.7 Predicting Molecular Shapes for Carbon

Molecular Shape	Number of Atoms Bonded Directly to Carbon
Tetrahedral	4
Trigonal Planar	3
Linear	2

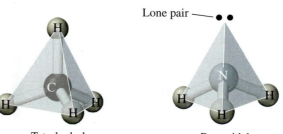

Tetrahedral   Pyramidal

FIGURE 3.12 Methane and ammonia. A nonbonded pair changes a tetrahedral shape to a pyramidal shape.

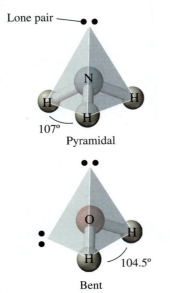

Lone pair

107°
Pyramidal

104.5°
Bent

FIGURE 3.13 Ammonia and water.
The bonding electron clouds are pushed
closer together by the extra space
requirements of the nonbonded electrons
making the bond angles different.

A lone pair, therefore, takes up more space in the valence shell than does a bonding pair. The nonbonded (lone) pair repels the bonded pairs more than the bonded pairs repel each other, resulting in a smaller angle between the bonded pairs (see **Figure 3.13**). The two nonbonded pairs in water push the bonded electron clouds closer to each other than the one nonbonded pair in ammonia, resulting in the reduced bond angle.

Molecular Shape of Larger Molecules

The majority of the molecules seen throughout the remainder of this text are not as structurally simple as CH_4 or NH_3. Most molecules contain many carbon atoms with several charge clouds. Can we determine the shape of a molecule that does not have a single clearly identifiable central atom?

Using VSEPR, the shape around any single atom that is bonded to at least two other atoms can be determined. A molecule of ethanol is shown. Ethanol is found in wine, beer, and other spirits as a product of fermentation and, more recently, has found new uses as a fuel additive. We can determine the shape around each of the carbons and the oxygen in ethanol, but cannot determine an overall shape for the molecule. Typically, we choose one or more atoms of interest in the molecule and determine the shape around those atoms.

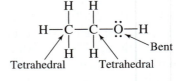

Tetrahedral   Tetrahedral   Bent

practice
problems

3.43 Determine the shape around the orange-colored atom (or atoms) in each of the following Lewis structures:

a. dimethylamine, an insect pheromone

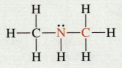

b. acrylonitrile, found in plastics

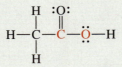

c. acetic acid, vinegar

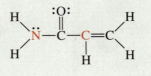

d. acrylamide, a byproduct of cooking starchy foods at high temperatures

3.44 Determine the shape around the orange-colored atom (or atoms) in each of the following Lewis structures:

a. dimethylether, found in wart treatments

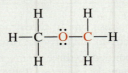

b. ethylene, a plant-ripening hormone

c. acetylene, a fuel used in welding

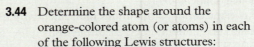

H—C≡C—H

d. formic acid, used as a preservative in livestock feed

Discovering the Concepts

❓ Inquiry Activity—Bond Polarity

Information

A *covalent bond* is formed when two atoms share electrons between them. If the two atoms are different, they will not share the electrons equally, and a *polar covalent bond* is formed. In this case, one of the atoms will have the electrons more of the time and be partially negatively charged (symbol δ^-) and the other atoms will have the electrons less of the time and be partially positively charged (symbol δ^+). The amount of charge present can be measured as a quantity called a *dipole moment*. This is represented by an arrow with a hash mark crossing the tail $\longmapsto$. The head of the arrow points toward the more negative atom.

Electronegativity

Fluorine is the most electronegative element. The closer an element is to fluorine on the periodic table, the more electronegative it is. For example, oxygen is more electronegative than nitrogen, which in turn is more electronegative than carbon.

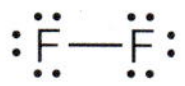

:F̈—F̈:   In a nonpolar covalent bond, the electrons are shared equally. This occurs if the atoms in the bond are identical.
One exception: C—H bonds are also nonpolar covalent.

FIGURE 1A A nonpolar covalent bond.

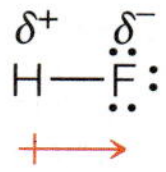

δ^+ δ^-
H—F̈:   In a polar covalent bond, the electrons are not shared equally. The more electronegative element is the negative pole of the dipole. This polarity can be represented with delta symbols (above) or the dipole moment arrow (below).
$\longmapsto$

FIGURE 2A A polar covalent bond.

Questions

Bond Polarity

1. Based on their ability to share electrons equally or unequally, are the following bonds considered polar covalent or nonpolar covalent? Explain your reasoning.
   a. C—N   b. C—Cl   c. O—H
   d. H—H   e. C—H   f. N—H
2. For the polar covalent bonds in question 1, indicate the polarity using the partial symbols (δ^+/δ^-) and the dipole moment arrow.

Molecule Polarity

3. Draw the Lewis structure for ammonia, NH_3. Show the polarity of the bonds between N and H on the structure you drew using the dipole moment arrow. What is the molecular shape (geometry) around the N in ammonia? Draw the overall dipole using the dipole moment arrow with the head pointing toward the negative side of your Lewis structure and the tail pointing toward the positive side. Draw this to the right of your molecule.
4. Draw the Lewis structure for H_2O. Show the polarity of the bonds on the structure you drew using the dipole moment arrow. Considering that the shape of H_2O is bent, do you think that H_2O has a dipole? (In other words, is there a side of the molecule that is more negative and a side more positive?) If so, indicate this to the right of your structure using the dipole moment arrow.
5. Draw the Lewis structure for carbon dioxide. What is the molecular shape (geometry) around the carbon in carbon dioxide? Show the polarity of the bonds between C and O on the structure you drew using the dipole moment arrow. Unlike NH_3 and H_2O, CO_2 overall has no dipole. Although CO_2 has polar bonds, it is a nonpolar molecule. Provide an explanation for this. [HINT: Consider its shape.]

6. Draw the Lewis structure for carbon tetrachloride, CCl_4. What is the molecular shape (geometry) around the carbon in carbon tetrachloride? Show the polarity of the bonds between C and Cl on the structure you drew using the dipole moment arrow. What is the overall (net) dipole for this molecule? Explain.

7. Why do some molecules have polar bonds and yet are nonpolar molecules whereas other molecules like water and ammonia have polar bonds and are polar molecules?

8. Using a dipole moment arrow ($\longmapsto$), indicate the overall molecular polarity for the following molecules (you may have to draw out Lewis structures to do this). If the molecule is nonpolar, state no dipole.

   a. HBr
   b. OCS
   c. CH_3F
   d.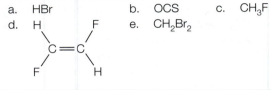
   e. CH_2Br_2

3.7 Electronegativity and Molecular Polarity

3.7 Inquiry Question:
How is the polarity of a molecule determined?

Electrons are shared in covalent bonds, but often the electrons are not shared equally. In this section, we explore a fundamental property of atoms that causes this inequality to occur.

Electronegativity

How can we tell which atoms in a covalent bond will attract the shared electrons more? The ability of an atom to attract the bonding electrons of a covalent bond to itself is known as **electronegativity.** Figure 3.14 shows the main-group elements from the periodic table with the electronegativities for each element below the symbol. Notice that the element with the greatest electronegativity on the periodic table is fluorine. Also note that the electronegativity increases as you move closer to fluorine. For example, oxygen is more electronegative than carbon, and chlorine is more electronegative than iodine.

An inequality occurs when two *different* atoms are involved in a covalent bond; for example, in H—Cl, the sharing of electrons is not equal. In the HCl molecule, the chlorine atom attracts the shared electrons more than the hydrogen atom, which means that the shared electrons in the bond between H and Cl spend more time near the chlorine atom.

Electronegativity increases →

	H 2.1					

Group 8A (18)

Group 1A (1)	Group 2A (2)		Group 3A (13)	Group 4A (14)	Group 5A (15)	Group 6A (16)	Group 7A (17)	
Li 1.0	Be 1.5		B 2.0	C 2.5	N 3.0	O 3.5	F 4.0	
Na 0.9	Mg 1.2		Al 1.5	Si 1.8	P 2.1	S 2.5	Cl 3.0	
K 0.8	Ca 1.0		Ga 1.6	Ge 1.8	As 2.0	Se 2.4	Br 2.8	
Rb 0.8	Sr 1.0		In 1.7	Sn 1.8	Sb 1.9	Te 2.1	I 2.5	
Cs 0.7	Ba 0.9		Tl 1.8	Pb 1.9	Bi 1.9	Po 2.0	At 2.1	

Electronegativity increases (vertical arrow)

FIGURE 3.14 Electronegativities of the main-group elements. Fluorine is the most electronegative element. The noble gases are not considered because they are not very reactive.

In other words, the electrons are shared unequally. When two identical atoms share electrons to form a covalent bond, as in H_2, the electrons are equally attracted to each atom and are shared equally. A covalent bond in which the electrons are not shared equally is called a **polar covalent bond.** When the electrons in a covalent bond are shared equally, the bond is called a **nonpolar covalent bond.**

We can denote the uneven sharing or distribution of electrons in a polar covalent bond using the symbol δ (lowercase Greek delta, meaning "partial"). **Figure 3.15** demonstrates how this notation is used in the case of HCl. We can also use an arrow with a hash mark ($\longmapsto$) to represent bond polarity. This arrow is commonly referred to by chemists as the dipole moment arrow. Both notations are shown in Figure 3.15. (Note that the arrow points toward the more electronegative element and the plus on the tail of the arrow is at the partial positive end.)

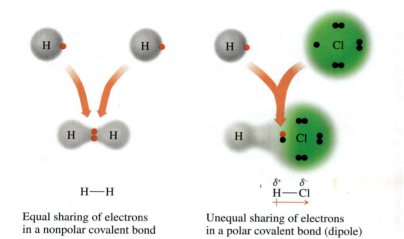

H—H

Equal sharing of electrons
in a nonpolar covalent bond

$\overset{\delta^+}{H}—\overset{\delta^-}{Cl}$
$\longmapsto$

Unequal sharing of electrons
in a polar covalent bond (dipole)

FIGURE 3.15 Covalent Bonding. In the nonpolar covalent bond of H_2, electrons are shared equally. In the polar covalent bond of HCl, electrons are shared unequally.

Like the opposite poles on a magnet, a covalent bond that does not share electrons equally has two distinct poles or ends—one that is partially negative and one that is partially positive. Because of this we refer to these bonds as *polar*. Of course, a bond that is "nonpolar" (sharing equally) has no ends or poles.

We can predict the type of bond likely to form by taking the difference between the electronegativities of the two elements, as shown in the table in **Figure 3.16**. Generally, elements with an electronegativity difference of 1.8 or more will form an ionic bond resulting from the *transfer* of electrons from the less electronegative element to the more electronegative element. For example, NaCl, which we know as an ionic compound, has an electronegativity difference of 2.1 [3.0 (Cl) − 0.9 (Na)] between chlorine and sodium.

In turn, two elements with an electronegativity difference of less than 1.8 will most likely *share* electrons to form a covalent bond. When two atoms are involved in a covalent bond, the greater the difference in electronegativity of the two elements, the more polar the bond.

Electonegativity Difference and Types of Bonds

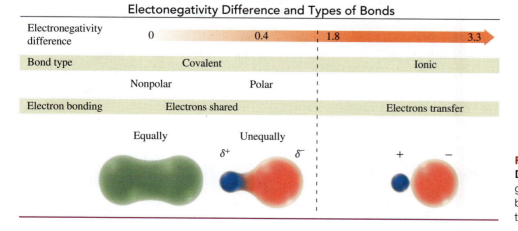

FIGURE 3.16 Electronegativity Difference and Types of Bonds. The greater the electronegativity difference between two atoms, the more polar the bond.

While it is possible to use the information in Figures 3.14 and 3.16 to calculate an electronegativity difference for any bond and determine where along the bond-type continuum it lies, without doing any calculations we can distinguish ionic bonds from covalent bonds by looking at the types of elements present (metals vs. nonmetals).

Let's see how we can distinguish between nonpolar bonds, slightly polar bonds, and strongly polar bonds for covalent compounds. We can do this without numbers by remembering two facts:

1. Fluorine is the most electronegative of all the elements and
2. electronegativities increase as you move toward fluorine on the periodic table.

With these rules in mind, let's consider the following carbon-containing bonds. Which of these bonds is the most polar?

$$C-N \qquad C-O \qquad C-F$$

Since each of the three bonds contains a carbon atom, we can compare the electronegativity of each of the other atoms relative to carbon. Looking at the periodic table, we see that carbon and nitrogen are side by side, which means that their electronegativities are more alike than those of carbon and oxygen or carbon and fluorine. Therefore, the $C-N$ bond, with the smallest difference in electronegativities, is the least polar. Applying the same reasoning, we can determine that the $C-O$ bond would fall next in the order and the $C-F$ bond would be the most polar. When we want to determine relative polarities of bonds, we can do it without electronegativity numbers. In fact, our example illustrates a rule of thumb we can use for comparing the polarity of covalent bonds: The *farther apart* two nonmetals are on the periodic table within a period, the *greater* the polarity of the bond between them. One major exception to this rule, which we will see again in Chapter 4, is that a $C-H$ bond is considered to be nonpolar, because the electronegativities of carbon and hydrogen are not that different.

---

sample problem

3.16 **Polarity of Covalent Bonds**

For each of the following covalent bonds, label the less electronegative atom with a δ^+ and the more electronegative atom with a δ^-. Then, label the bond using the dipole moment arrow ($\longmapsto$).

a. $N-C$ b. $N-O$ c. $P-F$

Solution

a. $\overset{\delta-}{N} \underset{}{\rightleftharpoons} \overset{\delta+}{C}$ On the periodic table, nitrogen is closer to fluorine than carbon, so nitrogen is more electronegative and attracts the bonding electrons more strongly, giving it a partial negative charge.

b. $\overset{\delta+}{N} \underset{}{\rightleftharpoons} \overset{\delta-}{O}$ Oxygen is closer to fluorine than carbon, so oxygen is more electronegative and attracts the bonding electrons more strongly, giving it a partial negative charge.

c. $\overset{\delta+}{P} \underset{}{\rightleftharpoons} \overset{\delta-}{F}$ Fluorine is the most electronegative of all elements, so it attracts the electrons in this bond more strongly, giving it a partial negative charge.

sample problem 3.17   Comparing Covalent Bonds

In each of the following bond pairs, circle the pair with the more polar bond:

a.   C—N and C—Cl   b.   N—Cl and S—N   c.   Si—F and Si—O

Solution

In general, the farther apart two atoms are within a period on the periodic table, the greater the difference in their electronegativities and, therefore, the more polar the bond between them.

a.   C—Cl is the more polar bond.

b.   N—Cl is the more polar bond.

c.   Si—F is the more polar bond.

Molecular Polarity

Molecules, like the bonds that hold them together, can be polar or nonpolar. A **polar molecule** is a molecule in which one end or area of the molecule is more negatively charged than the rest of the molecule. In other words, the electrons in a polar molecule are unevenly distributed over the molecule. In contrast, a **nonpolar molecule** has an even distribution of electrons over the entire molecule.

For simple molecules containing only two atoms connected by a covalent bond, we can determine the polarity of the molecule simply by determining the polarity of the bond. In our previous examples, the bond between the chlorines in Cl_2 is nonpolar, so Cl_2 is a nonpolar molecule. Likewise, since the bond in HCl is a polar bond, HCl is a polar molecule.

What happens when a molecule is composed of three or more atoms connected by several bonds? How do we determine the polarity of a molecule if it has more than one bond dipole? We *can* determine the polarity of larger molecules, but when we do, we must consider both the electronegativity of the atoms involved *and* the shape of the molecule.

As a first example, consider carbon dioxide and water. Their Lewis structures are shown.

$$:\ddot{O}=C=\ddot{O}: \qquad H-\ddot{O}-H$$

First, determine the bond polarity for each bond in each molecule:

$$\overset{\delta^-}{:\ddot{O}}\!\!=\!\!\underset{\delta^+\ \ \delta^-}{\overset{\delta^+}{C}}\!\!=\!\!\overset{}{\ddot{O}}: \qquad \underset{\delta^-\ \ \delta^+}{\overset{\delta^+\ \ \delta^-}{H-\ddot{O}-H}}$$

Each molecule has two polar bonds. Despite this apparent similarity, we know from experimental evidence that carbon dioxide is a nonpolar molecule whereas water is a polar molecule. How can two molecules that seem so similar actually be so different? Remember, these two molecules have different shapes.

Applying the principles of VSEPR from Section 3.6, we know that carbon dioxide is a linear molecule, but water is bent. To see how shape affects the polarities of these molecules, let's use the dipole moment arrow to represent the bond dipoles and draw the molecules to more closely represent their true shape.

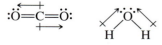

Notice that in CO_2 the bonding electrons are being pulled equally toward each of the oxygen atoms, but in opposite directions. So, like a tug of war that is equally matched, the electrons that are being pulled away from carbon by one of the oxygen atoms are also being pulled away from the carbon atom in the opposite direction by the other oxygen atom. In other words, the bond dipoles cancel each other.

Because of the bent shape of the water molecule, instead of canceling each other as they do in carbon dioxide, the bond dipoles actually add together to give an uneven distribution of electrons. The area around the oxygen atom where the lone pairs of electrons are positioned has more negative charge than the rest of the molecule. This results in an overall molecular dipole for water, shown as a orange arrow.

The interaction of electronegativity and molecular shape is the basis for determining whether a molecule is polar or nonpolar.

---

sample problem 3.18   **Determining the Polarity of Molecules**

Determine whether each of the following molecules is polar or nonpolar. If polar, show the direction of the molecular dipole using a dipole moment arrow.

a. CCl_4 b. NH_3 c. CH_3F

Solution

To see all the bonds in the molecule, it is useful to draw the Lewis structure first.

a.

$:\overset{..}{\underset{..}{Cl}}-C-\overset{..}{\underset{..}{Cl}}:$ No overall dipole

Each C—Cl bond is polar. Like CO_2, they are pulling equally and oppositely in all directions, so the charge is evenly distributed in this tetrahedral molecule. The individual bond dipoles are shown in black. Despite the many nonbonded pairs on the chlorines in this molecule, it has no overall dipole and is *nonpolar*.

b.

$H-\overset{..}{N}-H$ Overall dipole in orange

Each N—H bond is polar and N has extra electrons on it, which unevenly distributes the electrons in this pyramidal molecule. This compound is *polar*. The individual bond dipoles are shown in black. The overall dipole (orange) is going to point toward the extra electrons on the central atom.

c.

$H-C-H$ Overall dipole in orange

Recall that C—H bonds are nonpolar. The only polar bond in this tetrahedral molecule is the C—F bond and its dipole is shown (black). The fluorine pulls the electrons toward itself more, causing an uneven distribution of charge, so the compound is *polar*. The overall dipole (orange) mirrors the bond dipole.

practice problems

3.45 For each of the following molecules, (1) draw the correct Lewis structure; (2) label each polar covalent bond with a dipole moment arrow; and (3) determine if the molecule is polar or nonpolar and draw the dipole moment arrow for the molecule:

a. HCl
b. CH_4
c. NCl_3
d. COS (C is the central atom)
e. H_2S
f. CH_2Br_2 (C is the central atom)

3.46 For each of the following molecules, (1) draw the correct Lewis structure; (2) label each polar covalent bond with a dipole moment arrow; and (3) determine if the molecule is polar or nonpolar and draw the dipole moment arrow for the molecule:

a. CH_3Br (C is the central atom)
b. HF
c. CH_2O (C is the central atom)
d. SiO_2
e. PBr_3
f. OF_2

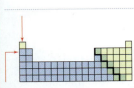

3.1 Electron Arrangements and the Octet Rule

3.1 Inquiry Question: How are electrons distributed in an atom?

In atoms, electrons are distributed only at distinct energy levels in an electron cloud. These energy levels can be designated as n, where $n = 1, 2, 3$, and so on. The electrons in the outermost energy level or shell are called the valence electrons. The noble gases (Group 8A) are stable atoms and each has an octet of electrons in its valence shell. An octet is eight electrons for all elements except hydrogen and helium, where a full valence shell is two electrons. Atoms, other than the noble gases, attain a stable valence octet by combining with each other.

3.2 In Search of an Octet, Part 1: Ion Formation

3.2 Inquiry Question: How are ions formed?

Ions are formed by gaining or losing electrons; this is one way that atoms can attain a valence octet. An atom that gains electrons becomes a negatively charged anion, and an atom that loses electrons becomes a positively charged cation. Nonmetal atoms form anions and metal atoms form cations. During ion formation, atoms gain or lose electrons to attain the same number of electrons as the noble gas with the closest atomic number. For the main-group elements, the atom's position on the periodic table determines its charge when it forms an ion. Polyatomic ions are groups of atoms that together have an ionic charge. Cations have the same name as the element with the word *ion* added to the end, but the name of an anion is derived by changing the ending of the elemental name to *ide*. Transition metals that form more than one cation are distinguished by using a Roman numeral in parentheses after the name of the metal to designate the charge.

3.3 Ionic Compounds—Electron Give and Take

3.3 Inquiry Question: How do ionic compounds form?

An ionic compound is formed when an ionic bond is created between a metal and nonmetal. Ionic bonds are formed by the attraction between an anion and a cation. Ionic compounds are neutral. The total charge (number of ions times the charge of each) of the cations must equal the total charge of the anions in the compound. To name an ionic compound, first name the cation (dropping the word *ion*) and then name the anion. If a transition metal with varying charge is present, a Roman numeral follows the name of the metal to designate its charge. In writing ionic formulas, the metal ion always goes first.

3.4 In Search of an Octet, Part 2: Covalent Bonding

3.4 Inquiry Question: How are covalent compounds formed?

Covalent compounds are formed when nonmetals share electrons to achieve valence octets. Sharing of electrons between two

atoms is called a covalent bond. The number of covalent bonds that an atom will form is determined by the number of electrons necessary to complete its valence octet and leads to the atom's preferred covalent bonding pattern. The molecular formula of a covalent compound gives the number of each type of atom in a molecule, and the Lewis structure shows how those atoms are bonded. Covalent compounds containing only two elements can be named by naming the first element, naming the second element and changing the ending to *ide*. Greek prefixes indicate the number of atoms of each in the compound.

3.5 The Mole: Counting Atoms and Compounds

3.5 Inquiry Question: How is the molar mass used to determine the number of moles or particles present in a known mass of a compound?

The molar mass relates the mass of a substance to a counting unit called the mole. Avogadro's number is a measure of the number of particles in a substance: 6.02×10^{23} per one mole. It is possible to calculate a substance's molar mass, which is equivalent to the atomic mass for an atom or the formula weight for a compound. By using Avogadro's number and molar mass as conversion factors, the number of moles or particles present in the mass of a compound can be determined.

3.6 Getting Covalent Compounds into Shape

3.6 Inquiry Question: How is the shape of a covalent compound determined?

The arrangement of the electrons in a molecule determines the three-dimensional shape of covalent compounds. A Lewis structure does not imply the three-dimensional shape of a molecule. Valence-shell electron-pair repulsion theory (VSEPR) explains that the electrons around any given atom arrange themselves to get as far away from each other as possible. For carbon atoms, this leads to shapes such as tetrahedral, trigonal planar, and linear. The presence of nonbonding pairs on an atom reduces the bond angles around that atom and alters the shape of the molecule.

3.7 Electronegativity and Molecular Polarity

3.7 Inquiry Question: How is the polarity of a molecule determined?

A molecule's polarity is based on the individual bond polarities in the molecule and molecular shape. The bond polarities are determined by comparing the electronegativities of the atoms in the bond. Nonpolar covalent bonds are formed when atoms share electrons equally (electronegativities are equal), and polar covalent bonds are formed when the sharing of electrons is unequal. Polarity can be represented by the delta (δ) symbol or the dipole moment arrow ($\longmapsto$).

The study guide will help you check your understanding of the main concepts in Chapter 3. You should be able to

3.1 Electron Arrangements and the Octet Rule

- Predict the number of valence electrons and energy levels for the main-group elements in the first four periods.
- Recognize the unique stability associated with a valence shell containing eight electrons.

3.2 In Search of an Octet, Part 1: Ion Formation

- Predict the ionic charge of a main-group element using the periodic table.
- Distinguish the name of an ion from its corresponding atom name.
- Gain familiarity with polyatomic ions and their charges.

3.3 Ionic Compounds—Electron Give and Take

- Predict the ionic charges present in an ionic compound.
- Predict the ionic charge of a transition metal using the compound's formula and the anionic charge.
- Name ionic compounds given the formula.
- Write the formula for ionic compounds given the name.

3.4 In Search of an Octet, Part 2: Covalent Bonding

- Distinguish between ionic and covalent compounds.
- Establish the relationship between the number of valence electrons present in the Period 1–3 nonmetals and Group7A elements and the number of bonds that the atom typically makes in a molecule.
- Draw Lewis structures for covalent compounds containing C, O, N, H, and the halogens (Group 7A).
- Name binary covalent compounds given the formula.
- Write the formula for a binary covalent compound given the name.

3.5 The Mole: Counting Atoms and Compounds

- Describe the mole unit and Avogadro's number.
- Calculate molar mass for a compound.
- Convert among the units of mole, number of particles, and gram.

3.6 Getting Covalent Compounds into Shape

- Predict the molecular shapes of small molecules using VSEPR.
- Determine the effect of lone pair electrons on molecular shape.

3.7 Electronegativity and Molecular Polarity

- Predict covalent bond polarity based on electronegativity.
- Predict molecular polarity from bond polarities and molecular shape.

Key Terms

anion—An ion with a net negative charge; formed when an atom gains electrons.

Avogadro's number—The number of particles present in a mole of a substance, 6.02×10^{23}.

binary compounds—Covalent compounds that are composed of only two elements.

bonding pair—A pair of electrons shared by two atoms to form a covalent bond.

cation—An ion with a net positive charge; formed when an atom gives up electrons.

charge clouds—Groupings of electrons in a molecule.

covalent bond—The result of atoms sharing electrons in order to achieve an octet in their valence shells.

covalent compound—A type of compound made up of atoms held together by covalent bonds.

dalton—A unit to measure the weight of a compound used in biochemistry. One dalton is equivalent to one-twelfth of a carbon-12 atom.

double bond—Covalent bonds formed when two atoms share two pairs of electrons.

electronegativity—A measure of the ability of an atom in a covalent bond to attract the bonding electrons.

ion—An atom with a charge due to an unequal number of protons and electrons.

ionic bond—The strong attraction between an anion and a cation.

ionic compound—The result of two or more ions attracted by ionic bonds to form a new electrically neutral species.

isoelectronic—Having the same number of electrons.

Lewis structure—A chemical formula used for covalent compounds to show the number, type, and connectivity of atoms in a molecule.

lone pair—A pair of electrons belonging to only one atom. A lone pair is also called a nonbonded pair.

molar mass—The mass in grams of one mole of a substance. For an element, this is numerically equivalent to atomic mass.

mole—The number of atoms in 12 g of carbon-12. This unit is used to count small objects like atoms and molecules. The number of objects in a mole of a substance is Avogadro's number.

molecular formula—The chemical formula of a molecular compound that represents all atoms in a molecule.

molecule—The smallest unit of a covalent compound; formed when two or more nonmetal atoms are joined by covalent bonds.

noble gases—The Group 8A (18) elements. These elements are chemically inert and satisfy the octet rule.

nonbonded pair—A pair of electrons belonging to only one atom. It is also referred to as a lone pair.

nonpolar covalent bond—A bond in which the electrons are shared equally by the two atoms.

nonpolar molecule—A molecule that has an even distribution of electrons over the entire molecule.

octet rule—The tendency of atoms to react with other atoms to obtain eight total electrons in their outer shell in a compound.

polar covalent bond—A bond in which the electrons are more strongly attracted by one atom than by the other.

polar molecule—A molecule that has an uneven distribution of electrons over the molecule.

polyatomic ion—A group of atoms that interact to form an ion.

single bond—A covalent bond formed when two atoms share one pair of electrons.

tetrahedral—A molecular shape with four atoms arranged around a central atom to give an angle of 109.5°.

triple bond—Covalent bonds formed when two atoms share three pairs of electrons.

valence electrons—Electrons found in the highest occupied energy level.

valence shell—The highest occupied energy level of an atom containing electrons.

valence-shell electron-pair repulsion model (VSEPR)—A method for determining molecular shape based on the repulsion of groups of electrons in the valence shell of the central atom.

Additional Problems

3.47 What is the number of electrons in the outer energy level and the group number for each of the following elements?

Example: Fluorine—7 electrons, Group 7A

  a. magnesium   **b.** chlorine   **c.** oxygen

  d. nitrogen   **e.** barium   **f.** bromine

3.48 What is the number of electrons in the outer energy level and the group number for each of the following elements?

Example: Fluorine—7 electrons, Group 7A

  a. lithium   **b.** silicon   **c.** neon

  d. argon   **e.** tin   **f.** cesium

3.49 How many valence electrons are present in the following atoms?

  a. C   **b.** Cl   **c.** K   **d.** Al

3.50 How many valence electrons are present in the following atoms?

  a. N   **b.** Mg   **c.** S   **d.** Si

3.51 For each of the atoms in Problem 3.49, determine the minimum number of electrons the atom must gain, lose, or share to become stable (isoelectronic with a noble gas).

3.52 For each of the atoms in Problem 3.50, determine the minimum number of electrons the atom must gain, lose, or share to become stable (isoelectronic with a noble gas).

3.53 Complete the following statements:

  a. An anion has more/fewer (*circle one*) protons than electrons.

  b. The electrons found in the outermost shell of an atom are called _____ electrons.

  c. Ions formed from metals have a positive/negative (*circle one*) charge.

3.54 Complete the following statements:

  a. An ion is charged because the numbers of _____ and electrons are not the same.

  b. An ion that has a positive charge is called a _____.

  c. The ion formed from a sulfur atom is called _____.

3.55 Determine the number of protons and electrons in the following ions:

  a. chloride   **b.** Fe^{3+} **c.** Cr^{6+}
  d. N^{3-} **e.** sodium ion   **f.** H^+

3.56 Determine the number of protons and electrons in the following ions:

  a. oxide   **b.** Ag^+ **c.** copper(I) ion
  d. H^- **e.** zinc(II) ion   **f.** I^-

3.57 Give the name of each of the unnamed ions in Problem 3.55.

3.58 Give the name of each of the unnamed ions in Problem 3.56.

3.59 Give the name and symbol of the ion that has the following number of protons and electrons:

  a. 3 protons, 2 electrons
  b. 35 protons, 36 electrons

3.60 Give the name and symbol of the ion that has the following number of protons and electrons:

  a. 9 protons, 10 electrons
  b. 38 protons, 36 electrons

3.61 Each of the following ions is isoelectronic with a noble gas. Give the symbol of the noble gas.

  a. Rb^+ **b.** F^- **c.** S^{2-} **d.** Ca^{2+}

3.62 Each of the following ions is isoelectronic with a noble gas. Give the symbol of the noble gas.

  a. Li^+ **b.** Mg^{2+} **c.** Al^{3+} **d.** O^{2-}

3.63 Complete the table of polyatomic ions by supplying the missing information.

Name	Formula
a. acetate	
b. bicarbonate	
c.	NO_3^-
d.	CN^-

3.64 Complete the table of polyatomic ions by supplying the missing information.

a.	NH_4^+
b.	CO_3^{2-}
c. phosphate	
d. hydroxide	

3.65 Give the formula for the ionic compound formed by the ammonium ion and each of the following anions:

  a. sulfide   **b.** chloride   **c.** sulfate   **d.** hydroxide

3.66 Give the formula for the ionic compound formed by the phosphate ion and each of the following cations:

  a. sodium   **b.** aluminum
  c. magnesium   **d.** iron(III)

3.67 Give the formula for each of the following ionic compounds:

  a. sodium hydroxide, drain cleaner
  b. aluminum oxide, used as an abrasive
  c. potassium nitrate, an ingredient in gunpowder

3.68 Give the formula for each of the following ionic compounds:

  a. potassium chlorate, used in fireworks displays
  b. magnesium sulfate, Epsom salts
  c. silver iodide, used to seed clouds in rainmaking

3.69 Name the following ionic compounds:

  a. Na_2O **b.** $BaSO_4$ **c.** $CuCl_2$
  d. $Mg(NO_3)_2$ **e.** Fe_2O_3 **f.** KF

3.70 Name the following ionic compounds:

  a. $AlCl_3$ **b.** $NH_4C_2H_3O_2$ **c.** AgBr
  d. K_3PO_4 **e.** CsF   **f.** MnO

3.71 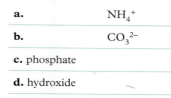 Magnesium salicylate is sold as an over-the-counter drug to treat pain and inflammation. If there are two salicylate ions for each magnesium ion in the compound, what is the charge on the salicylate ion?

3.72 Two common additives to bottled water are the ionic compounds potassium chloride and magnesium sulfate. Write the formulas for these ionic compounds.

3.73 Sodium metabisulfite is used in processed foods as a preservative and antioxidant. The metabisulfate anion has the formula $S_2O_5^{2-}$. Provide the formula for this compound.

3.74 Stannous or tin(II) fluoride is a compound found in many toothpastes. Write the formula for this compound.

3.75 How many covalent bonds does each of the following atoms tend to form?

a. C   b. F   c. P   d. O

3.76 How many covalent bonds does each of the following atoms tend to form?

a. Cl   b. Si   c. S   d. N

3.77 For each of the following compounds, indicate whether it is covalent or ionic and give its correct name:

a. CBr_4 b. SiO_2 c. $MgBr_2$ d. NCl_3
e. $CrCl_3$

3.78 For each of the following compounds, indicate whether it is covalent or ionic and give its correct name:

a. Cu_2O b. $SiCl_4$ c. H_2O d. BaS
e. CS_2

3.79 Explain the difference between an ionic formula and a molecular formula.

3.80 Explain the difference between a Lewis structure and a molecular formula.

3.81 Draw a Lewis structure for each of the following covalent compounds:

a. HCN   b. CS_2 c. PCl_3 d. CH_5N

3.82 Draw a Lewis structure for each of the following covalent compounds:

a. CH_4O b. N_2H_4 c. Br_2 d. C_2H_3Cl

3.83 Give the name of each of the following covalent compounds:

a. SeO_2 b. SiF_4 c. P_4S_3 d. OF_2

3.84 Give the name of each of the following covalent compounds:

a. SO_2 b. CI_4 c. SiS_2 d. SCl_2

3.85 Explain the difference between an ionic bond and a covalent bond.

3.86 What are the units of Avogadro's number?

3.87 What is the mass of 4.00 moles of the following?

a. H   b. Li   c. KNO_3 d. C_2H_6O

3.88 How many particles are in 5.0 moles of the following?

a. O   b. N   c. $MgCl_2$ d. C_2H_3NO

3.89 A pencil mark (made with graphite, a form of carbon) on paper has a mass of about 0.000005 g (5×10^{-6} g). How many atoms of graphite are present in the pencil mark?

3.90 The recommended daily allowance of the niacin (vitamin B3) is 14 mg per day for women. The molecular formula for niacin is $C_6H_5NO_2$. Calculate the number of molecules of niacin in 14 mg.

3.91 Aspartic acid, a naturally occurring amino acid, is a component of aspartame (Nutrasweet™), an artificial sweetener widely used in diet soft drinks.

a. Determine the shape around each orange-colored atom in the Lewis structure of aspartic acid.

b. Give the bond angle around each of the orange-colored carbon atoms.

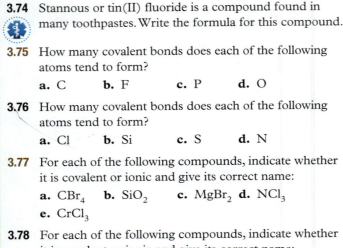

3.92 Cyanoacrylic acid is one of the compounds used to make Super Glue®. Determine the shape of the molecule around each orange-colored atom in its Lewis structure.

a. Determine the shape around each orange-colored atom in the Lewis structure of cyanoacrylic acid.

b. Give the bond angle of each of the orange-colored carbon atoms.

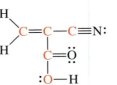

3.93 Methyl isocyanate is used in the manufacturing of pesticides and was the deadly gas released from a chemical plant in the 1984 disaster in Bhopal, India, that killed thousands. Determine the shape around each of the carbon and nitrogen atoms in the Lewis structure of methyl isocyanate.

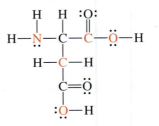

3.94 Vinyl acetate is used in the production of safety glass for the automotive industry. Determine the shape around each of the carbon atoms in the Lewis structure of vinyl acetate.

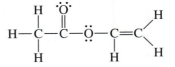

3.95 Identify the more electronegative atom in each of the following pairs:

a. S and Se   b. P and N
c. C and Cl   d. N and I

3.96 Identify the more electronegative atom in each of the following pairs:

a. H and Cl   **b.** N and O

c. C and N   **d.** F and Br

3.97 Using the Lewis structures provided,

a. label each polar covalent bond with a dipole moment arrow.

b. determine if the molecule is polar or nonpolar and draw the molecular dipole.

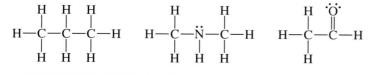

3.98 Using the Lewis structures provided,

a. label each polar covalent bond with a dipole moment arrow.

b. determine if the molecule is polar or nonpolar and draw the molecular dipole.

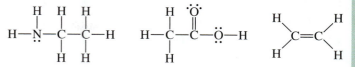

Challenge Problems

3.99 Tooth enamel contains a hard mineral called hydroxyapatite composed of calcium, phosphate, and hydroxide ions. If the formula for the compound contains six phosphate ions and two hydroxide ions, provide the complete formula for the ionic compound.

3.100 Two elements, E and X, are combined and react to form a compound containing E^{2+} and X^{2-} ions. Answer the following questions about E and X based on this information.

a. Which of the elements, E or X, is more likely to be a metal?

b. Which of the elements, E or X, is more likely to be a nonmetal?

c. What is the formula of the compound formed from E and X?

d. To what group on the periodic table does each element belong?

3.101 Precious metals are commonly measured out in troy ounces. One troy ounce of platinum is equivalent to 31.10 g of platinum. How many atoms of platinum are in 1.00 troy ounce of platinum?

3.102 Vinyl chloride, C_2H_3Cl, is used in the production of polyvinyl chloride (PVC) a plastic used to make tubing and pipes for a variety of medical applications.

a. Draw the Lewis structure of vinyl chloride.

b. Determine the shape and the bond angle around each of the carbon atoms.

c. Label each of the polar covalent bonds with a dipole moment arrow.

d. Determine if this compound is polar or nonpolar.

3.103 One of the most common compounds used in commercial dry cleaning is tetrachloroethylene, which has the molecular formula C_2Cl_4.

a. Draw the Lewis structure of tetrachloroethylene.

b. Determine the shape and the bond angle around each of the carbon atoms.

c. Label each of the polar covalent bonds with a dipole moment arrow.

d. Determine if this compound is polar or nonpolar.

Answers to Odd-Numbered Problems

Practice Problems

3.1 2 electrons, 8 electrons

3.3 **a.** 2 e⁻ in first energy level

b. 2 e⁻ in first energy level, 4 e⁻ in second energy level

c. 2 e⁻ in first energy level, 8 e⁻ in second energy level, 1 e⁻ in third energy level

d. 2 e⁻ in first energy level, 8 e⁻ in second energy level

3.5 **a.** 6   **b.** 4   **c.** 5   **d.** 1

3.7 In atoms, the number of protons and electrons are the same and there is no net charge. In cations, there are

more protons than electrons present, giving the cation a positive charge.

3.9 In naming an anion, the last several letters of the nonmetal element name are dropped and the suffix *ide* is applied.

3.11 **a.** 20 p, 18 e⁻   **b.** 53 p, 54 e⁻

c. 16 p, 18 e⁻   **d.** 30 p, 28 e⁻

3.13 **a.** calcium ion   **b.** iodide

c. sulfide   **d.** zinc ion

3.15 **a.** fluoride, F^- **b.** chromium(III) ion, Cr^{3+}

3.17 **a.** ammonium ion **b.** acetate **c.** cyanide

3.19 **a.** gold (III) chloride, $AuCl_3$

b. calcium carbonate, $CaCO_3$

c. magnesium hydroxide, $Mg(OH)_2$

3.21 **a.** one Au^{3+}, three Cl^- **b.** one Ca^{2+}, one CO_3^{2-}

c. one Mg^{2+}, two OH^-

3.23 **a.** sodium carbonate, Na_2CO_3

b. iron (II) carbonate, $FeCO_3$

c. aluminum carbonate, $Al_2(CO_3)_3$

3.25 **a.** two $Na+$, one CO_3^{2-} **b.** one Fe^{2+}, one CO_3^{2-}

c. two Al^{3+}, three CO_3^{2-}

3.27 (a)   (b)

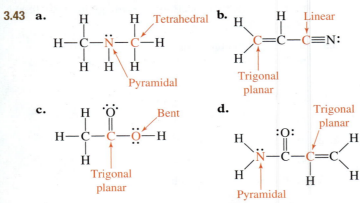

(c)   (d)

3.29 **a.** Carbon prefers four bonds, hydrogen only one each, so CH_2 is unlikely to exist.

b. Nitrogen prefers three bonds, hydrogen only one each, so NH_2 is unlikely to exist.

c. These halogens could form one covalent bond like Cl_2, so BrCl could exist.

d. Sulfur will form two bonds, hydrogen one each, so H_2S could exist.

3.31 **a.** covalent **b.** ionic **c.** ionic **d.** covalent

e. covalent **f.** ionic **g.** covalent **h.** ionic

i. covalent **j.** covalent

3.33 SO_3, diphosphorus pentoxide, SeF_4, carbon monoxide, N_2O_3

3.35 **a.** They are equal; 6.02×10^{23} atoms.

b. Gold weighs more, 197.0 g versus 107.9 g.

3.37 **a.** 1.51×10^{23} atoms

b. 0.019 mole **c.** 2.57×10^{23} atoms

3.39 **a.** 120.38 g/mole **b.** 180.16 g/mole **c.** 44.01 g/mole

3.41 1.09×10^{21} molecules

3.43 **a.**   Tetrahedral **b.**   Linear

       Pyramidal   Trigonal planar

c.   Bent   **d.**   Trigonal planar

    Trigonal planar   Pyramidal

3.45 **a.** polar

H—Cl

Polar

Because this molecule contains one bond, the bond and molecular dipoles are the same.

b. nonpolar

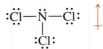

Nonpolar

C–H bonds are considered nonpolar.

c. polar

Polar

N–Cl bonds are nonpolar (same electronegativity), but the pair of electrons on the N gives the molecule polarity.

d. polar

Polar

C–O bonds are polar and C–S bonds are not. The molecular dipole is shown in orange.

e. polar

Polar

S–H bonds are polar. This molecule has a bent shape. The lone pairs of electrons are on one side of the molecule in three dimensions. The molecular dipole is shown in orange.

f. polar

Polar

C–Br bonds are polar. Considering the tetrahedral shape, the bromine side of the molecule is the negative side.

Additional Problems

3.47 **a.** two extrons, Group 2A

b. seven electrons, Group 7A

c. six electrons, Group 6A

d. five electrons, Group 5A

e. two electrons, Group 2A

f. seven electrons, Group 7A

3.49 a. 4 **b.** 7 **c.** 1 **d.** 3

3.51 a. 4 **b.** 1 **c.** 1 **d.** 3

3.53 a. fewer **b.** valence **c.** positive

3.55 a. 17 p, 18 e⁻ **b.** 26 p, 23e⁻ **c.** 24 p, 18e⁻
d. 7 p, 10e⁻ **e.** 11 p, 10e⁻ **f.** 1 p, 0e⁻

3.57 a. iron(III) **b.** chromium(VI)
c. nitride **d.** hydrogen ion, also proton

3.59 a. lithium ion, Li^+ **b.** bromide, Br^-

3.61 a. Kr **b.** Ne **c.** Ar **d.** Ar

3.63 a. $C_2H_3O_2^-$ **b.** HCO_3^-
c. nitrate **d.** cyanide

3.65 a. $(NH_4)_2S$ **b.** NH_4Cl
c. $(NH_4)_2SO_4$ **d.** NH_4OH

3.67 a. NaOH **b.** Al_2O_3 **c.** KNO_3

3.69 a. sodium oxide **b.** barium sulfate
c. copper(II) chloride **d.** magnesium nitrate
e. iron(III) oxide **f.** potassium fluoride

3.71 1⁻

3.73 $Na_2S_2O_5$

3.75 a. 4 **b.** 1 **c.** 3 **d.** 2

3.77 a. covalent, carbon tetrabromide
b. covalent, silicon dioxide
c. ionic, magnesium bromide
d. covalent, nitrogen trichloride
e. ionic, chromium(III) chloride

3.79 Ionic formulas are written for ionic compounds and use the smallest ratio of ions; molecular formulas are written for covalent compounds and indicate the exact number of atoms in the molecule.

3.81 a. **b.**

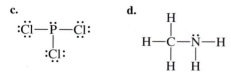

c. **d.**

3.83 a. selenium dioxide
b. silicon tetrafluoride
c. tetraphosphorus trisulfide
d. oxygen difluoride

3.85 A covalent bond results when two atoms share one or more pairs of electrons. An ionic bond is the attraction between two or more ions that are formed by the loss and gain of electrons.

3.87 a. 4.03 g **b.** 27.8 g **c.** 404 g **d.** 184 g

3.89 3×10^{17} atoms of carbon

3.91

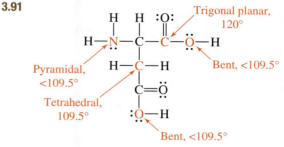

3.93

3.95 a. S **b.** N **c.** Cl **d.** N

3.97

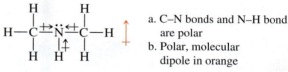

a. All bonds are nonpolar
b. Nonpolar

a. C–N bonds and N–H bond are polar
b. Polar, molecular dipole in orange

a. C=O bond is polar
b. Polar, molecular dipole in orange

3.99 $Ca_{10}(PO_4)_6(OH)_2$

3.101 9.60×10^{22} atoms of Pt

3.103

a. Lewis structure shown at left.
b. Both carbons are trigonal planar and the bond angle is 120 degrees.
c. C—Cl bonds are polar, dipoles shown at left.
d. The polar bonds are pulling outward equally and oppositely so the molecule has no molecular dipole and is nonpolar.

When you think of the word *organic* you may think of pesticide-free, naturally grown produce and grains, but in chemistry the term has a different meaning—like vitamins, medications, pesticides, and even plastics. Vitamin C is an organic compound that can be found naturally in foods or synthesized and formed into tablets. Whether natural or synthetic, the vitamin C molecule behaves the same way in the body. Many prescription drugs like Ritalin® are organic compounds. In Chapter 4, we unravel the mysterious stick drawings that are used to represent organic compounds and the unique properties of carbon-containing compounds.

4

Introduction to Organic Compounds

THE WORD *ORGANIC* can describe a set of environmentally friendly farming practices or imply that a product is in some way more natural, but chemists use *organic* to identify covalent compounds containing carbon. **Organic compounds** are composed primarily of carbon and hydrogen but may also include oxygen, nitrogen, sulfur, phosphorus, and a few other elements. The molecules of life, or **biomolecules,** such as proteins, carbohydrates, lipids, and DNA, are also organic compounds. Chapter 4 combines the concepts about atoms (Chapter 2) and how they combine to form molecules (Chapter 3) to begin the study of the organic compounds that make up living things and the chemistry of life.

Why are carbon-containing compounds singled out in this special category? Organic compounds, whether naturally occurring or synthesized in the laboratory, are found in familiar substances such as gasoline, cotton, plastics, vitamins, medicines, and cosmetics. Besides the importance of these compounds to life itself, carbon is unique in its ability to form bonds with other atoms of carbon to create chains and rings of various sizes and shapes. In fact, of the over 60 million known chemical compounds, most of these compounds are organic.

Carbon's ability to form such a tremendous variety of compounds and our ability to manipulate and react these compounds in the laboratory contribute to an ever-increasing interest in the chemistry subdiscipline called **organic chemistry.** Organic chemistry is dedicated to the study of the structure, properties, and reactivity of carbon-containing compounds. Compounds that do not contain carbon and hydrogen are classified as **inorganic compounds.**

Luckily, it is not necessary to study the millions of organic compounds to understand organic chemistry. Chemists group organic compounds into distinct families based on their molecular structure and composition. Furthermore, members of the same family behave similarly in chemical reactions. By recognizing a few key structural features, it is possible to gain a basic understanding of organic structure and reactivity.

? **What's an Inquiry Question?**
Inquiry Questions are designed to focus your reading on the main concepts by section. An Inquiry Question appears at the beginning of each section.

We start our exploration in Chapter 4 with the simplest family of organic compounds, the hydrocarbons. From this starting point, we can form a structural foundation that applies to the other families of organic compounds.

4.1 Alkanes: The Simplest Organic Compounds

4.1 Inquiry Question: How are the structure and polarity of alkanes and cycloalkanes characterized?

We use fossil fuels for heating, transportation, and generating electricity. Most of these fossil fuels belong to the first family of organic compounds that we will consider. This family of compounds is known as the **alkanes.** These structurally simple organic compounds are made up solely of carbon and hydrogen. In fact, the alkanes are typically referred to as **saturated hydrocarbons.** The term *hydrocarbon* indicates that alkanes are made up entirely of hydrogen and carbon, and *saturated* indicates that these compounds contain only single bonds. Each carbon atom, in addition to being bonded to other carbon atoms, is bonded to the maximum number of hydrogen atoms (which saturate it).

Straight-Chain Alkanes

A portable form of heating fuel often used in remote locations, in gas grills, and in recreational vehicles is the alkane called propane. Compounds like propane belong to the simplest group of alkanes, known as **straight-chain alkanes,** and are made up of carbon atoms joined to one another to form continuous, unbranched chains of varying length. To distinguish individual alkanes from one another, each compound is given a name that is based on the number of carbon atoms in its chain. For example, propane is the name for an alkane containing three carbon atoms connected in a straight chain and saturated with hydrogen.

The names and Lewis structures of the first 10 alkanes are shown in Table 4.1. The names of the first four alkanes may seem odd, but they are in fact historical names still used today. The names of the straight-chain alkanes with five or more carbons are derived from a Greek numerical prefix (for example, *pent* = 5) followed by the ending *ane* (indicating the alkane family). It is important to recognize the names of these first 10 alkanes because the names of the majority of organic compounds are derived from them.

As you look down the molecular formulas in Table 4.1, notice that the formulas systematically increase. For every carbon present in an alkane, the number of hydrogens is two times the number of carbons plus two. For propane, this means that for the three carbons in propane there are $(2 \times 3) + 2$ or a total of eight hydrogens, giving a molecular formula C_3H_8. We can write an alkane general formula as C_nH_{2n+2} where n is the number of carbons in the alkane.

Propane, C_3H_8

Propane is one of the straight-chain alkanes. Notice in the ball-and-stick model that the chain is not literally "straight." The carbons have a tetrahedral geometry bonded end to end.

TABLE 4.1 Names of the First 10 Straight-Chain Alkanes

Number of Carbon Atoms	Prefix	Name of Alkane	Molecular Formula	Lewis Structure
1	meth	Methane	CH_4	
2	eth	Ethane	C_2H_6	
3	prop	Propane	C_3H_8	
4	but	Butane	C_4H_{10}	
5	pent	Pentane	C_5H_{12}	
6	hex	Hexane	C_6H_{14}	
7	hept	Heptane	C_7H_{16}	
8	oct	Octane	C_8H_{18}	
9	non	Nonane	C_9H_{20}	
10	dec	Decane	$C_{10}H_{22}$	

Cycloalkanes

In addition to forming long chains, carbon can also form rings. These compounds are called **cycloalkanes** (*cyclo* implies circle). Because they are saturated hydrocarbons, they are still alkanes. The names of the cycloalkanes are formed by adding the prefix *cyclo* to the alkane name for the compound containing the same number of carbon atoms. For example, the simplest cycloalkane is cyclopropane. Notice that the name indicates that this compound is a ring compound (cyclo) of three carbon atoms (propane).

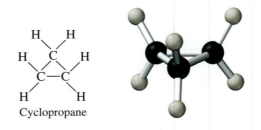

Cyclopropane

In theory, cycloalkanes (and similar ring-containing organic compounds) can exist in many different ring sizes. However, rings of five and six carbon atoms are the most common in nature. Here we focus on the more common rings of five and six carbon atoms. Table 4.2 gives the names and Lewis structures of cycloalkanes with three to six carbons.

TABLE 4.2 Cycloalkanes: Names and Common Structures

Name			
Cyclopropane	Cyclobutane	Cyclopentane	Cyclohexane

Ball-and-Stick Models

Lewis Structures

4.1   Cycloalkanes

The general molecular formula for a straight-chain alkane is C_nH_{2n+2}. What is the general molecular formula for a cycloalkane? Explain.

Solution

Inspection of Table 4.2 shows the molecular formulas C_3H_6, C_4H_8, C_5H_{10}, and C_6H_{12} for cyclopropane, cyclobutane, cyclopentane, and cyclohexane respectively. Because the carbons in a cycloalkane connect as a ring, there are two fewer hydrogens present as opposed to the corresponding straight-chain alkane. The general molecular formula is therefore C_nH_{2n}.

Alkanes Are Nonpolar Compounds

In Chapter 3, the polarity of molecules was determined by examining both the electronegativity of the atoms in the bonds and the overall shape of the molecule. Nonpolar molecules have an even distribution of electrons over their entire surface, while polar molecules have a partially negative part and a partially positive part (poles) due to an uneven distribution of electrons in the molecule.

The electronegativities of carbon and hydrogen are so similar that when these two elements form covalent bonds, the electrons are shared equally and the bond is nonpolar. Alkanes are composed solely of carbon and hydrogen, so regardless of their shape, alkanes are nonpolar. The nonpolar nature of alkanes affects the behavior of these compounds in aqueous systems. (Remember that water is polar.) This is discussed in more detail in Chapter 7.

4.2   Alkanes and Cycloalkanes

Using Tables 4.1 and 4.2, name the alkanes or cycloalkanes with the molecular formulas shown:

a. C_2H_6 b. C_6H_{12} c. C_9H_{20}

Solution

Compare each formula with the general formulas for straight-chain and cycloalkanes. Use Table 4.1 or 4.2 to find the name of the appropriate alkane.

a.   ethane   b.   cyclohexane   c.   nonane

practice problems

Use Tables 4.1 and 4.2 to help you answer these practice problems.

4.1   Name the alkanes or cycloalkanes with the molecular formulas shown:

    a. C_4H_8 b. C_5H_{12} c. C_7H_{16}

4.2   Name the alkanes or cycloalkanes with the molecular formulas shown:

    a. CH_4 b. C_6H_{14} c. C_7H_{14}

4.3   Write the molecular formulas for the alkanes shown:

    a. butane   b. cyclopentane   c. octane

4.4   Write the molecular formulas for the alkanes shown:

    a. cyclooctane   b. decane

    c. cyclobutane

Discovering the Concepts

? Inquiry Activity—Representing Molecules on Paper

Information

The following structures are representations of two covalent compounds that are *structural isomers*: isopropyl alcohol (rubbing alcohol) and *n*-propyl alcohol.

Isopropyl alcohol

C_3H_8O
Molecular formula

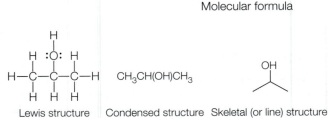

Lewis structure   Condensed structure   Skeletal (or line) structure

Ball-and-stick model

n-Propyl alcohol

C_3H_8O
Molecular formula

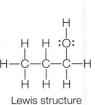
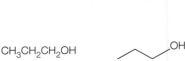
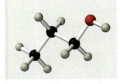

Lewis structure   Condensed structure   Skeletal (or line) structure

Ball-and-stick model

Questions

1. How are *n*-propyl alcohol and isopropyl alcohol the same? How are they different?
2. How do a molecular formula and a condensed structure differ?
3. How do a Lewis structure and a condensed structure differ?
4. a.   What atoms do the black spheres in the ball-and-stick models represent?
    b.   What two atoms do the red sphere and the small gray sphere to the right of it represent?
5. What atom is always at the intersection of two line segments in the skeletal structure?
6. Provide a definition for structural isomer. (HINT: What is the same about the *n*-propyl alcohol and isopropyl alcohol molecules? What is different?)
7. Provide the corresponding condensed and skeletal structures for the Lewis structures of isobutane and *n*-butane.

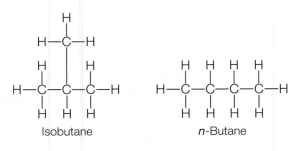

Isobutane   *n*-Butane

8. Are *n*-butane and isobutane structural isomers? How do you know?
9. Provide the corresponding skeletal structure for each of the following hydrocarbons:

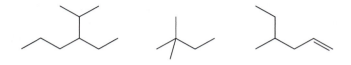

$$CH_3CH_2CH_2\overset{\overset{\displaystyle CH_3}{|}}{C}HCH_2CH_2CH_3$$

$$CH_3\overset{\overset{\displaystyle CH_2CH_2CH_3}{|}}{C}HCH_2CH_2\overset{\overset{\displaystyle |}{}}{C}HCH_2CH_3 \\ \overset{\overset{\displaystyle |}{CH_3}}{}$$

10. Explain how the number of hydrogens written after each carbon in the condensed structures in Question 9 was determined.
11. Provide the condensed structure for the following skeletal structures.

4.2 Representing the Structures of Organic Compounds

In Chapter 3, molecules were drawn using Lewis structures. Lewis structures show the connectivity of all the atoms in a molecule by showing all atoms, bonds, and lone pairs. In Tables 4.1 and 4.2, the Lewis structures of some common alkanes and cycloalkanes are shown.

All the information provided by a Lewis structure is important, but as molecules get larger, chemists simplify the representations by showing the whole molecule but omitting some of the details. For example, because we have established that an uncharged oxygen will always have two bonds and two lone pairs of electrons in a covalent compound, the lone pairs of electrons are often omitted (yet implied) when drawing oxygen in an organic compound.

4.2 Inquiry Question:
How are organic compounds represented?

Condensed Structural Formulas

The first representation considered is the **condensed structural formula,** also called the condensed structure. Like Lewis structures, these structures show *all the atoms* in a molecule, but they show *as few bonds as possible.* Condensed structures may or may not show lone pairs.

When drawing or interpreting condensed structures, it is helpful to remember the preferred covalent bonding patterns shown in Chapter 3 (see Table 3.4). For example, carbon always makes four bonds to other atoms. Because few bonds are shown between atoms in condensed structures, knowing these patterns will help you quickly interpret condensed structural formulas.

The condensed structural formula for propane is shown between its Lewis structure and its molecular formula. Note the differences. The molecular formula shows only the number of each atom in the molecule, while the Lewis structure shows the molecule's complete connectivity—all atoms and all bonds. With regard to information provided, the condensed structural formula is somewhere between the two. The bonding of the hydrogens to the carbons is expanded in the condensed structure versus the molecular formula, but the bonds to hydrogen are not shown as they are in the Lewis structure.

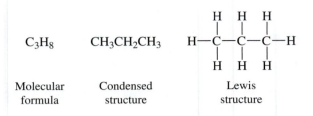

C_3H_8	$CH_3CH_2CH_3$	
Molecular formula	Condensed structure	Lewis structure

In the condensed structural formula, the atoms are listed from left to right in the order in which they are connected in the molecule. It is understood that the octet of each atom is completed according to its preferred pattern of covalent bonding.

Notice that a carbon bonded to three hydrogen atoms in the Lewis structure is written as CH_3 (or sometimes H_3C), in the condensed structure and a carbon bonded to two hydrogen atoms is written as CH_2. In each case, the carbon atom's preferred bonding pattern of four bonds is completed by bonding to other carbon atoms.

---

sample problem 4.3   Drawing Condensed Structural Formulas

Draw a condensed structural formula for each of the Lewis structures shown:

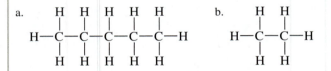

a.

b.

Solution

The condensed structural formula shows all atoms but omits the bonds between atoms. To write a condensed formula, begin with the carbon atom farthest to the left in the structure. List that carbon and then the hydrogens bonded to it, using a subscript to indicate the number of hydrogens. Proceed in the same manner for each carbon atom in the Lewis structure.

a. The correct condensed structure is $CH_3CH_2CH_2CH_2CH_3$. To further condense this structure, it could be written as $CH_3(CH_2)_3CH_3$.

b. The correct condensed structure is CH_3CH_3.

---

sample problem 4.4   Converting Condensed Structures to Lewis Structures

Draw a Lewis structure for each of the condensed structures shown:

a. $CH_3CH_2CH_2CH_3$

b. $CH_3CH_2CH_2CH_2CH_2CH_2CH_3$

Solution

Reviewing the preferred bonding patterns from Table 3.4, notice that hydrogen can make only one bond, while carbon makes four. The H following the C implies that those hydrogens are bonded to the preceding carbon. Bond the hydrogens to their respective carbons to draw the Lewis structures.

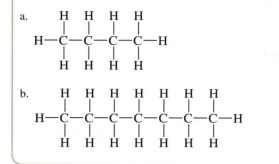

Skeletal Structures

The other representation that we will consider is commonly used on pharmaceutical inserts and on the labels of pesticides and herbicides. It is the main representation used in drawing cycloalkanes. This type of structure, called a **skeletal structure,** is truly a "bare bones" structure because it shows the bonds between carbon atoms as lines. Carbon and hydrogen atoms are not shown explicitly. The carbon atoms in skeletal structures are understood to exist at the end of the bond (line) if no other atom is shown.

The skeletal structure of propane is shown in **Figure 4.1** along with other representations. Propane contains three carbons, so it is drawn with just two lines. All bonds to carbon are drawn in a skeletal structure, *except* those to hydrogen. In propane, the middle carbon atom shows two bonds to other carbons, implying that two hydrogens are also bonded to it. The end carbons show one bond. Therefore, they would have three hydrogens bonded to them. Unless otherwise specified, each carbon atom is assumed to be saturated (see Section 4.1) with hydrogen, but hydrogens bonded to carbon are not shown, only implied. Because only bonds between carbon atoms are shown, it is not possible to draw a skeletal structure for CH_4, methane, because it has only one carbon. In fact, we typically do not draw skeletal structures for alkanes with fewer than three carbon atoms.

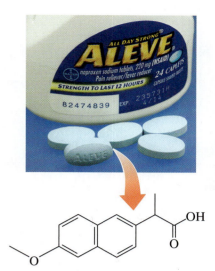

The information about a pharmaceutical often shows the active ingredient in the skeletal structure.

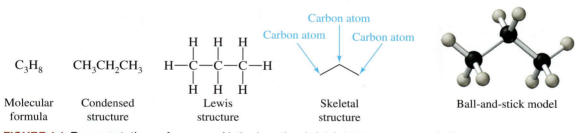

FIGURE 4.1 Representations of propane. Notice how the skeletal structure compares to the others. In the skeletal structure, the carbon and hydrogen atoms are not shown.

Skeletal structures are particularly useful for representing the cycloalkanes. When these molecules are represented using skeletal structures, they appear as familiar geometric figures. The three-carbon ring of cyclopropane looks like a triangle, the four-carbon ring of cyclobutane looks like a square, the five-carbon ring of cyclopentane looks like a pentagon, and the six-carbon ring looks like a hexagon. **Figure 4.2** shows the skeletal structures of the common cycloalkanes.

Cyclopropane   Cyclobutane   Cyclopentane   Cyclohexane   Cycloheptane

FIGURE 4.2 Skeletal Structures of Some Common Cycloalkanes. The cycloalkane structures look like geometric figures. Each corner represents a carbon atom.

Drawing skeletal structures takes some practice, but can be accomplished by keeping in mind a few simple rules as you draw the structures.

Rules for drawing skeletal structures

- Bonds to carbon are shown.
- Bonds between carbon and hydrogen are not shown, but are implied.
- Other elements bonded to carbon are drawn at the end of the bond using their symbol.
  - If these atoms have hydrogens bonded to them, these hydrogens are shown.
  - Lone pairs of electrons on these elements are not shown.

Solving a Problem

Drawing Skeletal Structures

Draw a skeletal structure for the following given their Lewis structure.

a.

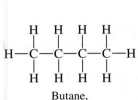

Butane,
used as a fuel in lighters

b.

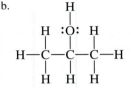

Isopropyl alcohol,
rubbing alcohol

STEP 1: Determine the number of carbons connected end to end.

a. In butane, there are four carbons.
b. In isopropyl alcohol, there are three carbons.

STEP 2: Draw the bonds between the carbons (the carbon skeleton). These are highlighted in the molecules that follow. Zigzag the lines so that you can see where one bond ends and the next begins. Sometimes it is useful to put a dot at the end of the bond so that you can see the position of the carbons.

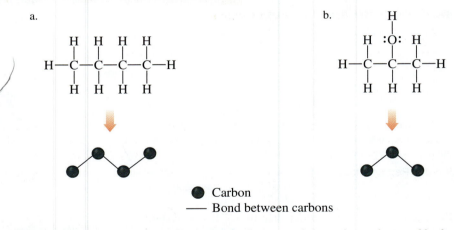

● Carbon
—— Bond between carbons

STEP 3: Draw bonds to noncarbon atoms. Butane only contains carbon and hydrogen. Isopropyl alcohol contains an oxygen with a hydrogen bonded to it. This is represented in skeletal structure as shown. Note that the lone pairs of electrons on the oxygen are also implied in the skeletal structure.

sample problem 4.5   Drawing Skeletal Structures

Draw a skeletal structure for each of the following compounds:

a.   Pentane

CH₃CH₂CH₂CH₂CH₃

b.   Cyclobutane

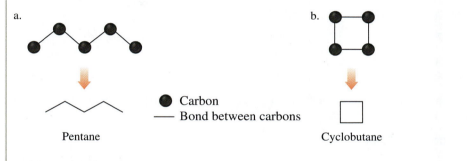

Solution

STEP 1: Determine the number of carbons connected end to end.

a.   In pentane, there are five carbons.
b.   In cyclobutane, there are four carbons.

STEP 2: Draw the carbon skeleton. Zigzag the five carbons of pentane. Sometimes it is useful to draw dots to keep track of the carbon atoms while assembling the skeletal structure. To draw the skeletal structure for a cycloalkane, you can count the number of carbon atoms in the ring and draw the corresponding geometric figure.

a.

Pentane

● Carbon
— Bond between carbons

b.

Cyclobutane

STEP 3: Draw bonds to noncarbon atoms. Because these two examples only have carbon and hydrogen, step 3 is not necessary in this problem.

---

practice problems

4.5   Describe the difference between a Lewis structure and a condensed structure in terms of atoms and bonds shown in the structures.

4.6   Describe the difference between a Lewis structure and a skeletal structure in terms of atoms and bonds shown in the structures.

4.7   Draw condensed structural formulas for the 10 straight-chain alkanes in Table 4.1.

4.8   Draw skeletal structures for the straight-chain alkanes with three carbons or more in Table 4.1.

4.9   Explain why it is not possible to draw a skeletal structure for methane.

4.10   Draw a skeletal structure for ethane, C_2H_6. Do you see why ethane is usually not represented in skeletal structure?

4.11   Cyclohexane, C_6H_{12}, is used to clean greasy machines and parts in many industrial settings. Draw the Lewis structure and the skeletal structure of cyclohexane without referring back to Table 4.2.

4.12   Octane, C_8H_{18}, is a component of gasoline. Draw the Lewis structure, the condensed structure, and the skeletal structure of octane without referring back to Table 4.1.

Discovering the Concepts

? Inquiry Activity—Meet the Hydrocarbons

Information

Organic Compounds
- contain carbon
- form covalent bonds
- are grouped into families
  - The families, with the exception of the alkanes, are classified by the *functional group(s)* present.
  - Functional groups are the reactive parts of molecules. Polar groups are more reactive than nonpolar groups.

The Hydrocarbon Family of Organic Compounds	
Family Member	**Functional Group**
Alkane	None, all C and H are single bonded
Alkene	C═C double bond
Alkyne	C≡C triple bond
Aromatic	Planar, ring structures based on benzene. Can contain *heteroatoms* such as N, O, or S. Benzene

Hydrocarbon Molecule Set

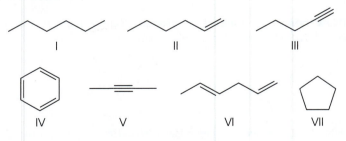

Questions

1. How are all the molecules in the molecule set the same? What do all hydrocarbons have in common?
2. Which molecules in the molecule set are polar? Which are nonpolar?
3. Based on the information, are hydrocarbons very reactive molecules?
4. Identify the molecules in the molecule set that belong to the following families.

Family Member	Functional Group
Alkane	
Alkene	
Aromatic	
Alkyne	

5. What is the molecular shape (geometry) of molecule IV in the molecule set?
6. Distinguish between an organic family of compounds and its functional group.

4.3 Families of Organic Compounds— Functional Groups

There are so many organic compounds, which means there must be a way of organizing all of them into manageable groups or families. So far we have met the saturated hydrocarbon family of organic compounds called alkanes. Yet organic compounds can also contain oxygen, nitrogen, sulfur, phosphorus, and other elements. Elements other than carbon and hydrogen that are present in an organic compound are called **heteroatoms.**

There are distinct patterns to carbon bonding in organic compounds. A group of atoms bonded in a particular way is called a **functional group.** Organic compounds are classified into families based on common functional groups (their carbon bonding pattern). Each functional group has specific properties and chemical reactivity. Organic compounds that contain the same functional group behave similarly. By identifying the properties and reactivity of fewer than 20 functional groups, we can understand much about the properties and reactivity of many of the millions of organic compounds.

Some common functional groups are listed in Table 4.3. First, we consider the families of compounds that contain just carbon and hydrogen (the first four groups listed in Table 4.3). Note that most of the functional groups listed contain the most common heteroatoms—oxygen, nitrogen and sulfur. Also notice that many of the functional groups contain a carbon double bonded to an oxygen. The C=O group found in several families is called a **carbonyl** (pronounced carbon-NEEL). We will encounter many of the functional groups listed here in later chapters as they appear in the structures of biomolecules.

The functional group is the reactive part of an organic molecule. To keep the focus on the functional group, an R is often used to represent the Rest of the molecule. The R part is usually bonded to the functional group through a carbon bond. For example, there are many molecules that fit into the family of **carboxylic acid.** This functional group contains a carbonyl bonded to an —OH group. Acetic acid or vinegar, and palmitic acid, a fatty acid discussed later in this section, are both part of the family of carboxylic acid because they both contain the functional group even though the rest of the molecule, R, is different.

? **4.3 Inquiry Question:** How are carbon compounds organized?

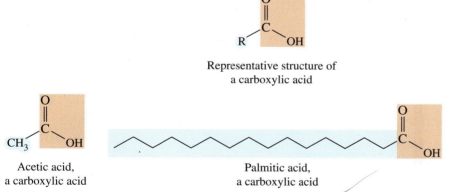

Representative structure of
a carboxylic acid

Acetic acid,
a carboxylic acid

Palmitic acid,
a carboxylic acid

Both acetic acid and palmitic acid contain the functional group carboxylic acid and are in the same family of organic molecules (R group highlighted in blue).

The use of R allows us to simplify a structure and highlight just the functional group of interest. Remember that the abbreviation R can represent anything from one carbon to a complex group containing many carbons.

TABLE 4.3 Families of Organic Compounds: Common Functional Groups

Family Name	Representative Structure of the Functional Group	Biological Example Containing Functional Group
Alkane (saturated hydrocarbon)	All C—C bonds are single	
Alkene (unsaturated hydrocarbon)	Contain C=C double bonds	
Alkyne (unsaturated hydrocarbon)	Contain C≡C triple bonds	
Aromatic	Planar, ring structures, based on benzene. Can contain heteroatoms.	
Alcohol	Primary (1°) / Secondary (2°) / Tertiary (3°)	
Phenol (aromatic alcohol)		
Ether	R—O—R	
Thiol / Sulfide (thioether) / Disulfide	R—S—H / R—S—R / R—S—S—R	

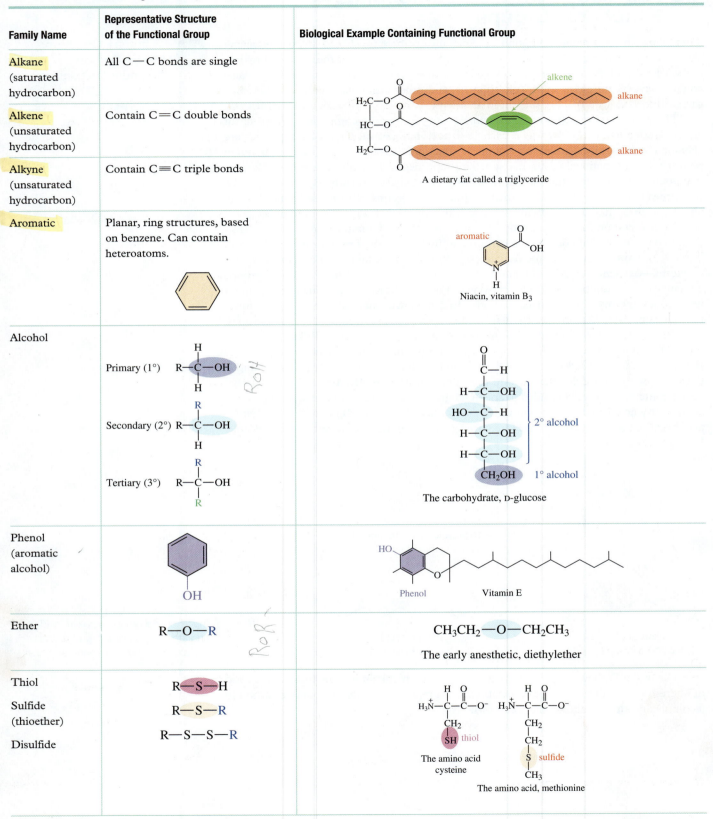

A dietary fat called a triglyceride

Niacin, vitamin B₃

The carbohydrate, D-glucose

Vitamin E

$CH_3CH_2—O—CH_2CH_3$

The early anesthetic, diethylether

The amino acid cysteine

The amino acid, methionine

Note: R represents the "Rest" of the molecule bonded through carbon. See text.

R = Carbon Group chain

TABLE 4.3 Families of Organic Compounds: Common Functional Groups (continued)

Family Name	Representative Structure of the Functional Group	Biological Example Containing Functional Group

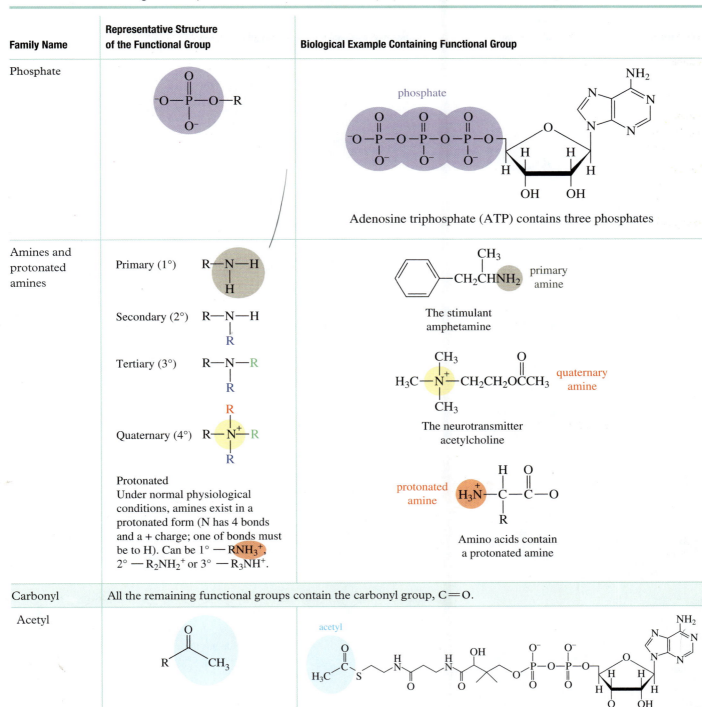

Phosphate — Adenosine triphosphate (ATP) contains three phosphates

Amines and protonated amines

Primary (1°)

Secondary (2°)

Tertiary (3°)

Quaternary (4°)

The stimulant amphetamine — primary amine

The neurotransmitter acetylcholine — quaternary amine

Protonated
Under normal physiological conditions, amines exist in a protonated form (N has 4 bonds and a + charge; one of bonds must be to H). Can be 1° — RNH_3^+, 2° — $R_2NH_2^+$ or 3° — R_3NH^+.

Amino acids contain a protonated amine — protonated amine

Carbonyl   All the remaining functional groups contain the carbonyl group, $C{=}O$.

Acetyl — acetyl

Acetyl coenzyme A

Note: R represents the "Rest" of the molecule bonded through carbon. See text.

(continued)

TABLE 4.3 Families of Organic Compounds: Common Functional Groups (continued)

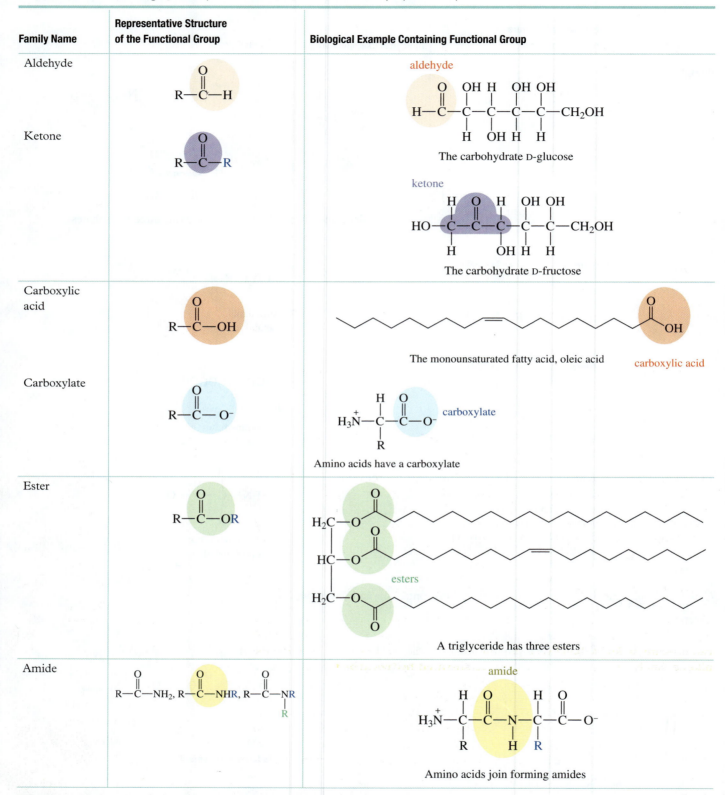

Family Name	Representative Structure of the Functional Group	Biological Example Containing Functional Group
Aldehyde		
Ketone		
Carboxylic acid		
Carboxylate		
Ester		
Amide		

Note: R represents the "Rest" of the molecule bonded through carbon. See text.

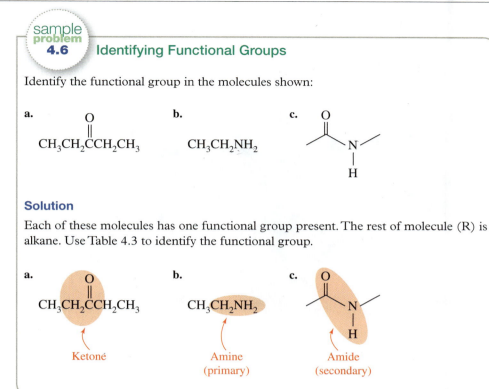

sample problem 4.6   **Identifying Functional Groups**

Identify the functional group in the molecules shown:

a.

$$\text{CH}_3\text{CH}_2\overset{\displaystyle O}{\overset{\displaystyle \|}{\text{C}}}\text{CH}_2\text{CH}_3$$

b.

$$\text{CH}_3\text{CH}_2\text{NH}_2$$

c.

Solution

Each of these molecules has one functional group present. The rest of molecule (R) is alkane. Use Table 4.3 to identify the functional group.

a.

$$\text{CH}_3\text{CH}_2\overset{\displaystyle O}{\overset{\displaystyle \|}{\text{C}}}\text{CH}_2\text{CH}_3$$

Ketone

b.

$$\text{CH}_3\text{CH}_2\text{NH}_2$$

Amine
(primary)

c.

Amide
(secondary)

Unsaturated Hydrocarbons—Alkenes, Alkynes, and Aromatics

In Section 4.1, we took a first look at alkanes, the simplest hydrocarbons. Here we look at the remaining hydrocarbon-only families and their functional groups. Organic compounds that contain at least one multiple bond are broadly characterized as unsaturated. There are three families of organic compounds that contain unsaturated hydrocarbons: alkenes, alkynes, and aromatics.

Alkenes

As tomatoes, bananas, and other fruits ripen on the vine or tree, the plant hormone ethene controls this process. Ethene (also known as ethylene) is also an important industrial chemical used in the production of the plastic polyethylene found in milk jugs and plastic bags. An ethene molecule contains a carbon–carbon double bond. A carbon–carbon double bond is an unsaturated hydrocarbon functional group in the **alkene** family. Alkenes are considered **unsaturated hydrocarbons** because they have more than one bond between two carbon atoms. In the skeletal structure, the double bond is represented with a second line between the two carbons.

Carbon atoms joined by a double bond are held closer together and more strongly than those joined by just a single bond. In other words, a double bond is shorter and stronger than a single bond. However, the second bond of a double bond is not as strong as the first bond. As we will see in Section 5.6, when alkenes react in an addition reaction, the second bond of the double bond is broken, but the carbon atoms remain joined together by the single bond. This means that alkenes are more reactive than alkanes, which do not have double bonds.

Alkenes are also found in a set of naturally occurring compounds called **terpenes.** These compounds all contain alkenes and contain a multiple of five carbons (5, 10, 15, 20, and so on). Some common terpenes are β-carotene, the alkene responsible for the orange pigment in carrots and sweet potatoes, used by our bodies to produce

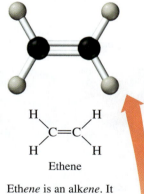

Ethene

Eth*ene* is an alk*ene*. It contains a C=C bond.

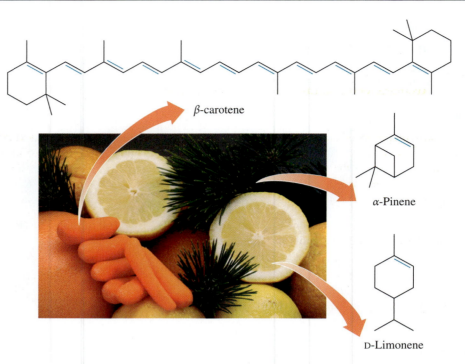

β-carotene

α-Pinene

D-Limonene

β-carotene, α-pinene, and D-limonene all contain the alkene functional group (blue).

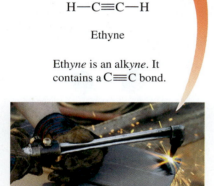

H—C≡C—H

Ethyne

Eth*yne* is an alk*yne*. It contains a C≡C bond.

vitamin A. Essential oils in plants also are terpenes. D-Limonene and α-pinene are essential oils found in citrus fruits and pine trees respectively. These oils give the plant their characteristic smell. Cholesterol and the hormones testosterone and estrogen also contain the alkene functional group and are synthesized from terpenes.

Alkynes

The fuel used in high-temperature welding torches to melt iron and steel is called ethyne or, more commonly, acetylene. Compounds like acetylene that contain one or more carbon–carbon triple bonds are members of the **alkyne** family. Alkynes are also unsaturated hydrocarbons because they have more than one bond between two carbon atoms.

As you might expect, the alkyne's triple bond is even shorter and stronger than the alkene's double bond. However, the two additional bonds of the triple bond mean that alkynes are even more reactive (and, therefore, less stable) than alkenes. Because they are so reactive, alkynes are rare in nature, occurring mainly in short-lived compounds. The poison dart frog secretes a poison that contains two alkyne functional groups.

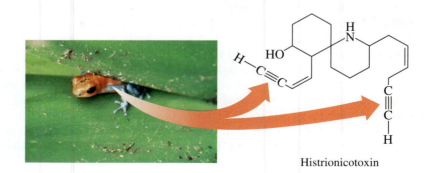

Histrionicotoxin

Aromatics

One of the more remarkable families of unsaturated hydrocarbons is the aromatic family. These compounds were originally dubbed "aromatic" because many of the first ones discovered had pleasant aromas. Oil of spearmint and peppermint are compounds in the aromatic family. **Aromatic compounds** have a cyclic structure like benzene. With so many double bonds, these compounds could be presumed to be very reactive. However, the members of this family are unusually stable and not very reactive at all.

Many members of the aromatic family contain the unsaturated six-membered ring that is called a benzene ring. (When it is part of a larger molecule, it is also called *phenyl*.) Benzene has the structure shown in Figure 4.3. Notice that benzene appears to have a series of alternating carbon–carbon single and double bonds. From the previous discussion of the lengths of double versus single bonds, we would expect benzene to have three longer single bonds and three shorter double bonds in an alternating pattern. We might also expect such a structure to have reactivity like an alkene.

However, experimental evidence shows that all the carbon–carbon bonds of benzene are the same length and that benzene is an exceptionally stable compound. In fact, benzene resists reactions that would break any (or all) double bonds. Unsaturated cyclic aromatic compounds like benzene, which are unusually and unexpectedly stable (more stable than normal alkenes), are said to exhibit **aromaticity.**

The properties and behavior of benzene and other aromatic compounds arise because the double bonds between adjacent carbon atoms are not static. In other words, the electrons shared to create those double bonds could be in either of the two arrangements shown in Figure 4.4.

In fact, the electrons in the double bonds are shared evenly by all six carbons. This is why benzene and benzene rings are often represented by the structure on the far right in the figure. This representation is called the **resonance hybrid** form and implies equal sharing of electrons in benzene instead of distinct double and single bonds.

Because these electrons are free to roam among the six carbons, they are much less likely to react with other molecules. This decreased reactivity translates to the enhanced stability exhibited by aromatic compounds.

There are many aromatic compounds containing two or more benzene rings present in tobacco smoke. These compounds are called polycyclic aromatic hydrocarbons or PAHs. Phenanthrene and benzo[a]pyrene are two such molecules. Polycyclic aromatic hydrocarbons are often formed when organic matter, such as the tobacco leaf, is burned. Some of these PAHs have been shown to be cancer-causing or carcinogenic. Aromatic compounds are also components of many plastics and pharmaceuticals.

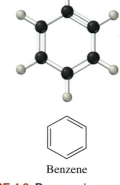

Benzene

FIGURE 4.3 Benzene is aromatic. Many aromatic compounds contain the six-membered ring seen in benzene.

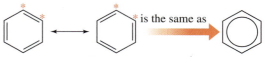

FIGURE 4.4 Benzene exhibits resonance. The electrons in the double bonds in benzene are distributed evenly among the six carbons in the ring. The starred carbon atoms are equivalent in both structures.

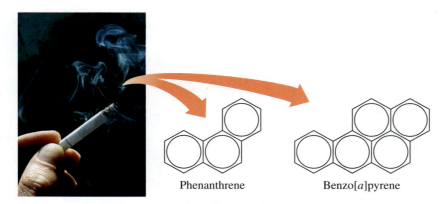

Phenanthrene   Benzo[a]pyrene

Tobacco smoke produces many polycyclic aromatic hydrocarbons

sample
problem
4.7 **Identifying Unsaturated Hydrocarbons**

Identify the unsaturated hydrocarbon functional group seen in the molecules shown.

a.   b.   c.

$CH_3C{\equiv}CH$

Solution

Recall the definitions of the three unsaturated hydrocarbon functional groups introduced in the section or refer to Table 4.3.

a.   **b.**   **c.**

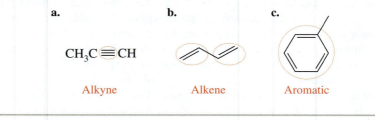

$CH_3C{\equiv}CH$

Alkyne   Alkene   Aromatic

Pharmaceuticals Are Organic Compounds

Open the package of any prescription medication and you find an insert of prescribing information that includes a description and chemical structure. Most organic molecules (pharmaceuticals included) contain more than one functional group. The structures are often shown as a hybrid of the condensed structural formula and a skeletal structure. Most often aromatic functional groups are shown using the resonance hybrid structure discussed previously. You can put your chemistry knowledge to work by identifying the functional groups and structures of the compounds. As we will see later in Chapter 7, correctly identifying the structure allows us to predict properties such as how well a substance dissolves in water and its chemical reactivity.

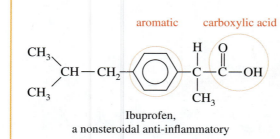

Ibuprofen,
a nonsteroidal anti-inflammatory

Epinephrine,
a stimulant and active ingredient in the
EpiPen®

Lipids Are Hydrocarbons — Fatty Acids

In the popular press, we often read about the health issues related to diets high in *saturated* fats. We are encouraged to include monounsaturated and polyunsaturated fats as part of a healthy diet. Earlier, alkanes were described as saturated hydrocarbons and alkenes, aromatics, and alkynes as unsaturated hydrocarbons. Whether hydrocarbons or fats, the terms *saturated* and *unsaturated* have the same meaning: carbon–carbon single bonds (saturated) or carbon–carbon multiple bonds (unsaturated). Fats with one double bond are called **monounsaturated** and those with two or more double bonds are called **polyunsaturated.** Here we take a brief look at the main component of fats, saturated fatty acids.

Fatty acids belong to a class of biomolecules called **lipids.** Although a structurally diverse class, lipids all share the common property of being mainly nonpolar biomolecules. This means most of a lipid molecule is hydrocarbon.

Saturated fatty acids are long, straight-chain alkane-like compounds that have a carboxylic acid group at one end (see **Figure 4.5**). The most common and biologically most important fatty acids are compounds containing from 12 to 22 carbon atoms. Most naturally occurring fatty acids have even numbers of carbon atoms.

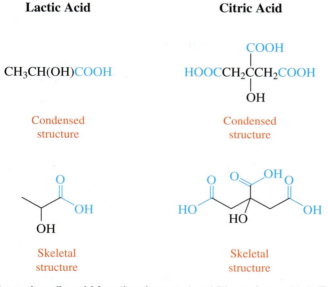

Lactic Acid

$CH_3CH(OH)COOH$

Condensed
structure

Citric Acid

$HOOCCH_2CCH_2COOH$
with COOH above and OH below the central C

Condensed
structure

Skeletal
structure

Skeletal
structure

FIGURE 4.5 The carboxylic acid functional group. In addition to fatty acids in Table 4.4, organic acids like citric acid found in fruits and lactic acid, a byproduct of strenuous muscular activity, also contain the carboxylic acid functional group (blue). This group is often represented as COOH in condensed formula.

The names and structures of even-numbered saturated fatty acids are given in **Table 4.4** in both condensed and skeletal structure. Compare the structures of these compounds with those of the straight-chain alkanes in Table 4.1. Notice that the structures of alkanes and saturated fatty acids both contain straight chains of carbon atoms, but the fatty acids contain the functional group carboxylic acid at one end. Organic acids that have the word *acid* as part of their name often contain the carboxylic acid functional group. We will explore unsaturated fatty acids in more detail in Section 4.5.

TABLE 4.4 Saturated Fatty Acids: Names and Structures

Name	Carbon Atoms	Source	Structures
Lauric acid	12	Coconut	
Myristic acid	14	Nutmeg	
Palmitic acid	16	Palm	
Stearic acid	18	Animal fat	
Arachidic acid	20	Peanut	
Behenic acid	22	Canola	

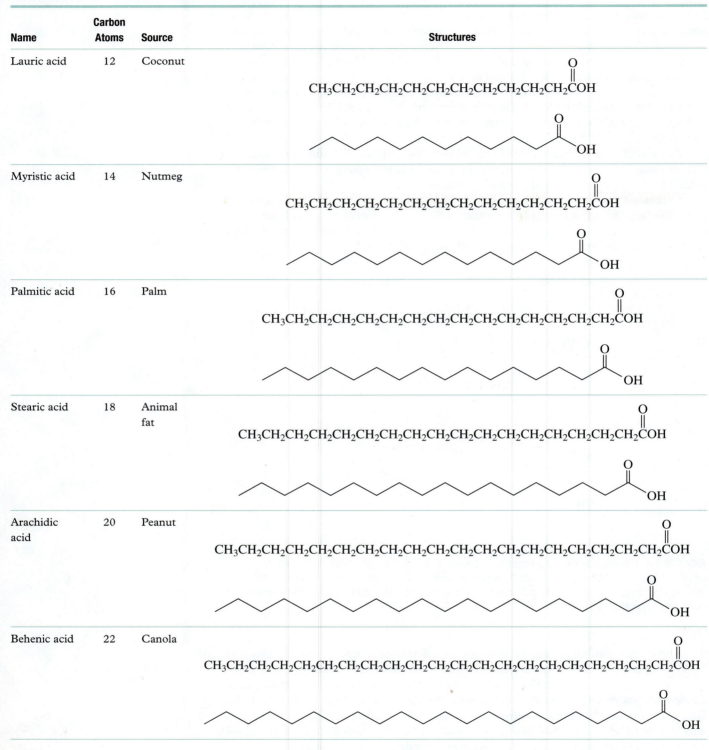

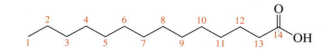

sample problem 4.8   **Drawing Saturated Fatty Acids**

Draw the skeletal structure for myristic acid, a saturated fatty acid with 14 carbons.

Solution

All fatty acids will have the carboxylic acid functional group at one end.

For myristic acid, 14 carbons are drawn in skeletal structure with a carboxylic acid group at one end. Keep in mind that the carboxylic acid group contains one of the carbons in the 14. The carbons are numbered in the solution for clarity.

Fatty Acids in Our Diets

Fats are important in a balanced diet because they play important roles as insulators and protective coverings for internal organs and nerve fibers. Mono- and polyunsaturated fats are part of a healthy diet. The Food and Drug Administration (FDA) recommends that a maximum of 30% of the calories in a normal diet come from such fatty acid-containing compounds. The FDA also recommends that the majority of our fat intake come from foods containing a higher percentage of mono- and polyunsaturated fatty acids. Table 4.5 shows the fatty acid composition of some common fats.

integrating chemistry

TABLE 4.5 Fatty Acid Composition of Common Dietary Fats

	Saturated	Monounsaturated	Polyunsaturated
	%[a]	%[a]	%[a]
Butter	68	28	4
Canola oil	7	61	32
Coconut oil	91	7	2
Corn oil	13	28	59
Lard	43	47	10
Olive oil	15	73	12
Palm oil	49	40	11
Soybean oil	15	25	60
Sunflower oil	13	21	66

[a]Approximate percentages by weight of total fatty acids

4.13 Identify the family of hydrocarbon present in the following:

a. b. c.

 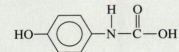

4.14 Identify the family of hydrocarbon present in the following:

a.

$CH_3CH_2CH=CH_2$

b. c.

CH_3
|
$CH_3CCH_2CHCH_3$
| |
CH_3 CH_3

Naphthalene, makes up moth balls

Isooctane, a component in gasoline

4.15 Identify all the functional groups present in the following:

a.

Acetaminophen, an analgesic

b.

Compound found in garlic

4.16 Identify all the functional groups present in the following:

a.

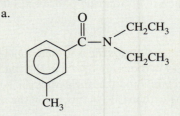

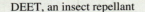

DEET, an insect repellant

b.

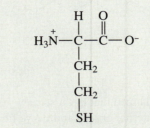

Homocysteine, an amino acid

4.17 The most prevalent fatty acid in coconut oil is lauric acid, a saturated fatty acid containing 12 carbons. Draw lauric acid in skeletal structure.

4.18 The most common fatty acid found in animals is stearic acid, a saturated fatty acid containing 18 carbons. Draw stearic acid in skeletal structure.

4.4 Nomenclature of Simple Alkanes

? **4.4 Inquiry Question:**
How are alkanes named?

The octane rating compares the compounds in gasoline to a branched-chain alkane with eight carbons.

Much as a limb on a tree branches off a tree's trunk, carbon atoms can branch off the straight chain of carbon atoms in an alkane. Alkanes that do not have all their carbon atoms connected in a single continuous chain are called **branched-chain alkanes.** Straight-chain octanes and branched-chain octanes are given different names to avoid confusion.

Here we explore the systematic naming (also called *nomenclature*) developed to precisely name organic compounds. This naming system was developed by the International Union of Pure and Applied Chemistry (IUPAC) and is used by chemists worldwide. The IUPAC nomenclature rules provide a unique name for any organic compound and ensure that every compound, like the two different octanes, has just one correct systematic name.

The names of organic compounds have three basic parts.

Substituents	+	Parent Name	+	Suffix
(Attachments)		*(Number of carbons in the longest continuous chain)*		*(Family name)*

Parts of the name of an organic compound.

The naming of any organic compound follows this outline. In this section, the naming of some simple alkanes is introduced to give you a working understanding of organic nomenclature.

Branched-Chain Alkanes

In Section 4.1, the names of the first 10 straight-chain alkanes were introduced. How is an alkane with carbons branching from the straight chain named?

Solving a Problem

Naming Branched-Chain Alkanes

Provide the IUPAC name for the compound shown.

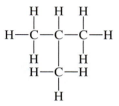

STEP 1: Find the longest continuous chain of carbon atoms. This is called the parent chain. Name the parent chain according to the alkane names given in Table 4.1. (HINT: Count from each end to every other end to make sure you find the longest chain.)

In this example, the name of the parent chain is *propane* since the longest chain has three carbons (circled at right).

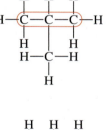

STEP 2: Identify the groups bonded to the main chain but not included in the main chain (ignore hydrogen). These groups are known as **substituents** and, in the case of branched-chain alkanes, they are called **alkyl groups.** The name of each alkyl group is derived from the alkane with the same number of carbon atoms by changing the *ane* ending of the name of the corresponding alkane to *yl*. For example, a one-carbon alkyl group is named as a *methyl* group. [Methane (CH_4) minus *ane* = meth and meth plus *yl* = methyl ($-CH_3$)]. Similarly, a two-carbon alkyl group is an ethyl group, a three-carbon alkyl group is propyl, and so forth (see **Table 4.6**). So in our example, the substituent is a methyl group (circled at right).

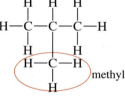

STEP 3: Number the parent chain starting at the end nearer to a substituent. In this example, the substituent is on carbon number 2 no matter which end you start numbering.

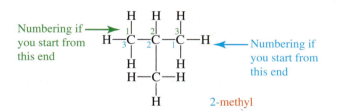

Numbering if you start from this end

Numbering if you start from this end

2-methyl

STEP 4: Assign a number to each substituent based on location, listing the substituents in alphabetical order at the beginning of the name. Separate numbers and words in the name by a dash.

The IUPAC name of this compound is 2-methylpropane.

(Note: If more than one of the same type of substituent is present in the compound, you would indicate this using the Greek prefixes *di*, *tri*, and *tetra* but ignore these prefixes when alphabetizing. For example, two methyl groups branching from a parent chain would be named *di*methyl.)

TABLE 4.6 Four Simplest Alkyl Substituents

	Methyl	Ethyl	Propyl	Isopropyl
Lewis structure				
Condensed structure	$-\xi-CH_3$	$-\xi-CH_2CH_3$	$-\xi-CH_2CH_2CH_3$	CH_3CHCH_3
Line structure	$-\xi-$			
Ball-and-stick model				

ξ represents the point where the group is bonded to the parent chain.

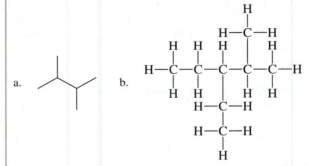

sample problem **4.9**   **Naming Branched-Chain Alkanes**

Determine the correct IUPAC name for each of the following branched-chain alkanes:

a.   b.

Solution

Carefully follow steps 1 to 4 to determine the correct name for each compound.

a.   **STEP 1: Find longest continuous chain of carbons.**

 butane

STEP 2: Identify the substituents.

 methyl

methyl

STEP 3: Number the parent chain.

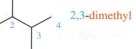

2,3-dimethyl

STEP 4: Assign numbers to the substituents and alphabetize them in the name. 2,3-dimethylpentane
Note that the two numbers are separated by a comma.

b.   **STEP 1: Find longest continuous chain of carbons.**

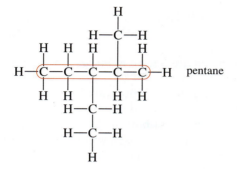

pentane

STEP 2: Identify the substituents.

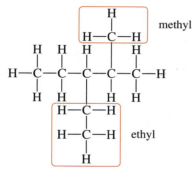

methyl

ethyl

STEP 3: Number the parent chain. Begin numbering from the end closest to a substituent.

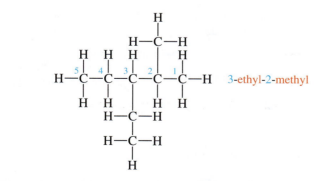

3-ethyl-2-methyl

STEP 4: Assign numbers to the substituents and alphabetize them in the name. 3-ethyl-2-methylpentane
Note that there is more than one chain of five carbons in length. All five-carbon chains will produce the same IUPAC name.

Haloalkanes

The members of Group 7A on the periodic table, collectively known as the halogens, are common substituents on alkane chains. Alkanes with a halogen substituent in place of a hydrogen atom are known as **haloalkanes** or alkyl halides. When a halogen is a substituent, the name is changed by replacing the *ine* ending of the name of the element with an *o*. The substituent names for the halogens become *fluoro, chloro, bromo,* and *iodo*. The rules for naming haloalkanes are the same as those for naming branched-chain alkanes with the halogen being the substituent.

Cycloalkanes

The rules for determining the IUPAC names for cycloalkanes are much the same as those for the branched-chain alkanes with a couple of modifications to the previous steps:

STEP 1: The ring serves as the parent name as long as it has more carbons than any substituent.
STEP 2: As before, identify the substituents.
STEP 3: Number the carbons in the ring. Carbon 1 will always have a substituent.
STEP 4: Assign numbers to the substituents. On a ring bearing a single substituent, that substituent is assumed to be on position 1 of the ring. If only one substituent is present, a 1 is implied and need not be listed. See Sample Problem 4.11c. When more than one substituent is present, the ring should be numbered to give the substituents the lowest possible combination of numbers.

---

sample problem 4.10 — Naming Haloalkanes and Cycloalkanes

Determine the correct IUPAC name for each of the following compounds:

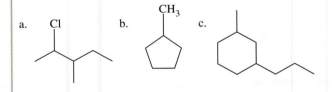

a.   Cl   b.   CH₃   c.

Solution

For haloalkanes, follow steps 1 to 4 given for the branched-chain alkanes to determine the correct name for each compound.

a. **STEP 1: Find longest continuous chain of carbons.**

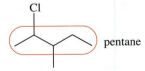

pentane

STEP 2: Identify the substituents.

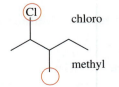

chloro

methyl

STEP 3: Number the parent chain.

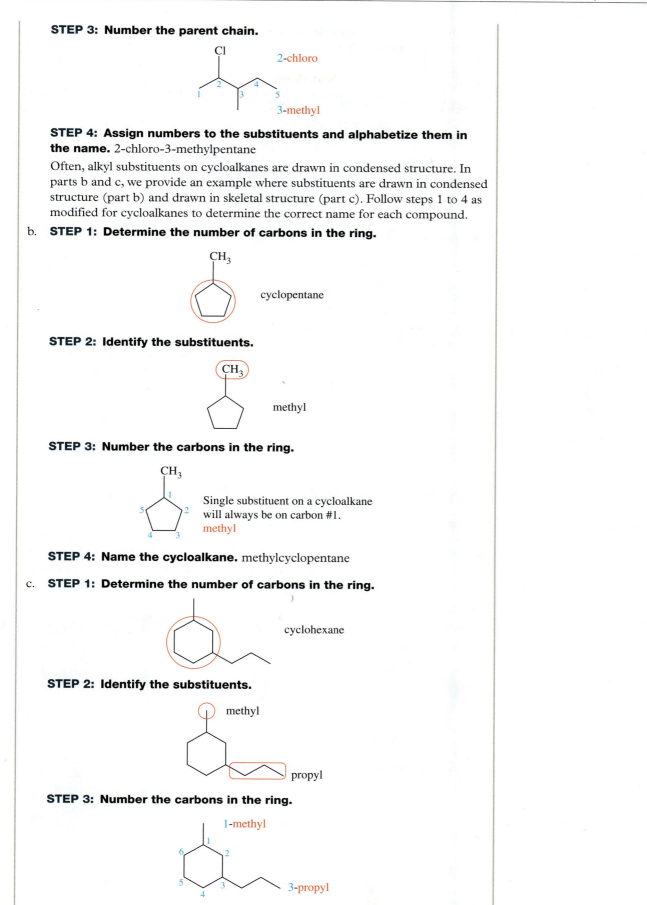

2-chloro

3-methyl

STEP 4: Assign numbers to the substituents and alphabetize them in the name. 2-chloro-3-methylpentane

Often, alkyl substituents on cycloalkanes are drawn in condensed structure. In parts b and c, we provide an example where substituents are drawn in condensed structure (part b) and drawn in skeletal structure (part c). Follow steps 1 to 4 as modified for cycloalkanes to determine the correct name for each compound.

b. **STEP 1: Determine the number of carbons in the ring.**

cyclopentane

STEP 2: Identify the substituents.

methyl

STEP 3: Number the carbons in the ring.

Single substituent on a cycloalkane will always be on carbon #1.
methyl

STEP 4: Name the cycloalkane. methylcyclopentane

c. **STEP 1: Determine the number of carbons in the ring.**

cyclohexane

STEP 2: Identify the substituents.

methyl

propyl

STEP 3: Number the carbons in the ring.

1-methyl

3-propyl

STEP 4: Assign numbers to the substituents and alphabetize them in the name.

1-methyl-3-propylcyclohexane

Notice that the numbering of the ring could have started at either the methyl or the propyl substituent to get the combination of 1 and 3. In such a case, give the lower number to the substituent that comes first alphabetically.

4.19 Draw the condensed structural formula for each of the following alkyl groups:

a. propyl   b. methyl

4.20 Give the correct name for each of the following substituents:

a. $CH_3CH_2 -$

b. $CH_3CH_2CH_2CH_2 -$

c. $I -$

4.21 Draw the skeletal structure for each of the following compounds:
a. 2,3-dimethylpentane
b. 2-ethyl-1,4-dimethylcyclohexane
c. 1,2-dichlorohexane
d. 3-ethylhexane
e. 1-bromo-3-iodocyclopentane
f. propylcyclopentane

4.22 Give the correct IUPAC name for each of the following compounds:

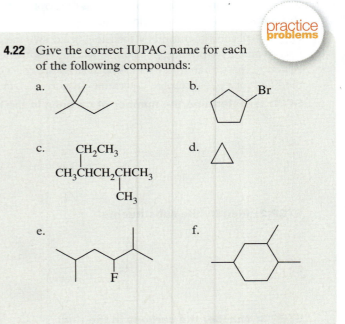

c.

$$CH_3CHCH_2CHCH_3$$

with CH_2CH_3 above the first CH and CH_3 below the second CH

4.5 Isomerism in Organic Compounds

4.5 Inquiry Question:
How are isomers the same and how do they differ?

Consider a cooking contest where the chefs are given the same basket of ingredients and asked to prepare their best dish from just those ingredients. Because each chef is unique and has personal taste, it is likely that each chef will prepare a different dish from those same ingredients. Some of the dishes may differ greatly from each other while others may be more alike.

Just as chefs can cook different dishes from the same ingredients, isomers have the same atoms, but they are arranged differently.

Not unlike the cooking contest, organic compounds with the same number of carbon atoms can have the carbons connected many different ways. Molecules with the same molecular formula but different connectivity or arrangements of the atoms are called **isomers** from the Greek *iso* meaning "same," and *meros* meaning "part." In this section, we compare and contrast different types of isomers possible in organic compounds.

Structural Isomers and Conformational Isomers

When an organic compound has four or more carbons, there is more than one way that the carbons can be connected. Let's look at the following two compounds both with the molecular formula C_4H_{10}:

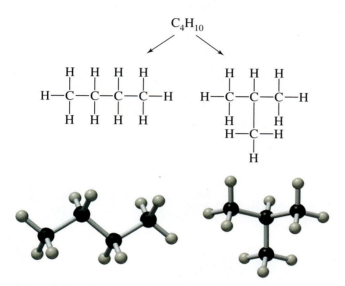

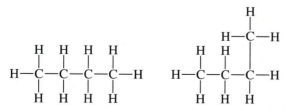

Butane (left) and isobutane (right) are structural isomers.

The compound on the left is the straight-chain alkane, butane. The compound on the right is a branched-chain alkane and has the IUPAC name 2-methylpropane, commonly called isobutane. The compounds are isomers, but the carbons are not bonded in the same way. Butane (left) has four carbons bonded in a row while 2-methylpropane (right) has three carbons bonded in a row with one carbon branching off the center carbon. When two molecules have the same molecular formula but a different connectivity of atoms, the molecules are related to each other as **structural isomers.**

Next, look at the following two Lewis structures. The structure on the left is the structure of straight-chain butane that you've seen before. The structure on the right looks different. Are these two representations of the same molecule, or are they structural isomers?

Are these two molecules structural isomers or the same molecule?

Because the connectivity of the carbons is the same, these are two representations of the same molecule. You can trace the chain from the first to the fourth carbon without picking up your finger for both structures. This means that these two compounds both have four carbon atoms connected in a continuous chain. Because both have a molecular formula of C_4H_{10}, although these two Lewis structures look different, they represent the same compound—butane.

Another way you can convince yourself that these are the same molecule is by naming each structure. Both molecules have the same name, butane, since the longest continuous chain in both cases is four carbons and there are no substituents.

Because both representations shown have the same parts, they are a type of isomer. This type of isomer is called a **conformational isomer.** These isomers, also called **conformers,** are not different compounds like structural isomers but are different arrangements of the same compound. Single bonds in a noncyclic compound can rotate and make molecules look different, but they are actually the same molecule.

Solving a Problem

Distinguishing Structural and Conformational Isomers

Indicate whether the pair of molecules shown are related as structural isomers (different molecules), conformational isomers (the same molecule), or not related as isomers.

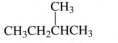

$$CH_3CH_2\overset{\overset{\displaystyle CH_3}{|}}{C}HCH_3$$

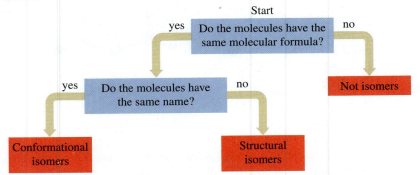

Use the following flow chart to distinguish between structural isomer and conformational isomer:

Flow chart for distinguishing structural isomers and conformational isomers.

In this case, the two molecules have the same molecular formula, C_5H_{12}. When correctly named, both compounds are 2-methylbutane, so the molecules are related as conformational isomers (conformers).

---

sample problem
4.11   Determining Structural and Conformational Isomers

For each of the following pairs, indicate whether they are related as structural isomers, conformational isomers, or not related.

a.

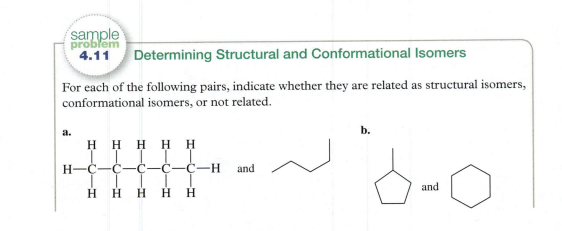

b.

Solution

Use the flow chart given previously in the Solving a Problem exercise to establish the relationship between the molecules shown.

a. The molecular formula of each of these compounds is C_5H_{12}, so they are isomers. Notice that both structures have a continuous chain of five carbon atoms and are named pentane. These are conformational isomers.

b. The molecular formula of each of these cycloalkanes is C_6H_{12}, so they are isomers. The name of the molecule on the left is methylcyclopentane and the name of the molecule on the right is cyclohexane. The carbons are bonded differently. These are structural isomers.

Cis–Trans Stereoisomers in Cycloalkanes and Alkenes

Look at the two molecules represented in **Figure 4.6**. These two compounds have the same molecular formula and the *same* connectivity of their atoms. So these two compounds are *not* structural isomers. Are they conformers? If they were conformers, we could change one into the other without breaking any bonds. This is not the case. The ring of carbons makes it impossible for one of the molecules to rotate and produce the other.

If we look at their three-dimensional shapes, they are different. The positioning of the methyl groups on the ring is different in the two compounds. The skeletal structure shows the compound on the left with a wedge-shaped bond (◢◣) and a dashed bond (·····ıll), while the compound on the right has two dashed bonds.

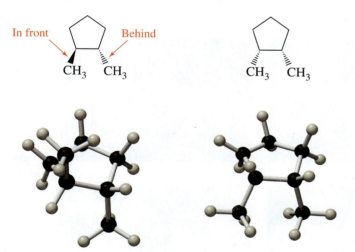

FIGURE 4.6 Stereoisomers of 1,2-dimethylcyclopentane. There are two ways that two methyl groups can be arranged on a cyclopentane. The ball-and-stick model shows two spatial arrangements. In the skeletal structures, this spatial arrangement is represented by wedges (in front of the paper page) and dashes (behind the paper page).

Chemists draw these types of bonds, known as wedge-and-dash bonds, to show the three-dimensional nature of a molecule on the flat page. The wedge bond is used to represent atoms that project out of the textbook page toward the viewer. The dash bond is used to represent atoms that project behind (or into) the textbook page away from the viewer.

Unlike the alkanes, in which bonds freely rotate, the cycloalkanes are limited in their bond rotations and flexibility. Because of this restricted rotation about the carbon–carbon bonds in cycloalkane rings, these compounds have two distinct sides or faces. To view this better, we can look at cyclopentane across one edge in a projection where the point at the top of the structure is behind the plane of the page and the bottom edge is in front of the plane of the page. Using this representation, we can see that the ring has a "top" face and a "bottom" face.

With this in mind, compare the skeletal structures of the two substituted cyclopentane compounds with the three-dimensional models. Notice

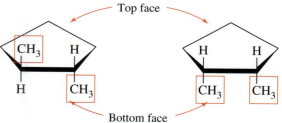

The methyl groups are connected to the same carbons but arranged differently on the two molecules.

how the methyl groups on the wedge bonds are on the top face of the ring while those on the dashed bonds are on the bottom face of the ring. The wedges and dashes allow us to show the three-dimensional nature of a molecule in the two-dimensional space of a page.

If we were to name either of the two molecules using the IUPAC rules developed previously, the name of both molecules would be 1,2-dimethylcyclopentane. Yet in the representations shown, the two molecules are clearly different. How can we distinguish the two by name?

When two molecules have the same molecular formula and the same attachments to the carbon skeleton, but have a different spatial arrangement, they are related to each other as **stereoisomers.** The two cyclopentane compounds shown in Figure 4.6 belong to a class of stereoisomers called **cis–trans stereoisomers.**

In this example, the stereoisomer with the methyls on the same face of the cyclopentane ring (both on top in this case) is called the cis stereoisomer. Use this memory device to remember cis: *same side is cis.* Notice all of the "s" sounds. The other isomer with one methyl on the top face and the second on the bottom face is called the trans stereoisomer. *Trans* comes from the Latin, meaning "across," as in the word *transatlantic.* In the trans stereoisomer, the methyls are across the ring from each other. The prefixes *cis* and *trans* in italics are included at the beginning of the compound names to denote the arrangement of the substituents in the compound and to give each compound a unique name. In Figure 4.6, the complete name of the isomer on the left is *trans*-1,2-dimethylcyclopentane and the isomer on the right is *cis*-1,2-dimethylcyclopentane.

Cis–trans isomerism also occurs in alkenes. Consider the carnival trinket called the Chinese handcuff shown in **Figure 4.7.** If you join the index fingers of each hand with one of these toys (and try not to pull your hands apart), your two hands can still rotate independently, despite the fact that they are "bonded" together. The same is true for two carbon atoms joined by a single bond.

FIGURE 4.7 Rotation in single versus double bonds. Like a Chinese handcuff, the rotation about a single bond (one handcuff) is allowed (a), but the rotation about a double bond (two handcuffs) is restricted (b).

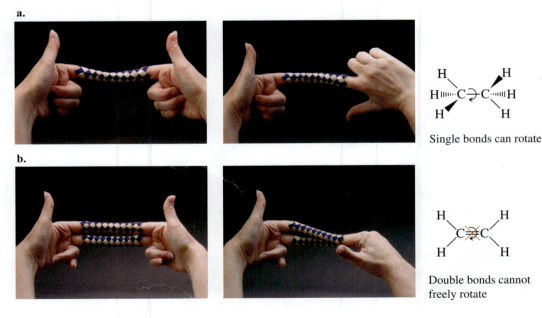

Single bonds can rotate

Double bonds cannot freely rotate

Now consider what would happen if you used two Chinese handcuffs and joined both of your index fingers *and* both of your middle fingers. With two "bonds" now joining your hands, they can no longer rotate independently of each other. The same is true for two carbon atoms joined by a double bond. The double bond between the two atoms will not allow them to rotate independently of each other. Organic chemists say that the carbon–carbon double bond is rigid or that it has **restricted rotation.**

Previously we saw that the rings of cycloalkanes have both a top face and a bottom face because of their inability to rotate freely. Because of the rigidity of the carbon–carbon double bond, alkenes also have two faces and some can have cis–trans stereoisomers.

Consider the two alkenes in **Figure 4.8**. The dashed line separates the two sides of the double bond into a top face (above the line) and a bottom face (below the line). Notice that the compound on the left has both similar groups (often carbon-containing groups) on the same side (face) of the double bond, while the one on the right has them on opposite sides of the double bond. Because there is restricted rotation around a double bond, these two compounds are not identical, nor are they conformers. (They cannot be changed into the other by rotating around a bond.) These two compounds are cis–trans stereoisomers. The compound with the similar groups on the same side of the double bond is the cis stereoisomer. The compound with the similar groups on the opposite sides of the double bond is the trans stereoisomer.

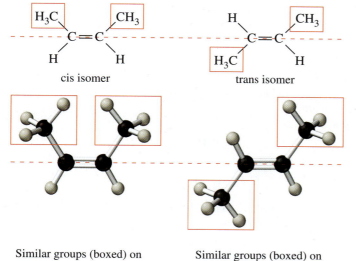

cis isomer   trans isomer

Similar groups (boxed) on   Similar groups (boxed) on
the same side of the line   opposite sides of the line
through the double bond.   through the double bond.

FIGURE 4.8 Cis—trans stereoisomers of alkenes. Alkenes have two sides to them due to restricted rotation around the double bond.

Figure 4.9 shows some other alkenes and their isomers, as well as the skeletal structures of each. Notice that not all alkenes exhibit cis-trans stereoisomerism. If one of the alkene carbons has two identical groups bonded to it (hydrogen in the case shown), this compound cannot have cis-trans stereoisomers.

a)

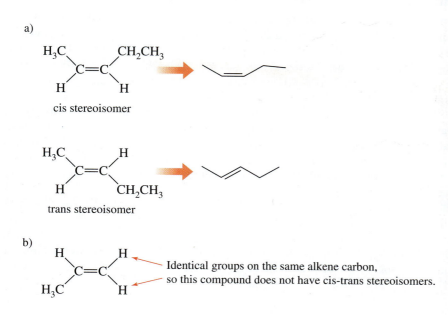

cis stereoisomer

trans stereoisomer

b)

Identical groups on the same alkene carbon, so this compound does not have cis-trans stereoisomers.

FIGURE 4.9 Structures of cis–trans stereoisomers. (a) Structures of cis–trans stereoisomers and (b) structure showing why some alkenes do not have cis–trans stereoisomers.

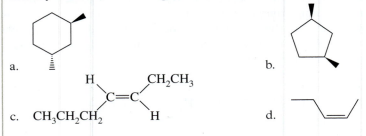

Identify each of the following as either the cis or trans stereoisomer:

a.

b.

c.

d.

Solution

a.   The wedge and dash indicate opposite sides of the ring. This is a trans stereoisomer.
b.   Both the substituents are on wedges and are on the same side of the ring. This is a cis stereoisomer.
c.   For alkenes, start by drawing a line along the double bond.

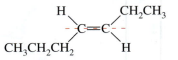

The carbon groups bonded to the alkene carbons are on opposite sides of the line. This is a trans stereoisomer.

d.   For alkenes, start by drawing a line along the double bond.

The carbon groups bonded to the alkene carbons are on the same side of the line. This is a cis stereoisomer.

Unsaturated Fatty Acids Contain Cis Alkenes

Now think back to the fatty acids from Section 4.3. There we examined saturated fatty acids, one type of lipid. Notice the difference in the structure of oleic acid shown in Figure 4.10 and the structure of the saturated fatty acids found in Table 4.4. Oleic acid and similar fatty acids are **unsaturated fatty acids** because they contain carbon–carbon double bonds (alkenes), which are unsaturated hydrocarbons. Naturally occurring unsaturated fatty acids contain cis alkenes.

Table 4.7 shows the structures and names of six common unsaturated fatty acids. Take particular note of the ω (omega) designations of the unsaturated fatty acids. These designations are commonly used in nutrition literature and in the popular press. In this system, the carbon farthest from the carboxylic acid is numbered 1, the next carbon is 2, and so forth. The **omega number** indicates the carbon positioning of a double bond in the structure of the fatty acid.

The unsaturated fatty acid α-linolenic acid commonly called Omega-3 (see Table 4.7) is given a designation of [18:3], ω-3. This notation indicates that the fatty acid has 18 carbons and 3 double bonds with the first double bond located between carbons 3 and 4. Two more double bonds are understood to occur at intervals of three carbons for a total of three unsaturations. This means that the other two double bonds of α-linolenic acid are located between carbons 6 and 7, and 9 and 10.

Oleic acid

CH₃CH₂CH₂CH₂CH₂CH₂CH₂CH₂CH=CHCH₂CH₂CH₂CH₂CH₂CH₂CH₂COOH

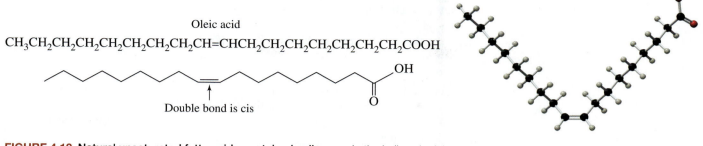

Double bond is cis

FIGURE 4.10 Natural unsaturated fatty acids contain cis alkenes. In the ball-and-stick model this changes the structure by putting a kink in the hydrocarbon chain.

The polyunsaturated fatty acids linoleic and α-linolenic are considered **essential fatty acids** because our bodies cannot produce these compounds so they must be obtained in the diet. Fatty acids are mainly taken in as a component of fats and oils in molecules called triglycerides. More on the properties of these molecules can be found in Chapter 7.

TABLE 4.7 Unsaturated Fatty Acids: Names and Structures

Name	Carbon Designation	Source	Structures
Monounsaturated Fatty Acids			
Palmitoleic acid	[16:1], ω-7	Butter	
Oleic acid	[18:1], ω-9	Olives, corn	
Polyunsaturated Fatty Acids			
Linoleic acid* (commonly called LA or Omega-6)	[18:2], ω-6	Soybean, safflower, corn	
α-Linolenic acid* (commonly called ALA or Omega-3)	[18:3], ω-3	Flaxseed, canola	
γ-Linolenic acid (commonly called GLA)	[18:3], ω-6	Formed from ALA, evening primrose, spirulina	
Arachidonic acid (commonly called AA)	[20:4], ω-6	Formed from LA, lard, eggs	

ᵃIndicates essential fatty acid.

Drawing Unsaturated Fatty Acids

Draw the skeletal structure for eicosenoic acid [20:1], ω-9.

Solution

Remember that the double bonds in naturally occurring fatty acids are cis double bonds. The omega number tells us that the double bond will be between carbons 9 and 10 counting from the hydrocarbon end. Be sure to include a carboxylic acid group at the other end.

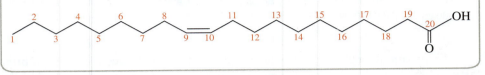

Stereoisomers—Chiral Molecules and Enantiomers

When you enjoy the refreshing taste of spearmint chewing gum or the pleasantly spicy, licorice-like flavor of caraway seeds on fresh rye bread, your taste buds are actually distinguishing between two stereoisomers. The flavors of spearmint and caraway come from a pair of stereoisomers called carvones. **Figure 4.11** shows that the carvone in spearmint is slightly different than the carvone in caraway seeds. Even though stereoisomers are identical except for the arrangement of groups in space, because they have different shapes our bodies can actually distinguish between them.

FIGURE 4.11 The carvone molecules. The two stereoisomers found in caraway seeds and spearmint are identical except for their spatial arrangement.

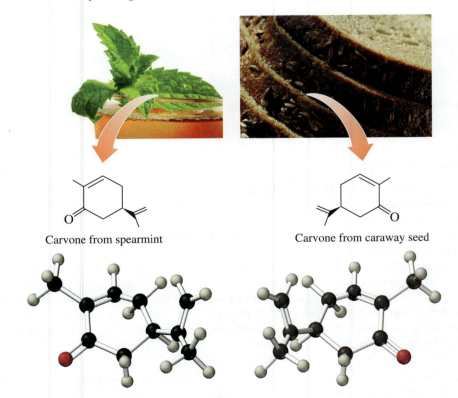

Carvone from spearmint

Carvone from caraway seed

From Figure 4.11, we can determine that both carvones have a molecular formula of $C_{10}H_{16}O$, so they are isomers. The connectivity of the atoms in the two structures is also the same, meaning that they are not structural isomers. However, the carvones have only one wedge- or dash-type bond present in each molecule. The two carvones cannot be cis–trans isomers. As we saw earlier with cycloalkanes, cis–trans isomers must have two such bonds. Are they conformational isomers and the same molecule?

In Section 4.4, we said that if two molecules had the same name, they are the same molecule. Besides their names, if molecules are identical, they would be superimposable. In other words, if we placed one directly on top of the other, every atom on each compound would align or stack on top of one another like two baseball caps. Let's try this with the carvones. First, take the carvone from spearmint and flip it over so that its ketone carbonyl group is oriented in the same direction as the carvone from caraway seeds.

Carvone from spearmint   "Flipped" carvone   Carvone from caraway
                                 from spearmint

Mentally slide one so that the molecules are stacked on top of each other and notice that the molecules are the same in all but one detail. Before we flipped the carvone from the spearmint, the group at the right of the structure was in front of the page, but when we flipped the carvone over, that group was then behind the page. The identical group on the carvone from caraway seeds was still in front of the page. Therefore, the two compounds are not superimposable because of the difference in the location in space of this group on each molecule.

In fact, if you look carefully, you can see that the molecules are mirror images of each other. Compounds like the two carvones, which have the same molecular formula and the same connectivity but are nonsuperimposable mirror images of each other, belong to a class of stereoisomers called **enantiomers.**

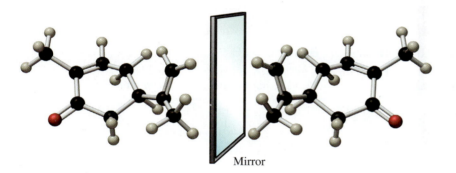

Mirror

In our everyday life, we encounter numerous common objects that have nonsuperimposable mirror images. Your shoes are a good example. Notice that your two shoes look very much the same, but they are not identical. Your left shoe is the mirror image of your right shoe. The same holds true for a right-handed and left-handed baseball glove. Objects such as these that have nonsuperimposable mirror images are termed **chiral** (from the Greek *cheir* meaning "the hand").

Molecules can also be chiral. The two carvones, which exist as a pair of enantiomers, can be thought of as a right-handed and left-handed form of the same molecule. Most molecules that exist as enantiomers contain a **chiral center**—a tetrahedral carbon atom bonded to four different atoms or groups of atoms. In **Figure 4.12**, the colored spheres around the atoms or groups of atoms highlight the four different groups on the chiral center in the molecule. Notice that the carbon atom (shown in black) is bonded to the four different groups: hydroxyl ($-OH$), hydrogen ($-H$), methyl ($-CH_3$), and ethyl ($-CH_2CH_3$).

There are two forms of this molecule, and they are mirror images that are nonsuperimposable on each other (see **Figure 4.13**). Like the carvones, this molecule can exist in two forms that are enantiomers of each other. On paper, we can represent the attachments to the chiral center with wedges and dashes and indicate the location of a chiral center with an asterisk.

FIGURE 4.12 A chiral carbon. Four different atoms or groups of atoms are bonded to the carbon (shown in black) forming a chiral center.

FIGURE 4.13 The mirror image enantiomers. Enantiomers are nonsuperimposable, so this is a chiral compound.

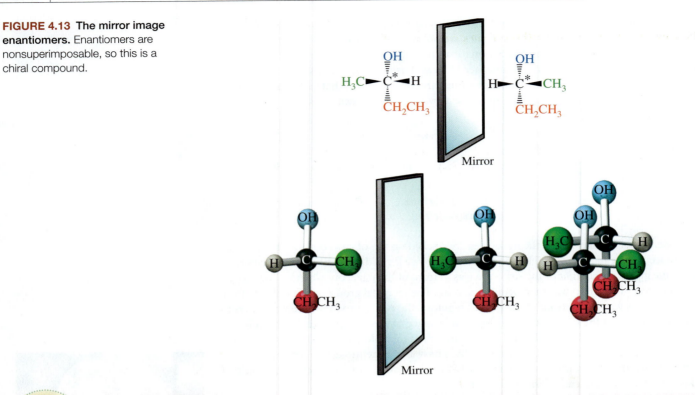

Solving a Problem

Identifying Chiral Carbons in a Molecule

Mark the chiral center(s) in the following molecule, if any, with an asterisk (★).

$$CH_3—CH_2—\overset{\overset{\displaystyle Br}{|}}{CH}—\underset{\underset{\displaystyle CH_3}{|}}{CH}—CH_3$$

STEP 1: Locate the tetrahedral carbons (carbons with four atoms bonded to them). Carbons with double or triple bonds are typically not chiral centers, since they do not have four groups attached. In this case, all the carbons have four atoms bonded to them (no multiple bonds). These are highlighted in blue.

$$CH_3—CH_2—\overset{\overset{\displaystyle Br}{|}}{CH}—\underset{\underset{\displaystyle CH_3}{|}}{CH}—CH_3$$

STEP 2: Inspect the tetrahedral carbons. Determine if the four groups attached to the tetrahedral carbons are different. To determine if the groups are different, you must examine the entire group of atoms bonded to the carbon, not just the atom directly bonded to it.

All carbons atoms bonded to two or more hydrogen atoms cannot be chiral centers because such carbons are not bonded to four different groups. These carbons can be removed from further consideration. In this case, that leaves only two carbon atoms (highlighted in blue) to look at more carefully.

$$\cancel{CH_3}—\cancel{CH_2}—\overset{\overset{\displaystyle Br}{|}}{CH}—\underset{\underset{\displaystyle \cancel{CH_3}}{|}}{CH}—\cancel{CH_3}$$

The blue-highlighted carbon atom on the right is bonded to two methyl groups. Since these two groups are the same, this carbon cannot be chiral.

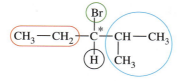

Expanding the structure to show the bond between the carbon and the hydrogen and then comparing the groups, we see that the carbon atom bonded to Br is bonded to four distinct groups and, therefore, is a chiral center.

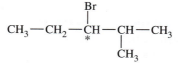

STEP 3: Assign the chiral centers. Typically an asterisk is drawn next to the chiral center.

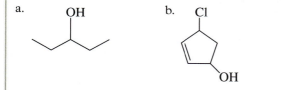

4.14 **Identifying Chiral Centers**

Mark the chiral centers in the following molecules, if any, with an asterisk:

a.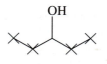

b.

c.

Solution

a. **STEP 1: Locate the tetrahedral carbons.** All the carbons in this molecule are tetrahedral.

   STEP 2: Inspect the tetrahedral carbons. Identifying chiral centers on skeletal structures may seem more difficult at first. With practice, looking for these types of structures can make the process quicker. Remember that, in a skeletal structure, a bend or corner is a —CH_2— if no other bonds exist at the corner. You can immediately rule out these carbons as well as those at the end of a line (—CH_3).

   This leaves only one carbon atom to consider further. At this point, you may find it useful to draw in the hydrogen atom that we typically omit in a skeletal structure and perhaps even add the symbol for the carbon itself.

Now we can see that the OH group and the H are not the same, but notice that the other two groups bonded to this carbon atom are identical: Both are ethyl groups.

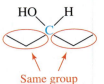

STEP 3: Assign the chiral centers. Answer: no chiral centers

b. **STEP 1: Locate the tetrahedral carbons.** Any carbons in double or triple bonds can be eliminated.

STEP 2: Inspect the tetrahedral carbons. As before, eliminate those carbon atoms that have two or more hydrogen atoms bonded to them. This also eliminates the carbon on the right that is a —CH_2.

Following the steps from the previous example, consider the remaining carbon atoms one at a time, adding the hydrogen and the C.

On this carbon atom, the H and Cl are different, but what about the other groups? To determine if these groups are the same or different, list the atoms and what is bonded to them moving clockwise around the ring, starting at the carbon that you are considering. Then do the same in the counterclockwise direction. For this carbon, in the clockwise direction, the atoms are CH_2, CH—OH, CH, and CH. In the counterclockwise direction the atoms are CH, CH, CH—OH, and CH_2. Notice that the lists are *not* identical. The CH—OH is second in the clockwise direction, but third in the counterclockwise direction. The two groups do not have the same connectivity of atoms and, therefore, are not the same. This is a chiral center.

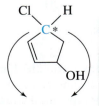

Groups are not identical
in the clockwise and
counterclockwise directions
around the ring.

Following the same procedure for the carbon atom bearing the OH group verifies that it is also a chiral center.

STEP 3: Assign the chiral centers. Answer:

c. **STEP 1: Locate the tetrahedral carbons.** All the carbons in this molecule are tetrahedral.

STEP 2: Inspect the tetrahedral carbons. Crossing out all the "corners" in this molecule, we quickly see that only one carbon remains for further consideration.

Don't let the presence of the wedge-and-dash bonds in this structure lead you to the quick conclusion that this carbon must be chiral. Instead, follow the procedure we used in Problem 4.15b. Doing this, we discover that in the clockwise direction, the group bonded to the carbon contains five consecutive CH$_2$'s. The same is true for the counterclockwise direction: five consecutive CH$_2$'s. Because the atoms are the same and in the same order in both the clockwise and counterclockwise directions around the ring, the groups are identical. This carbon is not a chiral center.

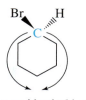

Groups are identical in the clockwise and counterclockwise directions around the ring.

STEP 3: Assign the chiral centers. Answer: no chiral centers

The Consequences of Chirality

Why does one carvone give the taste of spearmint and the other the taste of caraway? The taste receptors on your tongue cells also are "handed." Much as you cannot put your right hand into a left-handed baseball glove, a chiral molecule can fit only into a complementary receptor—one that accommodates its shape. This is also the case with many pharmaceuticals where only a single enantiomer has biological activity. One of the enantiomers of a drug called L-dopa is effective in the treatment of Parkinson's disease, while its mirror image has no biological effect. The anti-inflammatory ibuprofen, found in Advil®, has one active enantiomer. The other enantiomer is readily converted to the active form by the body.

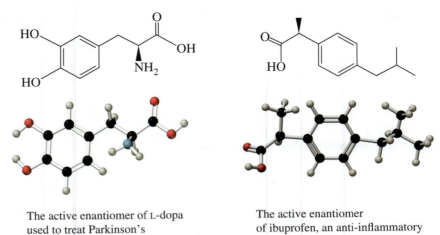

The active enantiomer of L-dopa
used to treat Parkinson's

The active enantiomer
of ibuprofen, an anti-inflammatory

Often only one enantiomer of a chiral pharmaceutical is biologically active.

In some cases, one enantiomer of a drug can be beneficial and the other harmful. This was the case with the drug thalidomide, which was marketed outside the United States in the late 1950s and early 1960s for use against morning sickness. One enantiomer of the drug was effective in alleviating the symptoms of morning sickness. The mirror image, as was tragically discovered later, was teratogenic—it caused birth defects. Because the drug was initially sold as a 50:50 mixture of the enantiomers, many mothers who took it later gave birth to babies with severe birth defects including shortened arms or legs and, sometimes, no limbs. Thalidomide was subsequently removed from markets worldwide. Recently, thalidomide has reemerged in the pharmaceutical market as a treatment for HIV, leprosy, and some forms of cancer, but it is prescribed under very tight restrictions.

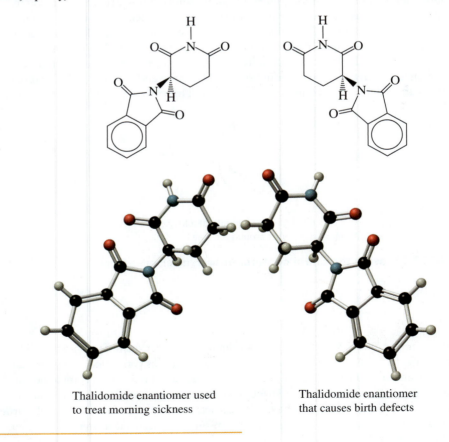

Thalidomide enantiomer used
to treat morning sickness

Thalidomide enantiomer
that causes birth defects

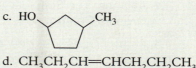

4.23 a. Describe what conformational isomers have
in common.
b. Describe how conformational isomers are different.

4.24 a. Describe what structural isomers have in common.
b. Describe how structural isomers are different.

4.25 Determine the relationship between each of the pairs
of the following compounds. Are they structural isomers (different molecules), conformational isomers
(the same molecule), or not related?

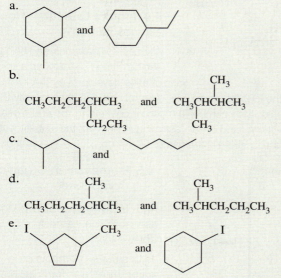

4.26 Determine the relationship between each of the pairs of
the following compounds. Are they structural isomers
(different molecule), conformational isomers (the same
molecule), or not related?

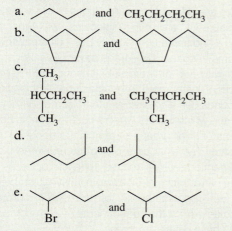

a. $CH_3CH_2CH_2CH_2CH_3$ and $CH_3CH_2CH_2CH_3$

e.
Br   Cl

4.27 Determine if each of the following cycloalkanes or
alkenes can exist as cis–trans stereoisomers. For those that
can, draw the two isomers. Label each of the isomers you
drew as the cis stereoisomer or the trans stereoisomer.
a. Cl   b. $CH_3CH_2CH_2CH_2CH=CH_2$

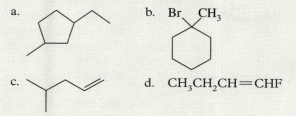

d. $CH_3CH_2CH=CHCH_2CH_2CH_3$

4.28 Determine if each of the following cycloalkanes or
alkenes can exist as cis–trans stereoisomers. For those
that can, draw the two isomers. Label each of the
isomers you drew as the cis stereoisomer or the
trans stereoisomer.

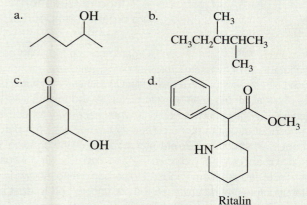

d. $CH_3CH_2CH=CHF$

4.29 Mark the chiral centers in the following molecules, if
any, with an asterisk (*):

4.30 Mark the chiral centers in the following molecules, if
any, with an asterisk (*):

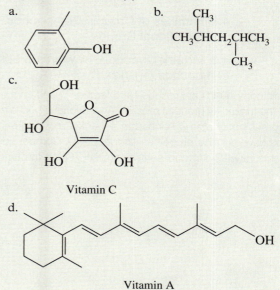

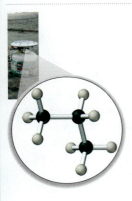

4.1 Alkanes: The Simplest Organic Compounds

4.1 Inquiry Question: How are the structure and polarity of alkanes and cycloalkanes characterized?

An organic compound is any compound that is composed primarily of carbon and hydrogen. The simplest of these are the alkanes, containing *only* single-bonded carbon and hydrogen. Alkanes can exist as straight-chain alkanes and as cycloalkanes. Because carbon—hydrogen bonds are nonpolar, alkanes are nonpolar molecules and not very reactive molecules.

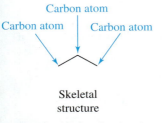

Skeletal structure

4.2 Representing the Structures of Organic Compounds

4.2 Inquiry Question: How are organic compounds represented?

Organic compounds can be represented with Lewis structures showing all atoms, bonds, and lone pairs of electrons in a molecule. Condensed structural formulas show all atoms in an organic molecule and their relative positioning, but they show bonds only when necessary for conveying the correct structure. Lone pairs may or may not be a part of a condensed structure. Skeletal structures show the bonding "skeleton" of an organic molecule by showing all carbon—carbon bonds. Hydrogen atoms bonded to carbon are not shown in skeletal structures, and carbon atoms are understood to exist at the corner formed when two bonds meet or at the termination of a bond. Atoms other than carbon and the hydrogens bonded to them are shown in skeletal structures.

4.3 Families of Organic Compounds — Functional Groups

4.3 Inquiry Question: How are carbon compounds organized?

Organic compounds are grouped into families based on the identity of the functional group(s) present. A functional group is a common grouping of atoms bonded in a particular way. Functional groups have specific properties and reactivity. Compounds with the same functional group behave similarly. Since the functional group is the part of the molecule that is of interest, we typically represent the hydrocarbon portion as *R* (the Rest of the molecule). The hydrocarbon families of alkene, alkyne, and aromatic are highlighted in this section. Fatty acids are alkanelike biomolecules that are the primary components of dietary fats. Fatty acids with a carbon–carbon double bond in their structure are referred to as unsaturated, while those without a double bond are referred to as saturated.

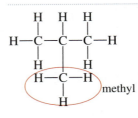

4.4 Nomenclature of Simple Alkanes

4.4 Inquiry Question: How are alkanes named?

Alkanes, cycloalkanes, and haloalkanes can be named by following a simple set of rules developed by the IUPAC. First find the longest continuous chain of carbons, then identify the substituents, then number the parent chain from the end closest to a substituent, and finally assign numbers to the substituents and alphabetize them in the name.

4.5 Isomerism in Organic Compounds

4.5 Inquiry Question: How are isomers the same and how do they differ?

Isomers are compounds with the same molecular formula, but a different arrangement of atoms. Structural isomers are two molecules with the same molecular formula, but a different connectivity. Conformational isomers are different representations of the same compound. Stereoisomers are two molecules with the same molecular formula and same connectivity but a different arrangement of the atoms in space. Cis—trans stereoisomers can exist in cycloalkanes with two or more substituents and in alkenes. Naturally occurring unsaturated fatty acids contain cis-alkenes. Organic compounds can also exist as stereoisomers if they contain a chiral center, which occurs when a carbon atom is bonded to four different atoms or groups of atoms. Compounds with a single chiral center can exist as a pair of stereoisomers called enantiomers. Enantiomers are related to each other as nonsuperimposable mirror images. One way to distinguish between isomers and two representations of the same compound (conformer) is to determine the correct name of each compound. Two isomers will have different names.

The study guide will help you check your understanding of the main concepts in Chapter 4. You should be able to

4.1   Alkanes: The Simplest Organic Compounds

- Distinguish organic and inorganic compounds.
- Define the terms saturated and unsaturated hydrocarbon.
- Compare the molecular formulas for straight-chain alkanes and cycloalkanes.

4.2   Representing the Structures of Organic Compounds

- Draw Lewis, condensed, and skeletal structures for organic compounds.

4.3   Families of Organic Compounds—Functional Groups

- Identify common functional groups in organic molecules.
- Characterize the unsaturated hydrocarbons alkenes, alkynes, and aromatics.
- Draw saturated fatty acids in skeletal structure.

4.4   Nomenclature of Simple Alkanes

- Name branched-chain alkanes, haloalkanes, and cycloalkanes using IUPAC naming rules.

4.5   Isomerism in Organic Compounds

- Distinguish structural isomers from conformational isomers.
- Identify cis and trans isomers in cycloalkanes and alkenes.
- Draw unsaturated fatty acids in skeletal structure.
- Locate chiral centers in organic molecules.

Key Terms

alkanes—A family of organic compounds whose members are composed of only singly bonded carbons and hydrogens; this family has no functional group.

alkenes—A family of organic compounds whose functional group is a carbon–carbon double bond ($C\!=\!C$).

alkyl group—An alkane group minus one hydrogen atom.

alkynes—A family of organic compounds whose functional group is a carbon–carbon triple bond ($C\!\equiv\!C$).

aromatic compounds—A family of cyclic organic compounds whose functional group is a benzene ring.

aromaticity—A term used to describe the unexpected stability of aromatic compounds.

biomolecules—The molecules of life, including carbohydrates, lipids, nucleic acids, and proteins.

branched-chain alkanes—Alkanes that have more than one chain of carbon atoms; the shorter chains are considered branches from the main, longer chain.

carbonyl—A $C\!=\!O$ functional group found as part of several other functional groups.

carboxylic acids—A family of organic compounds whose functional group is a carbonyl ($C\!=\!O$) bonded to an OH; the functional group is a carboxyl group ($-C\!\!\begin{smallmatrix}O\\\\OH\end{smallmatrix}$) abbreviated as COOH.

chiral—Objects or molecules that have nonsuperimposable mirror images.

chiral center—A carbon atom with four different atoms or groups of atoms bonded to it.

cis–trans stereoisomers—A type of isomerism in which compounds differ only in the position of two substituents on a ring or a double bond; the cis isomer has the two substituents on the same face of the ring or double bond and the trans isomer has them on opposite faces.

condensed structural formula—A representation of organic compounds that shows all atoms and their arrangement, using as few bonds as necessary to convey the correct arrangement of atoms.

conformational isomer—An isomer that differs only by rotation about one or more bonds.

conformer—See *conformational isomer*.

cycloalkanes—Hydrocarbon compounds containing a ring of carbon atoms.

enantiomers—Compounds containing at least one chiral center that are nonsuperimposable mirror images.

essential fatty acids—Simple lipid compounds that must be obtained from our diet.

fatty acids—Simple lipid compounds with a long alkane-like hydrocarbon chain bonded to a carboxyl group.

functional group—A group of atoms bonded in a particular way that is used to classify organic compounds into the various families; each functional group has specific properties and reactivities.

haloalkanes—Hydrocarbon compounds in which a hydrogen atom has been replaced by a halogen atom (F, Cl, Br, or I).

heteroatoms—Atoms in an organic compound other than carbon and hydrogen.

inorganic compounds—Compounds composed of elements other than carbon and hydrogen.

isomer—A compound with the same molecular formula as another compound, but a different arrangement of the atoms.

lipid—A class of biomolecules whose structures are mainly nonpolar.

monounsaturated—A term used to describe organic compounds that have one double bond in their structures.

omega number—This number specifies the position of the first double bond in a fatty acid chain when numbering from the methyl end of the fatty acid.

organic chemistry—The field of chemistry dedicated to studying the structure, characteristics, and reactivity of carbon-containing compounds.

organic compounds—Compounds composed primarily of carbon and hydrogen but that may also include oxygen, nitrogen, sulfur, phosphorus, and a few other elements in their structures.

polyunsaturated—A term used to describe organic compounds that have more than one double or triple bond.

resonance hybrid—A representation used to indicate electron sharing among several atoms.

saturated fatty acids—Fatty acids that have no carbon–carbon double bonds; each carbon of the alkyl portion of a fatty acid is bonded to the maximum number of hydrogen atoms.

saturated hydrocarbons—Organic compounds containing only carbon and hydrogen in which each carbon is bonded to the maximum number of hydrogen atoms.

skeletal structure—A representation of organic compounds that shows only the bonding of the carbon framework.

stereoisomers—Compounds differing only in the arrangement of their atoms in space.

straight-chain alkanes—Alkanes that have all their carbon atoms connected in a single continuous chain.

structural isomers—Compounds with the same molecular formula (number and type of atoms) but a different connectivity of the atoms.

substituent—Atom or group of atoms that are not part of the main chain or ring in an organic compound but are bonded to it.

terpene—An naturally occurring alkene that contains $5n$ carbons.

unsaturated fatty acids—Fatty acids that contain one or more carbon–carbon double bonds.

unsaturated hydrocarbon—A hydrocarbon with more than one bond between carbon atoms.

Additional Problems

4.31 Identify the following formulas as organic or inorganic compounds:

    a. KCl   **b.** C_4H_{10} **c.** CH_3CH_2OH

    d. H_2SO_4 **e.** $CaCl_2$ **f.** CH_3CH_2Cl

4.32 Identify the following formulas as organic or inorganic compounds:

    a. $C_6H_{12}O_6$ **b.** Na_2SO_4 **c.** I_2

    d. C_2H_5Cl **e.** $C_{10}H_{22}$ **f.** CH_4

4.33 Alkanes are also referred to as *saturated hydrocarbons.* Explain the meaning of the term *hydrocarbon.* Why are alkanes called saturated hydrocarbons?

4.34 Are alkanes considered polar or nonpolar compounds? Explain.

4.35 Give the molecular formula and draw the Lewis structure of cyclopentane.

4.36 Give the molecular formula and draw the Lewis structure of pentane.

4.37 Give the name of the alkane with 10 carbons.

4.38 Give the name of the cycloalkane with seven carbons.

4.39 Convert each of the Lewis structures shown into a condensed structural formula:

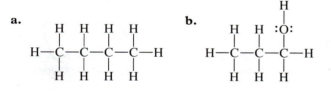

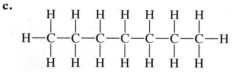

4.40 Convert each of the Lewis structures in Problem 4.39 into a skeletal structure.

4.41 Convert the condensed structures shown to skeletal structures.

    a. $CH_3CH(CH_3)CH_2CH_3$ **b.**

$$CH_3CH_2\overset{\displaystyle O}{\overset{\displaystyle \|}{C}}CH_2CH_3$$

    c. $CH_3CH_2CH(CH_2CH_3)CH_2CH_2CH_3$

4.42 Convert the condensed structures shown to skeletal structures.

    a. $CH_3CH_2CH{=}CH_2$ **b.** $CH_3CH_2CH_2CH_2OH$

    CH_3CHCH_3

    c. $CH_3CH_2CHCH_2CH_2CH_3$

4.43 Convert the skeletal structures shown to condensed structures.

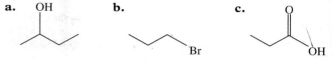

4.44 Convert the skeletal structures shown to condensed structures.

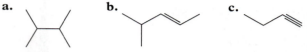

4.45 Lewis structures, condensed structural formulas, and skeletal structures are used to represent the structure of an organic compound. Each of the following compounds is shown in one of these representations. Convert each compound into the other two structural representations not shown.

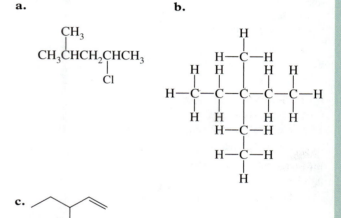

4.46 The molecular formula of an organic compound gives the least detail about the compound's structure. Give the molecular formula of each of the compounds in Problem 4.45.

4.47 Explain the structural difference between a saturated and an unsaturated fatty acid.

4.48 Draw the condensed structural formula and skeletal structure of the saturated fatty acid with 16 carbon atoms. What is the name of this fatty acid?

4.49 Identify all of the functional groups in each of the following molecules:

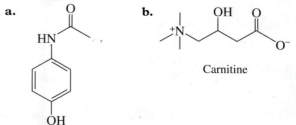

4.50 Identify all of the functional groups in each of the following molecules:

a.

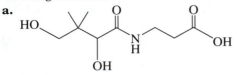

Pantothenic acid (vitamin B5)

b.

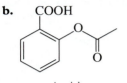

Aspirin

4.51 Zantac™ is used to treat the overproduction of stomach acid for patients with chronic heartburn or gastric ulcers. Identify the four functional groups circled in the structure of Zantac.

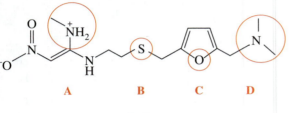

Zantac

4.52 Name the four functional groups circled in the following molecule:

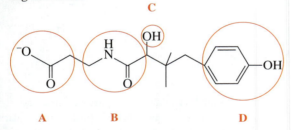

4.53 Write the condensed formula for each of the following molecules:

   a. 3-ethylhexane

   b. 1,3-dichloro-3-methylheptane

   c. 4-isopropyl-2,3-dimethylnonane

4.54 Draw skeletal structures for each of the following molecules:

   a. ethylcyclopropane

   b. *cis*-1-chloro-3-methylcyclohexane

   c. isopropylcyclohexane

4.55 A widely used general anesthetic is called halothane or Fluothane. Its IUPAC name is 2-bromo-2-chloro-1,1,1-trifluoroethane. Draw the Lewis structure for this compound.

4.56 In Section 4.4, we noted that octane ratings are based on an eight-carbon branched-alkane that is a structural isomer of octane called 2,2,4-trimethylpentane. Draw the skeletal structure for this compound.

4.57 Using condensed structural formulas, draw three conformers of hexane.

4.58 Using skeletal structures, draw two conformers of butane.

4.59 Draw the skeletal structure and give the correct IUPAC name for three cycloalkane structural isomers with the molecular formula C_5H_{10}.

4.60 Draw a condensed structural formula and give the correct IUPAC name for the three alkane structural isomers with the molecular formula C_5H_{12}.

4.61 Draw the two alkane structural isomers with molecular formula C_4H_{10}. Give the IUPAC name of each compound.

4.62 Draw the two cycloalkane structural isomers with the molecular formula C_4H_8. Give the IUPAC name of each compound.

4.63 How many structural isomers are possible for the molecular formula $C_5H_{11}F$? Draw the skeletal structure and give the IUPAC name of each compound.

4.64 Draw the skeletal structure and give the correct IUPAC name for three haloalkane structural isomers with the molecular formula C_4H_9Br.

4.65 Using wedge-and-dash bonds, draw both the cis and trans stereoisomers for each of the following compounds:

   a. 1,3-dimethylcyclohexane

   b. 1-bromo-2-ethylcyclopentane

4.66 Using wedge-and-dash bonds, draw both the cis and trans stereoisomers for each of the following compounds:

   a. 1-chloro-4-fluorocyclohexane

   b. 1,3-diethylcyclobutane

4.67 For each of the following compounds, indicate whether or not it can exist as cis–trans stereoisomers. If it can exist as the two isomers, draw both as a condensed structure.

   a. $H_2C{=}CHCH_2CH_3$

   b.

   c. $CH_3CH_2CH{=}CHCH_2CH_3$

   d.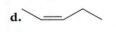

4.68 For each of the following compounds, indicate whether or not it can exist as cis–trans stereoisomers. If it can exist as the two isomers, draw both as a condensed structure.

   a.

   b.

   c. $CH_3CH{=}CHCHCH_3$ with CH_3

   d. $H_3C{-}C{=}CHCH_2CH_3$ with CH_3

4.69 Determine whether each of the following is the cis or the trans stereoisomer:

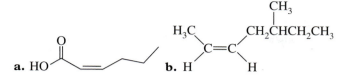

a. HO—

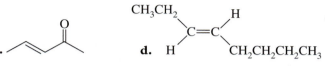

b.

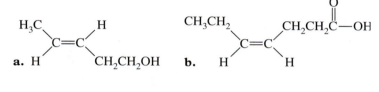

c. d.

4.70 Determine whether each of the following is the cis or the trans stereoisomer:

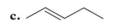

a. b.

c. d.

4.71 Give the name and omega number of the following fatty acid:

$CH_3CH_2CH_2CH_2CH_2HC{=}CHCH_2HC{=}$
$CHCH_2CH_2CH_2CH_2CH_2CH_2COOH$

4.72 Draw the structure of γ-linolenic acid [18:3] and give its omega number.

4.73 Mark the chiral centers in the following molecules, if any, with an asterisk (★).

a. b.

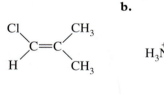

The amino acid cysteine

c.

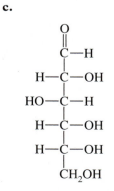

The carbohydrate D-glucose

d.

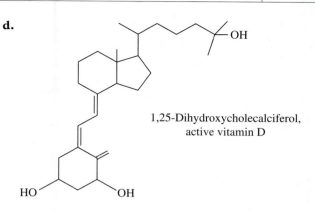

1,25-Dihydroxycholecalciferol, active vitamin D

4.74 Mark the chiral centers in the following molecules, if any, with an asterisk (★).

a.

The amino acid phenylalanine

b.

Epinephrine (adrenaline)

c.

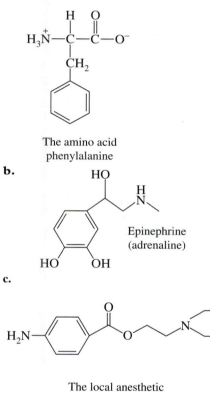

The local anesthetic novocaine

d.

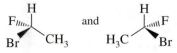

The anti-inflammatory naproxen (Aleve, Naprosyn)

4.75 Are the following compounds structural isomers, cis-trans isomers, or enantiomers?

a.

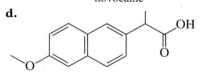 and

b.

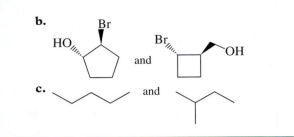

c. 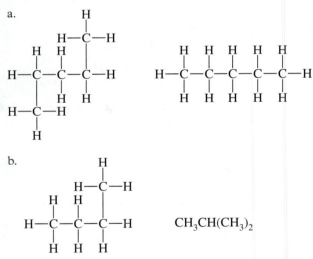 and

4.76 Draw the enantiomer of each of the following compounds. If the compound is not chiral, state that fact.

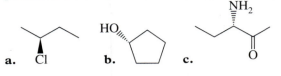

a.   b.   c.

Challenge Problems

4.77 Certain omega-3 fatty acids can be found only in animal sources, such as fatty fish. Two of these are eicosapentaenoic acid (EPA) [20:5] and docosahexaenoic acid (DHA) [22:6], both of which are ω-3 fatty acids. DHA has been shown to be important in healthy brain development, so it has recently been added to infant formulas. Breast milk is rich in DHA as long as the mother maintains a healthy diet that includes fish. Draw skeletal structures of the fatty acids EPA and DHA.

4.78 For each pair of molecules, identify the pair as:

A. structural isomers.

B. the same molecule (conformational isomers).

C. cis-trans stereoisomers.

D. different molecules.

a.

b.

CH₃CH(CH₃)₂

c.

H—C—C—C—C—H

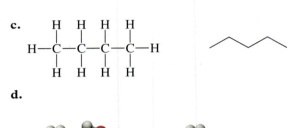

d.

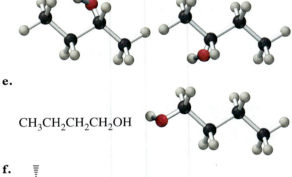

e.

$CH_3CH_2CH_2CH_2OH$

f.

Answers to Odd-Numbered Problems

Practice Problems

4.1   **a.** cyclobutane   **b.** pentane   **c.** heptane

4.3   **a.** C_4H_{10} **b.** C_5H_{10} **c.** C_8H_{18}

4.5   A Lewis structure shows all atoms, bonds, and nonbonding electrons. A condensed structure shows all atoms, but as few bonds as possible.

4.7 CH_4
CH_3CH_3
$CH_3CH_2CH_3$
$CH_3CH_2CH_2CH_3$
$CH_3CH_2CH_2CH_2CH_3$
$CH_3CH_2CH_2CH_2CH_2CH_3$
$CH_3CH_2CH_2CH_2CH_2CH_2CH_3$
$CH_3CH_2CH_2CH_2CH_2CH_2CH_2CH_3$
$CH_3CH_2CH_2CH_2CH_2CH_2CH_2CH_2CH_3$
$CH_3CH_2CH_2CH_2CH_2CH_2CH_2CH_2CH_2CH_3$

4.9 Skeletal structures show mainly bonds between carbon atoms. Since methane has only one carbon, it is not possible to draw its skeletal structure.

4.11

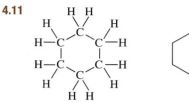

4.13 a. alkyne   **b.** alkene   **c.** alkene (a cycloalkene)

4.15 a.

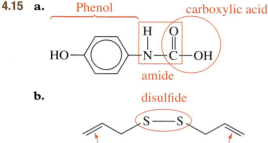

b.

4.17

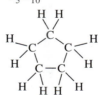

4.19 a. $CH_3CH_2CH_2-$ **b.** CH_3-

4.21 a.   **d.**

b.   **e.** Br

c. Cl
Cl   **f.**

4.23 a. Conformational isomers have the same molecular formula and connectivity.

b. Conformational isomers differ by rotation about one or more single bonds.

4.25 a. structural isomers

b. not related

c. not related

d. conformational isomers (the same molecule)

e. structural isomers

4.27 a. Cis–trans isomers not possible

b. Cis–trans isomers not possible

c. HO $\diagdown$ $\diagup$ CH_3 HO $\diagdown$ $\diagup$ CH_3
        trans   cis

d.
CH_3CH_2 $\diagdown$ $\diagup$ $CH_2CH_2CH_3$ H $\diagdown$ $\diagup$ $CH_2CH_2CH_3$
        C=C   C=C
        H   H CH_3CH_2 H
        cis   trans

4.29

a.   OH   **b.** CH_3
                          $CH_3CH_2CHCHCH_3$
                                      CH_3

c.   O   **d.**
        OH   HN   OCH$_3$
                          Ritalin

Additional Problems

4.31 a. inorganic   **b.** organic   **c.** organic
d. inorganic   **e.** inorganic   **f.** organic

4.33 Hydrocarbons are organic compounds composed only of carbon and hydrogen. Saturated refers to the fact that each carbon is bonded to the maximum number of hydrogens.

4.35 C_5H_{10}

4.37 Decane

4.39 a. $CH_3CH_2CH_2CH_3$ **b.** $CH_3CH_2CH_2OH$
c. $CH_3CH_2CH_2CH_2CH_2CH_2CH_3$

4.41 a.   **b.**   O   **c.**

4.43 a. $CH_3CH(OH)CH_2CH_3$ **b.** $CH_3CH_2CH_2Br$
c. CH_3CH_2COOH

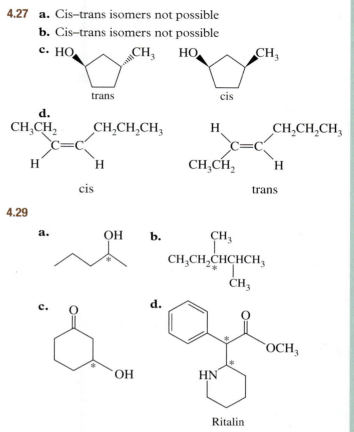

4.45 **a.**

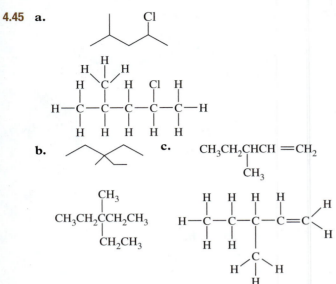

b. **c.** $CH_3CH_2CHCH = CH_2$
$\quad\quad\quad\quad\quad\quad\quad\quad$ CH_3

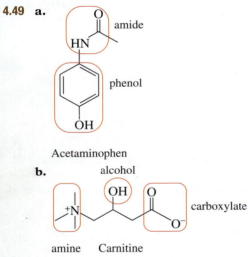

4.47 An unsaturated fatty acid contains one or more carbon–carbon double bonds, but a saturated fatty acid contains no double bonds.

4.49 **a.**

amide
phenol

Acetaminophen

b. alcohol

OH
amine carboxylate
O^-

amine   Carnitine

4.51 **A** = protonated amine, **B** = sulfide (thioether),
C = ether, **D** = amine

4.53 $CH_3CH_2CHCH_2CH_2CH_3$
a. $\quad\quad\quad CH_2CH_3$

$\quad\quad\quad\quad CH_3$
$CH_2CH_2CCH_2CH_2CH_2CH_3$
b. Cl$\quad$Cl

$\quad\quad\quad CH_3$
$CH_3CHCHCHCH_2CH_2CH_2CH_2CH_3$
c. $\quad CH_3\quad CHCH_3$
$\quad\quad\quad\quad\quad CH_3$

4.55

$$:\!\overset{\displaystyle ..}{\underset{\displaystyle ..}{F}}:\!\overset{\displaystyle ..}{\underset{\displaystyle ..}{Cl}}:$$

$$:\!\overset{..}{\underset{..}{F}}\!-\!C\!-\!\overset{..}{\underset{..}{C}}\!-\!\overset{..}{\underset{..}{Br}}:$$

$$:\!\overset{..}{\underset{..}{F}}:\ H$$

4.57
$CH_3CH_2CH_2CH_2CH_2CH_3$ $\quad$ CH_3CH_2 $\quad\quad\quad$ CH_3CH_2
$\quad\quad\quad\quad\quad\quad\quad\quad\quad\quad$ $CH_2CH_2CH_2CH_3$ $\quad$ $CH_2CH_2CH_2$
$\quad\quad\quad\quad\quad\quad\quad\quad\quad\quad\quad\quad\quad\quad\quad\quad\quad\quad\quad\quad$ CH_3

4.59 $\quad\quad\quad$ There are four possible isomers

Cyclopentane $\quad$ Methylcyclobutane $\quad$ 1,2-Dimethylcylopropane

1,1-Dimethylcylopropane

4.61

Butane $\quad$ 2-Methylpropane

4.63

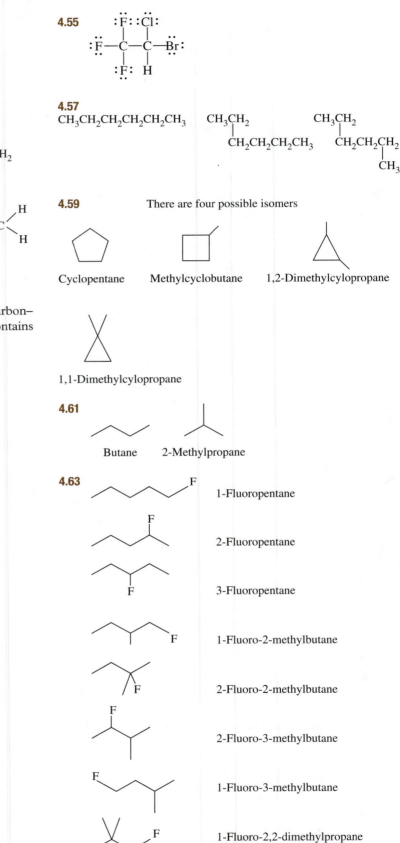

F $\quad$ 1-Fluoropentane

2-Fluoropentane

3-Fluoropentane

1-Fluoro-2-methylbutane

2-Fluoro-2-methylbutane

2-Fluoro-3-methylbutane

1-Fluoro-3-methylbutane

1-Fluoro-2,2-dimethylpropane

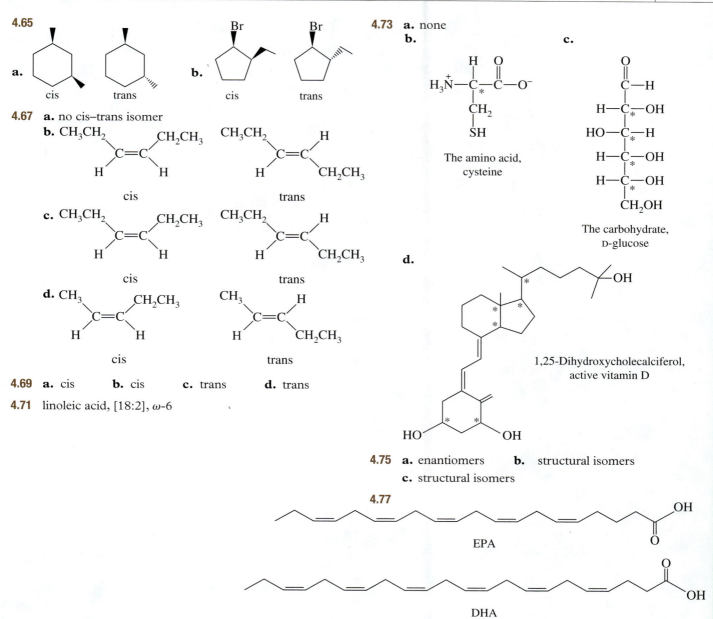

4.65

a. cis   trans

b. cis   trans

4.67 **a.** no cis–trans isomer

b. cis   trans

c. cis   trans

d. cis   trans

4.69 **a.** cis   **b.** cis   **c.** trans   **d.** trans

4.71 linoleic acid, [18:2], ω-6

4.73 **a.** none

b. The amino acid, cysteine

c. The carbohydrate, D-glucose

d. 1,25-Dihydroxycholecalciferol, active vitamin D

4.75 **a.** enantiomers   **b.** structural isomers
c. structural isomers

4.77

EPA

DHA

If you have ever spent hours outside on a cold winter day, a pocket hand warmer can help keep you warm. Within a few minutes the hand warmer begins to feel warm. Heat is generated through a chemical reaction. Why does it feel hot and not cold? How long will it work? To learn more about chemical reactions and their properties, read Chapter 5.

5   Chemical Reactions

WHAT IS HAPPENING inside a hand warmer?
A chemical reaction occurs when oxygen in the air
combines with a mixture of iron shavings, activated
charcoal, salt, and a little water to form—of all things—rust
(Fe_2O_3). When chemical reactions occur, energy, often in
the form of heat, gets exchanged. Understanding the heat
and energy exchanged in a chemical reaction describes an
area of chemistry called **thermodynamics.**

We know that rust damages cars and other metal parts
over time. How is it possible that these reactions, the hand
warmer and metal rusting, occur on such different time
scales? The secret is the addition of the activated charcoal
in the hand warmer. It acts as a *catalyst* to speed up the
chemical reaction. Chemical reactions in the body use
catalysts called *enzymes.* The
pace or rate of a chemical
reaction describes the study
of **reaction kinetics.** In this
chapter we explore both the
thermodynamics and kinetics
of chemical reactions, and then
we introduce several common
organic reactions and their
biochemical application.

A chemical reaction occurs in this hand
warmer when it is shaken.

5.1 Thermodynamics

? **5.1 Inquiry Question:** How are energy and heat transferred and measured
in a chemical reaction?

In Chapter 1, energy was defined as the ability to do work. Chemical reactions involve an
exchange of energy between substances. As we saw with the hand warmer, energy can be
dissipated in the form of heat. In other reactions, energy can be transformed from potential
energy into kinetic energy or vice versa. Energy, heat, and chemical reactions were first
investigated in Chapter 1. Remember that when reading chemical reactions, the reactants
always appear on the left, the products on the right, and the arrow connecting them means
"yields." Here, we take a deeper look at these topics.

Heat of Reaction

Some chemical reactions produce energy while others require energy in the form of heat.
A reaction like the hand warmer that gives off heat is considered an **exothermic reaction**
(*exo* means out of, *thermic* means heat; heat flows out).

The exothermic part of the hand warmer reaction can be represented in a chemical
equation simply by showing heat as a product.

$$4Fe(s) + 3O_2(g). \longrightarrow 2Fe_2O_3(s) + Heat$$

? **What's an Inquiry
Question?**
Inquiry Questions are
designed to focus your reading
on the main concepts by section.
An Inquiry Question appears at
the beginning of each section.

A cold pack produces an endothermic reaction.

An instant cold pack has the opposite effect of a hand warmer. We use a cold pack for injuries to slow blood flow to the injured area. These packs typically contain two compartments, one with water and the other containing a white solid (often ammonium nitrate). When the pack is squeezed, the water compartment breaks and water combines with the solid. The pack gets very cold because the mixing of the two substances requires energy in the form of heat. Reactions requiring heat energy are called **endothermic reactions** (*endo* means inside, *thermic* means heat; heat flows in).

The heat required in the instant cold pack can be represented in a chemical reaction by showing heat as a reactant.

$$\text{Heat} + NH_4NO_3(s) \longrightarrow NH_4NO_3(aq)$$

The difference in heat between the reactants and products in a chemical reaction is called the **heat of reaction** and can be exothermic or endothermic.

Randomness

Why do some reactions give off heat and some reactions require heat? Something else must be happening in the reactions at the molecular level. Consider what happens when you take an ice cube out of the refrigerator. It spontaneously melts, right? It is absorbing heat from its surroundings (endothermic) as it melts. The water molecules are also becoming less organized or more randomly arranged as the ice moves from a solid to a liquid. This demonstrates another key feature of chemical reactions. The reactants and products in a chemical reaction have a certain amount of randomness associated with them.

Consider the hand warmer and cold pack discussed earlier. In the case of the hand warmer, the iron and oxygen combined when rust was formed. This made the product less random than the reactants. This combined with the formation of an ionic compound gave off heat. In the case of the cold pack, the ammonium nitrate didn't actually chemically react, but the state changed when the solid dissolved in water. When it dissolved, the ammonium and nitrate ions became more disordered which caused the reaction to spontaneously get cold instead of hot. As seen in Chapter 1, the states of matter have differing amounts of randomness. In a solid, the particles are the least random and in a gas the particles are the most random (See Table 1.5).

Free Energy and ΔG

For any chemical reaction, both the heat of the reaction and the randomness of the reactants versus the products are taken into consideration when determining the energy exchanged during a chemical reaction. This energy is called the **free energy, (G),** or Gibbs free energy, and is defined as the amount of energy present in molecules available to do work whether due to heat or randomness. The **free energy change, ΔG** is the energy difference between the free energy present in the reactant and the product molecules. The free energy change (ΔG) includes contributions from both the heat of reaction and randomness. Because we are considering both heat and randomness when discussing free energy exchanged, the more general term **exergonic** (gives off energy) is used instead of exothermic. If energy is required for a reaction, the term **endergonic** is used.

For an exergonic reaction, the free energy of the reactants is greater than the free energy of the products and the value of ΔG is negative. Reactions with a negative ΔG value do not require energy input and occur spontaneously. **Spontaneous processes** will continue to occur once started and do not require energy from the surroundings. For example, a canoe will travel downstream spontaneously with the current. In an endergonic reaction, the free energy of the products is greater than that of the reactants, indicating that energy is absorbed and the value of ΔG is positive. Reactions with a positive ΔG value do require energy input from their surroundings and are considered nonspontaneous. **Nonspontaneous processes** do not occur naturally and require energy input. A canoe can only go upstream if the paddler inputs energy by paddling.

It should be noted that the spontaneity of a reaction does not tell us how fast a reaction occurs. For example, both graphite and diamond are solid forms of carbon. One can convert into the other.

$$C(s, \textit{diamond}) \longrightarrow C(s, \textit{graphite})$$

a.

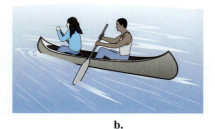

b.

(a) A canoe traveling downstream with the current is a spontaneous process. (b) A canoe traveling upstream is a nonspontaneous process that requires energy input.

The free energy change (ΔG) for this reaction has a negative value, meaning that diamonds will spontaneously convert to graphite. We do not see diamonds turning into graphite before our eyes, or even over our lifetime. This spontaneous change takes place very slowly. Knowing the value of ΔG can predict spontaneity, yet it does not imply the rate of a chemical reaction. We explore chemical reaction rates in Section 5.2.

Activation Energy

In addition to the energy difference between reactants and products, chemical reactions need a little "push" to get started. Reactants have to collide with enough energy and the correct orientation for the reaction to start.

The energy necessary to align the reactant molecules and to cause them to collide with enough energy to form products is known as the **activation energy.** If the energy in the reactant molecules is less than the activation energy, the molecules will bounce off each other without forming any products.

The conversion of carbon from diamond to graphite is a spontaneous, albeit slow, process.

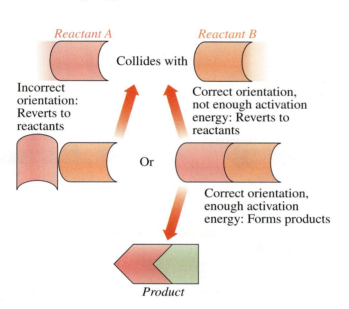

Activation energy is a property of a chemical reaction. We can think of activation energy as the energy hill that molecules must overcome before they change into products. If enough energy is available to allow the molecules to get over the energy hill (also known as the activation barrier), they can react to form products.

A reaction energy diagram is a way to visually examine the energy changes that occur in a chemical reaction. **Figure 5.1** shows reaction energy diagrams for an exergonic and an endergonic reaction. Energy appears on the y-axis and the reaction progress from reactants to products appears on the x-axis. Notice that each reaction has an activation energy hill to climb before proceeding to products.

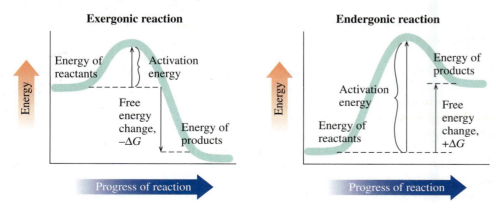

FIGURE 5.1 Reaction energy diagrams for an exergonic and an endergonic reaction. Reaction energy diagrams are a compact way of showing how the energy changes in a chemical reaction over time.

Each reaction also shows the difference between the energy for reactants and products giving the overall free energy (ΔG) of the reaction. In summary, a reaction energy diagram gives information about both the activation energy and free energy.

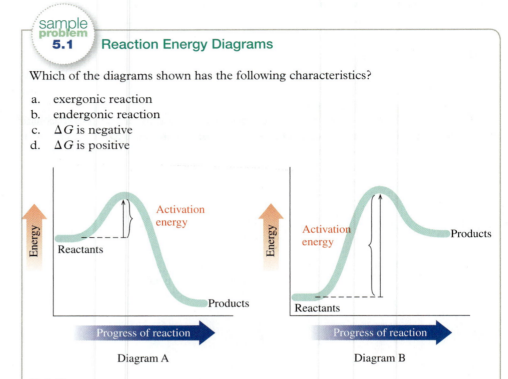

sample problem 5.1 **Reaction Energy Diagrams**

Which of the diagrams shown has the following characteristics?

a. exergonic reaction
b. endergonic reaction
c. ΔG is negative
d. ΔG is positive

Diagram A

Diagram B

Solution

a. The energy of the reactants is greater than the energy of the products in Diagram A. This represents an exergonic reaction.
b. The energy of the reactants is less than the energy of the products in Diagram B. This represents an endergonic reaction.
c. The energy of the reactants is greater than the energy of the products in Diagram A. Releasing energy represents a negative ΔG.
d. The energy of the reactants is less than the energy of the products in Diagram B. Absorbing energy represents a positive ΔG.

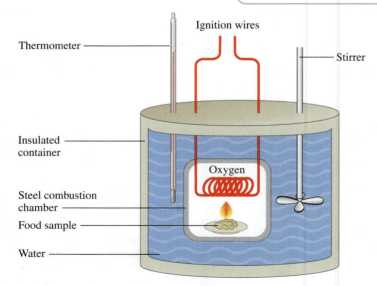

FIGURE 5.2 A cross section of a calorimeter. The number of Calories in a food sample can be determined from the amount of heat released when food is combusted in a calorimeter.

Calorimetry: An Application of Thermodynamics

If you look at the word *calorimetry,* what does it make you think of? Calories. Food molecules contain chemical potential energy (represented on food packages as Calories), which is transformed through chemical reactions to energy for our bodies.

Can you guess what an instrument called a calorimeter might do? A **calorimeter** measures the amount of energy in food by burning the food in the presence of oxygen. Chemical reactions with oxygen that produce heat are called **combustion** reactions. In a calorimeter, the food is placed inside a closed container surrounded by water and the food is burned (see **Figure 5.2**). The heat released during combustion is transferred to the water surrounding the container. (Combustion is explored in more detail in Section 5.3.)

The change in the temperature of the water is directly related to the amount of energy released as heat during the combustion reaction. The energy is commonly reported in units of nutritional Calories or kilojoules. Remember that a nutritional Calorie (Cal) is an energy unit equal to 1000 calories (cal). (For a review of the energy units nutritional Calorie (Cal), and kilojoule (kJ), refer to Section 1.2.) Whether burning food in a calorimeter or burning it in the body to produce energy, the result is the same. Energy is released.

Energy Content in Food

When trying to lose weight, we talk about *burning* calories although we are not lighting a match to ourselves. Whether burning outside the body (where we can see burning with oxygen in the air) or inside the body (where food combines with oxygen) the reaction is the same: the molecules in food react with oxygen producing carbon dioxide, water, and energy. The food we eat contains different amounts of energy. Energy (caloric) values for different nutrient molecules in food are given in Table 5.1.

The Nutrition Facts labels on packaged food list the total grams of each nutrient present per serving. The caloric content can be calculated by summing the Calories present for each nutrient. Values for some common foods are listed in Table 5.2.

TABLE 5.1 Caloric Values in Food

Nutrient Molecule in Food	Example	Cal/g	kJ/g
Carbohydrate	Table sugar, potatoes, flour	4	17
Protein	Meats, fish, beans	4	17
Fat	Oil, butter	9	38

How to calculate the number of calories from a nutritional label for peanut butter. Notice that the manufacturer has rounded the final value down to 180. To one significant figure, the Calories per serving should be 200.

$$12 \text{ g Fat} \quad \times \quad \frac{9 \text{ Cal}}{1 \text{ g Fat}} \quad = \quad 108 \text{ Cal}$$

$$15 \text{ g Carb} \quad \times \quad \frac{4 \text{ Cal}}{1 \text{ g Carb}} \quad = \quad 60 \text{ Cal}$$

$$7 \text{ g Protein} \quad \times \quad \frac{4 \text{ Cal}}{1 \text{ g Protein}} \quad = \quad 28 \text{ Cal}$$

$$\text{Total} = 196 \text{ Cal}$$

TABLE 5.2 Composition and Energy Content of Some Foods

Food	Carbohydrate (g)	Protein (g)	Fat (g)	Calories (kJ)
Apple	21	0.27	0.49	89 (370)
Bread, 1 slice white	14.1	2.4	0.9	74 (310)
Butter, 1 Tbs	0.008	0.12	11.5	104 (435)
Corn, ½ cup	20.6	2.72	1.05	103 (430)
Egg, chicken's, 1 large	0.6	6.07	5.58	77 (320)
Orange juice, 1 cup	27	2	0	116 (485)
Peanut butter, 1 Tbs	3.55	3.98	7.80	100 (420)
Rice, brown, ½ cup cooked	24.8	2.45	0.78	116 (485)
Steak, 3 oz	0	19	27	320 (1300)

sample problem 5.2   Calculating the Energy Content in Food

Calculate the number of Calories present in a slice of pepperoni pizza that contains 34 g carbohydrate, 13 g protein, and 12 g fat.

Solution

Use Table 5.1 to determine the number of Calories for each molecule type and sum them together.

$$\text{Carbohydrate: } 34 \ g \times \frac{4 \text{ Calories}}{1 \ g} = 136 \text{ Calories}$$

$$\text{Protein: } 13 \ g \times \frac{4 \text{ Calories}}{1 \ g} = 52 \text{ Calories}$$

$$\text{Fat: } 12 \ g \times \frac{9 \text{ Calories}}{1 \ g} = 108 \text{ Calories}$$

Total: 296 which rounds to 300 Calories (1 significant figure)

The measured calories per gram for each have only one significant figure present. The values must be rounded to one significant figure at the conclusion of the problem. If you spend much time reading nutrition labels, you may have observed that most contain rounded values.

Integrating chemistry

Low-Calorie Foods

Why does a regular soft drink have more Calories than a diet soft drink? What is in low-fat foods?

Most of the Calories in a regular soft drink come from the sugar in it. Sugar is a carbohydrate, which can be converted to energy in the body. Diet soft drinks contain substances that are not sugar but taste sweet. These sweet-tasting molecules, or sweeteners, undergo very few chemical reactions in the body and so do not produce much energy (Calories). Molecules that are unreactive in the body are not converted to energy.

Fats produce the highest amount of energy per gram when they undergo chemical reactions in the body (9 Cal/g). Manufacturers of baked goods have introduced fat substitutes into their products to produce low-fat alternatives. Substances like dextrins, celluloses, and gums are carbohydrate-based molecules that produce less energy (4 Cal/g) when eaten. The consumer gets a product that tastes similar but contains fewer Calories because less fat is present.

Diet soft drinks and low-fat foods produce less energy (Calories) when consumed by the body.

5.1 When vinegar (CH_3COOH) and baking soda ($NaHCO_3$) are combined, the mixture spontaneously undergoes an endothermic reaction while releasing CO_2 gas. Would the test tube feel hot or cold? Do you think this reaction would have a + or − value for ΔG? Explain your reasoning.

5.2 In your own words, define free energy change, ΔG.
  a. How does the ΔG differ in exergonic and endergonic reactions?
  b. Which is spontaneous, a reaction with a + or − value for ΔG?

5.3 Classify the following as exothermic or endothermic reactions:
  a. When two solids are combined in a test tube, the test tube gets hot.
  b. A reaction must be heated for it to continue.

5.4 Classify the following as exergonic or endergonic reactions:
  a. On a reaction energy diagram, the reactants show lower energy than the products.
  b. On a reaction energy diagram, the products are lower in energy than the reactants.

5.5 Classify the following as spontaneous or nonspontaneous processes:
  a. a hot bowl of oatmeal cooling on the table
  b. a chemical reaction that gives off free energy

5.6 Classify the following as spontaneous or nonspontaneous processes:
  a. working out on a treadmill at the gym
  b. a chemical reaction that requires free energy to occur

5.7 Calculate the number of Calories present in a chocolate chip cookie that contains 17 g carbohydrate, 1 g protein, and 7 g fat.

5.8 Use Table 5.2 to determine how many Calories would be found in a breakfast consisting of two eggs, two pieces of toast buttered with one tablespoon of butter, and one 6 oz glass of orange juice.

Discovering the Concepts

? Inquiry Activity—Reaction Energy Diagrams

Information

The diagrams shown are called reaction energy diagrams and graphically show the progress of two different chemical reactions on the *x*-axis and the amount of energy required as the reaction moves forward on the *y*-axis. The *activation energy* is the amount of energy required to get the reactants into position for collision with enough energy so that they actually react. Reactions with larger activation energies occur more slowly than reactions with smaller activation energies.

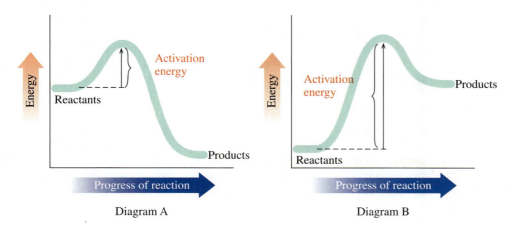

Diagram A   Diagram B

Questions

1. Which reaction has the larger activation energy?
2. Based on the diagrams, which reaction can form products more quickly?
3. A *catalyst* speeds up a chemical reaction by lowering the activation energy. Sketch

Diagram A and draw a second line on the same diagram for the reaction if a catalyst is present. Label the two traces as "uncatalyzed" and "catalyzed."

4. As we explored in Section 5.1, chemical reactions that release heat energy upon forming product are called *exergonic* and reactions that require heat energy are called *endergonic*.
    a. Which reaction (A or B) has more energy in the reactants than in the products?
    b. Which reaction is exergonic?
5. Examine your sketch from Question 3. Does a catalyst change the amount of energy produced or required in a chemical reaction? How did you decide?
6. Draw a reaction energy diagram for a slow, exergonic reaction and contrast it to a diagram for a fast, endergonic reaction.

5.2 Chemical Reactions: Kinetics

5.2 Inquiry Question:
How is the rate of a chemical reaction controlled?

To begin to understand reaction rates, consider this scenario. On an average day, a retailer will sell a certain amount of goods at a retail price to a certain number of shoppers. However, at the start of the holiday shopping season, retailers offer lots of sales on those same items, lowering the price. For this reason, more shoppers rush to the store to make their purchases, and the total number of shoppers and sales increase that day.

Now, think back to the reaction energy diagrams and activation energy hills that we examined in Section 5.1. Just as more shoppers make purchases more quickly at lower prices during a sale, chemical reactions with a lower activation energy proceed more quickly (have a faster rate) than reactions with a higher activation energy.

Factors Affecting Reaction Rates

We can measure the **rate of reaction** by determining the amount of product formed (or reactant used up) in a certain period of time. We know that for a chemical reaction to occur, the molecules of the reactant(s) must collide with each other with enough energy and the proper orientation. Any factor that affects either of these affects the rate of the reaction. The rate of a reaction is affected by several factors, three of which will be discussed here: temperature, amount of reactants, and the presence of a catalyst.

For a visual summary of the factors affecting chemical reaction rates, see **Figure 5.3**.

Temperature

The faster that cars drive on the highway, the more likely an accident is to occur. If you think of chemical reactants as the cars on the highway and a collision as the chemical reaction, this example implies that the rate of chemical reactions increases with temperature. Increasing temperature increases the kinetic energy (motion) of the reacting molecules, so they collide with each other more often, causing more reactions to occur. (See **Figure 5.3b**). In biological systems, if temperatures cannot be maintained, hypo- or hyperthermia results in part due to changing reaction rates. Under normal conditions, the temperature remains constant in a biological system and the reaction rate is not affected by temperature in this case.

Amount of Reactant

The more cars on the highway, the more likely an accident will occur. If the cars are the reactants and the accident is the chemical reaction, this example implies that the rate of chemical reactions increases with the amount of reactants. Having more reactant available allows the molecules to collide with each other more often, causing more reactions to occur. As reactants are consumed and form products, chemical reactions slow down due to a decrease in the amount of reactants.

Catalyst

Next consider driving either over a mountain road, or through a tunnel carved in the mountain, to get to the other side. Either route will get you there, but you arrive sooner if you drive through the tunnel. Similarly, a **catalyst** speeds up a reaction by *lowering* the activation energy. With a lower energy hill to climb, the reaction progresses to form products more

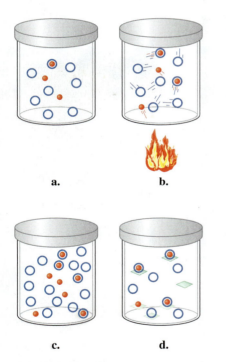

a.   **b.**

c.   **d.**

FIGURE 5.3 Factors affecting reaction rates. At the molecular level, the rate of a chemical reaction increases because the probability of a collision between the reactants increases. (a) Reactants in a chemical reaction (red fits in blue to form product). (b) Increasing the temperature increases the movement of the reactants and their collision frequency. (c) Doubling the amount of reactants increases the collision frequency. (d) A catalyst (green diamond) can increase the likelihood of collision by providing a surface with optimal orientation for the reaction to occur.

quickly. A catalyst participates in a chemical reaction but remains unchanged at the end of the chemical reaction; in other words, it is regenerated and can be used again. Whether a reaction is exergonic or endergonic, the energy of products or reactants is not affected by a catalyst—a catalyst only speeds up formation of the products. A catalyst is neither a product nor a reactant. Acids, bases, and metal ions often act as catalysts. Many catalysts immobilize reactants that are near to each other, increasing the likelihood of collision and therefore the rate.

a.   b.

Just as driving through the tunnel speeds up the time of a trip to the other side of a mountain, a catalyst speeds up the rate of a chemical reaction.

As an example, consider hydrogen peroxide, H_2O_2, often used to clean cuts. When hydrogen peroxide is put on a cut, the area bubbles because oxygen gas is produced by the reaction of hydrogen peroxide with a catalyst called *catalase* found in the blood. Catalase is a protein that acts as a biological catalyst called an **enzyme.**

The balanced chemical reaction is shown

$$2H_2O_2(l) \xrightarrow{\text{catalase}} 2H_2O(l) + O_2(g)$$

Because catalysts are not changed during a chemical reaction, they are written above or below the arrow in the reaction equation. In the absence of the catalyst, the hydrogen peroxide reaction occurs very slowly. In fact, if you pour hydrogen peroxide on your uncut skin, nothing will happen. Hydrogen peroxide cleans a wounded area by reacting with catalase to produce oxygen quickly at the wound site.

---

sample problem

5.3   **Factors Affecting Reaction Rate**

Determine if the following changes in condition would increase or decrease the rate of the chemical reactions in which they are involved:

a. baking cookies at 400° F instead of 325° F
b. taking medication every other day instead of every day
c. spraying an enzyme treatment on stained carpet fibers to clean them

Solution

a. Increase. Raising the temperature of baking will decrease the cooking time.
b. Decrease. By taking half the recommended medication, the amount of medication in the body is lowered and any chemical reactions involved will receive less impact.
c. Increase. Adding an enzyme to a stain catalyzes breaking up the stain from the carpet fibers.

The rate of shopping increases when catalyzed by a sale.

Enzymes Are Biological Catalysts

Because the body maintains a constant temperature and the amount of reactant present in cells does not fluctuate greatly, most biochemical reactions use enzymes to increase reaction rates. Most enzymes are proteins. Enzymes can enhance the rates of a chemical reaction by a factor of more than ten million (1×10^7). This rate is much faster than most chemical catalysts. Enzymes speed up reactions by immobilizing the reactants at a site on the enzyme called the **active site** and orienting them correctly for the reaction to occur.

Enzymes are highly specific to the reactions that they catalyze in the body. They are so important that when they are defective, it often results in an altered or diseased state in the organism. For example, albinism, the lack of pigment in the skin, hair, and eyes is caused by a defect in tyrosinase, an enzyme in the pathway that produces the pigment melanin. **Table 5.3** shows several diseases caused by enzyme deficiency. More information on proteins, enzymes and their catalysis is discussed in Chapter 10.

TABLE 5.3 Some Enzyme-Based Diseases

Disease	Enzyme	Clinical Symptoms
Albinism	Tyrosinase	Absence of pigment in skin, hair, and eyes
Gout	PRPP synthetase	Renal problems, joint problems, high levels of uric acid in the bloodstream
Lactose intolerance	Lactase	Diarrhea, bloating, flatulence, nausea, abdominal cramping
Glycogen storage deficiency	Glycogen synthase	Enlarged fatty liver, low sugar levels during fasting
Phenylketonuria (PKU-I)	Phenylalanine hydroxylase	Neurologic symptoms, mental retardation
Tay-Sachs disease	Hexosaminidase A	Mental retardation, blindness, muscular weakness, death

practice problems

5.9 a. How does increasing the temperature increase the rate of a chemical reaction?
  b. How does increasing the amount of reactants increase the rate of a chemical reaction?

5.10 a. Describe activation energy for a chemical reaction.
  b. How does adding a catalyst increase the rate of a chemical reaction?

5.11 Why does the rate of a chemical reaction decrease as the reaction progresses?

5.12 _____ are biochemical catalysts.

5.13 Determine if the following changes in condition would increase or decrease the rate of the chemical reactions in which they are involved:
  a. storing leftovers in the refrigerator

  b. increasing the kinetic energy of the reactant molecules
  c. increasing the frequency of collision

5.14 Determine if the following changes in condition would increase or decrease the rate of the chemical reactions in which they are involved:
  a. adding a metal surface for two reactants to locate each other
  b. lowering the number of one of the reactant molecules
  c. caramelizing onions on a medium-high stove setting versus a low setting

Discovering the Concepts

? Inquiry Activity—Types of Chemical Reactions

Data Set

Reaction Type

Synthesis

Decomposition

Exchange

Sample Chemical Reactions

$HCl(aq) + NaOH(aq) \longrightarrow NaCl(aq) + H_2O(l)$

$H_2(g) + Br_2(g) \longrightarrow 2HBr(g)$

$Br_2(g) + BaI_2(s) \longrightarrow BaBr_2(s) + I_2(g)$

$2H_2O(l) \longrightarrow 2H_2(g) + O_2(g)$

Questions

1. Match the sample reactions with the reaction type. The types can be used more than once. Support your choices.
2. How are a synthesis and a decomposition reaction the same and how are they different?
3. If the general reaction for a synthesis can be written as A + B $\longrightarrow$ AB, how would you write a general reaction for a decomposition?
4. One of the reactions in the data set is a single exchange, one is a double exchange. How are these the same, how are they different?
5. Complete the following general reactions for a single exchange and a double exchange:
   Single exchange: AB + C $\longrightarrow$
   Double exchange: AB + CD $\longrightarrow$
6. Categorize the following reactions as a synthesis, decomposition, or exchange reaction:
   a. $N_2(g) + O_2(g) \longrightarrow 2NO(g)$
   b. $C_4H_9Br(g) \longrightarrow C_4H_8(g) + HBr(g)$
   c. $Mg(s) + 2HCl(aq) \longrightarrow MgCl_2(aq) + H_2(g)$

5.3 Overview of Chemical Reactions

Now that we have explored the thermodynamics and kinetics of chemical reactions, we look next at some different types of chemical reactions. Most chemical reactions fit into one of three general types of reactions—decomposition, synthesis, or exchange reactions. These reactions can either be reversible or irreversible. Any conditions for the reaction like temperature or a catalyst will appear above or below the arrow.

5.3 Inquiry Question:
What are some general categories of chemical reactions?

Types of Chemical Reactions

Figure 5.4 shows a schematic of the types of chemical reactions. **Synthesis reactions** are like adding letters together to make words, except that in chemistry, smaller molecules or atoms combine to create larger molecules. An example is the combination of amino acids to form a protein. Synthesis reactions often require energy because products often have higher free energy (less random) than reactants (more random) when smaller molecules join to build larger ones. These reactions have only one product.

Decomposition reactions are just the opposite. These reactions are just what they sound like. Decomposition reactions break apart larger molecules into smaller molecules. As an example, plants store starch, which human beings can break down into the sugar glucose. Decomposition reactions often give off energy because product compounds form stronger bonds than the reactants. These reactions typically have only one reactant.

Exchange reactions, sometimes called *displacement* reactions, involve both synthesis and decomposition. Bonds are broken and new bonds are formed as substances decompose and then swap partners, synthesizing different product compounds. When you use an antacid like Milk of Magnesia® to combat heartburn, you break down the excess acid (HCl) in the stomach with magnesium hydroxide ($Mg(OH)_2$). The two substances exchange ions with each other, forming $MgCl_2$ and HOH (H_2O).

These three basic reaction types summarized in Table 5.4 allow us to begin to detect patterns as we look at chemical reactions more closely. It should be noted that not all reactions clearly fit into one of these general categories. These reaction types represent a first step in understanding the more specific chemical reactions later in the chapter.

Types of Chemical Reactions

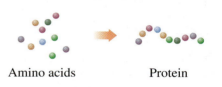

Amino acids   Protein

a. Synthesis—The formation of a protein from component amino acids is a synthesis reaction.

Starch   Glucose

b. Decomposition—The breakdown of starch is a decomposition reaction.

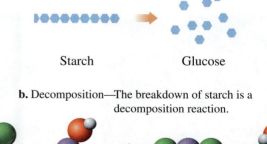

2HCl   + $Mg(OH)_2$ $MgCl_2$ + $2H_2O$

c. Exchange—Using an antacid to combat heartburn involves an exchange reaction.

FIGURE 5.4 Types of chemical reactions. (a) In a synthesis reaction, larger compounds are synthesized from smaller ones. (b) In a decomposition reaction, larger compounds are broken down into smaller ones. (c) In an exchange reaction, the reactants change partners.

TABLE 5.4 Types of Chemical Reactions

Reaction Type	General Reaction Scheme
Synthesis	$A + B \longrightarrow AB$
Decomposition	$AB \longrightarrow A + B$
Exchange	$AB + C \longrightarrow AC + B$ (single)
	$AB + CD \longrightarrow AD + CB$ (double)

sample problem 5.4 · Distinguishing Reaction Types

Categorize the following reactions as synthesis, decomposition, or exchange reactions:

a. $2Fe_2O_3(s) + 3C(s) \longrightarrow 3CO_2(g) + 4Fe(s)$

b. $2Al_2O_3(s) \longrightarrow 4Al(s) + 3O_2(g)$

c. $H_2(g) + I_2(g) \longrightarrow 2HI(g)$

Solution

To answer this problem, apply the patterns in Table 5.4 to the chemical equations.

a. Exchange. Two reactants (Fe_2O_3 and C) form two products (CO_2 and Fe). The element oxygen changes places.

b. Decomposition. A single reactant (Al_2O_3) breaks into two smaller parts (Al and O_2).

c. Synthesis. Two reactants (H_2 and I_2) form one product (HI).

Reversible and Irreversible Reactions

If you go to see a movie, you and other people enter a theater and become the audience. When the movie is over, the audience exits and everyone goes separate ways.

Just like entering and exiting a movie theater, chemical reactions are sometimes reversible. **Reversible reactions** can react and form products in either direction. If a chemical bond can be formed, with enough input of energy it can also be broken. Reversible chemical reactions are indicated with a pair of arrows between the reactants and products pointing in opposite directions.

$$A + B \rightleftharpoons AB$$

In this example, the forward reaction occurs when A and B form AB. When a sufficient amount of AB builds up, the reverse reaction occurs and A and B re-form. At the point when both the forward and the reverse reaction are occurring at the same rate, and neither reaction dominates, the reaction is said to be in a state of **chemical equilibrium.** This does not mean that at equilibrium there are equal amounts of reactants and products; it means that the reaction reaches a point where the amount of products remains constant. The topic of chemical equilibrium is discussed in more depth in Chapter 9.

Yet if you think about chemical reactions such as a candle burning, you know that it is not easy to reverse this reaction and bring the candle back to its original state. Reactions that are highly exothermic are for practical purposes considered **irreversible reactions.** Rejoining the combusted molecules from the candle wax to re-form the candle once they are dissipated into the atmosphere requires too much energy. When we break down food, these chemical reactions are never directly reversed to re-form the actual food. We use the energy produced to form other molecules and maintain bodily functions.

Reversible chemical reactions are like a set of individuals entering a movie theater and becoming an audience (the forward reaction—left to right). After the movie, the audience exits and returns to being the individuals (the reverse reaction—right to left). Reversible reactions can react and form products in either direction.

5.5 Reversible and Irreversible Reactions

Determine whether the following reactions are most likely reversible or irreversible:

a. $2N_2O(g) \rightleftharpoons N_2O_4(g)$
b. $HBr(aq) + NaOH(aq) \longrightarrow NaBr(aq) + H_2O(l)$
c. burning gasoline in an automobile engine

Solution

By inspecting these reactions, we can determine if they are reversible or irreversible.

a. Reversible. In this reaction, the formation of two gases is indicated by the equilibrium arrow, N_2O_4 in the forward direction and NO_2 in the reverse direction. The reaction is reversible.
b. Irreversible. In this reaction, water is formed as a product. Notice the reaction arrow only goes in the forward direction. The reaction is irreversible.
c. Irreversible. In the burning of gasoline, it is impossible to capture the exhaust from the reaction and re-form gasoline from it.

Combustion

Next we take a more in-depth look at one of the reactions we encountered earlier, combustion. Combustion involves a carbon-containing molecule reacting with oxygen to produce carbon dioxide and water. Earlier, we noted that combustion reactions tend to be highly exothermic. Because of the large amount of energy as heat produced, they are considered irreversible reactions. Combustion reactions are a type of decomposition reaction because larger molecules are broken down into smaller ones. They are also considered a special type of exchange reaction called an oxidation–reduction reaction because oxygen is reacted in a combustion reaction, something that we will explore further in Section 5.4.

Combustion reactions use oxygen and produce energy whether it dissipates as heat or performs work. They are an important part of understanding fuels, including fossil fuels used to heat homes and drive cars, and foods that fuel our bodies with energy. Because the products are always the same for complete combustion (carbon dioxide and water), these reactions are a useful place to start an exploration of organic chemical reactions.

Alkanes

Although alkanes are generally considered unreactive, they do undergo combustion. Alkanes are excellent fuel sources because they give off a lot of energy as heat when they are burned. Unfortunately, internal combustion engines like those found in gasoline-powered cars do not completely combust hydrocarbons. Such engines emit partially reacted hydrocarbons back to the environment in their exhaust. Other hydrocarbons are not completely oxidized, and forms of carbon soot result. The term *clean* burning refers to the efficiency of the combustion. The cleaner a fuel burns the more complete the combustion.

The combustion of butane (C_4H_{10}) is an exothermic, irreversible reaction.

5.6 Combustion of Alkanes

Provide the products and balance the following reaction for the complete combustion of propane, C_3H_8.

$$C_3H_8(g) + O_2(g) \longrightarrow ?$$

Solution

In combustion, the products are CO_2 and H_2O, so we complete the equation and then balance the equation as demonstrated previously in Section 1.6.

$$C_3H_8(g) + O_2(g) \longrightarrow CO_2(g) + H_2O(g)$$

STEP 1: Examine.

Element	Number in Reactants	Number in Products
C	3	1
H	8	2
O	2	3

STEP 2: Balance. Because they are found in both products, it is best to balance the oxygen atoms last. Starting with carbon, there are 3 carbons on the reactant side, so placing a 3 in front of the CO_2 will balance the carbon atoms. The 8 hydrogen atoms in propane can be balanced in the products by placing a 4 in front of the H_2O.

$$C_3H_8(g) + O_2(g) \longrightarrow 3CO_2(g) + 4H_2O(g)$$

At this point, we examine the number of oxygen atoms present and notice that there are 2 on the reactant side and 10 on the product side. Ten atoms of oxygen are necessary on the reactant side, so a coefficient of 5 is added in front of the O_2.

$$C_3H_8(g) + 5O_2(g) \longrightarrow 3CO_2(g) + 4H_2O(g)$$

STEP 3: Check.

Element	Number in Reactants	Number in Products
C	3	3
H	8	8
O	10	10

The equation is now balanced. In this equation, the propane is oxidized (oxygen added) to form CO_2, and the oxygen is reduced (hydrogen added) to form water.

Distinguishing Chemical Reactions

Up to this point, our understanding of chemical reactions has been limited to examining the formulas and the arrows, and categorizing and balancing the reaction. Recognizing the type of chemical reaction and balancing a chemical equation are important actions in determining how much product might be obtained and reactant required in a chemical reaction.

In contrast, for many organic reactions, the interesting part of the reaction is *how* the functional groups change as reactants form products, or what organic chemists call the **reaction mechanism,** so more details are necessary. When writing an organic chemical equation, the structure of organic molecules of interest are written out because there are many structural isomers for the same molecular formula. Often, the organic reactant is the only reactant shown. Small molecules that participate in the reaction are shown above the reaction arrow, and reaction conditions like a catalyst are typically shown below the arrow. Here is an example of an organic reaction called *hydrogenation* that will be discussed a little later in this chapter.

Organic reactions focus on changes in the functional groups. Here, an alkene is changed to an alkane.

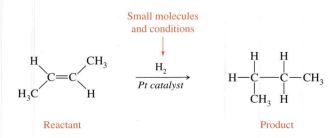

Reactant   Product

Biochemical reactions are written like organic reactions, and the reactant, product, and reaction conditions are shown. Sometimes biochemical reactions are *coupled* to each other through energy transfer; that is, two reactions occur at the same time, and energy is transferred from one reaction to fuel the other. The two reactions are often written together with the energy transfer reaction written at the arrow. **A**denosine **trip**hosphate, ATP, is commonly used to transfer energy in a biochemical reaction. Typically, this common energy-providing molecule does not have its structure shown.

For example, in the first reaction of **glycolysis,** the breakdown of the sugar molecule glucose in the body, glucose is converted to the molecule glucose-6-phosphate. The phosphate comes from the energy molecule ATP, which transfers a phosphate group and becomes **a**denosine **dip**hosphate, ADP, in the process. The energy transfer reaction is drawn at the arrow. The enzyme that catalyzes the reaction is shown below the arrow.

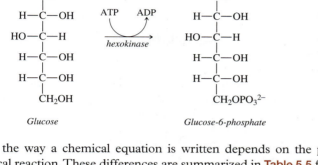

Glucose   Glucose-6-phosphate

Biochemical reactions show changes in functional groups coupled with a transfer of energy. Here, the energy is transferred through ATP.

In summary, the way a chemical equation is written depends on the purpose for writing the chemical reaction. These differences are summarized in Table 5.5 for a general reaction, an organic reaction, and a biochemical reaction.

In this section, we began by categorizing some general types of reactions and balanced some combustion reactions. In the next sections, we examine several organic reactions and provide some biochemical examples. More examples of biochemical reactions occur throughout the book as we examine biological molecules.

TABLE 5.5 Differences in Notation in Chemical Equations

Type of Chemical Reaction	Notation of Reactants and Products	Use of "Yields" Arrow	Purpose of Reaction Notation
General reaction	Chemical formula only	Indicate reversibility and conditions	Determine amounts (moles) required
Organic reaction	Drawn structure of organic molecule of interest	Indicate reversibility, small reactants, and conditions	Demonstrate change in a functional group
Biochemical reaction	Drawn structure of organic molecule of interest	Indicate reversibility, enzyme, and any common reactions coupled to reaction of interest, for example, $ATP \longrightarrow ADP$	Demonstrate change in a functional group and possibly energy-coupled reactions

5.15 Categorize the following reactions as synthesis, decomposition, or exchange reactions in the forward direction:
a. $CuO(s) + 2HCl(aq) \longrightarrow CuCl_2(aq) + H_2O(l)$
b. $C_6H_{12}O_6(aq) \longrightarrow 2C_2H_6O(aq) + 2CO_2(g)$
c. $2H_2(g) + O_2(g) \rightleftharpoons 2H_2O(g)$

5.16 Categorize the following reactions as synthesis, decomposition, or exchange reactions in the forward direction:
a. $N_2(g) + 3H_2(g) \rightleftharpoons 2NH_3(g)$
b. $CH_4(g) + 2O_2(g) \longrightarrow CO_2(g) + 2H_2O(g)$
c. $Al_2(SO_4)_3(aq) + 6KOH(aq) \longrightarrow 2Al(OH)_3(s) + 3K_2SO_4(aq)$

5.17 Determine if the reactions in Problem 5.15 are reversible or irreversible.

5.18 Determine if the reactions in Problem 5.16 are reversible or irreversible.

5.19 Write the products and balance the following reaction for the complete combustion of ethane, C_2H_6.
$$C_2H_6(g) + O_2(g) \longrightarrow ?$$

5.20 Write the products and balance the following reaction for the complete combustion of butane, C_4H_{10}.
$$C_4H_{10}(g) + O_2(g) \longrightarrow ?$$

5.21 List the similarities between chemical equations used in organic chemistry and chemical equations used in biochemistry.

5.22 List the differences between general chemical equations and organic chemical equations.

Discovering the Concepts

? Inquiry Activity—Oxidation–Reduction Reactions

Information
Oxidation and reduction reactions always occur together. One substance is oxidized at the same time that the other substance is reduced in the reaction. Oxidation–reduction reactions, also called *redox* reactions, can be identified in one of two ways: (1) they contain metal ions and involve the exchange of electrons (inorganic) or (2) they can involve the addition or removal of oxygen or hydrogen (organic). Sometimes, the reactants can be mixed (a metal and an organic molecule).

TABLE 1. Characteristics of Oxidation and Reduction (Redox) Reactions

Oxidation	
Always Involves	**May Involve**
Loss of electrons	Addition of oxygen
	Loss of hydrogen

Reduction	
Always Involves	**May Involve**
Gain of electrons	Loss of oxygen
	Gain of hydrogen

Questions
Sample Reaction 1: Inorganic, Redox Reaction Containing Metals

$$4Fe(s) + 3O_2(g) \longrightarrow 2Fe_2O_3(s)$$
$$\text{Rust}$$

1. Based on your previous knowledge of ionic compounds, assign charges to Fe(s), $O_2(g)$, and the cation and anion in $Fe_2O_3(s)$.
2. Complete the following table based on Table 1 and your answers to Question 1.

Substance	Did reactant gain or lose electrons to form product?	Oxidation or reduction?
Fe(s) forms iron ion		
$O_2(g)$ forms oxide		

Sample Reaction 2: Biological Redox Reaction

$$CH_3CH_2OH \quad + \quad NAD^+ \quad \xrightarrow[\text{Enzyme}]{} \quad CH_3\overset{\overset{\displaystyle O}{\|}}{C}H \quad + \quad NADH \quad + \quad H^+$$

3. Complete the following table based on Table 1 and Sample Reaction 2.

Substance	Did the reactant gain O (or lose H) or gain H (or lose O) to form product?	Oxidation or Reduction?
CH_3CH_2OH		
NAD^+		

Sample Reaction 3: Organic and Inorganic Mixed

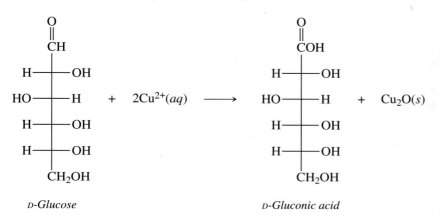

D-Glucose   D-Gluconic acid

4. Complete the following table based on Table 1 and Sample Reaction 3. *Note that the "always involves" column in Table 1 takes precedence over the "may involve" column.*

Substance	Oxidation or Reduction?
D-Glucose forming D-gluconic acid	
Cu^{2+} forming Cu_2O	

5. In each of the following reactions, determine which reactant is undergoing oxidation and which is undergoing reduction.

Reaction	Oxidation	Reduction
Formation of Salt: $2Na(s) + Cl_2(g) \longrightarrow 2NaCl(s)$		
Burning of Coal: $C(s) + O_2(g) \longrightarrow CO_2(g)$		
Complete Combustion of Glucose: $C_6H_{12}O_6 + 6O_2 \rightarrow 6CO_2 + 6H_2O$ D-glucose		

Reaction	Oxidation	Reduction

Reaction 8 of the Citric Acid Cycle:

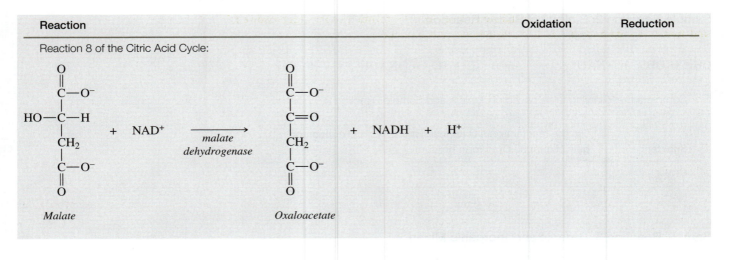

Malate + NAD$^+$ $\xrightarrow[\text{dehydrogenase}]{\text{malate}}$ *Oxaloacetate* + NADH + H$^+$

5.4 Oxidation and Reduction

5.4 Inquiry Question:
How are oxidation and reduction reactions identified?

Two chemical reactions called *oxidation* and *reduction* are essential to energy production and transfer in living systems. Oxidation–reduction or *redox* reactions can be identified in one of two ways:

(1) they contain metal ions and involve the movement of electrons (inorganic) or
(2) they can involve the addition or removal of oxygen and hydrogen (organic).

Both inorganic and organic redox reactions are present in biological systems, so we examine both here.

Inorganic Oxidation and Reduction

Rusting is caused by the oxidation of iron metal atoms to iron(III) ions in the presence of oxygen. The net reaction is shown.

$$4Fe(s) + 3O_2(g) \longrightarrow 2Fe_2O_3(s)$$
$$\text{Rust}$$

Let's examine the reactants and product more closely to see how the reactants (solid iron and oxygen) changed into products during this reaction.

The iron atoms (no charge) in the reactant formed the Fe^{3+} ions of the product. The iron atoms *lost electrons*. Atoms that lose electrons during a chemical reaction are said to undergo **oxidation.** Where did the electrons go? In this case, the O$_2$ (no charge) in the reactant formed the O^{2-} anions of the product Fe$_2$O$_3$. The reactant oxygen atoms *gained electrons*. Atoms that gain electrons during a chemical reaction are said to undergo **reduction.**

In other words, while the iron was being oxidized, it was reducing the oxygen. In this sense, iron is the **reducing agent** (undergoing oxidation itself), and oxygen is the **oxidizing agent** (undergoing reduction).

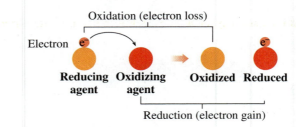

In oxidation–reduction reactions, metal atoms lose electrons forming cations (get oxidized) and nonmetals gain electrons forming anions (get reduced). Oxidation and reduction reactions are always coupled to each other because if a reactant is oxidized, losing one or more electrons, another reactant is simultaneously reduced, gaining one or more of those electrons. A mnemonic device that is helpful for remembering what is happening to the electrons in a redox reaction is to remember the letters in the words "OIL RIG," which stand for <u>O</u>xidation <u>I</u>s <u>L</u>oss (of electrons), <u>R</u>eduction <u>I</u>s <u>G</u>ain (of electrons).

In biological systems, cells oxidize and reduce metals, too. For example, a protein called cytochrome *c* plays an important role in the electron transport chain in the cell during ATP production. This protein contains an Fe^{2+} that undergoes oxidation to Fe^{3+} followed by reduction back to Fe^{2+} as it transports single electrons through the mitochondrial membrane. The iron ion is held in place on the protein by nitrogen atoms on a small organic molecule called a *heme* that is associated with the protein, as shown in **Figure 5.5**.

O	I	L		R	I	G
x	s	o		e	s	a
i		s		d		i
d		s		u		n
a				c		
t				t		
i				i		
o				o		
n				n		

Mnemonic (mind-jogging) device for remembering oxidation–reduction

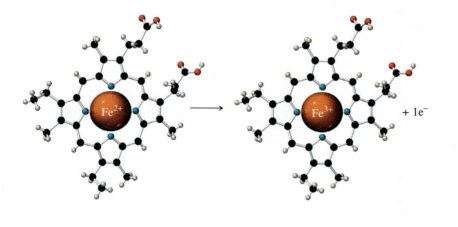

FIGURE 5.5 Biochemical redox of a metal ion. The iron (shown in orange) of cytochrome *c* is surrounded by an organic heme group (rest of structure). The iron is oxidized to Fe^{3+} when an electron is lost from Fe^{2+}.

sample problem 5.7   Inorganic Redox

In the following reaction, determine which of the reactants is undergoing oxidation and which is undergoing reduction.

$$Ca(s) + S(s) \longrightarrow CaS(s)$$

Solution

In this reaction, the ionic compound calcium sulfide is formed. The calcium atom (no charge) formed the cation, Ca^{2+}. The calcium atom lost electrons and underwent oxidation. The sulfur atom gained electrons to form the anion sulfide, S^{2-}. The sulfur underwent reduction.

Organic Oxidation and Reduction

In organic redox reactions, it is easier to see how functional groups change by looking at hydrogen and oxygen rather than to see where electrons are being gained or lost. An organic molecule is oxidized if it gains oxygen *or* loses hydrogen and is reduced if it gains hydrogen *or* loses oxygen (see **Table 5.6**). One organic group that readily undergoes oxidation and reduction is the carbonyl, $\overset{\text{O}}{\underset{\|}{\text{L}\quad\text{J}}}$. Carbonyls in functional groups like aldehydes can be reduced to alcohols and can be oxidized to carboxylic acids.

TABLE 5.6 Characteristics of Oxidation and Reduction Reactions

	Oxidation	
Always Involves	**May Involve**	
Loss of electrons	Addition of oxygen Loss of hydrogen	
	Reduction	
Always Involves	**May Involve**	
Gain of electrons	Loss of oxygen Gain of hydrogen	

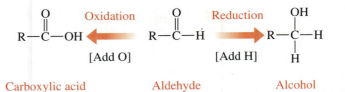

Aldehydes can be reduced to alcohols and oxidized to carboxylic acids. Notice that only the major organic product is shown.

As we move through the textbook, we will see examples of oxidation and reduction in reducing sugars (Chapter 6) and metabolic reactions (Chapter 12).

---

sample problem 5.8   **Organic Redox**

In the following reaction, determine if the organic reactant (benzoic acid) is undergoing oxidation or reduction.

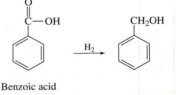

Benzoic acid

Solution

The carboxylic acid in the reactant is changed to an alcohol in the product. It lost one oxygen and gained two hydrogen atoms, so the carboxylic acid in benzoic acid underwent reduction.

Oxidation in Cells

We fuel our bodies with nutrients that are broken down through oxidation. In the body, the process of combustion (forming CO_2 and H_2O from reduced carbon) does not occur in one chemical reaction but through many steps in a chemical pathway, transferring energy as it goes. For example, one molecule of glucose ($C_6H_{12}O_6$) undergoes complete combustion in the body, but this process takes several steps and metabolic pathways to accomplish (see Chapter 12). We can write the overall reaction as

$$C_6H_{12}O_6(s) + 6O_2(g) \longrightarrow 6CO_2(g) + 6H_2O(l)$$

Collectively, the series of reactions through which glucose is combusted yielding carbon dioxide, water, and energy is a type of **cellular respiration.** We breathe in oxygen and exhale carbon dioxide as a product.

In addition to breaking down nutrients and producing energy, the body performs many other organic oxidation–reduction reactions. For example, when we metabolize alcohol (ethanol) from alcoholic beverages, an oxidation–reduction reaction occurs. An enzyme called liver alcohol dehydrogenase (LAD) catalyzes the reaction. Ethanol is oxidized to the aldehyde, ethanal, also called acetaldehyde. What is reduced? We noted earlier that oxidation–reduction reactions are always coupled, so if ethanol is oxidized, another substance is reduced. Remember that we also said that most biochemical reactions are coupled so that the energy is transferred efficiently. In the case of ethanol oxidation, the molecule undergoing reduction is nicotinamide adenine dinucleotide abbreviated NAD^+.

NAD^+ and flavin adenine dinucleotide (FAD) are two important molecules used in metabolic oxidation–reduction reactions. Each of these has an oxidized and reduced form and participates in many biological redox reactions (see **Table 5.7**) as energy transfer molecules. Their structures are shown in Chapter 12. Because most of the molecule stays the same, it is useful to use its acronym in a reaction.

The coupled redox reaction for the biological oxidation of ethanol is shown at left.

Reduction

CH_3CH_2OH $\xrightarrow[LAD]{NAD^+ \ NADH}$ H_3CCH (with O double-bonded)

Oxidation

NAD^+ undergoes reduction to NADH while ethanol is oxidized to ethanal.

TABLE 5.7 Common Biological Redox Molecules and Their Abbreviations

Name	Oxidized form	Reduced form
Nicotinamide adenine dinucleotide	NAD^+	NADH
Flavin adenine dinucleotide	FAD	FADH

5.23 Are the substances shown in italics undergoing oxidation or reduction?

a. *Copper(II) ions* in solution form solid copper metal on a surface.

b. *Aluminum* metal forms aluminum oxide (Al_2O_3).

c. *Wine* (containing ethanol, CH_3CH_2OH) sours to vinegar (CH_3COOH).

5.24 Are the substances shown in italics undergoing oxidation or reduction?

a. *Silver ions (Ag^+)* are electroplated into silver atoms on flatware.

b. *D-glucose* reacts with hydrogen to form the sugar alcohol D-sorbitol.

c. The biomolecule *FADH$_2$* loses hydrogen becoming FAD.

5.25 The reaction shown occurs in fuel cells and produces great quantities of energy. Identify the reactant that is oxidized and the reactant that is reduced.

$$H_2(g) + O_2(g) \longrightarrow H_2O(g)$$

5.26 Although we mine most sodium chloride, it can be synthesized by the reaction shown. Identify the substance oxidized and the substance reduced.

$$2Na(s) + Cl_2(g) \longrightarrow 2NaCl(s)$$

5.5 Organic Reactions: Condensation and Hydrolysis

Water forming on the outside of a glass of ice water displays the physical process called condensation. Similarly, in a **condensation** reaction, water is produced as two organic molecules are joined. Because water is produced, this type of reaction can also be called a **dehydration** reaction; however, we will use the more general term *condensation* in this textbook. Condensation reactions can occur among a number of functional groups that contain an —H in a polar bond (like O—H or N—H) and an —OH group that can be removed to form water.

A condensation reaction operating in reverse is a **hydrolysis** reaction (*hydro* means "water", *lysis* means "break"), where water is consumed as a reactant and the reactant molecule is split into two smaller molecules.

? 5.5 Inquiry Question: What are the characteristics of the organic reactions called condensation and hydrolysis?

Are condensation and hydrolysis reactions synthesis, decomposition, or exchange reactions?

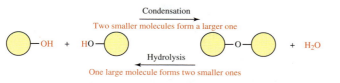

Condensation
Two smaller molecules form a larger one
—OH + HO— → —O— + H$_2$O
Hydrolysis
One large molecule forms two smaller ones

Condensation and hydrolysis reactions are common biochemical reactions. The energy molecule ATP is hydrolyzed to ADP through a hydrolysis.

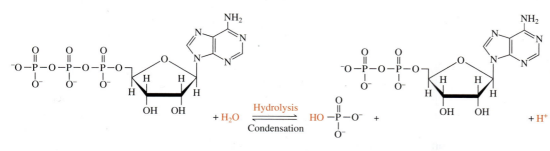

Adenosine triphosphate (ATP)

Adenosine diphosphate (ADP) + Energy

The formation of ADP from ATP is a biochemical hydrolysis.

Because ATP is so large and is often shown in biochemical reactions only to indicate energy transfer, the reaction is often written symbolically as

$$\text{ATP} \underset{\text{Condensation reaction}}{\overset{\text{Hydrolysis reaction}}{\rightleftharpoons}} \text{ADP} + P_i$$

where P_i symbolizes the hydrogen phosphate anion, HPO_4^{2-}.

Many functional groups are added and removed from molecules through condensation and hydrolysis reactions, respectively. Two common examples are the carboxyl (or carboxylic acid) group and the phosphate group.

Carboxylation reactions are condensation reactions. They involve, just as it sounds, the addition of a carboxyl, or carboxylic acid, group. As carbon dioxide moves through the cell, it is added to and removed from small molecules by two enzymes called carboxylase and decarboxylase, respectively. As an example, the small molecule pyruvate gets carboxylated to oxaloacetate, one of the steps in the pathway called *gluconeogenesis* ("the creation of new glucose") in the cell. The energy for this reaction is provided by the hydrolysis of ATP.

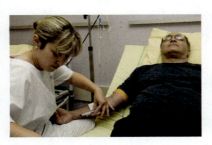

FIGURE 5.6 Enzymes in diagnosing disease. The enzyme alkaline phosphatase (ALP) can be screened in a routine blood test. Patients suspected of bone or liver disease have higher than normal values for this enzyme.

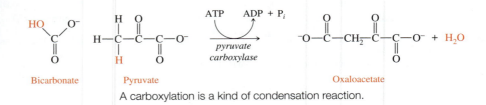

Bicarbonate   Pyruvate   Oxaloacetate

A carboxylation is a kind of condensation reaction.

The functional group, phosphate, symbolized as P_i, also moves around the cell through ATP. The addition and removal of phosphate from a molecule is one way cells regulate chemical pathways. These reactions are called **phosphorylation** and **dephosphorylation** and are condensation and hydrolysis reactions, respectively. The enzymes that catalyze the condensation are called phosphorylases, and the enzymes that hydrolyze phosphate are called phosphatases. One important phosphatase, alkaline phosphatase (ALP) is routinely screened in patients susceptible to liver or bone disease. When these cells are damaged, ALP levels will rise in the bloodstream. This enzyme catalyzes phosphate hydrolysis.

$$H_2O + R-O-\overset{\overset{O}{\|}}{\underset{\underset{O^-}{|}}{P}}-O^- \xrightarrow[\text{phosphatase}]{\text{alkaline}} ROH + HO-\overset{\overset{O}{\|}}{\underset{\underset{O^-}{|}}{P}}-O^-$$

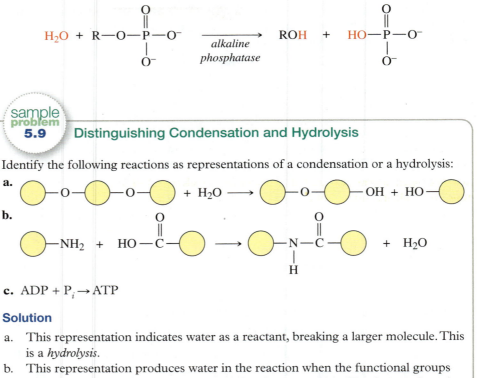

sample problem 5.9   Distinguishing Condensation and Hydrolysis

Identify the following reactions as representations of a condensation or a hydrolysis:

a.

b.

c. $ADP + P_i \rightarrow ATP$

Solution

a. This representation indicates water as a reactant, breaking a larger molecule. This is a *hydrolysis*.

b. This representation produces water in the reaction when the functional groups amine and carboxylic acid are combined to form an amide. This is a *condensation*.

c. This reaction connects ADP and a phosphate, P_i, the reverse reaction of ATP hydrolysis. This is a *condensation*.

5.27 Draw the products for the ester formed through the condensation reaction shown.

$$H_3C-\overset{\overset{\displaystyle O}{\|}}{C}-OH \ + \ HOCH_2CH_3 \ \xrightarrow[\text{reaction}]{\textit{Condensation}}$$

5.28 Draw the products formed from the ester hydrolysis reaction shown.

$$H_3CO-\overset{\overset{\displaystyle O}{\|}}{C}-CH_2CH_3 \ \xrightarrow[\textit{Hydrolysis}]{H_2O}$$

practice problems

5.6 Organic Addition Reactions to Alkenes

So far we have seen exchange reactions that transfer electrons (oxidation–reduction), synthesis reactions that combine smaller organic molecules (condensation), and decomposition reactions that break larger molecules apart (hydrolysis). Often, the name of the reaction can tell us something about what the reaction does. In an **addition** reaction to an alkene, an atom or group of atoms is *added* to a double bond in an organic molecule.

5.6 Inquiry Question:
How are small molecules added to alkenes?

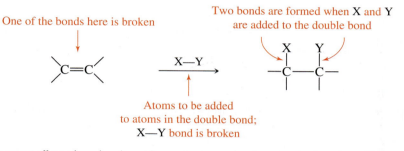

One of the bonds here is broken

Two bonds are formed when X and Y are added to the double bond

Atoms to be added to atoms in the double bond; X—Y bond is broken

In an addition reaction, the second bond in a double bond and the bond between the atoms added are broken and two single bonds are formed. Atoms are added to the carbons that were in the double bond. Are additions synthesis or decomposition?

Addition to an alkene is a simple yet important reaction in organic chemistry. Two addition reactions that commonly occur in biological molecules are discussed here: hydrogenation (hydrogen is added) and hydration (water is added).

Hydrogenation

If you read labels on processed foods, you may see partially hydrogenated soybean oil as an ingredient. This is soybean oil that has undergone a hydrogenation reaction. Oils derived from plants have long been considered a healthier alternative for human dietary requirements. They have a higher percent of unsaturated fatty acids that contain carbon–carbon double bonds (alkene functional groups). During a hydrogenation reaction, ==two hydrogen atoms are added, converting an alkene double bond to an alkane single bond.== A catalyst such as platinum (Pt), nickel (Ni), or palladium (Pd) is used.

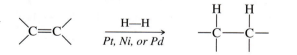

In a fatty acid, one of the components in oils, a complete hydrogenation changes an unsaturated fatty acid into a saturated one. As we will see in Chapter 7, this changes the physical properties of an oil or fat. Remember from Chapter 4 that fatty acids also contain a carboxylic acid functional group. The carbonyl double bond is not affected in a hydrogenation.

Because the reaction is difficult to control, hydrogenation of fatty acids often does not affect every alkene double bond. Some of the double bonds that are not completely hydrogenated will re-form, but when they do, they switch from the cis form to the

Palmitoleic acid, an unsaturated fatty acid

H_2 | platinum catalyst

Skeletal structure usually written as

Palmitic acid, a saturated fatty acid

Hydrogenation of an unsaturated fatty acid.

FIGURE 5.7 Partially hydrogenated soybean oil is found in many processed foods. The fatty acids in soybean oil undergo partial hydrogenation forming trans fats.

energetically more stable trans form, resulting in compounds known as *trans fats*. The reaction is called *partial* hydrogenation. Such partially hydrogenated oils are found in processed foods (see **Figure 5.7**).

Some studies have shown that trans fats have deleterious health effects. This has forced many manufacturers of partially hydrogenated oils to consider alternatives. Today, all nutritional labeling must give the amounts of trans fat present in food.

sample problem 5.10 **Predicting Products of Hydrogenation**

Provide the products of the complete hydrogenation for each of the reactions shown,

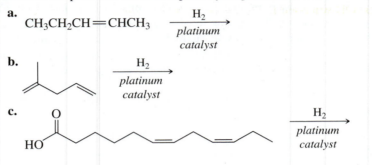

Solution

In each case, if the alkenes are completely hydrogenated, they will form saturated hydrocarbons, or alkanes. The products are

a. $CH_3CH_2CH_2CH_2CH_3$

b.

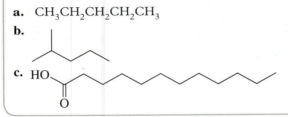

c. HO

Hydration

Hydration is the addition of water to the double bond in an alkene. The water gets added as —H and —OH. This reaction often requires acid as a catalyst.

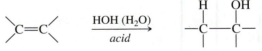

Hydration reactions occur in biochemistry, too. One example occurs at Reaction 7 in the citric acid cycle (also called the Krebs cycle, see Chapter 12). The reaction of fumarate to malate is a hydration reaction. This reaction is catalyzed by the enzyme fumarase.

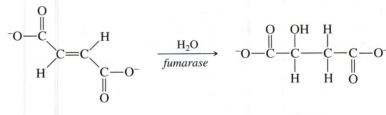

Fumarate   Malate

In contrast to hydrogenation, this reaction has two *different* groups attaching to the carbons that were in the double bond. What products might be produced when the alkene does not have the same groups attached to the double bond?

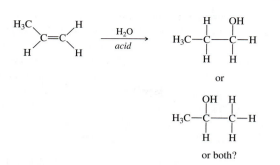

or both?

If the alkene double bond is asymmetric—that is, it does not have the same substituents on the carbons at the double bond—the H will usually bond to the carbon with more hydrogen atoms. The OH will bond to the carbon with more carbon groups attached. A Russian chemist named Vladimir Markovnikov was the first chemist to notice this in addition reactions of alkenes. This observation is called **Markovnikov's rule.**

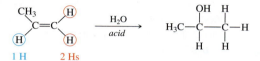

1 H   2 Hs

5.11 **Predicting Products of Hydration**

Provide the products for the hydration of each of the alkenes shown.

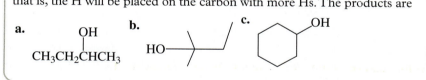

Solution

Remember that if the carbons in the alkene are unequally substituted, Markovnikov's rule applies; that is, the H will be placed on the carbon with more Hs. The products are

a.

$CH_3CH_2\overset{\displaystyle OH}{\underset{\displaystyle |}{C}}HCH_3$

b.

HO——

c.

OH

5.29 Write the products for the following hydrogenation reactions:

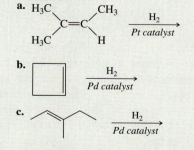

5.30 Write the products for the following hydrogenation reactions:

a.

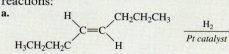

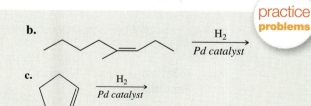

5.31 Write the products of hydration for the alkenes shown in Problem 5.29.

5.32 Write the products of hydration for the alkenes shown in Problem 5.30.

SUMMARY

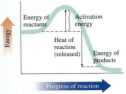

5.1 Thermodynamics

5.1 Inquiry Question: How are energy and heat transferred and measured in a chemical reaction?

Chemical reactions involve energy exchange. Reactions that give off heat are exothermic, and those that absorb heat are endothermic. For a chemical reaction to occur, the reactants must collide with each other with enough energy to react. This initial energy required for a reaction to occur is called the activation energy. The thermodynamics of a chemical reaction can be represented on a reaction energy diagram showing the energy of the reactants, products, the activation energy, and the free energy change, $\triangle G$. If $\triangle G$ is a negative value, a given chemical reaction is spontaneous, and if $\triangle G$ is positive, the reaction is nonspontaneous. We can measure the heat given off in a chemical reaction by calorimetry. This method is routinely done to determine the number of joules in a chemical reaction and calories (Cal) in food.

5.2 Chemical Reactions: Kinetics

5.2 Inquiry Question: How is the rate of a chemical reaction controlled?

Chemical reactions occur when reactants collide. Several factors control the rate (how fast reactants form products) of a chemical reaction. Increasing the rate of collision, or collision frequency, of chemical reactants increases the rate of the reaction. The more reactants present, the higher the likelihood of collision, so increasing the amount of reactants increases the rate. Temperature increases the rate by increasing the kinetic energy of the reactants. Because the reactants are moving more quickly, they are more likely to collide with enough energy to react. Catalysts also speed up a chemical reaction. Catalysts participate in a chemical reaction but remain unchanged at its completion. Catalysts increase the rate by lowering the activation energy. Enzymes are biological catalysts. Enzymes lower the activation energy by providing a site, called the active site, where the reactants are close and in the correct orientation to react instead of being positioned randomly in a reaction mixture.

5.3 Overview of Chemical Reactions

5.3 Inquiry Question: What are some general categories of chemical reactions?

There are three basic types of chemical reactions: synthesis, decomposition, and exchange. Depending upon the amount of energy released during a chemical reaction, these reactions can be further classified as either reversible or irreversible. Reversible reactions can reach a point called *chemical equilibrium* where the rates of the forward and reverse reactions are constant and no net products are formed in either direction. Organic hydrocarbons like alkanes can undergo combustion (reaction with oxygen, O_2) to form the products carbon dioxide (CO_2) and water (H_2O). Chemical equations are written for different purposes. General reactions are written to determine amounts, so balancing them is important. Organic reactions are usually written to show how functional groups change during the reaction, so their structures are written out. Biochemical reactions show organic structures and also can show energy coupling and enzymes used to catalyze the reactions.

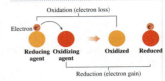

5.4 Oxidation and Reduction

5.4 Inquiry Question: How are oxidation and reduction reactions identified?

Oxidation and reduction (redox) reactions always occur simultaneously. Oxidation always involves a loss of electrons and may involve the addition of oxygen or removal of hydrogen. Reduction always involves a gain of electrons and may involve the addition of hydrogen or a removal of oxygen. If one substance in a reaction is oxidized, another substance in the reaction is reduced. If metals are involved, it is easier to determine which substance is oxidized or reduced by determining the charge on the metal. In organic reactions, it is often easier to look for the movement of oxygen and hydrogen. Combustion (reaction with oxygen) is an example of a redox reaction.

5.5 Organic Reactions: Condensation and Hydrolysis

5.5 Inquiry Question: What are the characteristics of the organic reactions called condensation and hydrolysis?

Organic condensation and hydrolysis reactions occur in opposite directions. A condensation reaction joins molecules and often produces water. Hydrolysis reactions break molecules and water is a reactant. Carboxylations and phosphorylations are examples of biological condensation reactions. Dephosphorylation is an example of a biological hydrolysis reaction.

5.6 Organic Addition Reactions to Alkenes

5.6 Inquiry Question: How are small molecules added to alkenes?

The addition reactions hydrogenation and hydration in alkenes are also very common reactions in biological molecules. In addition reactions, two atoms or groups of atoms are added to the alkene double bond, forming a carbon–carbon single bond. In a complete hydrogenation, one hydrogen atom is added to each carbon of the double bond. In hydration, an H and an OH are added to the carbons. In a hydration of an alkene with different groups attached to the alkene, the hydrogen will bond to the carbon in the double bond with more hydrogens bonded directly to the alkene. This demonstrates Markovnikov's rule.

The study guide will help you check your understanding of the main concepts in Chapter 5. You should be able to

5.1 Thermodynamics

- Draw reaction energy diagrams for exergonic and endergonic reactions.
- Predict spontaneity of a reaction based on the $\triangle G$ value.
- Describe how a calorimeter works.
- Calculate the energy content in foods from its nutrient molecules.

5.2 Chemical Reactions: Kinetics

- Predict relative activation energies and speed of reactions using a reaction energy diagram.
- Determine the effect that temperature, amount of reactants, and a catalyst have on the rate of a reaction.
- Describe how an enzyme catalyzes a biochemical reaction.

5.3 Overview of Chemical Reactions

- Classify reactions as synthesis, decomposition, or exchange reactions.
- Distinguish reversible and irreversible reactions.
- Predict the products and balance the chemical equation for a hydrocarbon undergoing combustion.
- Contrast a general chemical equation and an organic chemical equation.

5.4 Oxidation and Reduction

- Identify the substance oxidized and the substance reduced in an inorganic oxidation–reduction reaction.
- Identify the substance oxidized and the substance reduced in an organic oxidation–reduction reaction.

5.5 Organic Reactions: Condensation and Hydrolysis

- Predict the products of an organic condensation reaction.
- Predict the products of an organic hydrolysis reaction.

5.6 Organic Addition Reactions to Alkenes

- Predict the products of a hydrogenation reaction, an addition reaction of an alkene.
- Predict the products of a hydration reaction, an addition reaction of an alkene.

Key Terms

activation energy—The energy in a chemical reaction necessary for the reactants to collide with enough energy to form products.

active site—The location on an enzyme where catalysis occurs.

addition to alkenes—A chemical reaction where atoms or groups of atoms are added to a double bond (alkene). Some important biochemical addition reactions include hydration (add H_2O) and hydrogenation (add H_2).

biochemical reactions—Chemical reactions that occur in living systems.

carboxylation—A type of condensation reaction where a carboxyl group is added to another molecule.

catalyst—A substance that speeds up a chemical reaction by lowering the activation energy of the reaction. A catalyst participates in a chemical reaction but is unchanged at its completion.

cellular respiration—The oxidation of organic substances through a series of chemical reactions in the cell, ultimately producing carbon dioxide, water, and energy.

chemical equilibrium—A state in a reversible chemical reaction where both the forward and reverse reactions are occurring at the same rate.

combustion—An exothermic chemical reaction reacting O_2 with a hydrocarbon. Complete combustion produces carbon dioxide and water.

condensation—A synthesis reaction involving the combination of two organic molecules. A small molecule such as H_2O is also produced. The reverse reaction is hydrolysis.

decomposition reaction—A type of reaction where one reactant breaks into two or more smaller substances.

dehydration—A condensation reaction in which H_2O is produced.

endergonic reaction— A reaction that requires energy.

endothermic reaction— A reaction that absorbs heat from its surroundings.

enzyme—A biological catalyst.

exchange reaction—A type of reaction where two reactants exchange one or more parts forming different product substances.

exergonic reaction—A reaction that gives off energy.

exothermic reaction—A reaction giving off heat.

free energy (G)—The amount of energy available to do work. In a chemical reaction, the free energy change, $\triangle G$, is the difference between the amount of energy in the products and the amount of energy in the reactants. A negative value is indicative of a spontaneous reaction.

glycolysis—A series of chemical reactions in the body that break down glucose, producing energy.

hydration—An organic addition reaction where H and OH are added to the carbons in a carbon–carbon double bond.

heat of reaction— The energy difference between the products and reactants in a chemical reaction.

hydrogenation—An organic addition reaction where H is added to each carbon in a carbon–carbon double bond.

hydrolysis—A decomposition reaction involving the breaking of one large organic molecule into two smaller molecules. H_2O is also a reactant. The reverse reaction is condensation.

irreversible reaction—A chemical reaction where products cannot revert back to reactants.

Markovnikov's rule—In the addition of H and OH to an asymmetric alkene, the hydrogen will add to the carbon with more hydrogen directly attached to it.

nonspontaneous process—A process that does not occur naturally and often requires energy input.

oxidation—The loss of electrons during a chemical reaction; in organic reactions, often appears as a gain of oxygen or a loss of hydrogen.

oxidizing agent—Responsible for oxidizing another reactant, the reactant itself undergoes reduction.

phosphorylation—A condensation reaction where a phosphate (P_i) is bonded to another molecule.

rate of reaction—A measure of how much product is formed (or reactant used up) in a certain period of time in a chemical reaction.

reaction kinetics—The study of the relationship between a chemical reaction and its rate.

reaction mechanism—The stepwise description of how a chemical reactant transforms into products.

reducing agent—Responsible for reducing another reactant, the reactant itself undergoes oxidation.

reduction—The gain of electrons during a chemical reaction; in organic reactions often appears as a gain of hydrogen or a loss of oxygen.

reversible reaction—A chemical reaction where products can revert back to reactants and equilibrium is established.

spontaneous process—A process that will continue to occur once started.

synthesis reaction—A type of reaction where reactants combine to form one product.

thermodynamics—The study of the relationship between the heat produced or consumed in a chemical reaction and the associated energy.

Summary of Reactions

Types of Chemical Reactions

Reaction Type	General Reaction Scheme
Synthesis	A + B $\longrightarrow$ AB
Decomposition	AB $\longrightarrow$ A + B
Exchange	AB + C $\longrightarrow$ AC + B (single) AB + CD $\longrightarrow$ AD + CB (double)

Combustion

$$C_xH_y + O_2 \longrightarrow CO_2 + H_2O$$

Condensation/Hydrolysis

Reactions are the reverse of each other.

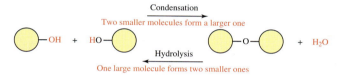

Oxidation–Reduction

Reactions occur simultaneously.

Oxidation	
Always Involves	**May Involve**
Loss of electrons	Addition of oxygen Loss of hydrogen

Reduction	
Always Involves	**May Involve**
Gain of electrons	Loss of oxygen Gain of hydrogen

Addition Reactions to Alkenes

Hydrogenation

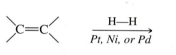

Hydration

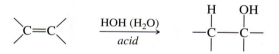

Additional Problems

5.33 Methane (a.k.a. natural gas) can react with oxygen in a combustion reaction as shown:

$$CH_4(g) + 2O_2(g) \rightarrow 2H_2O(g) + CO_2(g) + heat$$

a. Is the reaction exothermic or endothermic?

b. Considering that heat is a form of energy, predict whether the reactants or products have more free energy.

c. Is the reaction spontaneous?

d. Sketch a reaction energy diagram for this reaction. Label the axes, the energy of reactants, the energy of products, and ΔG.

5.34 In the following reaction energy diagram, label 1, 2, 3, and 4 as one of the following: (a) energy of reactants; (b) energy of products; (c) activation energy; or (d) free energy change.

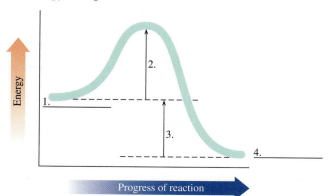

5.35 Which reaction occurs at a faster rate, an exergonic reaction with a low activation energy or an endergonic reaction with a high activation energy? Explain.

5.36 What is measured by the free energy change?

5.37 Use the following reaction energy diagram to answer the questions:

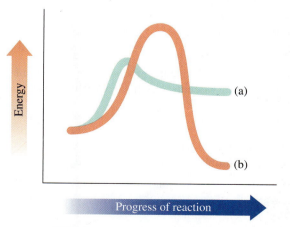

a. Which curve represents the faster reaction, and which represents the slower?

b. Which curve represents an endergonic reaction, and which curve represents an exergonic reaction?

c. Which curve represents a spontaneous reaction?

d. Which reaction has a positive ΔG value?

5.38 Two curves for the same reaction are shown in the following reaction energy diagram. Which curve represents the uncatalyzed reaction, and which curve represents the reaction with catalyst present?

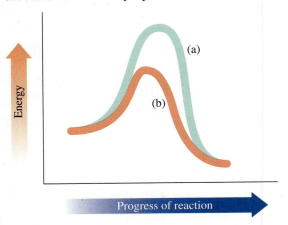

5.39 Draw and label a reaction energy diagram for an endergonic reaction in which the activation energy is two times greater than its free energy change (ΔG).

5.40 Draw and label a reaction energy diagram for an exergonic reaction in which the free energy change (ΔG) is three times greater than its activation energy.

5.41 Calculate the number of Calories present in a Burger King Whopper with Cheese® that contains 53 g carbohydrate, 35 g of protein, and 48 g of fat.

5.42 One gram of alcohol provides 7 Calories. Calculate the number of Calories in a 5 oz glass of red wine that contains 2.5 g carbohydrate and 13.7 g alcohol.

5.43 Try your hand at predicting the products that would result from the following reactions, and balance each equation:
  a. synthesis: $Mg(s) + Cl_2(g) \longrightarrow$
  b. decomposition: $HI(g) \longrightarrow$
  c. exchange (single): $Ca(s) + Zn(NO_3)_2(aq) \longrightarrow$
  d. exchange (double): $K_2S(aq) + Pb(NO_3)_2(aq) \longrightarrow$

5.44 Try your hand at predicting the products that would result from the following reactions, and balance each equation:
  a. synthesis: $Mg(s) + O_2(g) \longrightarrow$
  b. decomposition: $PbO_2(s) \longrightarrow$
  c. exchange (single): $KI(s) + Br_2(g) \longrightarrow$
  d. exchange (double): $CuCl_2(aq) + Na_2S(aq)$

5.45 Write the products of the following reactions:

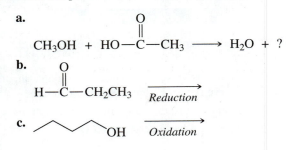

5.46 Write the products of the following reactions:

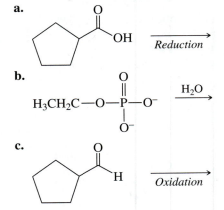

5.47 Identify the main organic reaction shown as either condensation, hydrolysis, oxidation, or reduction:

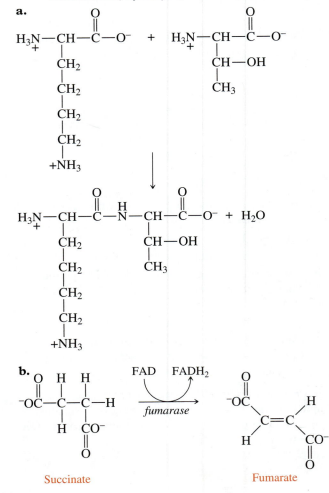

Succinate   Fumarate

5.48 Identify the main organic reaction shown as either condensation, hydrolysis, oxidation, or reduction:

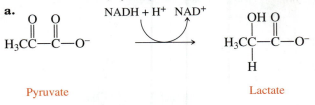

Pyruvate   Lactate

b.

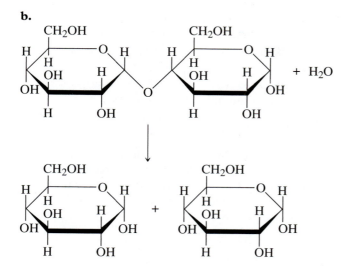

$$\downarrow$$

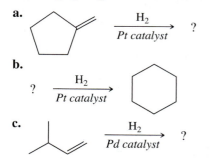

5.49 Provide the products and balance the following reaction below for the complete combustion of pentane, C_5H_{12}. Identify the reactant that is oxidized and the reactant that is reduced.

$$C_5H_{12}(g) + O_2(g) \longrightarrow ?$$

5.50 Provide the products and balance the following reaction below for the complete combustion of the fatty acid, octanoic acid, $C_8H_{16}O_2$. Identify the reactant that is oxidized and the reactant that is reduced.

$$C_8H_{16}O_2(s) + O_2(g) \longrightarrow ?$$

5.51 Identify the reactant that is oxidized and the reactant that is reduced in the following reactions:
a. $Cu(s) + AgNO_3(aq) \longrightarrow Cu(NO_3)_2(aq) + 2Ag(s)$
b. $4Al(s) + 3O_2(g) \longrightarrow 2Al_2O_3(s)$
c. $2AgBr(s) \longrightarrow 2Ag(s) + Br_2(g)$

5.52 Identify the reactant that is oxidized and the reactant that is reduced in the following reactions:
a. $Mg(s) + FeCl_2(aq) \longrightarrow MgCl_2(aq) + Fe(s)$
b. $2PbO(s) \longrightarrow 2Pb(s) + O_2(g)$
c. $2Li(s) + F_2(g) \longrightarrow LiF(s)$

5.53 Fill in the missing organic products or reactants for the following hydrogenation reactions:

a.

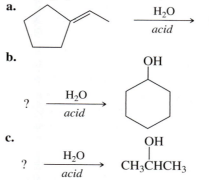

5.54 Fill in the missing organic products for the complete hydrogenation of the following:

a.

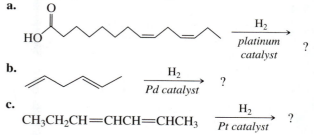

b.

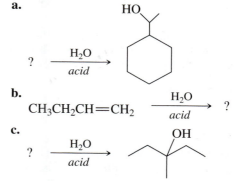

c.

CH_3CH_2CH=CHCH=CHCH_3 $\xrightarrow[\text{Pt catalyst}]{H_2}$?

5.55 Fill in the missing organic products or reactants for the following hydration reactions:

a.

$\xrightarrow[\text{acid}]{H_2O}$?

b.

? $\xrightarrow[\text{acid}]{H_2O}$ (cyclohexanol, OH)

c.

? $\xrightarrow[\text{acid}]{H_2O}$ OH / CH_3CHCH_3

5.56 Fill in the missing organic products or reactants for the following hydration reactions:

a.

? $\xrightarrow[\text{acid}]{H_2O}$ (HO cyclohexane)

b.

CH_3CH_2CH=CH_2 $\xrightarrow[\text{acid}]{H_2O}$?

c.

? $\xrightarrow[\text{acid}]{H_2O}$ OH (branched alcohol)

5.57 How do low-carb diets work? We store glucose molecules in our muscles and liver as glycogen, which consists of thousands of glucose molecules linked together. During periods of fasting, we can activate glycogen to provide glucose.

a. Determine which of the following reactions below would be a condensation and which would be a hydrolysis.

Thousands of glucose $\underset{?}{\overset{?}{\rightleftarrows}}$ Glycogen + H_2O

b. Individuals who do not eat carbohydrates do not store the same levels of glycogen as people who do. Explain the weight loss associated with storing less glycogen.

Challenge Problems

5.58 If you begin an exercise program and burn 5 Calories per minute by walking, how many hours of walking per day would you have to do in a two weeks to lose 5 pounds? You must burn an extra 3500 Calories to lose 1 pound of body weight.

5.59 Your daily latte contains 290 Calories. You want to lose weight by giving up your morning latte and drinking water instead. Assuming the rest of your diet remains constant, how many days will it take you to lose 5 pounds? You must burn an extra 3500 Calories to lose 1 pound of body weight.

5.60 Basal Metabolic Rate (BMR) is the amount of energy your body uses to function at rest in a day. This includes the energy required to keep the heart beating, the lungs breathing, the kidneys functioning, and the body temperature stabilized. Approximately 70 percent of an individual's energy expenditure goes into basal metabolism. The other 30 percent goes into digestion and absorption of nutrients (10%) and physical movement (about 20%).

a. Calculate your basal metabolic rate in Calories using the appropriate equation here:

Females: BMR = 655 + (9.6 × wt(kg)) + (1.8 × ht(cm)) − (4.7 × age(yrs))

Males: BMR = 66 + (13.7 × wt(kg)) + (5 × ht(cm)) − (6.8 × age(yrs))

b. Calculate how many Calories you burn overall in a day.

5.61 Several neurotransmitters are made through decarboxylation of amino acids. Serotonin, a neurotransmitter involved in mood response, is made in the brain when 5-hydroxytryptophan is decarboxylated. Provide the structure of serotonin.

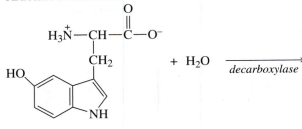

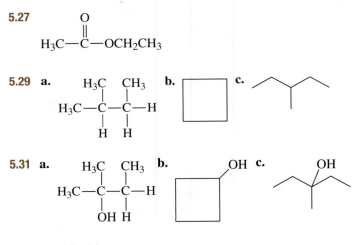

5-Hydroxytryptophan

Answers to Odd-Numbered Problems

Practice Problems

5.1 Cold. If the reaction is endothermic it absorbs heat from its surroundings which cools the reaction. This reaction occurs spontaneously, so it would have a $-\Delta G$.

5.3 **a.** exothermic **b.** endothermic

5.5 **a.** spontaneous **b.** spontaneous

5.7 135 Calories

5.9 **a.** Increasing the temperature increases the rate by increasing the movement of the reactants. The faster they move, the more likely they are to collide and react.
b. Increasing the amount of reactant introduces more reactant molecules, which increases the likelihood of collision and reacting.

5.11 The amount of reactants decreases as the reaction progresses, so the rate slows.

5.13 **a.** decrease **b.** increase **c.** increase

5.15 **a.** exchange **b.** decomposition **c.** synthesis

5.17 **a.** irreversible **b.** irreversible **c.** reversible

5.19 $2C_2H_6(g) + 7O_2(g) \rightarrow 4CO_2(g) + 6H_2O(g)$

5.21 In both general and organic reactions, the reactants appear on the left, the products appear on the right, and an arrow separates them.

5.23 **a.** reduction **b.** oxidation **c.** oxidation

5.25 Hydrogen is oxidized, and oxygen is reduced.

5.27

5.29 **a.** **b.** **c.**

5.31 **a.** **b.** **c.**

Additional Problems

5.33 **a.** exothermic **b.** reactants
c. yes, the reaction is spontaneous
d.

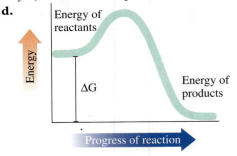

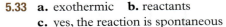

5.35 An exothermic reaction with a low activation energy occurs faster. The low activation energy allows the reactants to react more quickly.

5.37 **a.** Curve (a) is faster, and curve (b) is slower.
b. Curve (a) is endergonic, and curve (b) is exergonic.
c. Curve (b) is spontaneous.
d. Curve (a) has a positive ΔG value.

5.39

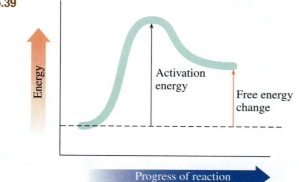

5.41 780 Calories

5.43 **a.** $Mg + Cl_2 \longrightarrow MgCl_2$
b. $2HI \longrightarrow H_2 + I_2$
c. $Ca + Zn(NO_3)_2 \longrightarrow Ca(NO_3)_2 + Zn$
d. $K_2S + Pb(NO_3)_2 \longrightarrow 2KNO_3 + PbS$

5.45 **a.** **b.** $HOCH_2CH_2CH_3$

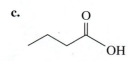

c.

5.47 **a.** condensation   **b.** oxidation

5.49 $C_5H_{12}(g) + 8O_2(g) \longrightarrow 5CO_2(g) + 6H_2O(g)$; pentane is oxidized and oxygen is reduced.

5.51 **a.** Copper is oxidized, and silver ion is reduced.
b. Aluminum is oxidized, and oxygen is reduced.
c. Bromide is oxidized, and silver ion is reduced.

5.53 **a.**   **b.**   **c.**

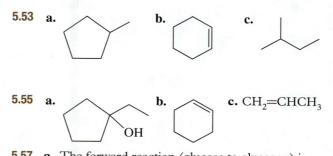

5.55 **a.**   **b.**   **c.** $CH_2{=}CHCH_3$

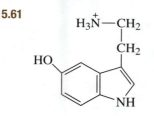

5.57 **a.** The forward reaction (glucose to glycogen) is condensation; the reverse reaction (glycogen to glucose) is hydrolysis.
b. A person who does not store as much glycogen does not produce water when forming glycogen from glucose. The initial weight lost is water weight.

Challenge Problems

5.59 60 days

5.61

Serotonin

Did you know that a tree is composed mainly of carbohydrates? From the cellulose in its woody trunk to the starch stored in its leaves to the sugars created in the fruit it bears, carbohydrates come in many forms and have a variety of functions. Chapter 6 explores the unique structures and functions of carbohydrates.

6

Carbohydrates—Life's Sweet Molecules

IF YOU WERE to ask a friend, "What do you know about carbohydrates?" you might get such varied answers as "carbohydrates are sugar," "they provide energy," or "there are *simple* carbohydrates and *complex* carbohydrates." Let's take a closer look at these common statements.

Carbohydrates *are* sugar, and they *do* provide energy. Most people associate carbohydrates with sugary sweet foods that give the body quick energy. Our bodies break down carbohydrates to extract energy. Carbon dioxide and water are released in the process. The carbohydrate glucose is the primary nutrient that our body uses to produce energy. Many carbohydrates can be recognized by their name, which often ends in *ose.* Carbohydrates are organic molecules essential to life, so they are classified as biomolecules.

Simple carbohydrates are referred to as *simple sugars* and are often sweet to the taste. We will start our exploration of carbohydrate structure with these smaller, simple sugars. If we eat more simple sugars than we need for energy, our body chemically converts them to fat. More *complex carbohydrates* include starches and insoluble fibers found in plants, the most common of which is cellulose. As we will see, starch and cellulose are larger molecules composed of glucose. We can digest and absorb starch, but we can not digest and absorb cellulose. This may seem strange, since they both contain glucose. Starch and cellulose combine glucose differently into their structures, accounting for this difference (see Section 6.5).

Something else that many people do not realize about carbohydrates is that they have functions in the body other than simply serving as an energy source. Carbohydrates are found on the surfaces of cells where they can act as "road signs," allowing molecules to distinguish one cell from another. For example, the ABO blood markers are carbohydrates found on the surface of red blood cells that allow us to distinguish our body's blood type from a foreign blood type. Our bodies also contain a carbohydrate called *heparin* that prevents blood clots. Carbohydrates are also found in larger molecules like our genetic material, DNA and RNA.

? What's an Inquiry Question?
Inquiry Questions are designed to focus your reading on the main concepts by section. An Inquiry Question appears at the beginning of each section.

6.1 Classes of Carbohydrates

6.1 Inquiry Question:
How do we classify
carbohydrates?

The simplest carbohydrates are **monosaccharides** (*mono* is Greek for "one," *sakkhari* is Greek for "sugar") These often sweet-tasting sugars cannot be broken down into smaller carbohydrates. The common carbohydrate glucose, $C_6H_{12}O_6$, is a monosaccharide. Monosaccharides contain the elements carbon, hydrogen, and oxygen, and have the general formula $C_n(H_2O)_n$, where n is a whole number 3 or higher.

Disaccharides consist of two monosaccharide units joined together. A disaccharide can be split into two monosaccharide units. Ordinary table sugar, sucrose, $C_{12}H_{22}O_{11}$, is a disaccharide that can be broken up through hydrolysis into the two monosaccharides glucose and fructose.

Oligosaccharides are carbohydrates containing anywhere from three to nine monosaccharide units. The blood-typing groups known as ABO are oligosaccharides.

When 10 or more monosaccharide units are joined together, the large molecules that result are termed **polysaccharides** (*poly* is Greek for "many"). In polysaccharides, the sugar units can be connected in one continuous chain or the chain can be branched. Starch, a polysaccharide in plants, contains large chains of glucose that can be broken down to produce energy.

The various classes of carbohydrates contain different numbers of monosaccharides.

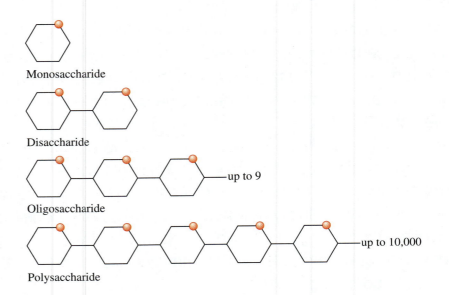

Monosaccharide

Disaccharide

Oligosaccharide — up to 9

Polysaccharide — up to 10,000

Fiber in Your Diet

Dietary fibers are carbohydrates that we cannot digest with our own enzymes. Dietary fiber is divided into two classes called soluble and insoluble fiber. **Soluble fiber** can mix with water, forming a gel-like substance that swells in the stomach and digestive tract. This gives a sense of fullness. This full feeling can slow sugar absorption into the bloodstream. Soluble fiber has also been shown to help lower blood cholesterol by interfering with cholesterol absorption. Some foods high in soluble fiber include oatmeal, legumes (peas, beans, and lentils), apples, psyllium husk, and carrots. Fruit pectins used in making jellies (giving a gel consistency) contain soluble fiber.

Dietary fiber consists of carbohydrates that we cannot digest. Yet they are important for a healthy digestive tract.

In contrast, **insoluble fibers** do not mix with water, although they play a critical role in the digestive tract. Insoluble fiber has a laxative effect and adds bulk to the diet, thus preventing constipation. The polysaccharide cellulose (Section 6.5) is an insoluble fiber. Some sources include whole grains, seeds, brown rice, cabbage, and vegetable skins.

6.1 Classify the following carbohydrates as a monosaccharide, disaccharide, oligosaccharide, or polysaccharide:

   a. carageenan, a seaweed extract containing up to 25,000 carbohydrate units

   b. levans, soluble fiber containing 3–6 carbohydrate units

   c. maltose, containing two glucose units

6.2 Classify the following carbohydrates as a monosaccharide, disaccharide, oligosaccharide, or polysaccharide

   a. raffinose, a soluble fiber containing three carbohydrate units

   b. starch, a storage carbohydrate in plants that contains thousands of glucose units

   c. fructose, a simple sugar found in fruit with the formula $C_6H_{12}O_6$

6.2 Functional Groups in Monosaccharides

? 6.2 Inquiry Question: What functional groups are present in monosaccharides?

Like most organic molecules, monosaccharides contain several functional groups. The functional groups of the monosaccharide glucose are shown in **Figure 6.1**. Note that glucose contains several alcohol (or hydroxyl) groups represented as —OH. Glucose also includes the carbonyl-containing aldehyde functional group. Some monosaccharides contain a ketone functional group instead of an aldehyde. Carbohydrates are considered polyhydroxyaldehydes or polyhydroxyketones because they contain several hydroxyl (alcohol) groups and either an aldehyde or ketone group. Before discussing monosaccharides further, let's look at these functional groups and some common alcohols, aldehydes, and ketones more closely.

Alcohol

Ethanol, a commercially important compound produced from the fermentation of the simple sugars in grains and fruits, is one of the simplest members of the family of organic compounds known as **alcohols.** Ethanol is the alcohol present in liquor, beer, and wine. It is also the main additive in the alternative fuel blends such as gasohol and E85, which is 85% ethanol and only 15% gasoline.

Alcohols are classified by the number of alkyl groups attached to the carbon atom that is bonded to the hydroxyl group. (This carbon atom is sometimes referred to as the alcoholic carbon.) The number of alkyl groups attached to the alcoholic carbon directly impacts the reactivity of the alcohol. A **primary (1°) alcohol** has *one* alkyl group attached to the alcoholic carbon, a **secondary (2°) alcohol** has *two* such alkyl groups, and a **tertiary (3°) alcohol** has *three*. Monosaccharides contain both primary and secondary alcohols.

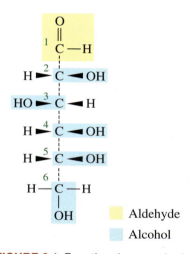

Aldehyde

Alcohol

FIGURE 6.1 Functional groups in the monosaccharide glucose. Glucose is a polyhydroxyaldehyde because it includes the carbonyl-containing functional group aldehyde and several hydroxy (alcohol) groups.

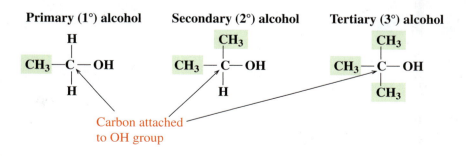

Primary (1°) alcohol   Secondary (2°) alcohol   Tertiary (3°) alcohol

Carbon attached to OH group

CH_3CH_2OH ◄— Alcohol

Ethanol, an alcohol

Classifying Alcohols

Classify each of the following alcohols as a primary (1°), secondary (2°), or tertiary (3°) alcohol:

a. $CH_3CH_2CH_2OH$

b.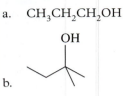

Solution

To determine if an alcohol is primary, secondary, or tertiary, answer the following question: How many carbons are directly bonded to the $C\!-\!OH$ carbon?

a. One carbon is bonded to the C with the OH; therefore, this is a primary (1°) alcohol.

b. Three carbons are bonded to the C with the OH; therefore, this is a tertiary (3°) alcohol.

Aldehyde

Benzaldehyde, the compound responsible for the aroma of almonds and cherries, is a member of the simplest family of carbonyl-containing organic compounds known as the **aldehydes.** The presence of the benzene ring in its structure further classifies benzaldehyde as an aromatic aldehyde.

Members of the aldehyde family always have a carbonyl group with a hydrogen atom bonded to one side of the carbonyl and an alkyl or aromatic group bonded to the other. The lone exception to this is formaldehyde, which has a hydrogen bonded to each side of the carbonyl. (Formaldehyde was once used as a preservative for biological specimens, but it is no longer used because it was found to cause cancer in some animals.) Monosaccharides can contain an aldehyde functional group at one end of the molecule (in addition to multiple hydroxyl groups).

The aldehyde functional group is found in many scented compounds. The simplest aldehyde is formaldehyde where the R group is a hydrogen.

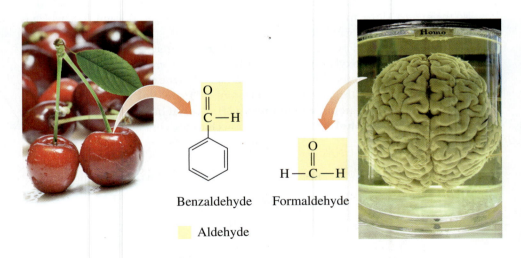

Benzaldehyde   Formaldehyde

Aldehyde

Ketone

The **ketone** family of organic compounds is structurally similar to the aldehydes. The difference is that ketones have an alkyl or aromatic group on *both* sides of the carbonyl. The simplest ketone is acetone, which was once the main component of fingernail polish remover. Because of its tendency to cause dry skin, acetone has now been largely replaced in many formulations of nail polish remover.

$$H_3C - \overset{\overset{\displaystyle O}{\|}}{C} - CH_3$$

Ketone   Acetone

Besides monosaccharides, ketones occur in a wide variety of biologically relevant compounds. For example, pyruvate is a ketone-containing compound formed during the metabolic breakdown of carbohydrates. Butanedione, the flavor of butter, contains two ketone groups. Note that in the structures of pyruvate and butanedione, the carbonyl has a carbon as the *first* atom on each side—this distinguishes the functional group as a ketone regardless of the other atoms that may be present in the structure.

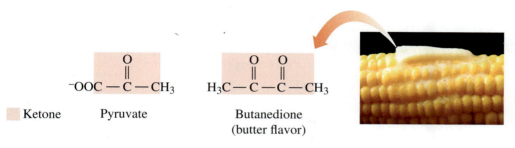

$$^-OOC - \overset{\overset{\displaystyle O}{\|}}{C} - CH_3 \qquad H_3C - \overset{\overset{\displaystyle O}{\|}}{C} - \overset{\overset{\displaystyle O}{\|}}{C} - CH_3$$

Ketone   Pyruvate   Butanedione
                                  (butter flavor)

The ketone functional group is found in many flavors like butter. It consists of a carbonyl bonded to two carbons.

A monosaccharide that contains an aldehyde functional group is referred to as an **aldose,** and one that contains a ketone functional group is called **ketose.** The most common monosaccharides contain three to six carbons. A monosaccharide with three carbons is a *triose,* one with four carbons is a *tetrose,* one with five carbons is a *pentose,* and one with six carbons is a *hexose.*

For instance, **glucose,** the most abundant monosaccharide found in nature, is an aldose containing six carbons and is therefore classified as an aldohexose, while **fructose,** a common monosaccharide found in fruits, is a ketose containing six carbons and is classified as a ketohexose. Look at the monosaccharides shown and convince yourself that these are examples of an aldohexose and ketopentose by counting the carbons and identifying the functional groups.

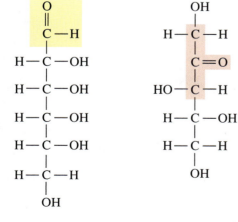

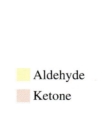

Aldehyde

Ketone

An aldohexose   A ketopentose

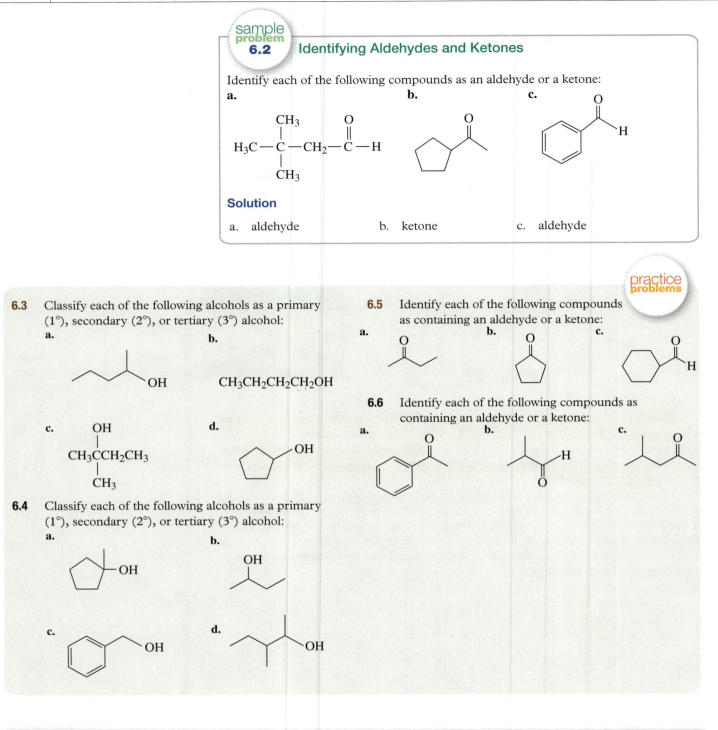

sample problem 6.2 **Identifying Aldehydes and Ketones**

Identify each of the following compounds as an aldehyde or a ketone:

a.

b.

c.

Solution

a. aldehyde

b. ketone

c. aldehyde

practice problems

6.3 Classify each of the following alcohols as a primary (1°), secondary (2°), or tertiary (3°) alcohol:

a.

b. $CH_3CH_2CH_2CH_2OH$

c.

d.

6.4 Classify each of the following alcohols as a primary (1°), secondary (2°), or tertiary (3°) alcohol:

a.

b.

c.

d.

6.5 Identify each of the following compounds as containing an aldehyde or a ketone:

a.

b.

c.

6.6 Identify each of the following compounds as containing an aldehyde or a ketone:

a.

b.

c.

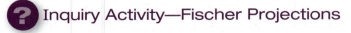

Discovering the Concepts

? Inquiry Activity—Fischer Projections

Information

In Chapter 4, we saw that a chiral center is a carbon in a molecule with four different groups bonded to it. When a chiral center appears in a molecule, there are two different ways that the four different groups can arrange themselves around the carbon. The two resulting molecules are related to each other as nonsuperimposable mirror images, a special type of stereoisomer called an *enantiomer.*

The simplest carbohydrate is glyceraldehyde, $C_3H_6O_3$. Glyceraldehyde contains an aldehyde functional group, so it is referred to as an *aldose.* In carbohydrates, a pair of enantiomers is designated by writing either D- or L- in front of the name. The D-isomer has the chiral center farthest from the carbonyl arranged like D-glyceraldehyde, and the L-isomer has the chiral center farthest from the carbonyl arranged like L-glyceraldehyde.

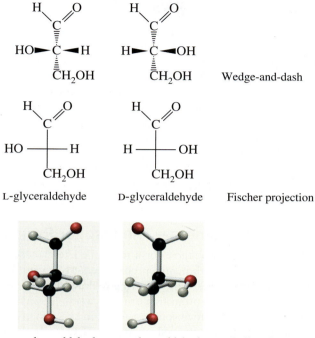

FIGURE 1 The enantiomers of glyceraldehyde (L- on left, D- on right) shown in three different representations: wedge-and-dash, Fischer projection, and ball-and-stick.

Questions

1. In D-glyceraldehyde **(Figure 1)**, list the groups attached to the chiral center that are in front of the plane of the paper. Which groups attached to the chiral center are behind the plane of the paper?
2. Where is the chiral center in the Fischer projection? Identify it on the glyceraldehyde molecules (Figure 1) with an asterisk.
3. What does the horizontal line in the center of the Fischer projections shown represent?
4. What does the vertical line in the center of the Fischer projections shown represent?
5. How are D- and L-glyceraldehyde different?
6. Identify the chiral centers on D-glucose **(Figure 2)** with an asterisk.
7. Draw the Fischer projection of D-glucose.
8. Draw the Fischer projection of L-glucose (the mirror image of D-glucose) next to D-glucose that you drew in question 7.
9. D-Galactose is an *epimer* of D-glucose: All the chiral centers are identical, except for C4, which is reversed. Draw the Fischer projection of D-galactose.

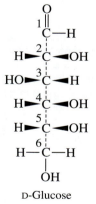

FIGURE 2 D-Glucose shown in wedge-and-dash.

6.3 Stereochemistry in Monosaccharides

Let's look at the structure of glucose a little more closely and see if we recognize any other properties of carbon-containing molecules seen earlier in Chapter 4. The carbons bonded to the alcohol (—OH) groups all have a tetrahedral geometry. Remember that a carbon atom with tetrahedral geometry that has four different atoms or groups of atoms attached

6.3 Inquiry Question:
What structural properties do monosaccharides display?

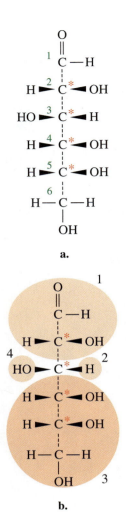

a.

b.

FIGURE 6.2 Monosaccharides have chiral centers. (a) The chiral centers in glucose are designated with an asterisk. (b) Do you see the four different groups (shaded circles) attached to carbon 3 making it a chiral center?

to it is a chiral center. As we also saw in Chapter 4, a compound with a single chiral center can exist in two different forms called *enantiomers*.

How many chiral centers does a glucose molecule contain? (See **Figure 6.2**.) Let's examine the carbons in glucose starting with carbon 1, which is at the top. Carbon 1, containing an aldehyde group, is not tetrahedral, so it cannot be a chiral center. Carbons 2 to 5 are tetrahedral and have four different atoms or groups of atoms attached, so they are chiral centers. Carbon 6 is tetrahedral but does not have four *different* groups of atoms attached to it, because two of the atoms attached are hydrogen, so it is not chiral. Therefore, glucose has a total of four chiral centers. The groups bonded to each chiral center of a glucose molecule could have two different arrangements (a pair of mirror images), so how many possible arrangements around the chiral centers, or stereoisomers, are possible?

The number of stereoisomers possible increases with the number of chiral centers present in a molecule. For molecules with one chiral center, there are only two different ways the attached atoms or groups can be arranged spatially (and these are mirror images). For molecules with two chiral centers, there are a total of four different ways to arrange the attached atoms or groups differently, and for molecules with three chiral centers, there are eight different ways to attach atoms or groups. The general formula for determining the number of stereoisomers is 2^n, where n is the number of chiral centers present in the molecule. Because glucose has four chiral centers, ($2^4 = 2 \times 2 \times 2 \times 2 = 16$), 16 stereoisomers are possible. It is amazing then that only one of these stereoisomers is our preferred energy source.

Representing Stereoisomers—The Fischer Projection

Considering all the stereoisomers possible for a molecule like glucose and the need for designating the position of the attachments on chiral centers, it would be convenient to have a way of representing molecules without having to draw all those wedges and dashes on all the chiral centers. The Fischer projection provides a simpler way of indicating chiral molecules by showing their three-dimensional structure in two dimensions.

Consider the simplest aldose, glyceraldehyde ($C_3H_6O_3$), shown in the following figure. Because it has one chiral center, it can exist as one of two enantiomers. These are designated as a D-enantiomer and an L-enantiomer. If we were to represent the enantiomers of glyceraldehyde with wedges and dashes to show their shape, we could represent the molecules as shown in the top part of the figure. In this representation, the only atom that is on the plane of the paper is the carbon at the chiral center, with all other atoms either in front of (wedges) or behind (dashes) the plane of the page.

Wedge-and-dash projections of glyceraldehyde

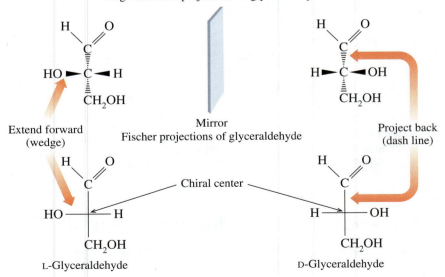

Mirror
Fischer projections of glyceraldehyde

L-Glyceraldehyde   D-Glyceraldehyde

In the **Fischer projection,** *horizontal lines on a chiral center represent wedges, and vertical lines on a chiral center represent dashes.* A chiral center is not shown as a "C" on a Fischer projection but is implied at the intersection of the lines. This gives the viewer a quick and easy way of identifying the number of chiral centers. The designation of D and L for glyceraldehyde and all other carbohydrates is based on the Fischer projection positioning in glyceraldehyde, used as a reference molecule for this designation.

All **D-sugars** have the —OH on the chiral center farthest from the carbonyl (C=O) on the *right side* of the molecule in the Fischer projection. The enantiomer of this is the **L-sugar**, which has the —OH group on the chiral center farthest from the C=O on the *left side* of the Fischer projection. Most of the carbohydrates commonly found in nature and the ones we use for energy are D-sugars. D- and L-glyceraldehyde are represented by the Fischer projections shown in the bottom part of the figure.

Similarly, **Figure 6.3** shows the molecule D-glucose transformed into a Fischer projection.

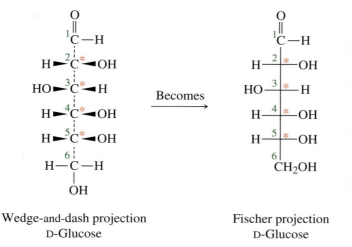

Wedge-and-dash projection
D-Glucose

Becomes

Fischer projection
D-Glucose

FIGURE 6.3 D-**Glucose represented in a Fischer projection.** Attachments to chiral centers on the horizontal point toward the viewer, and attachments to chiral centers on the vertical point away from the viewer. Note: The dashes in the wedge-and-dash projection at left are drawn as vertical lines for clarity.

When we draw enantiomers in a Fischer projection, they are written as if there is a mirror placed between the two molecules. The reflection of one molecule is identical to the second molecule, its enantiomer. All chiral centers have their horizontal groups switched in their enantiomer when viewed in a Fischer projection. Attached atoms or groups on the right in one enantiomer appear on the left side of the other. The enantiomers D- and L-glucose are shown in the following figure:

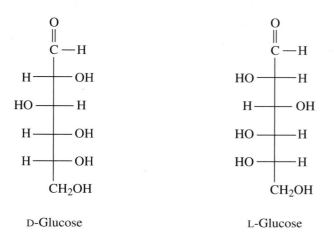

D-Glucose

L-Glucose

Solving a Problem

Drawing an Enantiomer in a Fischer Projection

Draw the Fischer projection for the enantiomer for D-Galactose.

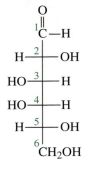

D-Galactose

STEP 1: Locate the chiral centers. In a Fischer projection, the chiral centers are located at the intersections of the vertical and horizontal lines. D-Galactose has four chiral centers at positions 2, 3, 4, and 5.

STEP 2: Switch horizontal groups on the chiral centers. If you imagine that a mirror exists between the pair of molecules so that one enantiomer's reflection will be the other enantiomer, what appears on the right of one enantiomer will appear on the left of the other. The positioning of the atoms attached to the achiral centers does not matter. The L-enantiomer is shown to the right of the original (blue).

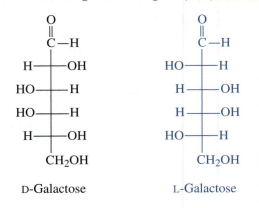

D-Galactose   L-Galactose

---

sample problem 6.3   Drawing an Enantiomer in a Fischer Projection

Draw the Fischer projection for the enantiomer for D-ribose.

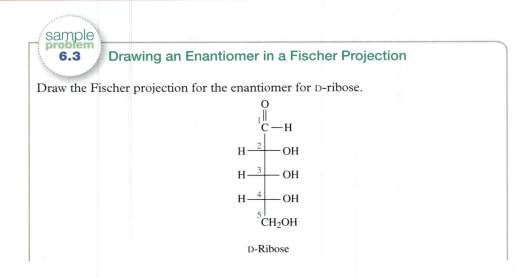

D-Ribose

Solution

STEP 1: Locate the chiral centers. D-Ribose has three chiral centers at carbons 2, 3, and 4.

STEP 2: Switch horizontal groups on the chiral centers. The L-enantiomer is shown to the right of the original (blue).

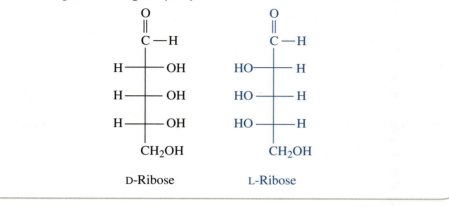

D-Ribose   L-Ribose

Stereoisomers That Are Not Enantiomers

So far we have seen one stereoisomer of D-glucose: its enantiomer L-glucose. Because there can be only one mirror image for any stereoisomer, how are the other 14 of 16 possible stereoisomers related to D-glucose? Stereoisomers that are not enantiomers are called **diastereomers.** Diastereomers are stereoisomers that are not exact mirror images. Figure 6.4 shows the monosaccharides D-galactose and D-talose in a Fischer projection. Both of them are diastereomers of D-glucose.

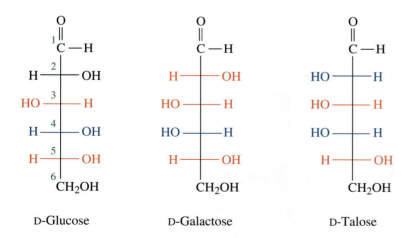

D-Glucose   D-Galactose   D-Talose

FIGURE 6.4 Diastereomers of D-glucose. The monosaccharides D-galactose and D-talose are classified as diastereomers of D-glucose because some chiral centers are oriented the same (shown in red) and some chiral centers are mirror images (shown in blue). Notice that all three monosaccharides are D-sugars.

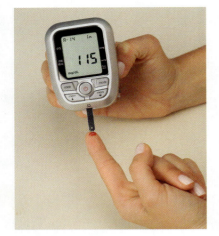

A glucose monitor.

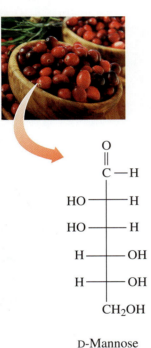

$$
\begin{array}{c}
\quad\;\; O \\
\quad\;\; \| \\
\quad\; C-H \\
HO-\!\!-\!\!-H \\
HO-\!\!-\!\!-H \\
H-\!\!-\!\!-OH \\
H-\!\!-\!\!-OH \\
\quad\; CH_2OH
\end{array}
$$

D-Mannose

Cranberries are a rich source of the monosaccharide D-mannose.

Important Monosaccharides

Several of the more common monosaccharides are hexoses (containing six carbons) produced and used by nature only as the D-isomers. The D-form is discussed here.

The most abundant monosaccharide found in nature is glucose. We commonly refer to D-glucose as dextrose, blood sugar, or grape sugar. It is found in fruits, vegetables, and corn syrup. As we noted in Chapter 5, glucose can be broken down inside cells to produce energy through glycolysis. Diabetics have difficulty getting glucose from the bloodstream into their cells so that glycolysis can occur. This is why they must regularly monitor their blood glucose levels. Glucose is also a sugar unit in the disaccharides sucrose (table sugar) and lactose (milk sugar) as well as the polysaccharides amylose, amylopectin, glycogen, and cellulose.

Galactose (Figure 6.4) is found in nature combined with glucose in the disaccharide lactose, which is present in milk and other dairy products. Galactose has one of its chiral centers (carbon 4) arranged opposite that of glucose. Diastereomers that differ in just one chiral center (as compared with more than one chiral center) are called **epimers.** The body can chemically convert galactose into glucose for use in glycolysis through help of an enzyme called an *epimerase*.

Mannose is a monosaccharide found in some fruits and vegetables. It is not easily absorbed by the body. The most notable fruit that contains high amounts of mannose is the cranberry. Mannose has been shown to be effective against urinary tract infections (UTIs). When the level of mannose builds up in the bladder, the bacteria causing the UTI will attach themselves to the mannose in the urine instead of the other carbohydrates found on the outside of the cells lining the urinary tract and will be eliminated. Mannose is also an epimer of glucose.

The ketose fructose is also commonly referred to as fruit sugar or levulose and is found in fruits, vegetables, and honey. In combination with glucose, it gives us the disaccharide sucrose (table sugar). Fructose is the sweetest monosaccharide, one and a half times sweeter than table sugar, making it popular with dieters who can get the same sweet taste with fewer calories. Even though it is not an epimer of glucose, fructose can be broken down for energy production in the body by the chemical reactions of glycolysis, as we will see in Chapter 12.

$$
\begin{array}{c}
\; CH_2OH \\
\; | \\
\; C=O \\
HO-\!\!-\!\!-H \\
H-\!\!-\!\!-OH \\
H-\!\!-\!\!-OH \\
\; CH_2OH
\end{array}
$$

D-Fructose

Apples and honey contain high amounts of the monosaccharide D-fructose.

The pentoses (five-carbon sugars) ribose and 2-deoxyribose are a part of larger bio-molecules called nucleic acids that make up our genetic material, a topic we will address in Chapter 11. The nucleic acids are distinguished in their name by the monosaccharide they contain. *Ribo*nucleic acid (RNA) contains the sugar ribose, and *deoxyribo*nucleic acid (DNA) contains the sugar deoxyribose. Structurally, the only difference between the two pentoses is the absence of an oxygen atom on carbon 2 of deoxyribose. Ribose is also found in the vitamin *ribo*flavin and other biologically important molecules, as we will see in Chapter 12.

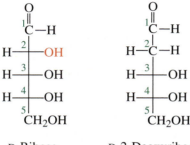

D-Ribose   D-2-Deoxyribose

---

sample problem

6.4   Distinguishing Stereoisomers

Classify structures A, B, and C in the figure as being either an enantiomer or a diaste-reomer of D-mannose.

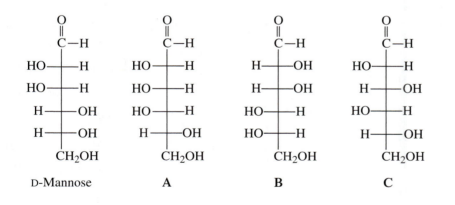

D-Mannose   **A**   **B**   **C**

Solution

Determine the differences between D-mannose and structures A, B, and C. Structure A has one chiral center oriented differently from D-mannose, so it is a diastereomer (epimer). Structure B is the mirror image enantiomer of D-mannose, L-mannose. Structure C has two chiral centers oriented differently from D-mannose, so it is a diastereomer.

6.7 Identify the following monosaccharides as the D- or the L-isomer:

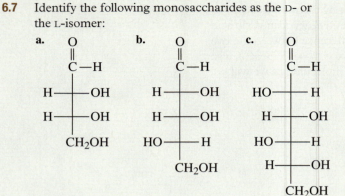

6.8 Identify the following monosaccharides as the D- or the L-isomer:

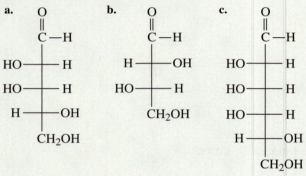

6.9 Draw the Fischer projection for the enantiomer (mirror image) of each of the following:

a.   **b.**

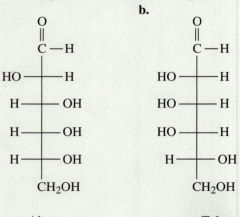

D-Altrose   D-Talose

6.10 Draw the Fischer projection for the enantiomer (mirror image) of each of the following:

a.   **b.**

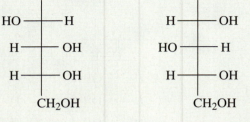

D-Fructose   D-Sorbose

6.11 Classify structures A, B, and C in the figure as being either an enantiomer or a diastereomer of D-galactose.

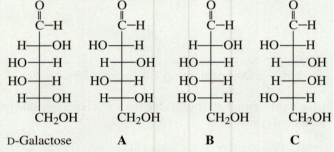

D-Galactose   A   B   C

6.12 Classify structures A, B, and C in the figure as being either an enantiomer or a diastereomer of D-glucose.

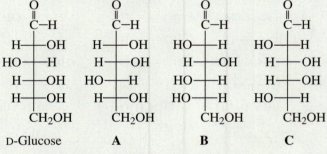

D-Glucose   A   B   C

6.13 Identify the monosaccharide that fits each of the following descriptions:

a. also referred to as dextrose
b. also called fruit sugar
c. used to treat urinary tract infections

6.14 Identify the monosaccharide that fits each of the following descriptions:

a. in combination with glucose produces the disaccharide lactose
b. also called blood sugar
c. also called levulose

Discovering the Concepts

? Inquiry Activity—Ring Formation

Part 1. Information

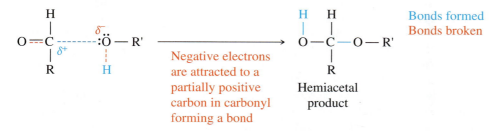

FIGURE 1 The formation of a hemiacetal from an aldehyde and an alcohol.

Questions

1. In **Figure 1**, name the functional groups of the compounds on the reactant side of the chemical equation.
2   Which oxygen in the hemiacetal product in Figure 1 (right or left) was the carbonyl oxygen from the aldehyde?
3. Which oxygen in the hemiacetal product in Figure 1 (right or left) was from the alcohol, R'OH?

Part 2. Information

The carbonyl carbon (C1) in a monosaccharide is referred to as the *anomeric* carbon.

Conventions for Drawing the Ring Form of a Six-Carbon D-Aldose

*   Carbon 6 will always be on the top side of the ring.
*   Hydroxyls that do not react and are on the right of a Fischer projection will be on the bottom side of the ring.
*   Hydroxyls that do not react and are on the left of a Fischer projection will be on the top side of the ring.

Questions

4. Monosaccharides chemically react with themselves (intramolecularly) due to polar opposites strongly attracting each other within the molecule.
    a. Considering the structure of D-glucose, if the O on carbon 5 of D-glucose (O5) is attracted to the carbonyl carbon 1 (C1), reacting to form a bond between them (making a hemiacetal functional group), how many atoms would be enclosed in a ring?
    b. How many of those atoms are carbon?
    c. How many atoms are oxygen? Recall from Section 4.5 that a carbon ring can be represented with a top and bottom face. Considering this, draw the ring form of D-glucose.
5. Where did you place the OH for C1 (top or bottom)?
6. If you placed it on the bottom, you drew *α*-D-glucose. If you placed the OH on C1 on the top, you drew *β*-D-glucose. Which one did you draw?
7. Can you devise a rule for identifying these two ring forms (*α* and *β* anomers) of D-glucose relative to the position of C6?

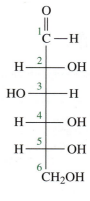

D-Glucose

6.4 Reactions of Monosaccharides

 6.4 Inquiry Question: How are the products of monosaccharide reactions drawn?

Now that we have established the structural characteristics of monosaccharides, we look at some of their common reactions: ring formation and oxidation–reduction. Both of these reactions are in part due to the highly polar nature of the carbonyl found in the aldehyde and ketone functional groups.

Ring Formation — The Truth about Monosaccharide Structure

The linear structures that helped us understand functional groups and chirality of monosaccharides do not show how most monosaccharides actually are structured. Pentoses and hexoses (5- and 6-carbon monosaccharides) bend back on themselves to form rings. When opposite charges on different functional groups within the monosaccharide attract strongly enough, new bonds are formed and other bonds are broken.

This reaction is illustrated for the functional groups in an aldose in **Figure 6.5**. The carbonyl group present in aldehydes and ketones is very polar and highly reactive. The partially positive (δ^+) carbon in the carbonyl attracts the lone pairs of electrons on the oxygen of a hydroxyl that is partially negative (δ^-). If a bond forms between this carbon and oxygen, the new pattern formed is called a **hemiacetal** (pronounced hem-ee-ass'-i-tal).

The carbonyl group in aldoses and ketoses is highly polar, making the group reactive.

FIGURE 6.5 Formation of the hemiacetal group. The electrons on the oxygen of the alcohol group (δ^-) are attracted to the partial positive charge (δ^+) of the carbon in the carbonyl (blue dashed line). To maintain an octet, the alcohol's hydrogen moves to the oxygen of the carbonyl. The joining of these two functional groups results in a hemiacetal.

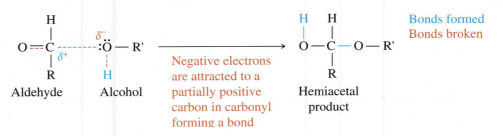

Monosaccharides contain both a carbonyl and several hydroxyl (alcohol) functional groups, and these two functional groups can react within the same molecule, or intramolecularly. Using D-glucose as an example, the hydroxyl on carbon 5 (C5) can bend around and react forming a new bond with the carbonyl at carbon 1 (C1), producing two possible ring structures as shown in **Figure 6.6**. When this occurs, five carbons and an oxygen form a ring, and one of the carbons (carbon 6) remains outside the ring. Monosaccharides exist in a ring form most of the time because the carbonyl group reacts readily with a hydroxyl to form this hemiacetal ring.

FIGURE 6.6 Ring formation in D-glucose. D-Glucose forms a ring when the alcohol group on C5 reacts with the carbonyl (C1), forming a new bond. α and β anomers are formed. C6 remains outside the ring. Dashed lines show bonds breaking and forming to create a ring.

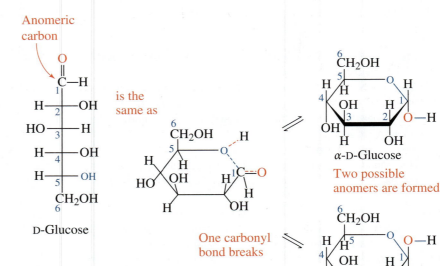

Two structures are possible during ring formation. Recall that a carbonyl group is trigonal planar (flat), so the oxygen in the hydroxyl group on carbon 5 can form its bond on either the top or the bottom side of the carbonyl. Two different ring arrangements can be produced from a single linear monosaccharide chain. These two interconvertible forms are termed **anomers.** In any monosaccharide, the carbonyl carbon that reacts to form the hemiacetal in the reaction is referred to as the **anomeric carbon.** (Note that in the ring form, the anomeric carbon is the only carbon bonded directly to two oxygen atoms.)

The two hemiacetal anomers of D-glucose are referred to as the alpha (α) and the beta (β) anomers. Anomers are distinguished by the positioning of the —OH group on the anomeric carbon relative to the position of the carbon outside the ring (for D-glucose carbon 6). In the six-member ring form of D-isomers called a **pyranose,** carbon 6 (C6) is always drawn on the top side of the ring. The following describes how to determine whether a monosaccharide is the alpha or the beta anomer:

- In the α anomer, the —OH on the anomeric carbon is trans to carbon 6 (the carbon outside the ring). They are on opposite sides of the ring.
- In the β anomer, the —OH on the anomeric carbon is cis to carbon 6 (the carbon outside the ring). They are on the same side of the ring.

---

sample problem 6.5 Alpha and Beta Forms

Identify the following carbohydrates as the α or β anomer:

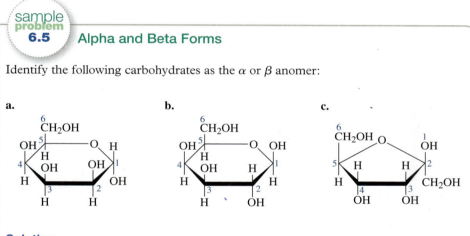

a. **b.** **c.**

Solution

a. The —OH on carbon 1 is on the opposite side of the ring (trans) from carbon 6, so this is the alpha (α) anomer.

b. The —OH on carbon 1 is on the same side of the ring (cis) as carbon 6, so this is the beta (β) anomer.

c. The —OH on carbon 1 is on the same side of the ring (cis) as carbon 6, so this is the beta (β) anomer.

Solving a Problem

Drawing Pyranose Rings from Linear Monosaccharides

Let's practice drawing monosaccharides in the pyranose ring form (six-member ring) from the linear Fischer projection.

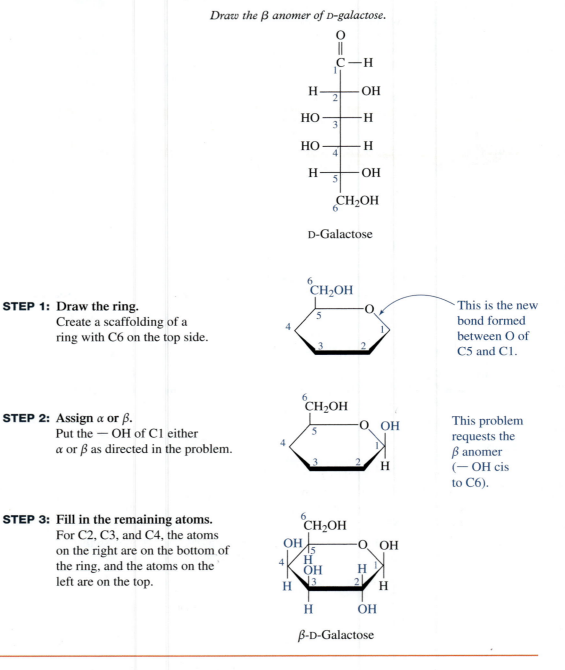

Draw the β anomer of D-galactose.

D-Galactose

STEP 1: Draw the ring.
Create a scaffolding of a ring with C6 on the top side.

This is the new bond formed between O of C5 and C1.

STEP 2: Assign α or β.
Put the — OH of C1 either α or β as directed in the problem.

This problem requests the β anomer (— OH cis to C6).

STEP 3: Fill in the remaining atoms.
For C2, C3, and C4, the atoms on the right are on the bottom of the ring, and the atoms on the left are on the top.

β-D-Galactose

Drawing Pyranose Rings from Linear Monosaccharides

Draw the alpha (α) anomer of D-mannose in pyranose ring form.

Solution

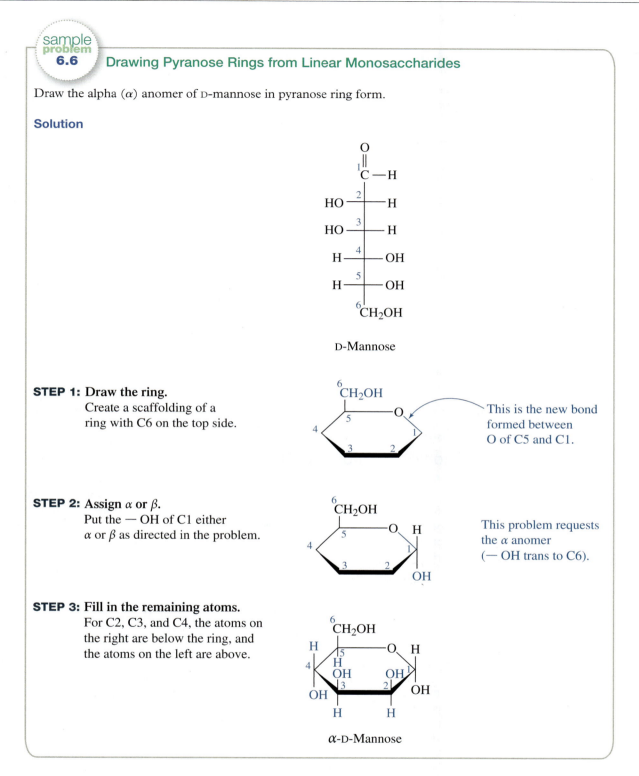

STEP 1: Draw the ring.
Create a scaffolding of a
ring with C6 on the top side.

This is the new bond
formed between
O of C5 and C1.

STEP 2: Assign α or β.
Put the —OH of C1 either
α or β as directed in the problem.

This problem requests
the α anomer
(—OH trans to C6).

STEP 3: Fill in the remaining atoms.
For C2, C3, and C4, the atoms on
the right are below the ring, and
the atoms on the left are above.

α-D-Mannose

D-Fructose contains *both* a ketone group and several hydroxyl groups. The —OH on
carbon 5 (C5) can curl around and react with the carbonyl positioned at carbon 2 (C2),
allowing two possible ring structures, as in D-glucose (see Figure 6.7). Four carbons and
an oxygen form the five-member ring called a **furanose.** Two of the carbons (C1 and
C6) remain outside the ring.

FIGURE 6.7 Ring formation in
D-fructose. D-Fructose forms a ring
when the alcohol group on C5 reacts
with the carbonyl C2. α and β anomers
are formed. C6 and C1 remain outside
the ring. Dashed lines show bonds
breaking and forming to create the ring.

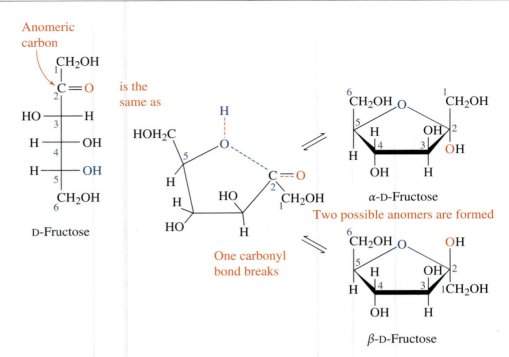

Alpha and beta anomers are determined by the positioning of the —OH group on the anomeric carbon (in D-fructose this is C2) relative to carbon 6 outside the ring.

- In the α anomer, the —OH on the anomeric carbon is trans C6 (on opposite sides).
- In the β anomer, the —OH on the anomeric carbon is cis to C6 (on the same side).

Oxidation–Reduction and Reducing Sugars

As well as forming rings, the carbonyl group (the anomeric carbon) in aldoses can also undergo organic oxidation and reduction as discussed in Chapter 5. This occurs by adding oxygen (or losing hydrogen) during oxidation and adding hydrogen (or losing oxygen) during reduction. Aldoses can undergo both oxidation *and* reduction. The aldehyde functional group can oxidize to a carboxylic acid or can reduce to an alcohol. In monosaccharides, an oxidation produces a sugar acid and a reduction produces a sugar alcohol.

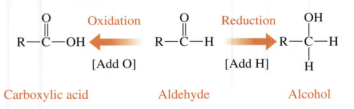

When molecules are oxidized, they act as a reducing agent and reduce a second reactant. One useful oxidation reaction for sugars occurs in **Benedict's test.** It tests for the presence of an aldose in a solution. Using Benedict's test, an aldehyde group undergoes oxidation while reducing copper ions from Cu^{2+} to Cu^+ (copper gained one electron). The Cu^{2+} ions are soluble, coloring the initial reaction solution blue. The aldehyde group in turn is oxidized by Cu^{2+}, forming a sugar acid, while Cu^{2+} undergoes reduction by the aldehyde, forming Cu^+ ions. The resulting copper(I) oxide (Cu_2O), is not soluble and forms a brick-red precipitate in solution (see **Figure 6.8**). Because aldoses are easily oxidized and can therefore serve as reducing agents, sugars capable of reducing substances like Cu^{2+} are referred to as **reducing sugars.** Fructose and other ketoses are also reducing sugars, even though they do not contain an aldehyde group, because in the presence of oxidizing agents, they can rearrange to aldoses.

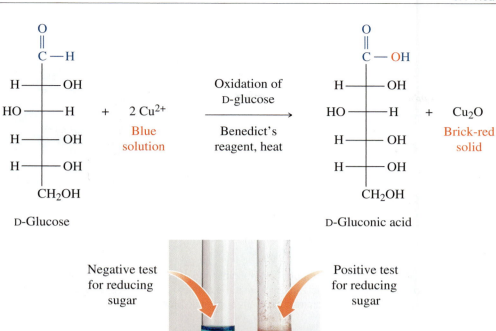

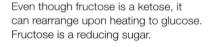

FIGURE 6.8 The oxidation of
D-glucose. In the presence of
Benedict's reagent (containing Cu^{2+}
ions), D-glucose reacts to produce the
sugar acid, D-gluconic acid, and a
brick-red precipitate, Cu_2O.

Negative test
for reducing
sugar

Positive test
for reducing
sugar

Brick-red precipitate

Cu^{2+} $Cu_2O(s)$

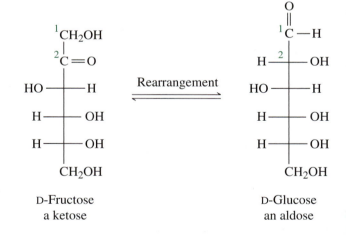

Even though fructose is a ketose, it
can rearrange upon heating to glucose.
Fructose is a reducing sugar.

Benedict's test can be used to monitor glucose levels in urine. Glucose test strips,
like the one shown at right, produce a range of color changes if glucose is present.
Excess glucose in urine suggests high levels of glucose in the bloodstream, an indicator
for diabetes.

An aldose or ketose can also be reduced to an alcohol when the carbonyl reacts with
hydrogen under the left conditions (see **Figure 6.9**). Sugar alcohols are produced
commercially as artificial sweeteners and are found in sugar-free foods.

Sugar alcohols can be produced in the body when glucose levels remain high in
the bloodstream. An enzyme called aldose reductase acts to reduce excessive glucose
to the sugar alcohol sorbitol, which at high concentration can contribute to cataracts
(clouding of the lens of the eye). These so-called sugar cataracts are commonly seen
in diabetics.

The color change in a glucose test
strip measures the amount of glucose
in the urine through an oxidation–
reduction reaction.

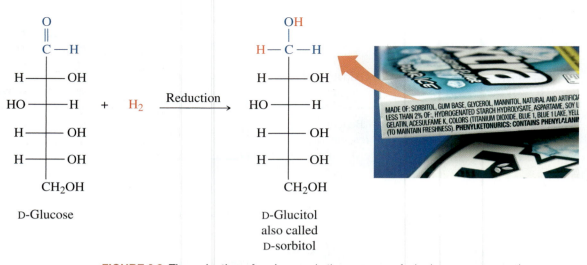

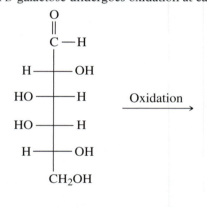

FIGURE 6.9 The reduction of D-glucose. In the presence of a hydrogen source, D-glucose reacts to produce the sugar alcohol D-glucitol, which is better known as D-sorbitol. Sugar alcohols like sorbitol, xylitol, and erythritol are used in many sugar-free products.

---

sample problem
6.7   **Redox Reactions in Monosaccharides**

Draw the product when D-galactose undergoes oxidation at carbon 1.

$$
\begin{array}{c}
\text{O} \\
\parallel \\
\text{C—H} \\
\text{H——OH} \\
\text{HO——H} \\
\text{HO——H} \\
\text{H——OH} \\
\text{CH}_2\text{OH}
\end{array}
\quad \xrightarrow{\text{Oxidation}}
$$

D-Galactose

Solution

The monosaccharides can be oxidized or reduced at the carbonyl group. An aldehyde functional group is oxidized to a carboxylic acid. When undergoing oxidation, D-galactose is oxidized to the following:

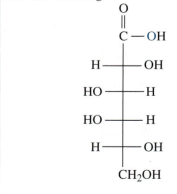

$$
\begin{array}{c}
\text{O} \\
\parallel \\
\text{C—OH} \\
\text{H——OH} \\
\text{HO——H} \\
\text{HO——H} \\
\text{H——OH} \\
\text{CH}_2\text{OH}
\end{array}
$$

6.15 Identify the following carbohydrates as the α or β anomer:

a.

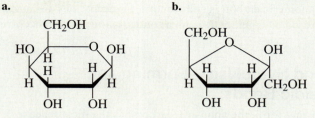

b.

6.16 Identify the following carbohydrates as the α or β anomer:

a.

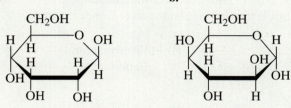

b.

6.17 Draw the α and β anomer of D-talose in pyranose ring form:

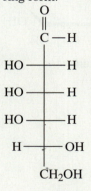

D-Talose

6.18 Draw the α and β anomer of D-altrose in pyranose ring form:

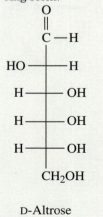

D-Altrose

6.19 When an aldehyde undergoes oxidation, the functional group _____ results.

6.20 When an aldehyde undergoes reduction, the functional group _____ results.

6.21 Write the products if (a) carbon 1 is oxidized and (b) if carbon 1 is reduced in D-ribose.

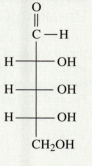

D-Ribose

6.22 Write the products if (a) carbon 1 is oxidized and (b) if carbon 1 is reduced in D-xylose.

D-Xylose

6.5 Disaccharides

6.5 Inquiry Question:
How are disaccharides
formed and identified?

So far, we have seen that the carbonyl at the anomeric carbon reacts to form a ring structure in monosaccharides and reacts in oxidation and reduction reactions. Even when
enclosed in a ring, the anomeric carbon is still reactive. Disaccharides are formed when
two monosaccharides are joined together through a condensation reaction occurring at an
anomeric carbon.

Condensation and Hydrolysis—Forming and Breaking Glycosidic Bonds

In a monosaccharide in ring form, the anomeric carbon has the most reactive —OH in
the molecule (C1 in an aldose). When this hydroxyl reacts with a hydroxyl on another
monosaccharide, a **glycosidic bond** forms. Glycosidic bonds join monosaccharides to
each other and, in a more general sense, connect monosaccharides to any alcohol. In
Figure 6.10, two glucose units are joined to form the disaccharide maltose.

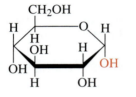
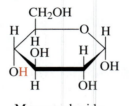
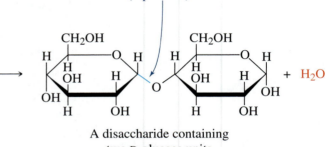

Monosaccharide
α-D-Glucose

Monosaccharide
α-D-Glucose

Glycosidic bond formed
(α position)

A disaccharide containing
two D-glucose units.
Maltose

+ H_2O

**FIGURE 6.10 Formation of the
disaccharide maltose.** Two molecules
of glucose are joined forming a
glycosidic bond. The loss of H and OH
from the glucose molecules produces a
molecule of water in this condensation
reaction.

sample problem 6.8   **Locating a Glycosidic Bond**

Find the glycosidic bond in the following carbohydrates:

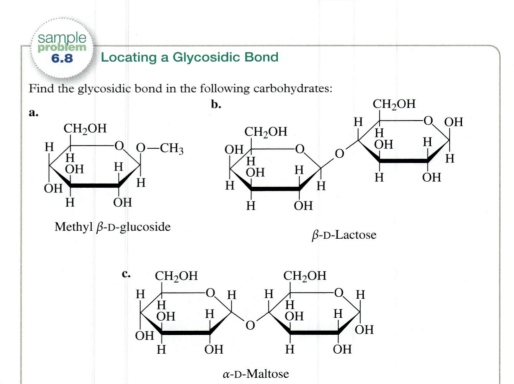

a.

Methyl β-D-glucoside

b.

β-D-Lactose

c.

α-D-Maltose

Solution

The glycosidic bond joins a monosaccharide through a condensation reaction to another alcohol group. One of the carbons bonded to the oxygen must be the anomeric carbon, and the glycosidic bond forms there. The glycosidic bonds are blue in the following figure.

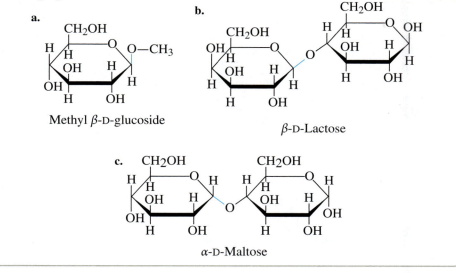

Methyl β-D-glucoside

β-D-Lactose

α-D-Maltose

The formation of a glycosidic bond or **glycoside** is an example of a condensation reaction. Recall from Section 5.5 that in a condensation reaction a molecule of water is eliminated as two molecules are joined. During digestion, we break up disaccharides at the glycosidic bond through the opposite reaction, hydrolysis, where water is consumed as a reactant.

sample problem 6.9 Identifying Condensation and Hydrolysis

Identify the carbohydrate reaction shown as a condensation or a hydrolysis.

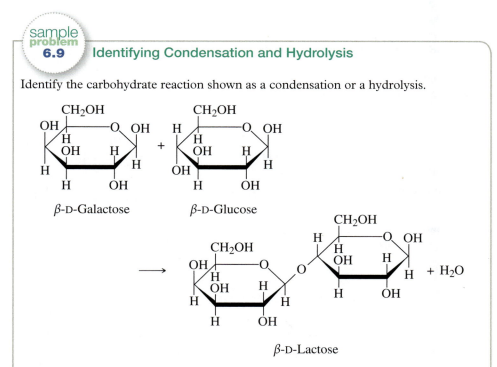

β-D-Galactose   β-D-Glucose

β-D-Lactose

Solution

This reaction connects two monosaccharides into a disaccharide through a glycosidic bond. This is a condensation reaction.

Naming Glycosidic Bonds

If we look back at the disaccharide maltose that was formed from two glucose units, we see that the glycosidic bond was formed in the α position. If instead the second monosaccharide had bonded to the β side (in this case the top) of the first glucose, the glycosidic bond would have created a different disaccharide molecule with a different shape (see **Figure 6.11**). Because of the two possible anomers, it becomes necessary to specify how the monosaccharides are bonded; that is, α or β. Besides the specific configuration bonded (α or β), the carbons that were joined must also be specified. Any —OH group on the second monosaccharide can, in principle, react with the first monosaccharide's anomeric carbon. Because a monosaccharide has many hydroxyl groups that can undergo condensation, it is necessary to specify the carbon atoms joined by the glycosidic bond. The convention for naming glycosidic bonds is to specify the anomer bonded and the carbons bonded. For maltose, the glycosidic bond is specified as $\alpha(1 \rightarrow 4)$ (stated "alpha-one-four").

FIGURE 6.11 Comparison of maltose and cellobiose. Maltose and cellobiose are disaccharides of D-glucose bonded $\alpha(1 \rightarrow 4)$ and $\beta(1 \rightarrow 4)$, respectively. The glycosidic bond positioning (alpha versus beta) makes the shape of these two molecules different.

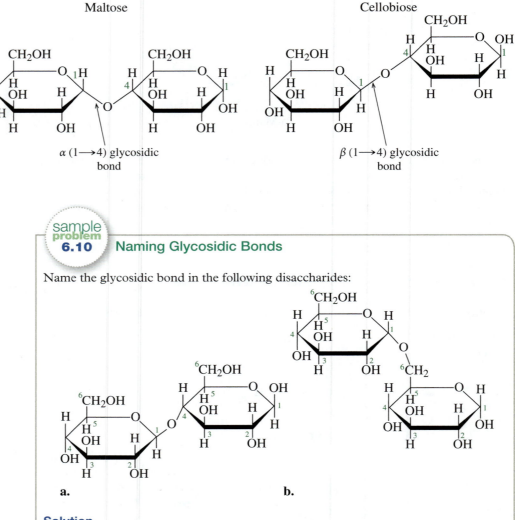

sample **problem**
6.10 Naming Glycosidic Bonds

Name the glycosidic bond in the following disaccharides:

a.

b.

Solution

a. The anomeric carbon (C1) in this glycosidic bond is in the β anomer form. The carbons that are connected are C1 on the left and C4 on the right. The glycosidic bond is designated $\beta(1 \rightarrow 4)$.

b. The anomeric carbon (C1) in this glycosidic bond is in the α anomer form. The carbons that are connected are C1 on the left and C6 on the right. The glycosidic bond is designated $\alpha(1 \rightarrow 6)$.

Three Important Disaccharides—Maltose, Lactose, and Sucrose

Three common disaccharides formed through the condensation of two monosaccharides are maltose, lactose, and sucrose. Their formation is outlined as follows:

α-D-glucose + D-glucose $\rightarrow$ maltose + H_2O [glycosidic bond $\alpha(1 \rightarrow 4)$]
β-D-galactose + D-glucose $\rightarrow$ lactose + H_2O [glycosidic bond $\beta(1 \rightarrow 4)$]
α-D-glucose + β-D-fructose $\rightarrow$ sucrose + H_2O [glycosidic bond $\alpha, \beta(1 \rightarrow 2)$]

In discussing saccharides larger than the monosaccharides, the "D" designation is often dropped from the name and assumed unless noted. For clarity, it is also common to see the ring form with the hydrogens not shown but implied. These shortcuts will be used from this point forward.

Maltose

Maltose, also known as malt sugar, is a disaccharide formed in the breakdown of starch. Malted barley, a key ingredient in beer, contains high levels of maltose. The malting process involves monitoring the germination of barley grains, during which the starch in the grain is converted to maltose through hydrolysis. This process is halted before the grains sprout by drying and roasting the grains. The glucose in the maltose of malted barley can then be converted to alcohol by yeast in the fermentation process.

The glycosidic bond between the glucose units in maltose is $\alpha(1 \rightarrow 4)$. Because one of the anomeric carbons (C1 on the second glucose unit) is not in a glycosidic bond and is considered a free anomeric carbon, maltose is a reducing sugar (see **Figure 6.12**).

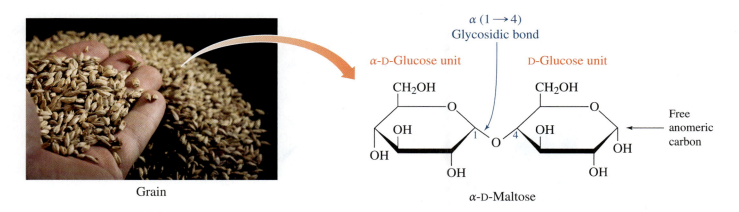

Grain

α-D-Maltose

FIGURE 6.12 The disaccharide maltose. One of the main sugars in malted barley (grains in photo) is maltose formed from hydrolyzed starch. The bonded anomeric carbon is in the α position, so this glycosidic bond is $\alpha(1 \rightarrow 4)$.

Lactose

Lactose, or milk sugar, is found in mammalian milk. Intolerance to lactose can occur in people who do not inherit or who lose the ability to produce the enzyme that hydrolyzes this disaccharide into its two monosaccharides. This hydrolytic enzyme, commonly known as lactase, is sometimes added to commercial products to assist in the digestion of milk products for lactose-intolerant people. When lactose remains undigested, intestinal bacteria break down undigested lactose and, in doing so, produce abdominal gas and cramping.

The glycosidic bond in lactose is $\beta(1 \rightarrow 4)$ because it occurs between C1 of a β-galactose and C4 of a glucose unit. Because the anomeric carbon on the glucose unit is free (not in a glycosidic bond), lactose is a reducing sugar (see **Figure 6.13**).

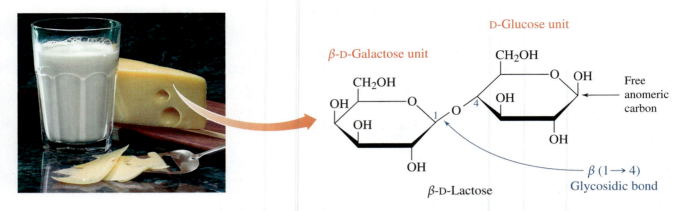

D-Glucose unit

β-D-Galactose unit

Free anomeric carbon

β (1 → 4) Glycosidic bond

β-D-Lactose

FIGURE 6.13 The disaccharide lactose. Lactose, found in the milk of mammals, contains galactose and glucose. The bonded anomeric carbon is in the β position, so this glycosidic bond is $\beta(1 \rightarrow 4)$.

Sucrose

Sucrose, ordinary table sugar, is the most abundant disaccharide in nature. High quantities of sucrose are found in sugar cane and sugar beets, which are the main commercial sources.

When glucose and fructose join in an $\alpha,\beta(1 \rightarrow 2)$ glycosidic bond, sucrose is formed. In sucrose, both anomeric carbons are bonded (carbon 1 of glucose and carbon 2 of fructose) so both are noted in the name of the glycosidic bond. Because there is no free anomeric carbon, sucrose does not react with Benedict's reagent and is *not* a reducing sugar (see **Figure 6.14**).

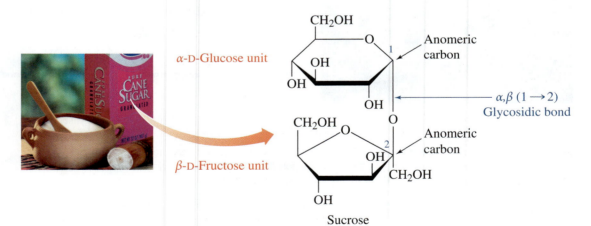

α-D-Glucose unit

Anomeric carbon

α,β (1 → 2) Glycosidic bond

β-D-Fructose unit

Anomeric carbon

Sucrose

FIGURE 6.14 The disaccharide sucrose. Sucrose, obtained from sugar beets and sugar cane, contains glucose and fructose. Both anomeric carbons are bonded in the glycosidic bond. Glucose is in the α position and fructose is in the β position, so this glycosidic bond is $\alpha, \beta(1 \rightarrow 2)$.

Relative Sweetness of Sugars and Artificial Sweeteners

The dieter looking for a sweet taste and fewer calories can use artificial sweeteners like Nutrasweet®, Splenda®, or Truvia™. Sweetness is perceived by our taste buds, where sugar molecules and other artificial sweeteners bind and register this taste in the brain.

Sweetness is referenced relative to sucrose (table sugar), which is given a value of 100. Fructose has a sweetness value at least one and a half times that of sucrose (table sugar). This means that it takes less fructose to achieve the same sweet taste as sucrose. Artificial sweeteners are hundreds of times sweeter than sucrose, so much less of them needs to be used. Further, many artificial sweeteners lack calories because they cannot produce energy, as many monosaccharides do through glycolysis. The sweetness of some sugars and other sweeteners compared to sucrose is shown in **Table 6.1**.

TABLE 6.1 Relative Sweetness of Sugars and Artificial Sweeteners

Sweetener	Sweetness Relative to Sucrose (= 100)	Description
Simple Sugars		
Fructose	140–175	Fruit sugar, a monosaccharide that is a component of sucrose
Invert sugar (Hydrolyzed sucrose)	120	Found in honey
Sucrose	100	Table sugar, a disaccharide containing glucose and fructose
Xylitol	100	A sugar alcohol, used in sugar-free products
Glucose	75	Dextrose, a monosaccharide that is a component of sucrose and lactose
Erythritol	70	A sugar alcohol used in sugar-free products
Sorbitol	36–55	A sugar alcohol used in sugar-free products
Maltose	32	A disaccharide of glucose
Galactose	30	A monosaccharide that is a component of the disaccharide lactose
Lactose	15	Milk sugar, a disaccharide containing galactose and glucose
Other Sweeteners		
Sucralose (Splenda®)	60,000	Chlorinated disaccharide
Saccharin	45,000	An organic substance initially discovered as a by product of research on dyes
Stevia (found in Truvia™)	25,000	Natural sweetener containing nonhydrolyzable glucosides found in the leaves of the plant *Stevia rebaudiana*
Aspartame (Nutrasweet®)	18,000	Contains two amino acids, aspartic acid and phenylalanine

6.23 Identify the following reactions as condensation or hydrolysis:
   a. two monosaccharides reacting to form a disaccharide
   b. the formation of a glycosidic bond
   c. a reaction in which one molecule breaks into two and an —H and —OH are added

6.24 Identify the following reactions as condensation or hydrolysis:
   a. a disaccharide breaking into two monosaccharides
   b. a reaction in which two molecules combine forming one and a molecule of water is produced
   c. the breaking of a glycosidic bond

6.25 Name the glycosidic bond present in mannobiose shown in the following figure:

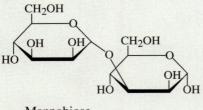

Mannobiose

6.26 Name the glycosidic bond present in melibiose, a disaccharide that has the sweetness of about 30 compared with sucrose.

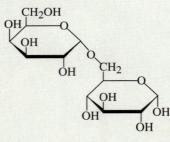

Melibiose

6.27 For each of the following disaccharides, name the glycosidic bond and draw the monosaccharide units produced by hydrolysis:

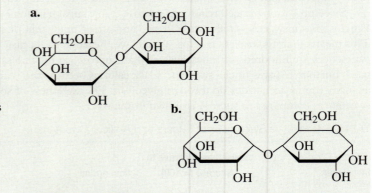

6.28 For each of the following disaccharides, name the glycosidic bond and draw the monosaccharide units produced by hydrolysis:

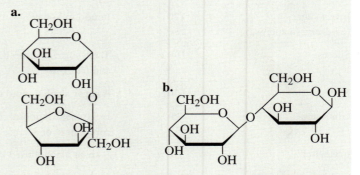

6.29 Identify a disaccharide that fits each of the following descriptions:
   a. ordinary table sugar
   b. found in milk and milk products
   c. also called malt sugar
   d. hydrolysis gives galactose and glucose

6.30 Identify a disaccharide that fits each of the following descriptions:
   a. not a reducing sugar
   b. composed of two glucose units
   c. also called milk sugar
   d. hydrolysis gives glucose and fructose

6.6 Polysaccharides

Some days there is more sunlight than other days—can plants store glucose for a rainy day? What happens if we eat more simple sugars (monosaccharides and disaccharides) than we need for energy—can we store that energy? Glucose can be stored in both plants and animals by connecting α-glucose units through glycosidic bonds as starch and glycogen respectively. Connecting many β-glucose units produces the molecule cellulose, a structural material in plants. These molecules, called polysaccharides, are very large. Because most of the monosaccharides in a polysaccharide are bonded through their anomeric carbon, polysaccharides do not contain a sufficient number of reducing ends to give a positive Benedict's test (see **Figure 6.15**).

a.

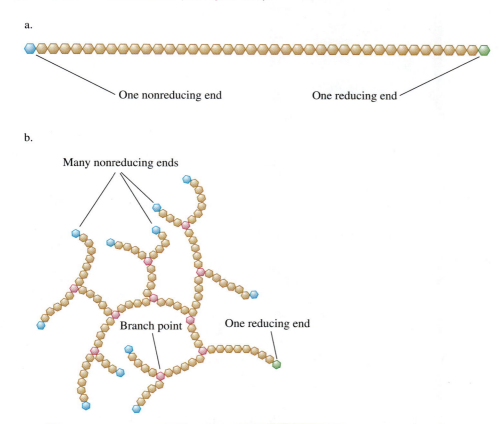

b.

FIGURE 6.15 Branching in polysaccharides. Polysaccharides can have all monosaccharides bonded in a straight chain (a) or can contain branching (b). Most polysaccharides contain a single reducing end and one (unbranched) or many (branched) nonreducing ends. They are considered nonreducing sugars.

Three important **storage polysaccharides** containing only α-glucose units are amylose, amylopectin, and glycogen. Two important **structural polysaccharides** containing β-glucose units are cellulose and chitin. Chitin contains a modified β-glucose unit. Even though these polysaccharides have a similar chemical makeup (glucose units), they differ structurally and functionally because of their glycosidic bonds and differences in branching.

Storage Polysaccharides

Amylose and Amylopectin—Starch

Have you ever eaten an unripe banana? Its taste is less sweet than a ripe banana, and the texture is rather mealy. The mealy or grainy texture is due to the stores of starch in the banana's fruit that have yet to be hydrolyzed to glucose as the fruit ripens. Starch is a glucose storage polysaccharide that accumulates in small granules in plant cells. Plant foods that we think of as starchy—like potatoes, grains, and beans—contain an abundance of these polysaccharides. Starch is a mixture of two polysaccharides, amylose and amylopectin. **Amylose,** which makes up about 20% of starch, is made up of anywhere from 250 to 4000 D-glucose units bonded $\alpha(1 \rightarrow 4)$ in a continuous chain. Long chains of D-glucose bonded $\alpha(1 \rightarrow 4)$ will tend to coil much like a telephone cord.

Amylopectin makes up about 80% of plant starch. It also contains D-glucose units connected by $\alpha(1 \rightarrow 4)$ glycosidic bonds. However, unlike amylose, about every 25 glucose units along a linear glucose chain, a second glucose chain branches off through an $\alpha(1 \rightarrow 6)$ glycosidic bond (see **Figure 6.16**). During ripening, starch granules undergo hydrolysis by the action of an enzyme called amylase that hydrolyzes the $\alpha(1 \rightarrow 4)$ glycosidic bonds in starch, producing glucose and maltose, which are sweet. When we eat starch, our digestive system breaks it down into glucose units for use by our bodies.

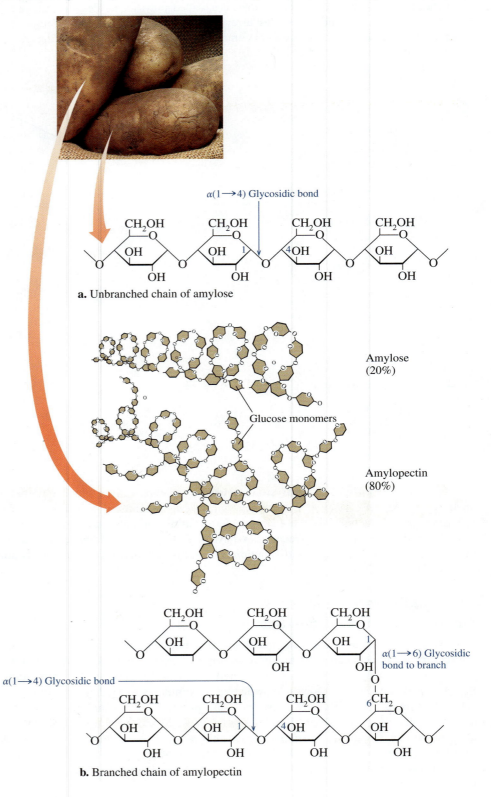

FIGURE 6.16 The storage polysaccharide starch. The structure of (a) amylose is a straight-chain polysaccharide of glucose units and (b) amylopectin is a branched chain of glucose.

Glycogen

Animals store glucose as a polysaccharide, too. **Glycogen** is the storage polysaccharide found in animals. Most glycogen stores are located in the liver and in muscles. Glycogen is identical in structure to amylopectin except that $\alpha(1 \rightarrow 6)$ branching occurs about every 12 glucose units. Glycogen is hydrolyzed to glucose in the liver and sent into the bloodstream to maintain constant glucose levels in the blood during fasting periods when sugars are not being consumed. The large amount of branching in this molecule allows for quick hydrolysis when glucose is needed.

Structural Polysaccharides

Cellulose

Trees stand tall because of the structural polysaccharide cellulose. **Cellulose** also contains glucose units, but they are D-glucose units bonded $\beta(1 \rightarrow 4)$. This single change in glycosidic bond configuration completely alters the overall structure of cellulose compared with that of amylose. Whereas amylose coils in an α-bonded glucose chain, the β-bonded chain of cellulose is straight. Many of these straight chains of cellulose align next to each other, forming a strong, rigid structure (see **Figure 6.17**). The straight chains of cellulose

Fibers formed from aligned cellulose

Cellulose aligns to form rigid structures

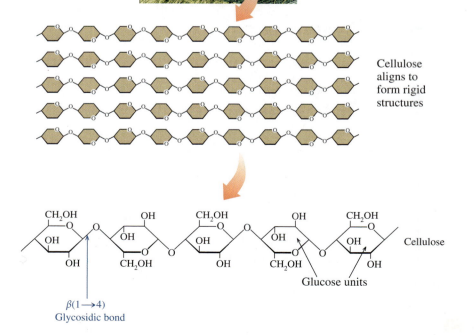

Cellulose

Glucose units

$\beta(1 \rightarrow 4)$
Glycosidic bond

FIGURE 6.17 The structural polysaccharide cellulose. Cellulose forms a rigid structure when the straight chain of the single polysaccharide formed from β-glucose units linking $\beta(1 \rightarrow 4)$ aligns with neighboring cellulose molecules. To easily show the straight chain, alternate glucose units are inverted.

are attracted through an attractive force called hydrogen bonding, which keeps the molecules rigidly aligned to one another. Hydrogen bonding is discussed in Chapter 7.

Nutritionally, cellulose is the most abundant insoluble fiber. We lack the enzyme called cellulase that hydrolyzes the $\beta(1 \to 4)$ glycosidic bonds in cellulose. Even though we do not eat wood, we get cellulose in our diet from whole-grain foods. We cannot digest cellulose, but it is still an important part of our diet because it assists with digestive movement in the small and large intestine. Some animals and insects—like sheep, cows, and termites—can digest cellulose because their digestive system contains bacteria and other microorganisms that produce cellulase.

Chitin

Chitin is a polysaccharide that makes up the exoskeleton of insects and crustaceans and the cell walls of most fungi. This polysaccharide is made up of a modified β-D-glucose called N-acetylglucosamine containing $\beta(1 \to 4)$ glycosidic bonds. An acetyl group bonded to the glucose ring through an amine replaces the hydroxyl group at carbon 2.

Like cellulose, chitin is a structurally strong material that has many uses, one of which is a surgical thread that biodegrades as a wound heals. Chitin is present in the exoskeletons of arthropods such as insects and spiders and serves to protect them from water. Because of this property, chitin can be used to waterproof paper. When ground up, chitin becomes a powder that holds in moisture, and it can be added to cosmetics and lotions. The structure of chitin is shown in **Figure 6.18**.

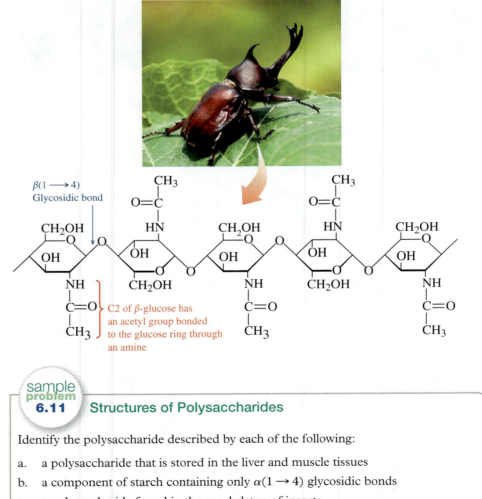

FIGURE 6.18 The structural polysaccharide chitin. Chitin is formed from β-N-acetylglucosamine units bonded $\beta(1 \to 4)$.

---

sample problem 6.11 **Structures of Polysaccharides**

Identify the polysaccharide described by each of the following:

a. a polysaccharide that is stored in the liver and muscle tissues

b. a component of starch containing only $\alpha(1 \to 4)$ glycosidic bonds

c. a polysaccharide found in the exoskeleton of insects

Solution

a. glycogen   b. amylose   c. chitin

6.31 Describe the similarities and differences of the following polysaccharides:
a. amylose and amylopectin
b. amylopectin and glycogen

6.32 Describe the similarities and differences of the following polysaccharides:
a. amylose and cellulose
b. cellulose and chitin

6.33 Give the name of one or more polysaccharides that matches each of the following descriptions:
a. not digestible by humans
b. the storage form of carbohydrates in plants
c. contains only $\alpha(1 \rightarrow 4)$ glycosidic bonds
d. glucose polysaccharide with the most branching

6.34 Give the name of one or more polysaccharides that matches each of the following descriptions:
a. the storage form of carbohydrates in animals
b. contains only $\beta(1 \rightarrow 4)$ glycosidic bonds
c. contains both $\alpha(1 \rightarrow 4)$ and $\alpha(1 \rightarrow 6)$ glycosidic bonds
d. produces maltose during digestion

6.7 Carbohydrates and Blood

We usually think of carbohydrates as food sources. However, they also play other roles in our body. They are found in the blood and other bodily fluids. In this section, we describe two other functions of carbohydrates as signals on the surface of blood cells and as anticoagulants.

6.7 Inquiry Question: What other roles can carbohydrates play in the body?

ABO Blood Types

What is your blood type? Many people know their blood type, but most do not know that the blood types of A, B, AB, and O refer to carbohydrates. Your red blood cells have a number of chemical markers bonded to the cell surface identifying the cells as yours as they circulate through the bloodstream. One set of those chemical markers are the oligosaccharides known as the ABO blood markers, which contain either three or four monosaccharides. Some of these are modified monosaccharides and contain other functional groups.

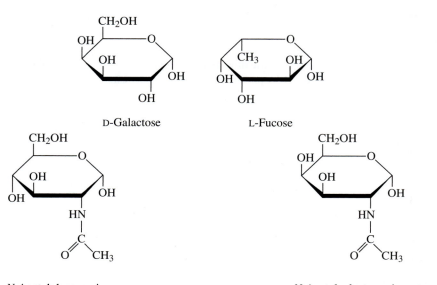

D-Galactose

L-Fucose

N-Acetylglucosamine

N-Acetylgalactosamine

Structures of the monosaccharides that are in the ABO blood groups.

All the blood types include the carbohydrates *N*-acetylglucosamine, galactose, and **fucose,** a carbohydrate that often appears in nature as its L-enantiomer. Type O blood contains only these three carbohydrates attached to the surface of the red blood cell through a glycosidic bond. Type A and type B blood contain these three plus a fourth carbohydrate. In type A, the fourth carbohydrate is an *N*-acetylgalactosamine bonded to the galactose unit, while in type B blood, the fourth carbohydrate is a second galactose bonded to the galactose. People with type AB blood have both the type A carbohydrate set and the type B carbohydrate set on their red blood cells.

Comparison of the sugar linking in the O, A, and B blood groups.

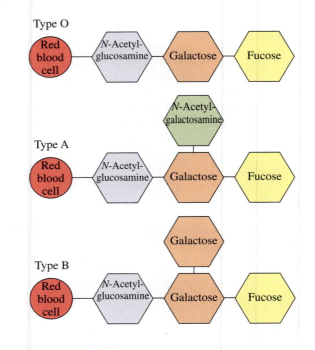

Universal Donors and Acceptors

Integrating chemistry

Everyone has a unique blood type. One of most important markers is the ABO blood marker group. The O blood type is considered the universal donor blood type, but why? The body can only recognize its own carbohydrate set (A, B, or O) and will try to destroy what it considers a foreign blood type by producing protein molecules called antibodies that act to destroy the foreign blood type. Antibodies are described in more detail in Chapter 10. Because the trisaccharide on the cells of O-type blood (*N*–acetylglucosamine—galactose—L-fucose) is present on cells of *all* blood types (A, B, and AB), O-type blood can be donated to any blood type and is considered the universal donor. No blood type recognizes the O carbohydrate set as foreign.

The AB blood type is considered the universal acceptor blood type for a similar reason. AB blood contains all possible ABO combination types, so any blood type transfused will be accepted by the body. The compatibility of blood types is summarized in **Table 6.2**.

TABLE 6.2 Compatibility of ABO Blood Groups

Blood Group	Can Receive Blood Types	Can Donate to Blood Types
A	A, O	A, O
B	B, O	B, O
AB[a]	A, B, AB, O	Cannot donate to other blood types
O[b]	O	A, B, AB, O (all types)

[a]AB universal acceptor.
[b]O universal donor.

Heparin

Heparin is a medically important polysaccharide that prevents clotting, acting as an anticoagulant in the bloodstream. Test tubes, tubing, and needles used for drawing blood are routinely coated with heparin to prevent the blood from clotting. Heparin is a highly ionic polysaccharide of many repeating disaccharide units. The disaccharide includes an oxidized monosaccharide and a D-glucosamine. This polysaccharide belongs to a group of polysaccharides called **glycosaminoglycans,** all of which have highly charged repeating disaccharide units. Negative charges are mainly due to the presence of sulfate groups.

Oxidized monosaccharide   D-Glucosamine monosaccharide

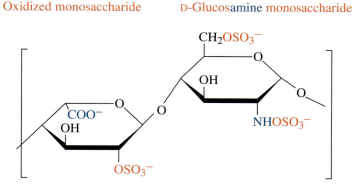

Heparin's repeating disaccharide

Heparin is a modified polysaccharide that serves as an anticoagulant in the bloodstream.

sample problem 6.12 | ABO Blood Types

Explain whether the following blood types could be donated to a person with type A blood:

a. A   b. B   c. AB   d. O

Solution

a. Two matched blood types can always be donated to each other; the transfused blood will not be recognized as foreign. Type A can be donated to type A.

b. Type B cannot be donated to type A because the person with type A blood would still recognize type B blood as foreign and try to eliminate it. Type B cannot be donated to type A.

c. Even though type AB contains the A carbohydrates, it also contains the B carbohydrates, which a person with type A blood would recognize as foreign. Type AB cannot be donated to type A.

d. Because the O carbohydrate set is present within the type A carbohydrates, type O blood will not be recognized as foreign and will be accepted. Type O can be donated to type A.

6.35 Explain whether the following blood types could be donated to a person with type B blood:
a. A   b. AB

6.36 Explain whether the following blood types could be donated to a person with type O blood:
a. B   b. AB

SUMMARY

6.1 Classes of Carbohydrates

6.1 Inquiry Question: How do we classify carbohydrates?

Carbohydrates are classified as monosaccharides (simple sugars), disaccharides (two monosaccharide units), oligosaccharides (three to nine monosaccharide units), and polysaccharides (many monosaccharide units). The simplest carbohydrates are the monosaccharides with a molecular formula of $C_n(H_2O)_n$ where $n = 3–6$. Edible carbohydrates that cannot be broken down by the body's enzymes are classified as either soluble or insoluble fibers based on their ability to mix with water.

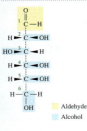

Aldehyde
Alcohol

6.2 Functional Groups in Monosaccharides

6.2 Inquiry Question: What functional groups are present in monosaccharides?

Monosaccharides contain several alcohol (hydroxyl) groups, and either an aldehyde or ketone functional group. Alcohols can be primary, secondary, or tertiary depending on the number of carbons attached to the alcohol carbon. Aldehydes and ketones are carbonyl-containing functional groups. Because of the functional groups present, monosaccharides can be called aldoses or ketoses and referred to as polyhydroxyaldehydes and polyhydroxyketones.

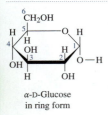

Fischer projection
D-Glucose

6.3 Stereochemistry of Monosaccharides

6.3 Inquiry Question: What structural properties do monosaccharides display?

Monosaccharides can be drawn linearly in a representation called a Fischer projection that highlights their chiral centers. Most carbohydrates found in nature are the D-isomer. Stereoisomers that have multiple chiral centers can either be related to each other as enantiomers (mirror images of each other) or diastereomers (not enantiomers). Some important monosaccharides are glucose, galactose, mannose, fructose, deoxyribose, and ribose.

6.4 Reactions of Monosaccharides

6.4 Inquiry Question: How are the products of monosaccharide reactions drawn?

α-D-Glucose
in ring form

A hydroxyl group and the carbonyl functional group of a linear monosaccharide can react to enclose the hydroxyl's oxygen in a ring. Because the carbonyl group is a planar structure, this reaction produces two possible ring arrangements about the anomeric carbon termed the α and the β anomer when a ring is formed. The anomeric carbon of carbohydrates (C1 of an aldose) is highly reactive and can undergo both oxidation to a carboxylic acid or reduction to an alcohol. Monosaccharides are considered reducing sugars because their anomeric carbon can react to reduce another compound in a redox reaction.

6.5 Disaccharides

6.5 Inquiry Question: How are disaccharides formed and identified?

Carbohydrates form glycosides when an anomeric carbon reacts with a hydroxyl on a second organic molecule. This condensation results in the formation of a glycosidic bond. Glycosidic bonds are named by designating the anomer of the reacting monosaccharide and the carbons that are bonded; for example, $\alpha(1 \rightarrow 4)$. Some important disaccharides formed through condensation reactions are maltose, lactose, and sucrose. Monosaccharides and disaccharides make up simple sugars, many of which are sweet to the taste. The sweetness of carbohydrates and other carbohydrate substitutes is indexed relative to the sweetness of sucrose (see Table 6.1).

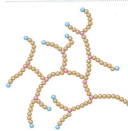

6.6 Polysaccharides

6.6 Inquiry Question: How are polysaccharides characterized?

Polysaccharides consist of many monosaccharide units bonded together through glycosidic bonds. Glucose can be stored as a polysaccharide called starch in plants and glycogen in animals. Starch consists of two polysaccharides: amylose, a linear chain of glucose, and amylopectin, a branched chain of glucose. Glycogen is also a branched polysaccharide of glucose, but it contains more branching than does amylopectin. Two polysaccharides that are structurally important in nature include cellulose in plants (wood) and chitin in arthropods (exoskeleton) and fungi (cell wall). Cellulose is a linear chain of glucose, but in cellulose the glycosidic bonds are β, whereas in starch they are bonded α. Chitin is also a linear chain of a modified glucose, N-acetylglucosamine bonded $\beta(1 \rightarrow 4)$. Both these structural polysaccharides form strong, water-resistant materials when the linear chains are aligned with each other.

6.7 Carbohydrates and Blood

6.7 Inquiry Question: What other roles can carbohydrates play in the body?

Carbohydrates are also used as recognition markers on the surfaces of cells and in other bodily fluids. The ABO blood groups are oligosaccharides of which one of the carbohydrate units is L-fucose, one of the few L-sugars found in nature. These ABO oligosaccharides are found on the surface of red blood cells. The A and B blood groups look like the O blood group except that they contain an additional monosaccharide. For this reason, the O blood type is considered the universal donor. Heparin is a polysaccharide consisting of a repeating disaccharide containing an oxidized monosaccharide and a glucosamine. Heparin functions in the blood as an anticoagulant and is commonly found as a coating on medical tubing and syringes used during blood transfusions.

The study guide will help you check your understanding of the main concepts in Chapter 6. You should be able to

6.1 Classes of Carbohydrates

- Classify carbohydrates as mono-, di-, oligo-, or polysaccharides.
- Distinguish soluble and insoluble fiber.

6.2 Functional Groups in Monosaccharides

- Distinguish primary, secondary, and tertiary alcohols.
- Recognize and draw the functional groups alcohol, aldehyde, and ketone.

6.3 Stereochemistry in Monosaccharides

- Distinguish D- and L- stereoisomers of monosaccharides.
- Draw Fischer projections of linear monosaccharides.
- Define enantiomer, epimer, and diastereomer.
- Draw enantiomers and diastereomers of linear monosaccharides.
- Characterize common monosaccharides.

6.4 Reactions of Monosaccharides

- Draw the cyclic α and β anomers from linear monosaccharide structures.
- Draw oxidiation and reduction products of aldoses.

6.5 Disaccharides

- Locate and name glycosidic bonds in disaccharides.
- Distinguish condensation and hydrolysis reactions of simple sugars.
- Characterize common dissacharides.

6.6 Polysaccharides

- Identify polysaccharides by glycosidic bond and sugar subunit.

6.7 Carbohydrates and Blood

- Predict ABO compatibility.
- Describe the structure and role of heparin.

Key Terms

alcohol—A family of organic compounds whose functional group is an OH (hydroxyl).

aldehyde—A family of organic compounds whose functional group is a carbonyl ($C=O$) bonded to at least one hydrogen atom.

aldose—A monosaccharide that contains the aldehyde functional group.

amylopectin—An α-D-glucose polysaccharide with $\alpha(1 \rightarrow 4)$ glycosidic bonds and $\alpha(1 \rightarrow 6)$ branch points approximately every 25 glucose units; a component of starch.

amylose—An unbranched α-D-glucose polysaccharide with $\alpha(1 \rightarrow 4)$ glycosidic bonds; a component of starch.

anomeric carbon—In a monosaccharide, the carbon that was the carbonyl in the linear structure; for example, carbon 1 of D-glucose.

anomer—A sugar diastereomer differing only in the position of the hydroxyl at the anomeric carbon.

Benedict's test—A common test for the presence of a reducing sugar. If a reducing sugar is present, a brick-red precipitate results.

carbohydrate—A simple or complex sugar composed of carbon, hydrogen, and oxygen; a primary source of energy in a normal diet.

carbonyl—A functional group that contains a carbon double bonded to an oxygen ($C=O$).

cellulose—An unbranched β-D-glucose polysaccharide with $\beta(1 \rightarrow 4)$ glycosidic bonds; the main component of wood and plants.

chitin—An unbranched β-D-acetylglucosamine polysaccharide with $\beta(1 \rightarrow 4)$ glycosidic bonds; the main component in the exoskeletons of arthropods and the cell walls of fungi.

diastereomer—A stereoisomer that is not a mirror image. Diastereomers exist for molecules with more than one chiral center.

disaccharide—A carbohydrate composed of two monosaccharides joined through a glycosidic bond.

epimer—A stereoisomer with multiple chiral centers that differs from another stereoisomer at a single chiral center; a type of diastereomer.

Fischer projection—A molecular representation that displays the three-dimensional shape of a chiral molecule on a two-dimensional space. Horizontal lines from a chiral center project groups in front of the plane (like wedges), and vertical lines from a chiral center project groups behind the plane (like dashes). The chiral center is at the intersection of the lines and is in the plane.

fructose—A monosaccharide found in honey and fruit juices; it is combined with glucose in sucrose; also called levulose and fruit sugar.

fucose—An L-monosaccharide found in the ABO blood groups.

furanose—The five-membered ring form of a monosaccharide containing four carbons and one oxygen atom in the ring.

galactose—A monosaccharide that occurs combined with glucose in lactose.

glucose—The most prevalent monosaccharide in the diet. An aldohexose found in fruits, vegetables, corn syrup, and honey; also known as blood sugar and dextrose.

glycogen—An α-D-glucose polysaccharide with $\alpha(1 \rightarrow 4)$ glycosidic bonds and $\alpha(1 \rightarrow 6)$ branch points approximately every 12 glucose units; storage polysaccharide found in animals.

glycolysis—A series of chemical reactions in the body that break down monosaccharides, producing energy.

glycosaminoglycan—A highly ionic polysaccharide that contains a repeating disaccharide unit consisting of an oxidized monosaccharide and a glucosamine.

glycoside—See glycosidic bond.

glycosidic bond—Formed when two hydroxyl groups join, one of which is on the anomeric carbon of a monosaccharide in ring form. When one of the hydroxyls is part of a second monosaccharide, a disaccharide is formed.

hemiacetal—(hem-ee-ass'-i-tal) The functional group formed when an alcohol and aldehyde functional group combine. It contains a tetrahedral carbon bonded to —OH and —OR.

heparin—A medically important polysaccharide that prevents clotting in the bloodstream.

insoluble fiber—Dietary carbohydrates that do not dissolve in water but move through the digestive tract unchanged. Important for a healthy digestive tract.

ketone—A family of organic compounds whose functional group is a carbonyl ($C=O$) bonded to two alkyl groups.

ketose—A monosaccharide that contains the ketone functional group.

lactose—A disaccharide consisting of glucose and galactose found in milk and milk products.

maltose—A disaccharide consisting of two glucose units; obtained from the hydrolysis of starch in germinating grains.

mannose—A monosaccharide that is not easily absorbed by the body but prevalent in some fruits like cranberries.

monosaccharide—A simple sugar unit containing three or more carbons with the general formula $C_n(H_2O)_n$.

oligosaccharide—A carbohydrate composed of three to nine monosaccharide units joined through glycosidic bonds.

polysaccharide—A carbohydrate composed of many monosaccharide units joined through glycosidic bonds. The type of monosaccharide, glycosidic bond, and branching vary.

primary (1°) alcohol—An alcohol that has one alkyl group bonded to the carbon atom with the —OH.

pyranose—The six-membered ring form of a monosaccharide that contains five carbons and one oxygen atom in the ring.

reducing sugar—A carbohydrate with a free aldehyde group capable of reducing another substance.

secondary (2°) alcohol—An alcohol that has two alkyl groups bonded to the carbon atom with the —OH.

soluble fiber—Dietary carbohydrates that mix with water and form a gel-like substance. When ingested, they give a feeling of fullness.

storage polysaccharide—Polysaccharides that are stored in cells as a glucose energy reserve; examples include starch in plants and glycogen in animals.

structural polysaccharide—Polysaccharides whose function is to provide structure for an organism; examples include cellulose, which

supports plant structure, and chitin in the exoskeleton of arthopods.

sucrose—A disaccharide composed of glucose and fructose; a nonreducing sugar, commonly called table sugar or "sugar."

D-sugar—An enantiomer whose configuration is based on the arrangement of the attachments on D-glyceraldehyde; these carbohydrate orientations are the most common in nature. In the Fischer projection, the hydroxyl group on the chiral center farthest from the carbonyl is on the right side of the molecule.

L-sugar—An enantiomer whose configuration is based on the arrangement of the attachments on L-glyceraldehyde; these carbohydrate orientations are rarely found in nature. In the Fischer projection, the hydroxyl group on the chiral center farthest from the carbonyl is on the left side of the molecule.

tertiary (3°) alcohol—An alcohol that has three alkyl groups bonded to the carbon atom with the —OH.

Summary of Reactions

Ring Formation (D-Glucose)

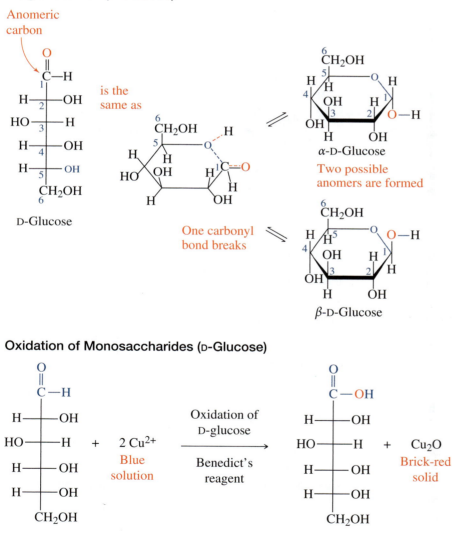

Oxidation of Monosaccharides (D-Glucose)

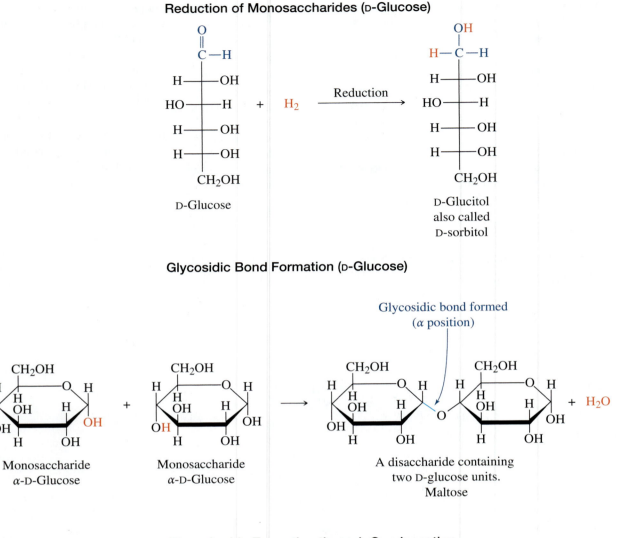

Reduction of Monosaccharides (D-Glucose)

D-Glucose + H$_2$ $\xrightarrow{\text{Reduction}}$ D-Glucitol also called D-sorbitol

Glycosidic Bond Formation (D-Glucose)

Glycosidic bond formed (α position)

Monosaccharide α-D-Glucose + Monosaccharide α-D-Glucose → A disaccharide containing two D-glucose units. Maltose + H$_2$O

Disaccharide Formation through Condensation

glucose + glucose $\longrightarrow$ maltose + H$_2$O

glucose + galactose $\longrightarrow$ lactose + H$_2$O

glucose + fructose $\longrightarrow$ sucrose + H$_2$O

Additional Problems

6.37 Write the molecular formula for a carbohydrate containing four carbons.

6.38 What would be the molecular formula of a monosaccharide characterized as an aldopentose?

6.39 Explain the difference between an oligosaccharide and a polysaccharide.

6.40 Explain the difference between a monosaccharide and a disaccharide.

6.41 Describe the properties of soluble fiber.

6.42 Describe the properties of insoluble fiber.

6.43 Name the functional groups present in aldoses.

6.44 Name the functional groups present in ketoses.

6.45 Classify each of the following as primary, secondary or tertiary alcohols:

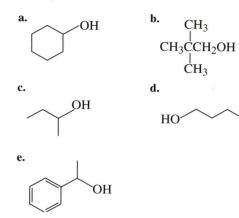

a.

b. CH₃CCH₂OH with CH₃ and CH₃

c.

d.

e.

6.46 Classify each of the following as primary, secondary or tertiary alcohols:

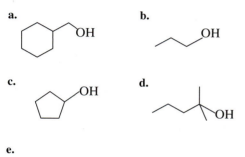

a.

b.

c.

d.

e.
CH₃CHCH₂CH₃
    |
   OH

6.47 How are the following pairs of carbohydrates, shown in a Fischer projection, related to each other? Are they structural isomers, enantiomers, diastereomers, or epimers?

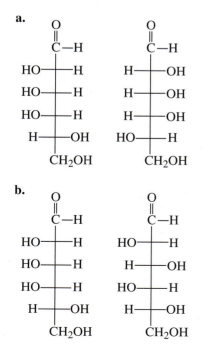

a.

b.

6.48 How are the following pairs of carbohydrates, shown in a Fischer projection, related to each other? Are they structural isomers, enantiomers, diastereomers, or epimers?

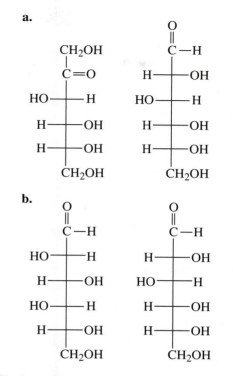

a.

b.

6.49 Identify the following carbohydrates as the α or β anomer:

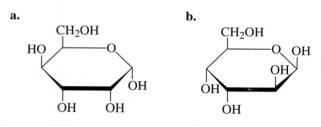

a.   b.

6.50 Identify the following carbohydrates as the α or β anomer:

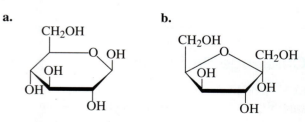

a.   b.

6.51 Draw the α and β anomer of D-mannose.

6.52 Draw the α and β anomer of D-fructose.

6.53 Draw the Fischer projection of the product of the oxidation of D-galactose at C1.

256   CHAPTER SIX Carbohydrates

6.54 Draw the Fischer projection of the product of the oxidation of the monosaccharide D-talose at C1.

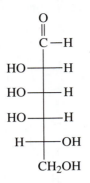

D-Talose

6.55 Draw the Fischer projection of the product of reduction reaction of D-galactose at C1.

6.56 Draw the Fischer projection of the product of the reduction reaction of D-talose at C1.

6.57 Will the following carbohydrates produce a positive Benedict's test?

a. D-glucose   **b.** lactose

c. sucrose   **d.** starch

6.58 Will the following carbohydrates produce a positive Benedict's test?

a. D-mannose   **b.** L-fucose

c. maltose   **d.** glycogen

6.59 Draw the product of the following 1 → 4 condensation and name the glycosidic bond:

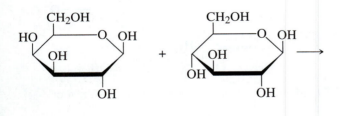

6.60 Draw the product of the following 1 → 4 condensation:

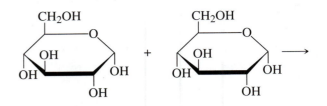

6.61 Isomaltose, a disaccharide formed during carmelization in cooking, contains two glucose units bonded α(1 → 6). Draw the structure of isomaltose.

6.62 Our bodies cannot digest cellulose because we lack the enzyme cellulase. Why is cellulose an important part of a healthy diet if we cannot digest it?

6.63 Give the name of one or more carbohydrates that match the following descriptions:

a. ordinary table sugar

b. dextrose

c. an L-sugar found in the ABO blood types

d. found in starch

e. found in the cell walls of fungi

6.64 Give the name of one or more carbohydrates that match the following descriptions:

a. a disaccharide of D-glucose bonded α(1 → 4)

b. fruit sugar

c. contains a sugar acid and sugar amine disaccharide repeat unit

d. storage polysaccharide in animals

e. insoluble fiber found in plants and trees

6.65 Explain whether the following blood types could be accepted by a person with type O blood:

a. B   **b.** A

6.66 Explain whether the following blood types could be accepted by a person with type A blood:

a. AB   **b.** O

Challenge Problems

6.67 Carbohydrates are also easily oxidized at C6. Provide the product of the oxidation of (a) D-galactose at C6 and (b) D-talose at C6.

6.68 Carbohydrates are abbreviated using a three-letter abbreviation followed by their glycosidic bond type. For example, maltose and sucrose can be written respectively as

Glcα(1 → 4)Glc   Glcα(1 → 2)βFru

Maltose   Sucrose

Provide the structure for the O-type blood carbohydrate set given the following abbreviation:

L- Fucα(1 → 2)Galβ(1 → 4)GlcNAc

6.69 The structure of sucralose, found in the artificial
sweetener Splenda®, is shown in the figure. It consists of
a chlorinated disaccharide made up of galactose and
fructose. In its structure shown,
   a. identify the galactose unit and the fructose unit.
   b. identify the type of glycosidic bond present.
   c. determine if Splenda® is a reducing sugar.

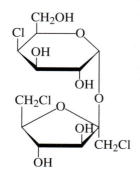

6.70 Which of the components in starch is more likely to be
broken down more quickly in plants, amylose or
amylopectin? Why?

6.71 How much energy is produced if a person eats 50 g of
digestible carbohydrate (not fiber) in a day? In this case,
what percent of a 2200 Calorie diet would be digestible
carbohydrate? Recall that carbohydrates provide
four Calories of energy per gram consumed.

6.72 D-Fructose can also form a six-membered ring. Draw the
β isomer of D-fructose in the six-membered ring form.

Answers to Odd-Numbered Problems

Practice Problems

6.1   **a.** polysaccharide
     b. oligosaccharide
     c. disaccharide

6.3   **a.** secondary
     b. primary
     c. tertiary
     d. secondary

6.5   **a.** ketone
     b. ketone
     c. aldehyde

6.7   **a.** D-isomer
     b. L-isomer
     c. D-isomer

6.9   **a.**   **b.**

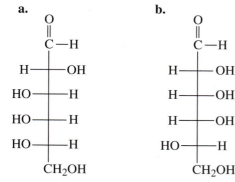

6.11 A—diastereomer; B—diastereomer; C—enantiomer

6.13   **a.** D-glucose
     b. D-fructose
     c. D-mannose

6.15   **a.** beta
     b. beta

6.17

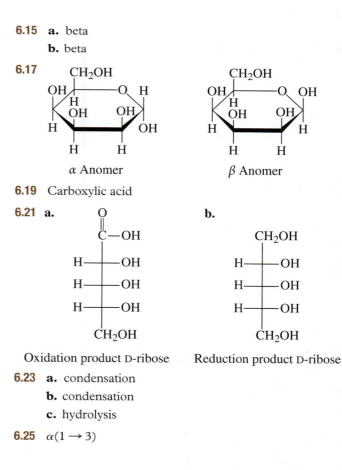

α Anomer   β Anomer

6.19 Carboxylic acid

6.21   **a.**   **b.**

Oxidation product D-ribose   Reduction product D-ribose

6.23   **a.** condensation
     b. condensation
     c. hydrolysis

6.25 $\alpha(1 \rightarrow 3)$

6.27 **a.** bond is β (1 → 4)

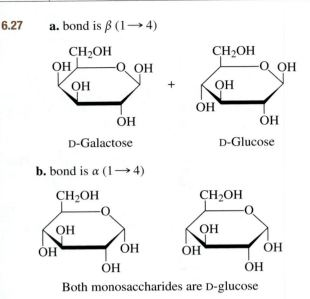

D-Galactose   +   D-Glucose

b. bond is α (1 → 4)

Both monosaccharides are D-glucose

6.29 **a.** sucrose
b. lactose
c. maltose
d. lactose

6.31 **a.** Both contain α(1 → 4) glycosidic bonds and only D-glucose. Amylopectin also contains branching α(1 → 6).
b. Both contain α(1 → 4) glycosidic bonds and branching α(1 → 6) and only D-glucose. Branching occurs more often in glycogen than amylopectin.

6.33 **a.** cellulose, chitin
b. amylose, amylopectin
c. amylose
d. glycogen

6.35 **a.** no
b. no

Additional Problems

6.37 $C_4H_8O_4$

6.39 An oligosaccharide is smaller, containing between 3 and 9 monosaccharide units, while a polysaccharide contains 10 or more monosaccharide units.

6.41 Soluble fibers mix with water and form a gel-like substance that gives a feeling of fullness when eaten.

6.43 aldehyde, hydroxyls (alcohol)

6.45 **a.** secondary
b. primary
c. secondary
d. primary
e. secondary

6.47 **a.** enantiomer
b. epimer

6.49 **a.** alpha
b. beta

6.51

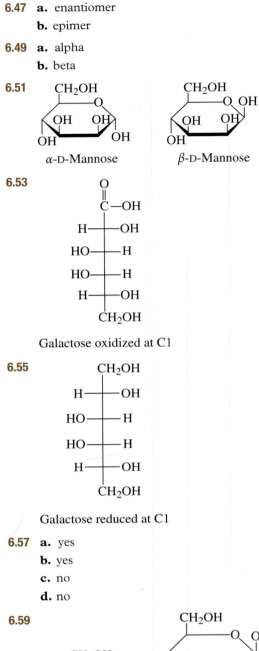

α-D-Mannose   β-D-Mannose

6.53

$$
\begin{array}{c}
O \\
\parallel \\
C-OH \\
H-\!\!-OH \\
HO-\!\!-H \\
HO-\!\!-H \\
H-\!\!-OH \\
CH_2OH
\end{array}
$$

Galactose oxidized at C1

6.55

$$
\begin{array}{c}
CH_2OH \\
H-\!\!-OH \\
HO-\!\!-H \\
HO-\!\!-H \\
H-\!\!-OH \\
CH_2OH
\end{array}
$$

Galactose reduced at C1

6.57 **a.** yes
b. yes
c. no
d. no

6.59

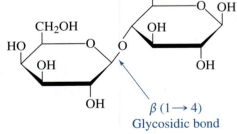

β (1 → 4)
Glycosidic bond

6.61

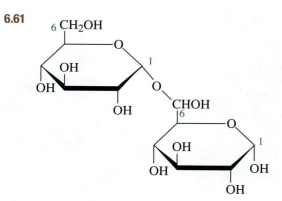

Isomaltose

6.63 **a.** sucrose

b. glucose

c. L-fucose

d. amylose, amylopectin

e. chitin

6.65 **a.** no

b. no

6.67 **a.**   **b.**

CHO			CHO
H	OH	HO	H
HO	H	HO	H
HO	H	HO	H
H	OH	H	OH
COOH			COOH

6.69 **a.**   **b.**

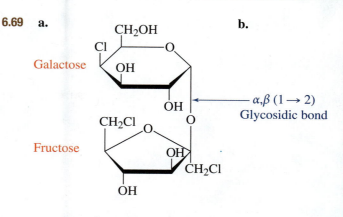

Galactose

Fructose

$\alpha,\beta\ (1 \rightarrow 2)$
Glycosidic bond

c. Splenda® is not a reducing sugar.

6.71 200 Calories, 9%

How does soap remove greasy dirt from our hands when we wash them? The answer can be found in the attractions between soap molecules, water, and greasy dirt. In this chapter, we explore attractive forces between compounds and how they affect the physical properties of compounds.

7

What's the Attraction? State Changes, Solubility, and Lipids

WE CAN STOP THE SPREAD OF DISEASE

through frequent hand washing. Why does soap and water eliminate microbes? The chemical structure of soap, water, and microbes affects how the substances behave when combined. In this case, the outer covering of a cell (the cell membrane) is attracted more to the soap than to plain tap water, allowing microbes to be washed down the drain when soap is used.

In Chapter 7, we explore the attractive forces that draw substances to each other. We take a more in-depth look at the states of matter and apply attractive forces to the physical properties of boiling, melting, and dissolving in water. From here, we apply our understanding of attractive forces to explore several lipids including soaps, dietary fats and oils, phospholipids, and cholesterol, finishing with the structure of a cell membrane.

Discovering the Concepts

? Inquiry Activity—The Attractive Forces

Information

Molecules "stick" together in the liquid and solid states due to attractions between the molecules. This accounts for some of their physical properties like boiling point, melting point, viscosity, and surface tension. In solids, each molecule is fixed (moves very little) relative to the neighboring molecules and the attractive forces are optimized.

Attractive forces are often referred to as "intermolecular forces" because they occur between (*inter* is a prefix meaning *between*) molecules, not within molecules. Technically, ions are not molecules, so the term *attractive force* is used here to encompass the five forces shown in Table 1.

Questions (Use Table 1 on the next page to answer the following questions)

1. What is common about the attractions between all the forces shown circled in Table 1?
2. Which is stronger, a formal +/− attraction (caused by a gain or loss of electrons), or a partial δ^+/δ^- attraction (caused by a shift in electron position)?
3. Which of the attractive forces occurs in nonpolar organic molecules?
4. Hydrogen bonds are a unique attractive force (not an actual covalent bond) present in some molecules. The attraction occurs between a δ^+ (also known as *polarized*) hydrogen in a polar bond with O, N, or F and a lone pair of electrons (:) on an O, N, or F. The hydrogen is called the *donor,* and the pair of electrons is referred to as the *acceptor*. This attraction is indicated with a dashed line between the polarized H and O, N, or F.
   a. How many polarized Hs are in water?
   b. How many lone pairs of electrons are in a molecule of H_2O?
   c. Based on your answers to a and b, how many H bonds can one water molecule make to neighboring water molecules?
   d. Carefully re-read the information given at the beginning of Question #4. Draw all possible hydrogen bonds to a water molecule.

? What's an Inquiry Question?

Inquiry Questions are designed to focus your reading on the main concepts by section. An Inquiry Question appears at the beginning of each section.

TABLE 1 Examples of the Attractive Forces Between Molecules (from weakest to strongest)

Force Name	Example

London forces

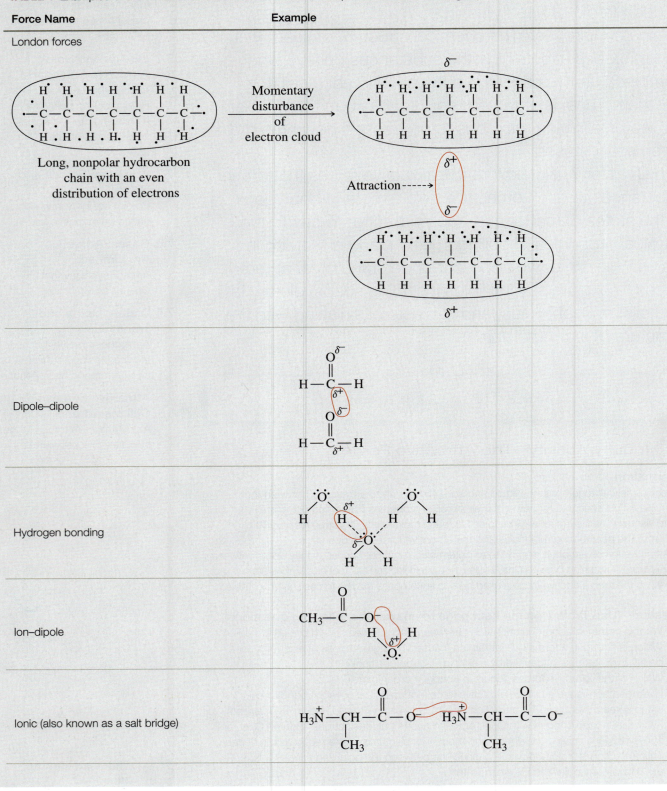

Dipole–dipole

Hydrogen bonding

Ion–dipole

Ionic (also known as a salt bridge)

5. How many hydrogen bonds can one methanol molecule (CH_3OH) make to neighboring methanol molecules? Draw them.
6. The boiling point of pure water is 100 °C and the boiling point of pure methanol is 65 °C. Provide an explanation for this difference based on the strength or number of attractive forces present in each.
7. Polar molecules can contain more than one type of attractive force. Fill in the following table with a yes or no answer to the question: "Does this attractive force operate between molecules of the given compound?"

Molecule	H Bond	Dipole–Dipole	London
CCl_4			
H_2O			
$CHCl_3$			

7.1 Types of Attractive Forces

In Chapter 3, we saw that atoms form compounds either by giving or taking electrons (ionic compounds) or by sharing electrons (covalent compounds) to form bonds. In both cases, electrons are distributed unevenly and a charge builds up within the compound. Because opposite charges attract, compounds with either an ionic charge (+/−) or a partial charge (δ^+/δ^-) can be attracted to each other.

Attractive forces are caused by the attraction of an electron-rich area of one compound (where electrons spend more of their time) to an electron-poor area of another molecule (where electrons spend less time.) Attractive forces can occur between two molecules, between two ions, or between an ion and a molecule. If the attraction is between two molecules, it is called an **intermolecular force.**

We will consider five attractive forces shown in order of strength in **Figure 7.1**. We will start with the weakest, London forces, and work up to the strongest, **ionic attraction.**

London Forces

If you have ever wrestled with a piece of plastic wrap that clings to itself rather than to the food you are trying to wrap, you have experienced **London forces.** These attractive forces occur momentarily in all molecules when electrons become unevenly distributed over a molecule's surface. When this happens, the partially positive side of this temporary dipole attracts the electrons of the second molecule, creating an attraction between these two molecules and inducing a temporary dipole in the second molecule. The result is the attractive force that you feel as you pull apart two sheets of plastic wrap (see **Figure 7.2**). While *all compounds exhibit London forces,* these forces are significant only in the case of nonpolar molecules because these are the only attractive force present in nonpolar molecules.

Many plastic wraps contain molecules which have long, nonpolar hydrocarbon chains that interact with each other, creating temporary dipoles throughout the material. The attraction between the temporary dipoles causes the wrap to stick to itself. Although this is the weakest attractive force, many of these weak forces acting together can become quite strong. The terms *induced dipole* and *dispersion force* describe the same attractive force as London forces.

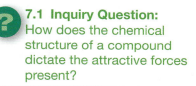

? 7.1 Inquiry Question: How does the chemical structure of a compound dictate the attractive forces present?

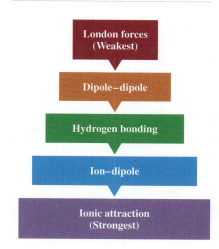

London forces (Weakest)

Dipole–dipole

Hydrogen bonding

Ion–dipole

Ionic attraction (Strongest)

FIGURE 7.1 Types of attractive forces. Attractive forces can occur between molecules, ionic compounds, or a combination of the two. Ionic attractions are the strongest attractive force, and London forces are the weakest.

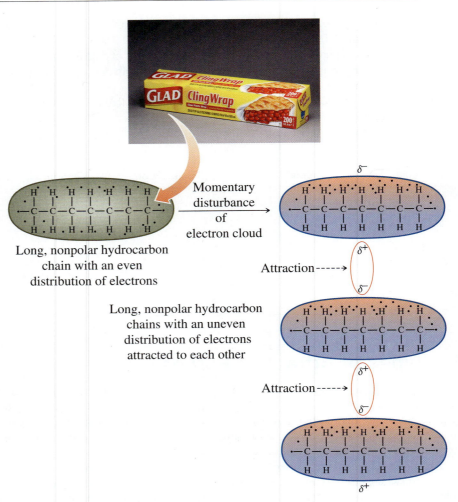

FIGURE 7.2 **An example of London forces.** When the normally even distribution of electrons around the long hydrocarbon chain of the plastic wrap is disturbed (top), the newly formed temporary dipole induces temporary dipoles in the other hydrocarbon chains and results in an attraction between the molecules (bottom), called London forces.

Dipole–Dipole Attractions

Some molecules—polar molecules—have a permanently uneven distribution of electrons caused by electronegativity differences in the atoms that make up the molecules. (See Section 3.7 for a review of polarity.) These molecules have a permanent separation of charge where one area of the molecule is partially positive and another partially negative. Such molecules have a **permanent dipole.**

Because the dipole in these molecules does not come and go as it does in the case of London forces, the attraction of the partially positive end of one molecule for the partially negative end of another molecule is stronger and more pronounced than London forces. This type of attraction involves the interaction of two dipoles and is called a **dipole–dipole attraction** (see Figure 7.3).

Dipole–dipole attractions do not exist between nonpolar molecules because these molecules do not have permanent dipoles. Molecules with permanent dipoles will also have London forces, but the attraction of the dipoles is much stronger, making the contribution from London forces negligible.

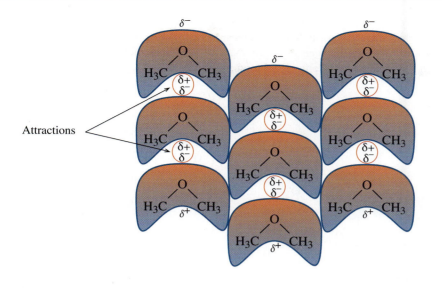

Attractions

FIGURE 7.3 An example of a dipole–dipole attraction. Molecules of dimethyl ether are polar. The partially negative end of the dipole of one molecule of dimethyl ether (shaded red) is attracted to the partially positive end of the dipole of the second molecule (shaded blue).

Hydrogen Bonding

The attractive force called hydrogen bonding is so prevalent in nature that it has been given its own name even though it is just a very strong dipole–dipole attraction. **Hydrogen bonding** involves a polarized hydrogen (hydrogen in a polar bond) and is much stronger than other dipole–dipole forces. It is important to understand that this attraction is not a bond like the covalent bonds discussed in Chapter 3 because there is no sharing of electrons. The hydrogen bond is simply a very strong dipole–dipole attraction.

Hydrogen bonding requires the interaction of two players, a donor hydrogen and an acceptor pair of electrons, as described in Table 7.1.

TABLE 7.1 Requirements for Hydrogen Bonding

Name	Description
Hydrogen-bond donor (δ^+)	A molecule with a hydrogen atom covalently bonded to an oxygen, nitrogen, or fluorine (O, N, or F)
Hydrogen-bond acceptor (δ^-)	A molecule with a nonbonding (lone) pair of electrons on an oxygen, nitrogen, or fluorine (O, N, or F)

The high electronegativity of O, N, and F polarizes the hydrogen atom of the donor, giving it a high partial positive charge (δ^+). This strongly attracts the high partial negative charge (δ^-) centered on the nonbonding electron pair of the O, N, or F of the acceptor.

Let's look at a glass of pure water as our first example of hydrogen bonding (see Figure 7.4). A molecule of water has two hydrogen atoms that are bonded to the oxygen. Both of these hydrogen atoms are attached to O by polar bonds (because O is more electronegative than H) polarizing the H so that it is partially positive (δ^+). Each of the two hydrogen atoms in water can act as a hydrogen-bond donor (d) in a hydrogen-bonding interaction. A water molecule also contains two nonbonding (lone) pairs of electrons on the oxygen. These nonbonding pairs mean that each water molecule can also accept two hydrogen bonds (a). So, one water molecule can make up to four hydrogen bonds to other water molecules at any one time as shown in Figure 7.4. Hydrogen bonds are always illustrated using a dashed line.

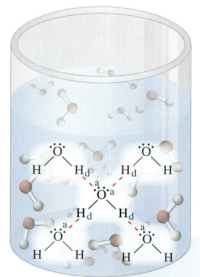

FIGURE 7.4 Hydrogen bonding in water. A single water molecule (center) can make up to four hydrogen-bonding interactions with neighboring water molecules. The hydrogen-bond donor hydrogen is designated d, and the hydrogen-bond acceptor electrons are designated a. Hydrogen bonds are indicated with red dashed lines.

Hydrogen bonds can occur between the same molecules as seen in pure water, between two different polar molecules as seen in the interactions of ethanol and acetone with water, or even between different parts of the same molecule (intramolecularly) as seen in the molecule ethylene glycol monomethyl ether.

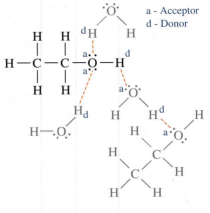

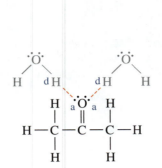

Intramolecular hydrogen bonding in a molecule containing an ether and alcohol group

Intermolecular hydrogen bonding between acetone and water

Intermolecular hydrogen bonding between ethanol and water

Hydrogen bonding can occur between any acceptor and donor hydrogen atoms.

Solving a Problem

Drawing Hydrogen Bonds

How many hydrogen bonds can the molecule shown make with water molecules? Draw them.

$$OH$$
$$|$$
$$CH_3CHCH_3$$

STEP 1: Determine the number of hydrogen bond donors and acceptors present in the molecule. Only hydrogen atoms bonded to an O, N, or F are polarized enough to form hydrogen bonds. Remember that C—H bonds are nonpolar, so these hydrogens do not participate in hydrogen bonding. This molecule is both a hydrogen-bond donor (O—H hydrogen) and hydrogen-bond acceptor (two electron pairs). This molecule can form three hydrogen bonds with water.

STEP 2: Draw hydrogen bonds. Connect donors and acceptors from the molecule to acceptors and donors on water respectively with a dashed line.

$$CH_3CHCH_3$$

Drawing Hydrogen Bonds

How many hydrogen bonds can the molecules shown make with water molecules? Draw them.

a. $CH_3CH_2NHCH_2CH_3$ b. CH_3CH_2Cl c. CH_3CN

Solution

To hydrogen bond to water, the molecule must have either a hydrogen that is bonded to a N, O, or F (donor) or a nonbonding pair of electrons on an O, N, or F (acceptor). Remember that C—H bonds are nonpolar, so these hydrogens do not participate in hydrogen bonding.

a.   **STEP 1: Determine the number of hydrogen-bond donors and acceptors present in the molecule.** This molecule has one N—H bond, so there is one donor hydrogen and one lone pair of electrons that can act as an acceptor. Two hydrogen bonds can be formed to water.

   STEP 2: Draw hydrogen bonds.

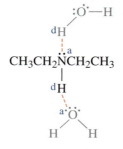

b.   **STEP 1: Determine the number of hydrogen-bond donors and acceptors present in the molecule.** This molecule has no O, N, or F atoms so there are no hydrogen-bond acceptors present. The H atoms in this molecule are bonded to carbon, making them nonpolar; therefore, there are no hydrogen-bonding donors present in this molecule. This molecule cannot form hydrogen bonds to water.

c.   **STEP 1: Determine the number of hydrogen-bond donors and acceptors present in the molecule.** This molecule has one nitrogen atom with one lone pair of electrons present that can act as a hydrogen-bond acceptor. The H atoms in this molecule are bonded to carbon, making them nonpolar and, therefore, no hydrogen-bonding donors are present in this molecule. This molecule can make one hydrogen bond to water.

   STEP 2: Draw hydrogen bonds.

$$CH_3CN \underset{a}{:}\text{---}\underset{d}{H} \overset{\cdot\cdot\cdot\cdot}{\underset{}{O}} \diagdown H$$

Ion–Dipole Attraction

When you add a teaspoon of salt to a glass of water and stir, you observe that the salt dissolves. What you are witnessing is another very strong attractive force in action. The **ion-dipole attraction** occurs between ionic charges like those found in salt and polar molecules such as water. Ion–dipole attractions are an important attractive force often seen in biological systems because water is present. This attractive force is stronger than hydrogen bonding.

The interaction between sodium or chloride ions and water molecules results from the attraction of an ion to the opposite partial charge on a polar molecule. As shown in **Figure 7.5**, a cation like sodium ion is attracted to the partially negative end of the dipole of a polar molecule like water. Similarly, an anion like chloride is attracted to the partially positive end of the dipole of water. As we will see later in the chapter, this interaction plays an important role in the solubility of ionic compounds.

Water   Sodium ion   Chloride ion   Water

FIGURE 7.5 An ion–dipole attraction. The cation Na$^+$ is attracted to the negative dipole of water and the anion Cl$^-$ is attracted to the positive dipole of water, allowing NaCl to dissolve in water.

Ionic Attraction

As we saw in Chapter 3, there are two types of ions in ionic compounds, anions and cations. Anions have more electrons than their neutral atom and possess a − charge. Cations have fewer electrons than their neutral atom and possess a + charge. When these opposite charges attract each other, an ionic attraction exists. Ionic attractions are also called *ionic bonds* because they occur between two ions with opposite charges. An ionic attraction is the strongest attractive force because it involves more than just an uneven distribution of electrons as seen in a dipole–dipole (δ^+/δ^-) attraction.

Ionic attractions are sometimes called *salt bridges* because **salts** are ionic compounds. (More information about how salts form is found in Section 9.2.) Ionic charges are also found in the organic functional groups carboxylate and protonated amine containing − and + charges respectively. These groups are found in the amino acids that form proteins and can form salt bridges (ionic bonds) when they come into contact with each other. For example, the oxygen transport protein hemoglobin forms salt bridges between different areas of the protein as it releases oxygen at the tissues. For more information on hemoglobin, go to Chapter 10.

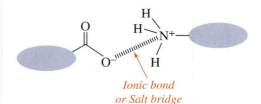

Ionic bond or Salt bridge

When two functional groups contain opposite charges, an ionic attraction can result.

Recognizing the attractive forces present in a compound from its structure will allow you to predict the physical properties of compounds. The flow chart in **Figure 7.6** will help you to predict the strongest attractive force present.

Questions

Attractive Force
(weakest to strongest)

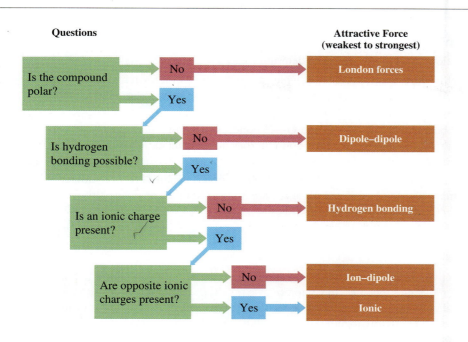

Solving a Problem

Identifying Attractive Forces in Compounds

Name the strongest attractive force present in the pure substances shown.
a. Ammonia, NH_3
b. Hexane, C_6H_{14}
c.

$$H_3\overset{+}{N}-\underset{\underset{H}{|}}{\overset{\overset{H}{|}}{C}}-\overset{\overset{O}{\|}}{C}-O^-$$

Glycine, an amino acid

Use the flow chart in Figure 7.6 to determine each.

a. Ammonia has N—H bonds, which are polar. N—H bonds can participate in hydrogen bonding. An ionic charge is not present. The strongest attractive force between NH_3 molecules is hydrogen bonding.
b. Hexane contains only carbon and hydrogen, so it is a nonpolar molecule. The strongest attractive force between hexane molecules is London forces.
c. Glycine contains opposite charges, so glycine can form ionic bonds. The strongest attractive force between glycines is ionic.

sample
problem
7.2   **Identifying Attractive Forces**

Identify the strongest attractive force present in the pure substances shown.

a.   one of the first anesthetics, diethyl ether, $CH_3CH_2OCH_2CH_3$
b.   rubbing alcohol, isopropanol, $CH_3CH(OH)CH_3$
c.   the ozone depleting substance, trichloroethane, CCl_3CH_3

Solution

Use the flow chart in Figure 7.6 to determine each.

a. Diethyl ether has C—O bonds, which are polar. It cannot form hydrogen bonds because no hydrogens are bonded to O, N, or F. The strongest attractive force in diethyl ether is dipole–dipole.

b. Rubbing alcohol has C—O and O—H bonds, which are polar. The O—H bonds allow isopropanol to hydrogen bond. It does not contain an ionic charge. The strongest attractive force is hydrogen bonding.

c. Trichloroethane has C—Cl bonds, which are polar. It cannot form hydrogen bonds because none of its hydrogens are bonded to O, N, or F. The strongest attractive force present is dipole–dipole.

Attractive Forces Keep Biomolecules in Shape

integrating chemistry

The attractive forces discussed in this section are used extensively in nature to hold biological molecules together. In Chapter 6, we noted the formation of cellulose fibers from the alignment of zigzagging $\beta(1 \rightarrow 4)$ bonded glucose units. The linear cellulose molecules are held tightly together through hydrogen bonding between the glucose units in neighboring molecules.

London forces hold cell membranes together (see Section 7.6). Hydrogen bonding holds a DNA double helix in its twist (see **Figure 7.7**). Protein structures are held together by combinations of all the attractive forces discussed. We explore attractive forces in DNA and in protein structures in more detail as they are discussed in later chapters.

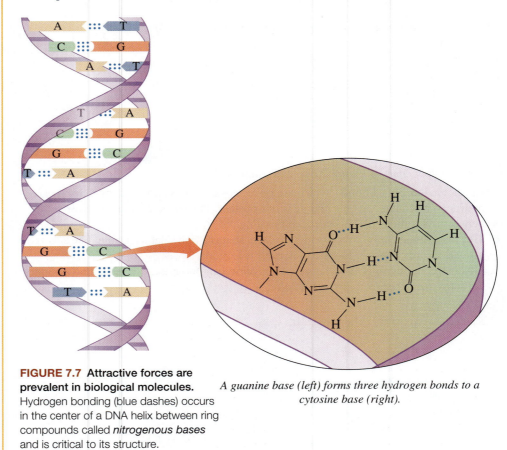

FIGURE 7.7 Attractive forces are prevalent in biological molecules. Hydrogen bonding (blue dashes) occurs in the center of a DNA helix between ring compounds called *nitrogenous bases* and is critical to its structure.

A guanine base (left) forms three hydrogen bonds to a cytosine base (right).

practice
problems

7.1 Explain the difference between a covalent bond and an attractive force.

7.2 Polar molecules can contain more than one type of attractive force. Fill in the table with a yes or no answer to the following question: "Does this attractive force operate between molecules of the given compound?"

Molecule	H-Bond	Dipole–Dipole	London
CF_4			
H_2O			
CH_3NH_2			

7.3 Polar molecules can contain more than one type of attractive force. Fill in the table with a yes or no answer to the following question: "Does this attractive force operate between molecules of the given compound?"

Molecule	H-Bond	Dipole–Dipole	London
CH_3OH			
NH_3			
CH_3CH_3			

7.4 What type(s) of attractive forces exist between all molecules?

7.5 Explain the difference between the dipole in London forces and the dipole in dipole–dipole attractions.

7.6 Given that only polar molecules can participate as donors in hydrogen bonding, is it true that all polar molecules can be hydrogen-bond donors? Explain.

7.7 Explain the requirements for a molecule to be a hydrogen-bond acceptor.

7.8 Considering your answers to Problems 7.6 and 7.7, can water and carbon dioxide form a hydrogen bond? Explain.

7.9 An ion–dipole attraction often occurs when ionic compounds mix with water. Why do you think this is so?

7.10 Name two functional groups that contain ionic charges.

7.11 Identify the strongest attractive force present in the pure substances shown: London forces, dipole–dipole, hydrogen bonding, ion–dipole, or ionic attractions. Refer to the flow chart in Figure 7.6.
a. NaF
b. $CH_3CH_2CH_3$
c. CH_3NH_2
d. CH_3F

7.12 Identify the strongest attractive force present in the pure substances shown: London forces, dipole–dipole, hydrogen bonding, ion–dipole, or ionic attractions.
a. $CH_3CH_2CH_2CH_2OH$
b. KCl
c. CH_2Cl_2
d.

$$\underset{H \qquad H}{\overset{O}{\underset{\diagdown\;\diagup}{\overset{\|}{\underset{C}{}}}}}$$

7.13 How many hydrogen bonds can CH_3NH_2 make to water? Draw them.

7.14 How many hydrogen bonds can CH_2O make to water? Draw them.

7.2 Liquids and Solids: Attractive Forces Are Everywhere

You may remember the scene in the children's Christmas classic *Frosty the Snowman* when Frosty takes his little friend Karen into the greenhouse to help her warm up. Once they get inside, Professor Hinkle, the evil magician, locks them in, and Frosty changes from a snowman into a puddle of water. In scientific terms, he went from being a solid (snow or ice) to a liquid (water).

How did this happen? How did the greenhouse cause this major change in Frosty's life? Of course, you know that it was the heat in the greenhouse that changed Frosty from a solid to a liquid, but how did the heat cause that change?

Did the heat in the greenhouse cause Frosty's water molecules to move faster? Yes, and by doing so it disrupted the attractive forces between the water molecules. The state in which a given substance exists depends largely on the motion of its molecules—molecules of a liquid move faster and with more types of motion than the molecules of a solid. Now that we can identify the attractive forces present by looking at the structure of a compound, we will spend much of the rest of Chapter 7 using this knowledge to predict some of the physical properties of substances, like boiling point and solubility.

? **7.2 Inquiry Question:** How can the boiling points and melting points of compounds be predicted?

Heat and Attractive Forces

Just as holding hands with your sweetie is much easier when you are sitting together on a couch than when you are dancing, attractive forces are stronger between two compounds moving slowly than between two compounds bouncing wildly about. When two compounds with low energy come near each other, there is a much greater possibility that an attraction will develop than when two compounds with high energy move quickly past each other. In other words, as a substance is heated, the particles (molecules or ions) begin to move faster and faster. In turn, the attractions between them start to become less important—they are no longer held together as tightly.

When a solid substance is heated, it can melt and become a liquid, and when a liquid is heated, it can vaporize and become a gas. These transitions are called **changes of state**. The changes of state between a solid, liquid, and gas are summarized in Figure 7.8. These include **freezing, melting,** (between liquids and solids), **vaporization** and **condensation** (between liquids and gases), and **sublimation** and **deposition** (between solids and gases).

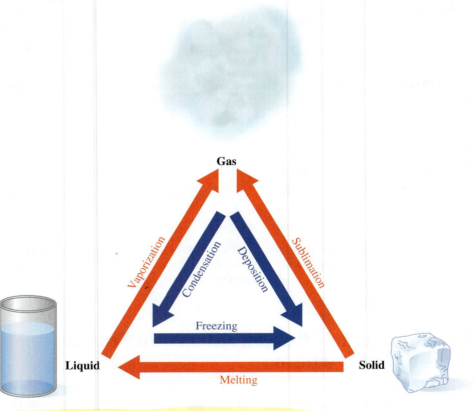

FIGURE 7.8 Changes in the states of matter. Changes to a more ordered state of matter (blue) have more attractive forces between particles than changes to a less ordered state of matter (red).

Adding heat to a substance causes its particles to move faster, creating less order among them, which, in turn, disrupts the attractive forces holding the matter together. Conversely, removing heat slows the movement of the particles in a substance down, and they attract other particles by allowing attractive forces to hold particles more firmly.

sample
problem
7.3   **Changes in States of Matter**

Name the change of state occurring for each.

a.   ice forming on a puddle of water during the winter
b.   dew forming on grass
c.   water boiling

Solution

Use Figure 7.8 as a guide for this problem. Establish the initial state and final state of matter—solid, liquid, or gas. The change of state will be on the arrow connecting the initial state to the final state.

a. Water in the liquid state (initial state) forms ice, a solid (final state). Freezing.

b. Water as a gas (initial state) deposits on blades of grass forming a liquid (final state). Condensation.

c. Water in the liquid state (initial state) boils forming water in the gaseous state (final state). Vaporization.

Boiling Points and Alkanes

Pentane and octane are both alkanes with simple straight-chain structures that exist as liquids at room temperature. When heated, pentane boils at 36 °C, but octane does not boil until the temperature reaches 125 °C. Why do two molecules that are both alkanes boil at such different temperatures? To understand this, we must understand what happens at the molecular level in the process of boiling.

For a liquid to boil, the molecules of the liquid must push back the gas molecules of the atmosphere at the surface of the liquid, allowing gas molecules of the liquid to escape. This also requires that the liquid molecules overcome their attractive forces to the other molecules in the liquid and that each moves into the gas phase *as a single molecule*. The heat supplied during boiling provides the energy necessary for each molecule to vaporize, moving individually from the liquid into the gas phase.

When the **boiling point** is reached, the molecules have enough energy to change from a liquid to a gas (vaporize). **Boiling** occurs when gas bubbles form in a liquid and escape at the surface.

As heating begins, gas bubbles are formed in the liquid

During boiling, gas bubbles rise to the surface and escape. (36 °C is boiling point for pentane)

Because the molecules of the atmosphere stay the same, the difference in the boiling points of two or more compounds must lie in their attractive forces. Let's see how this works for our alkane example.

The structures of pentane and octane are shown Figure 7.9. Like all alkanes, both are nonpolar compounds and, therefore, exhibit only London forces. When interacting in the liquid phase, note that molecules of octane have more opportunities to interact than do molecules of pentane simply because a single octane has more carbon atoms and, therefore, a greater surface area.

Recall that London forces between molecules are a result of a disturbance in the distribution of electrons over the surface of a molecule. A molecule with a larger surface area has more surface contact with other molecules as well as more electrons to disturb. In other words, straight-chain alkanes with more carbons have stronger attractions between molecules. Therefore, because these attractions must be overcome for the compound to boil, more heat is necessary to disrupt these attractions, which means the boiling point is higher.

In the liquid state

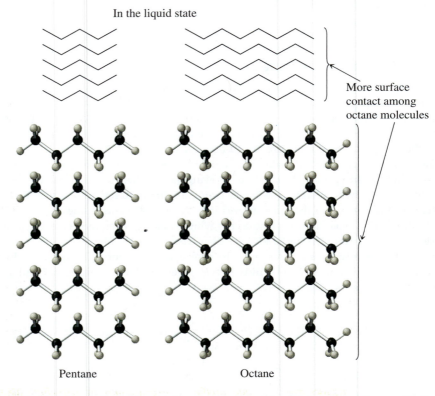

More surface contact among octane molecules

Pentane   Octane

FIGURE 7.9 More surface contact between molecules raises the boiling point. Octane has a larger surface area and more potential surface for contact than pentane, so its boiling point is higher.

Table 7.2 shows the boiling points for the first 10 straight-chain alkanes. Notice that as the number of carbon atoms in the chain (and therefore the surface area of the molecule) increases, the boiling point also increases. This trend to higher boiling points with increasing carbon chain length is also true for **melting points,** the temperature at which molecules move from the solid to the liquid phase.

TABLE 7.2   Boiling Points of Common Straight-Chain Alkanes

Name	Structure	Boiling Point °C
Methane	CH_4	−161
Ethane	CH_3CH_3	−89
Propane		−42
Butane		−0.5
Pentane		36
Hexane		69
Heptane		98
Octane		125
Nonane		151
Decane		174

Next let's consider two alkanes with the same number of carbon atoms, but with structures that differ—hexane and 2,3-dimethylbutane. These two compounds are structural isomers, meaning that they have the same molecular formula, but their connectivity is different. As shown in **Figure 7.10**, hexane is a straight-chain alkane, but 2,3-dimethylbutane is a branched-chain alkane.

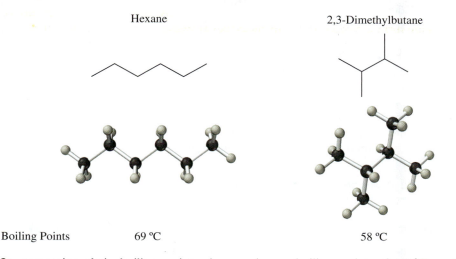

Hexane	2,3-Dimethylbutane
Boiling Points 69 °C	58 °C

FIGURE 7.10 Two structural isomers of C_6H_{14}. Hexane and 2,3-dimethylbutane are two alkanes with the same molecular formula but different connectivity, which affects their boiling points.

In comparing their boiling points, hexane has a boiling point of 69 °C and 2,3-dimethylbutane has a boiling point of 58 °C. If two molecules have the same number of carbon atoms, why do their boiling points have an 11 °C difference? The answer lies in attractive forces and in the surface area of the molecules.

Consider a plate of spaghetti and an accompanying bowl of meatballs. The long noodles have more points of contact with each other and can line up with each other if you stretch them out on your plate. Similarly, the molecules of hexane, a straight-chain alkane, can "stack" close together like the spaghetti. In contrast, molecules of 2,3-dimethylbutane, are, like meatballs, more spherical in their overall shape. When next to each other on your plate, two meatballs touch at only one small point. Likewise, the molecules of 2,3-dimethylbutane have less surface contact than do the hexane molecules. As mentioned earlier, the more contact between two molecules, the greater the attraction caused by the London forces between them. In turn, the greater London forces attraction means that the boiling point of hexane is higher than that of 2,3-dimethylbutane. In fact, as a general rule, *for* ==*alkanes with the same number of carbon atoms, straight-chain alkanes have higher boiling points than do branched alkanes.*==

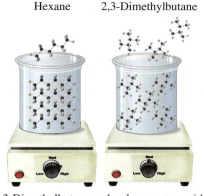

Hexane   2,3-Dimethylbutane

2,3-Dimethylbutane makes less contact with surrounding molecules in the liquid than hexane and requires less heat to boil.

The strands of spaghetti have more surface contacts between them than the meatballs do, just as straight-chain alkanes have more contacts between them than branched ones.

The Unique Behavior of Water

Now, let's consider the case of water, H_2O. From its molecular formula, it is clear that water is a small molecule—it only has 3 atoms. Compare it to propane that has 11 atoms (C_3H_8). In terms of contact with neighboring molecules, you might guess that water has a lower boiling point than propane. Yet, from Table 7.2 we can see that propane boils at $-42\ ^\circ$C, while we know that water boils at 100 $^\circ$C.

Earlier in the chapter, we discussed the fact that water is a polar molecule. Water's electrons are always unevenly distributed, unlike the electrons of alkanes. Water molecules strongly attract each other through hydrogen bonding, a much stronger attractive force than London forces. For water to boil, that is, to move from the liquid to the gas phase, there must be enough heat energy to disrupt the hydrogen-bonding interactions. This requires more heat energy than it does to disrupt the London forces in compounds like propane, leading to the unusually high boiling point of water.

Melting points follow the same trends as boiling points in that the stronger and more numerous the forces between molecules, the higher the melting point. For example, hydrogen sulfide, H_2S, has a similar structure to H_2O but cannot form hydrogen bonds. The melting points of the two compounds are extremely different with hydrogen sulfide having a melting point of $-80\ ^\circ$C to water's 0 $^\circ$C.

Table 7.3 summarizes the trends in boiling and melting points previously discussed.

TABLE 7.3 Boiling and Melting Point Trends

Nonpolar Molecules	Similar-Sized Molecules, Different Forces	Molecules with Same Forces
The greater the surface area, the higher the boiling or melting point.	The stronger the forces, the higher the boiling or melting point.	The stronger and more numerous the forces, the higher the boiling or melting point.

Solving a Problem

Predicting Boiling Points

Predict which substance in the pair would have the higher boiling point. Justify your prediction using attractive forces.

a. ⌇⌇⌇ or ⌇⌇⌇   b. H_2O or CH_3OCH_3

Solution

a. **STEP 1: Determine the strongest attractive force present in each.** Using the flow chart in Figure 7.6, we see that the alkanes shown are nonpolar and both have the same attractive force present, London forces.

   STEP 2: Predict the higher boiling point based on the strength and number of forces present. The trends in Table 7.3 indicate that if two molecules have the same forces present, the one with more of that force will have a higher boiling point. Because both have London forces only, we must consider the number of forces present. The straight-chain alkane, butane, makes more contacts with its neighboring molecules in the liquid state, so more London forces would be present. The straight-chain alkane, butane, is predicted to have the higher boiling point.

b. **STEP 1: Determine the strongest attractive force present in each.** Using the flow chart in Figure 7.6, the strongest attractive force present in H_2O is hydrogen bonding and the strongest attractive force present in CH_3OCH_3 is the dipole–dipole attraction.

   STEP 2: Predict the higher boiling point based on the strength and number of forces present. Because hydrogen bonding is a stronger attractive force than dipole–dipole, H_2O is predicted to have the higher boiling point.

sample problem
7.4   **Predicting Boiling Points**

Predict which substance in the pair would have the higher boiling point. Justify your prediction using attractive forces.

a. CH_3OH or CH_3CH_3 b. HF or HCl   c. CH_3NHCH_3 or CH_3CH_2OH

Solution

a. **STEP 1: Determine the strongest attractive force present in each.**
The strongest attractive force present in CH_3OH is hydrogen bonding, and the strongest attractive force present in CH_3CH_3 is London forces.

STEP 2: Predict the higher boiling point based on the strength and number of forces present. Because hydrogen bonding is a stronger attractive force than the London force, CH_3OH is predicted to have the higher boiling point.

b. **STEP 1: Determine the strongest attractive force present in each.** The strongest attractive force present in HF is hydrogen bonding, and the strongest attractive force in HCl is dipole–dipole.

STEP 2: Predict the higher boiling point based on the strength and number of forces present. Because hydrogen bonding is a stronger attractive force than dipole–dipole, HF is predicted to have the higher boiling point.

c. **STEP 1: Determine the strongest attractive force present in each.** The strongest attractive force present in CH_3NHCH_3 is hydrogen bonding, and the strongest attractive force in CH_3CH_2OH is hydrogen bonding.

STEP 2: Predict the higher boiling point based on the strength and number of forces present. The trends in Table 7.3 indicate that if two molecules have the same forces present, the one with more of that force will have a higher boiling point. CH_3NHCH_3 has one hydrogen-bond donor and one hydrogen-bond acceptor present forming up to two hydrogen bonds. CH_3CH_2OH has one donor and two acceptors present forming up to three hydrogen bonds. Because CH_3CH_2OH can form more hydrogen bonds per molecule than CH_3NHCH_3, CH_3CH_2OH is predicted to have the higher boiling point. (For a review of hydrogen bonding, see Section 7.1.)

practice problems

7.15   In the following pairs of molecules, predict which one of the molecules in the pair would have the higher boiling point by naming the strongest attractive force present in each molecule.

a. CH_4 or H_2O b. NH_3 or CO_2

c.

$$\underset{CH_3}{\overset{\displaystyle O}{\overset{\displaystyle \|}{\underset{\diagdown}{C}}}}\overset{\diagup}{}CH_3 \quad \text{or} \quad \underset{CH_3}{\overset{\displaystyle O}{\overset{\displaystyle \|}{\underset{\diagdown}{C}}}}\overset{\diagup}{}OH$$

  Acetone   Acetic acid

d. KCl or CH_3OCH_3

7.16   In the following pairs of molecules, predict which one of the molecules in the pair would have the higher melting point by naming the strongest attractive force present in each molecule.

a. CH_3OH or $CH_3CH_2OCH_2CH_3$

b. $CH_3CH_2CH_2CH_2CH_2CH_3$ or $CH_3CH_2CH_2CH_2CH_2CH_2CH_2CH_3$

c. $CH_3-\overset{\displaystyle }{\underset{\displaystyle H}{N}}-CH_3$ or $CH_3CH_2NH_2$

d. $CH_3CH_2CH_2CH_3$ or KBr

7.17 The formulas and boiling points of several similar compounds are listed in the following table. Determine the strongest attractive force present in each compound and explain the difference in their boiling points.

Formula	Boiling Point °C
CH_4	−164
CH_3F	−78
CH_3OH	65

7.18 The formulas and boiling points of several similar compounds are listed in the following table. Determine the strongest attractive force present in each compound and explain the difference in their boiling points.

Formula	Boiling Point °C
CH_3CH_3	−89
CH_3Cl	−24
CH_3NH_2	−6

7.19 Rank the following compounds in order of increasing boiling point (lowest to highest) and explain your ranking:

7.20 Rank the following compounds in order of increasing boiling point (lowest to highest) and explain your ranking:

Discovering the Concepts

? Inquiry Activity—Solubility in Water

Common Substances and Their Characteristics

Substance	Characteristics
NaCl (table salt)	Ionic compound
Sucrose (table sugar)	Polar covalent compound
Vegetable oil	Nonpolar covalent compound
Soap	Contains a highly polar part (often ionic) and a nonpolar part

Questions

1. Based on your experience, which of the compounds in the table will dissolve in water? To dissolve is to separate a substance into single particles in solution.
2. Which of the characteristics listed applies to water?
3. The *golden rule of solubility* states that "like dissolves like." What is *alike* about the compounds you listed in question 1 and water?

4. Based on their characteristics, predict whether each of the following compounds might be soluble in water.

    a. acetone,

    $$CH_3 - \overset{\overset{\displaystyle O}{\|}}{C} - CH_3$$

    b. ethanol, CH_3CH_2OH
    c. $NaHCO_3$
    d. cyclohexane,

    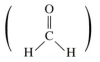

    e. octanol,

    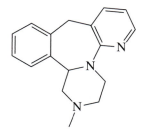

5. Ionic compounds containing sodium will separate into cations and anions in solution. Draw and describe how an ionic compound like NaCl might interact with water through ion–dipole attractions.

6. Draw and describe how a polar covalent compound like formaldehyde

    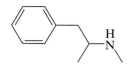

    might interact with water through dipole–dipole attractions.

7. Many pharmaceuticals contain aromatic rings or large areas of hydrocarbon (nonpolar). Quite often, pharmaceuticals are synthesized in an ionic form to increase their solubility in aqueous solution. Ionic charges on organic molecules dramatically increase the compound's solubility. Predict whether the following pharmaceuticals would be likely soluble in water based on their polarity and the distribution of polarity throughout the compound.

    Pentolinium—an antihypertensive

    Methamphetamine—mainly recreational usage yet sometimes prescribed for ADHD

    Remeron™—an antidepressant

8. Devise a rule to predict the solubility of an organic compound in water.

7.3 Attractive Forces and Solubility

? 7.3 Inquiry Question:
How is the solubility of a
compound in water
determined?

Before adding oil-and-vinegar-based dressing to a salad, you must vigorously shake the
mixture and then quickly pour it. Why? Because oil and water (vinegar is mostly water)
don't mix. Have you ever wondered why oil and water don't mix? The answer lies in the
attractive forces between the molecules.

The Golden Rule of Solubility

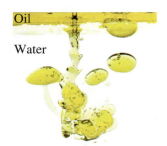

Oil

Water

Oil and water don't mix, but sugar and water do. In both cases, we are witnessing the solubility
(or lack thereof) of one substance in another. The maximum amount of a substance that can
dissolve in a specified amount of water at a given temperature defines a substance's **solubility
in water.** As we discussed in Section 7.1, the polarity of molecules (polar or ionic versus nonpolar) dictates the types of attractive forces displayed between molecules.

The **golden rule of solubility**—*like dissolves like*—means that molecules that are similar
will dissolve in each other. The "similarity" mentioned here relates to the character of the
molecules—are they polar or nonpolar? In short, the golden rule of solubility says that *molecules
that have similar polarity and participate in the same types of attractive forces will dissolve each other.*

Predicting Solubility

Nonpolar Compounds

Let's look more closely at oil and water. Dietary oils, known as **triglycerides,** are nonpolar
organic compounds formed through the condensation reaction of three fatty acids (to
review their structure see Section 4.3) with a compound called glycerol (see **Figure 7.11**).

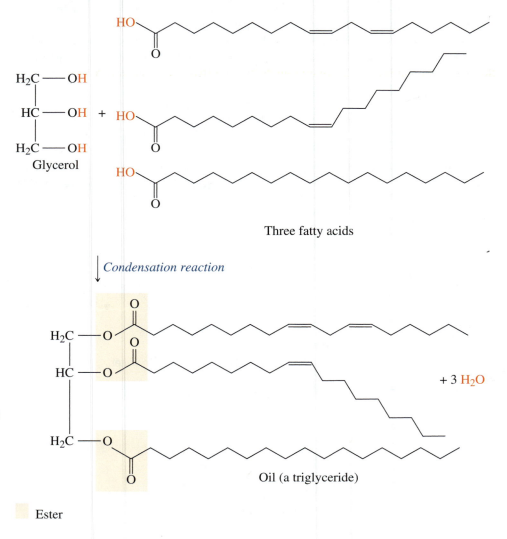

Three fatty acids

Condensation reaction

+ 3 H$_2$O

Oil (a triglyceride)

Ester

FIGURE 7.11 Esterification is a
type of condensation reaction. Here, it
occurs between three fatty acids and
a glycerol forming a triglyceride.
Although the triglyceride contains some
polar oxygen bonds in the ester functional groups, the nonpolar areas of the
molecule dominate the polarity, making
the molecule overall nonpolar.

Organic chemists refer to this condensation reaction more specifically as an **esterification** because an ester functional group is formed during the reaction (actually the triglyceride product contains three ester groups).

As nonpolar compounds, oils are attracted to neighboring molecules through London forces. Water, on the other hand, is a polar molecule and interacts with other substances through dipole–dipole, hydrogen bonding, and ion–dipole attractions. This means that, in terms of attractive forces, oil and water are very *unlike* each other. Yet for one molecule to dissolve another, the molecules must interact with each other. Oil and water do not interact with each other, so they do not dissolve in each other. The attractions among the water molecules are much greater than the attraction between a water molecule and an oil molecule. Even after a bottle of oil and vinegar salad dressing is shaken vigorously, the contents of the bottle will separate back into two layers—a watery layer (vinegar) and an oil layer.

Polar Compounds

Why does table sugar, also an organic compound, dissolve in water, but oil does not? In Chapter 6, we learned that the substance that we call table sugar is the disaccharide sucrose. Like all other carbohydrates, sucrose has multiple hydroxyl(—OH) groups. The key to the interaction of sucrose and water lies in these functional groups.

The hydroxyl groups of sucrose make it a polar compound and give it the ability to interact with water not only through dipole–dipole attractions, but also through hydrogen-bonding interactions as shown in **Figure 7.12**. Because table sugar and water are both polar and share these attractive forces, table sugar is an organic compound that *is* soluble in water.

Ionic Compounds

What about the solubility of ionic compounds in water? In Section 7.1, we saw that ion–dipole attractions can exist between water and ions. These attractions are much stronger than the partial dipole attractions that exist between water and polar covalent compounds. So ionic compounds are often more soluble in water than are covalent compounds.

In fact, for most ionic compounds, the ion–dipole attractions between the ions of an ionic compound and water are so strong that when enough water molecules interact with each ion, the ionic bond between the two ions is disrupted.

Individual ion–dipole attractions are not stronger than an ionic bond, but when multiple water molecules interact with an ion, the sum of these attractive forces is greater than the strength of the ionic bonds. The result, as shown in **Figure 7.13** for sodium chloride, is that in water many ionic compounds break apart into their component ions, and each ion is surrounded by water molecules. This process is called **hydration.**

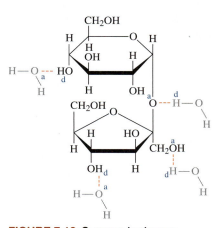

FIGURE 7.12 Sucrose hydrogen bonding with water. A few of the many hydrogen bonds (shown in red) possible between water and sucrose (acceptors and donors labeled). The lone pair electrons on O are implied, not shown, for the sake of clarity.

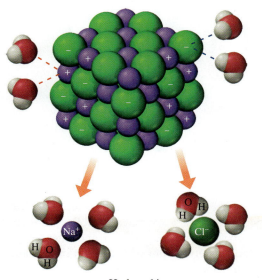

Hydrated ions

FIGURE 7.13 Hydration of NaCl. When NaCl dissolves in water, the multiple ion–dipole attractions between water and the Na⁺ and Cl⁻ ions cause the compound to break apart into its component ions.

Amphipathic Compounds

In Chapter 4, we saw that fatty acids are mostly hydrocarbons. Despite the presence of a carboxylic acid group containing two electronegative oxygen atoms, these compounds behave similarly to nonpolar alkanes. The carboxylic acid part of a fatty acid is polar, but relative to the large nonpolar hydrocarbon chain, it does not contribute as much to the overall polarity of the molecule. The nonpolar part of the fatty acid is so much larger than the polar carboxylic acid group that it dominates the character of the compound, making fatty acids mainly nonpolar. Molecules like fatty acids that have both polar and nonpolar parts are called **amphipathic** (from the Greek *amphi* meaning "both" and *pathic* meaning "condition") compounds.

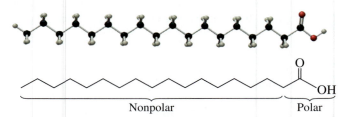

Fatty acids are amphipathic. They have a polar part and a nonpolar part.

One relevant application of these attractive forces to solubility occurs in soap. Soap is composed of fatty acid salts. Remember that salts are ionic compounds. Unlike fatty acids (Chapter 4), which contain a carboxylic acid functional group $\left(\begin{matrix} O \\ \| \\ R-C-OH \end{matrix} \right)$ and a long hydrocarbon chain making them largely nonpolar, fatty acid salts are ionic because they contain the carboxylate form $\left(\begin{matrix} O \\ \| \\ R-C-O^- \end{matrix} \right)$ of the functional group (hydrogen removed) at one end. The charge on the carboxylate makes this end of the molecule ionic. This end is commonly called the polar *head*. The rest of the molecule behaves like a nonpolar compound. Fatty acid salts have long nonpolar hydrocarbon *tails* and extremely polar (ionic) *heads*, so they are amphipathic.

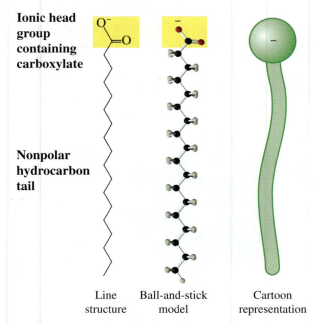

Ionic head group containing carboxylate

Nonpolar hydrocarbon tail

Line structure   Ball-and-stick model   Cartoon representation

An amphipathic compound like a fatty acid can be represented in cartoon with a polar head and nonpolar tails.

Amphipathic compounds like soaps will not dissolve in water because of the large nonpolar tails present. The nonpolar tails are **hydrophobic** (water fearing) and will be excluded from the water, associating with each other and interacting through London forces. The ionic heads, which are **hydrophilic** (water loving), interact with the water mainly through ion–dipole attractions.

Because the water interacts only with the ionic heads, the tails associate with each other, creating the core of a spherical structure in the water called a **micelle** (see Figure 7.14). The polar heads are attracted to the water and form the shell of the micelle. Micelles form because the ion–dipole and hydrogen-bonding attractions between water molecules and the ionic heads are stronger (and preferred) to water interactions with the hydrocarbon tails.

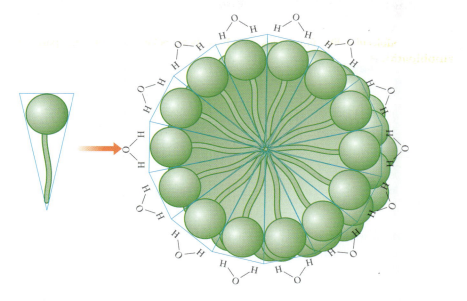

FIGURE 7.14 A spherical micelle.
A fatty acid salt has a conical shape so that when the nonpolar hydrocarbon tails are pushed to the center by the strong ion–dipole attraction between the water and the ionic charge on the head group, a spherical-shaped micelle is formed.

How Soap Works

How does soap do its job of removing greasy dirt? Most stains, including grease and dirt, are nonpolar. Based on the golden rule, greasy dirt is not soluble in water. When skin or clothing with a greasy dirt stain is washed with soapy water, the stain is attracted to the nonpolar hydrocarbon tails of the soap and is dissolved in the interior of the micelle formed by the soap molecules (see Figure 7.15). The ionic head groups surround the exterior surface of the micelle and form ion–dipole interactions with the water. Because the surface of the micelle is covered with the polar head groups, the entire micelle, with the greasy stain molecules now dissolved and "hidden" in its interior, is soluble in water and is washed down the drain. Amphipathic compounds like soaps are called **emulsifiers** because they allow nonpolar and polar compounds to be suspended in the same mixture.

integrating chemistry

Soap molecule

Fatty acid tail

H₂O

Hydrophilic

Oil and grease

Hydrophobic

FIGURE 7.15 The cleaning action of soap. Soaps emulsify oil and dirt in water by putting these nonpolar substances in the hydrophobic interior of a micelle, which are then rinsed away with water.

Many organic substances are soluble in water to some extent, making them fall into a gray area of solubility based on how polar (or like water) they are. Let's try to predict the solubility of some molecules in water.

Solving a Problem

Predicting Solubility in Water

Predict which of the following compounds is likely to be more soluble in water. Justify your answer.

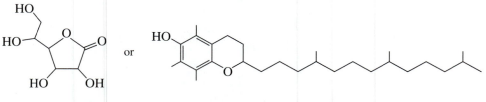

Vitamin C, ascorbic acid   or   Vitamin E, alpha-tocopherol

Solution

STEP 1: Determine the number of polar attractive forces in each. Because we are considering solubility in water, a polar compound, we examine the number of polar interactions present in the compounds. Vitamin C contains several oxygens and hydrogens available for hydrogen bonding and is therefore considered a highly polar molecule. Vitamin E has a few polar areas but overall is mostly hydrocarbon and mainly nonpolar.

STEP 2: Predict the more soluble substance based on the overall polarity of the entire molecule. Because water must surround the compound, molecules that are more soluble will have polar functional groups spread throughout. Vitamin C has polar bonds distributed throughout its structure whereas the polar areas in vitamin E are limited. Vitamin C is more soluble in water.

---

sample problem 7.5 — Predicting Solubility in Water

Predict which of the following compounds is likely to be more soluble in water. Justify your answer.

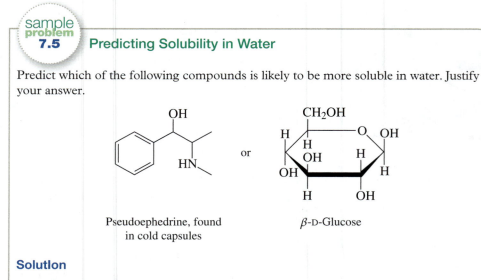

Pseudoephedrine, found in cold capsules   or   β-D-Glucose

Solution

STEP 1: Determine the number of polar attractive forces in each.
Pseudoephedrine has a few polar atoms OH and NH that can hydrogen bond, but the molecule is mainly nonpolar. Glucose has many OH groups that can hydrogen bond.

STEP 2: Predict the more soluble substance based on the overall polarity of the entire molecule. Because water must surround the compound, molecules that are more soluble will have polar functional groups spread throughout. Glucose is more soluble in water than pseudoephedrine.

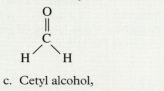

7.21 Based on their attractive forces, predict whether the following compounds are likely soluble, insoluble, or form a micelle in water.

　　a. Hexane, $CH_3CH_2CH_2CH_2CH_3$

　　b. Formaldehyde,

$$\underset{H}{\overset{\displaystyle \overset{O}{\parallel}}{\underset{}{C}}}\underset{H}{}$$

　　c. Cetyl alcohol,

~~~~~~~~~~OH

7.22 Based on their attractive forces, predict whether the following compounds are likely soluble, insoluble, or form a micelle in water.

　　a. NaCl

　　b. stearic acid, [18:0], a saturated fatty acid

　　c. isopropyl alcohol,

OH

7.23 Predict which of the following compounds is likely to be more soluble in water. Justify your answer.

　　a. fatty acid or fatty acid salt

　　b. KCl or $(CH_3)_3N$

7.24 Predict which of the following compounds is likely to be more soluble in water. Justify your answer.

　　a. CH_3CH_2OH or CH_3CH_3

　　b. CH_3NH_2 or H_2S

7.4 Gases: Attractive Forces Are Limited

We just saw that attractive forces between molecules affect properties in the liquid and solid states of matter. In liquids and solids (also called condensed states), particles of matter are close together—close enough to attract each other. In gases, particles of matter are extremely far apart. In common situations, there is a negligible attraction between gas particles. In this section, we take a brief look at some unique properties of gases due to the lack of attractive forces.

Gases and Pressure

When considering gases, one of the most important quantities to consider is pressure. To illustrate the concept of pressure, imagine an empty syringe (with no needle attached) like those used to dispense medicine orally to an infant.

If the plunger of the syringe is drawn all the way out, the syringe will fill with air. If you place your finger firmly over the open tip of the syringe and depress the plunger, what happens? The plunger moves in, and the sample of air is "squeezed." Remember that the particles of a gas are usually far apart and not closely associated with each other. In other words, a sample of gas is mostly empty space. When the air in the syringe is squeezed, the space between the particles is decreased, and the particles of the air are forced closer together and have less room to move about. By depressing the plunger, you are compressing the gas by applying pressure to it. **Pressure** is a force exerted against a given area.

Pressure measurements are expressed in a variety of different units, but for our discussion we will use two of the more common and familiar units, pounds per square inch (psi) and millimeters of mercury (mmHg). The latter unit is the one used when measuring blood pressure.

Suppose you placed a 5-pound bag of sugar on top of your thumb. You would be feeling pressure on your thumb. Your thumbnail area is about 1 square inch, so the pressure would be about 5 pounds per square inch or 5 psi. The unit of **pounds per square inch (psi)** is a measure of force (measured in pounds) applied to an area of 1 square inch. The pressure of the atmosphere at sea level is about 14.7 pounds per square inch, 14.7 psi.

The unit **millimeters of mercury (mmHg)** can also be used to measure the pressure exerted by the atmosphere surrounding Earth.

The pressure of the atmosphere can be measured using a device known as a mercury barometer, which can be constructed by filling a long, sealed glass tube with liquid mercury and inverting the tube into a dish of mercury without letting air into the tube

? 7.4 Inquiry Question: How are Boyle's law and Charles's law applied to gases?

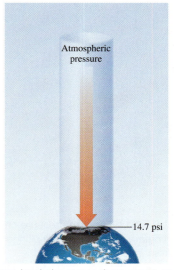

At sea level, the atmosphere exerts an average pressure of 14.7 pounds per square inch on Earth's surface.

(see Figure 7.16). The force of the atmosphere pushing down on the mercury in the dish prevents most of the mercury in the tube from draining out. In fact, if the tube is long enough, a column of mercury 760 mm high (29.92 in.) will remain inside the tube at sea level. Because the pressure exerted by the atmosphere determines the height of the column of mercury in the tube, measuring the height of the mercury column, in millimeters, gives us the pressure in mmHg.

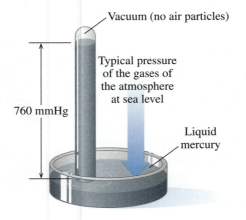

Vacuum (no air particles)

Typical pressure of the gases of the atmosphere at sea level

760 mmHg

Liquid mercury

FIGURE 7.16 **The mercury barometer.** The pressure of the gases of Earth's atmosphere is able to support a column of mercury 760 mm high.

As mentioned, the typical pressure exerted by the atmosphere at sea level supports a column of mercury 760 mm high. This pressure is equivalent to 14.7 psi but varies with altitude and changes in the weather.

Pressure and Volume—Boyle's Law

In the example of the syringe, as the plunger was depressed, the applied pressure increased and the volume of the air decreased (see Figure 7.17).

In the mid-1600s, the Irish chemist Robert Boyle noted these changes and began to experiment with the effect of pressure on the volume of a gas. In his experiments with gases, Boyle discovered that the volume of a gas and the pressure applied to it are related if the temperature and amount of the gas are not allowed to change. He found that when the pressure on a gas was doubled, the volume of the gas was reduced to one-half of its initial volume. Similarly, if he tripled the pressure on the gas, the volume was reduced to one-third of its initial volume. Figure 7.18 shows how these changes in pressure affect the volume of a gas.

Boyle found, and his law states, *that the volume of a fixed amount of gas at constant temperature is inversely proportional to the pressure.* Simply stated, Boyle's law says that an increase in the pressure on a gas will result in a decrease of its volume and vice versa.

Mathematically, Boyle's law allows us to determine what will happen to a sample of gas of known volume and pressure if we make changes to either the volume or the pressure

(a)

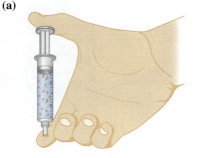

(b)

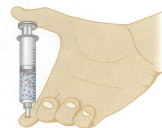

FIGURE 7.17 **Effect of pressure on the volume of a gas.** As the plunger is depressed (going from a to b), the pressure on the gas increases and the volume of the gas decreases.

Initial   Twice the pressure   Triple the pressure

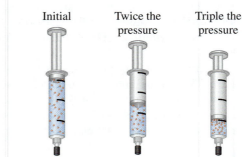

FIGURE 7.18 **Pressure versus volume of a gas.** When the pressure on the gas is doubled, the volume is halved. When the pressure is tripled, the volume decreases to one-third of the initial volume.

while the temperature stays the same. The following formula gives the relationship for such a determination:

$$P_i \times V_i = P_f \times V_f$$

where P_i = the initial or starting pressure

V_i = the initial or starting volume

P_f = the final or ending pressure

V_f = the final or ending volume

Solving a Problem

Boyle's Law

Suppose you have a syringe with one end sealed containing 10 cc of air at a pressure of 14.7 psi. If you double the pressure to 29.4 psi—by pressing on the plunger—what will the volume of the air be?

STEP 1: Determine the given information. In this problem, we are given

P_i = 14.7 psi, V_i = 10 cc, and P_f = 29.4 psi.

STEP 2: Solve for the missing variable using the Boyle's law relationship. In this case, V_f.

$$P_i \times V_i = P_f \times V_f$$

Divide both sides by P_f.

$$\frac{P_i \times V_i}{P_f} = \frac{P_f \times V_f}{P_f}$$

This gives us

$$\frac{P_i \times V_i}{P_f} = V_f$$

STEP 3: Substitute the given information into the equation and solve.

$$\frac{14.7\cancel{psi} \times 10cc}{29.4\cancel{psi}} = V_f$$

$$V_f = 5cc$$

When you complete the calculation (taking care to cancel the units on your numbers), check your answer by seeing if Boyle's law holds. Did the volume go down when the pressure went up? Yes, just as you may have predicted.

Boyle's Law and Breathing

Do you know that you use Boyle's law every day? Breathing is a very practical application of Boyle's law. When you breathe in, you are actually not forcibly drawing air into your lungs. Instead, the muscles of your rib cage and your diaphragm contract to cause the volume of your chest cavity to increase (see **Figure 7.19**). When this happens, the air pressure inside your lungs decreases (volume increase results in pressure decrease), and the pressure of the atmosphere causes air to rush into your lungs to equalize the internal and external pressures. When the muscles relax, the volume of your chest cavity decreases, increasing the pressure in your lungs above that of the outside, and air flows out to the lower pressure (atmosphere).

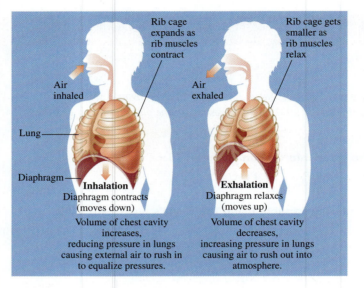

FIGURE 7.19 **Boyle's law and breathing.** Breathing is controlled by contraction and relaxation of the diaphragm muscles.

---

sample problem 7.6 — Boyle's Law Calculations

The lungs of a normal adult can hold 5.0 L of air under typical atmospheric pressure (760 mmHg). If a diver dives to a depth where his lungs compress to a volume of 4.0 L, what is the pressure surrounding her?

Solution

Changes in the volume and pressure of a gas are related using Boyle's law. First, check the problem to determine what is given and what information is requested.

STEP 1: Determine the given information. We are given $P_i = 760$ mmHg, $V_i = 5.0$ L, and $V_f = 4.0$ L.

STEP 2: Solve for the missing variable using the Boyle's law relationship. In this case, P_f.

$$\frac{P_i \times V_i}{V_f} = P_f$$

STEP 3: Substitute the given information into the equation and solve.

$$\frac{760 \text{ mmHg} \times 5.0 \text{ L}}{4.0 \text{ L}} = P_f$$

$$P_f = 950 \text{ mmHg}$$

Temperature and Volume—Charles's Law

Changing the temperature of a gas directly affects the motion of the particles. When the temperature of a gas is increased, heat is added to the gas, and the motion of the particles (speed) increases. Likewise, removing heat energy decreases the temperature of the gas and causes the gas particles to slow down. In each case, the change in the temperature results in a change in the motion of the particles of the gas.

To see how the temperature affects the volume of a gas, consider the photos in Figure 7.20. In the first picture, you see a fully inflated balloon just before it was placed in a cooler at very low temperature. In the second picture, you see the same balloon just after it was taken out of the cooler. Notice the difference? After being in the cooler, the balloon is smaller—its volume has decreased. This relationship between the temperature of a gas and its volume was discovered in the late 1700s by French scientist Jacques Charles.

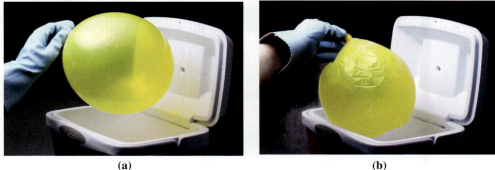

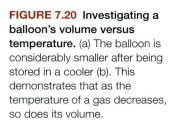

FIGURE 7.20 Investigating a balloon's volume versus temperature. (a) The balloon is considerably smaller after being stored in a cooler (b). This demonstrates that as the temperature of a gas decreases, so does its volume.

In his experiments with gases, Charles found that if the pressure and amount of a gas are not allowed to change, the volume of the gas is directly proportional to its absolute temperature. (We'll discover why he used absolute temperature a little later.) Charles discovered that when the absolute temperature of a gas was doubled, the volume of the gas also doubled.

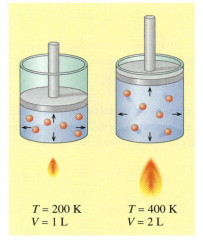

$T = 200$ K
$V = 1$ L

$T = 400$ K
$V = 2$ L

The effect of temperature on the volume of a gas can be shown in an expandable container. When the temperature is doubled, its volume is also doubled. The increased motion of the gas particles causes the volume to increase.

Charles's law states that *the volume of a fixed amount of gas at constant pressure is directly proportional to its absolute temperature.* In other words, an increase in the temperature of the gas will result in an increase in its volume, while a decrease in temperature will result in a decrease in its volume.

We can use Charles's law to determine what will happen to a sample of gas of known volume and temperature if we make changes to either its volume or its temperature. Charles's law can be applied as shown for making such determinations:

$$\frac{V_i}{T_i} = \frac{V_f}{T_f}$$

where T is the absolute temperature in units of Kelvin, V is volume, and the subscripts i and f stand for "initial" and "final," respectively (as they did in the Boyle's law discussion).

Solving a Problem

Charles's Law

If you remove a balloon with a volume of 105 mL of air from a freezer at 0 °C and allow the balloon to warm to room temperature (25 °C) what would its final volume be?

STEP 1: Determine the given information. We are given

$$V_i = 105 \text{ mL}, \ T_i = 0 \text{ °C, and } T_f = 25 \text{ °C.}$$

Temperatures of 0 °C cannot be put into Charles's law because mathematically division by zero is not allowed. In gas law problems, temperatures must be converted to the absolute temperature scale, Kelvin. (Refer to Chapter 1 to review how to convert temperature units.) So $T_i = 273$ K and $T_f = 298$ K.

STEP 2: Solve for the missing variable using the Charles's law relationship. In this case, V_f.

$$\frac{V_i \times T_f}{T_i} = V_f$$

STEP 3: Substitute the given information into the equation and solve.

$$\frac{105 \text{ mL} \times 298 \text{ K}}{273 \text{ K}} = V_f$$

$$115 \text{ mL} = V_f$$

Doing the math (taking care to cancel units) results in the answer $V_f = 115$ mL. Does the final answer make sense? Should the volume of the balloon increase as the temperature increases? Yes. That is exactly what Charles's law predicts.

---

sample problem 7.7 Using Charles's Law

Determine the volume of the balloon we just discussed (105 mL in the freezer) if it is removed from the freezer at 0 °C and warmed to 145 °F.

Solution

STEP 1: Determine the given information. We cannot complete our calculation with the temperature in degrees Fahrenheit. The temperature must be converted to the Kelvin scale. We convert the temperature first to °C and then to K as follows.

$$°C = (145 \text{ °F} - 32) \times \frac{1 \text{ °C}}{1.8 \text{ °F}}$$

$$°C = 63$$

$$K = 63 + 273 = 336 \text{ K}$$

The final temperature T_f is 336 K. From the information in the problem, we are also given $V_i = 105$ mL and $T_i = 273$ K (after converting from Celsius).

STEP 2: Solve for the missing variable using the Charles's law relationship. In this case, V_f.

$$\frac{V_i \times T_f}{T_i} = V_f$$

STEP 3: Substitute the given information into the equation and solve.

$$\frac{105 \text{ mL} \times 336 \text{ K}}{273 \text{ K}} = V_f$$

$$V_f = 129 \text{ mL}$$

This answer makes sense: The volume increased because the temperature went up.

In the following problems, you may assume that the temperature and amount of the gas remain unchanged.

7.25 A balloon is filled with helium gas, which is lighter than air. When the following changes are made at a constant temperature, which of the diagrams (A, B, or C) shows the new volume of the balloon?

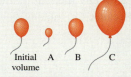

Initial   A   B   C
volume

a. The balloon floats to a higher altitude where the outside pressure is lower.

b. The balloon is taken inside a house, but the atmospheric pressure remains the same.

c. The balloon is placed in a hyperbaric chamber where the pressure is increased.

7.26 When you swim underwater, the pressure of the water pushes against your chest cavity and subsequently reduces the size of your lungs. At 10 m below the water surface, the pressure exerted by the water is 14.7 psi. If a swimmer has a lung volume of 6 L at sea level (atmospheric pressure), what would the volume of her lungs be when she is at the bottom of a pool that is 5 m deep?

7.27 A child's balloon containing 8.5 L of helium gas at sea level is released and floats away. What will the volume of the helium be when the balloon rises to the point where the atmospheric pressure is 380 mmHg?

7.28 Assuming no air can move in or out, what happens to the volume of a hot air balloon as the air is heated?

7.29 Select the diagram that shows the new volume of a balloon when the following changes are made at constant pressure.

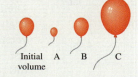

Initial   A   B   C
volume

a. If the temperature is changed from 150 K to 300 K.

b. If the balloon is placed in a freezer.

c. If the balloon is warmed and then cooled to its initial temperature.

7.30 A gas has a volume of 10 L at 0 °C. What is the final temperature of the gas (in °C) if its volume increases to 25 L?

7.5 Dietary Lipids and Trans Fats

Up to this point, we have examined a compound's state of matter and attractive forces present. Establishing this information allows us to correctly predict some of the compound's physical properties. Now we leave gases and turn our attention to applying these principles to some biologically relevant liquids and solids, lipids. Recall that in Section 7.3 we identified the lipid called a triglyceride as a nonpolar (hydrophobic) molecule that is not soluble in water. Triglycerides exist as solids or liquids. In this section we examine structural changes that affect their melting points.

? 7.5 Inquiry Question:
How are fats and oils converted into one another?

Fats Are Solids

A juicy steak, a tub of lard, or a stick of butter all contain animal fats that you can find at your local grocery store. Animal **fat** is a solid or semisolid material at room temperature that, like the oils we saw earlier in the chapter, is a triglyceride made up of three fatty acids joined to a glycerol backbone. But if both fats and oils are triglycerides, why is one solid and the other liquid at room temperature? Attractive forces are key to the explanation.

In Section 7.3, we saw the formation of a triglyceride from the condensation reaction of three fatty acids with glycerol. When the hydrocarbon chains (tails) of the fatty acids are mostly saturated, the triglyceride product is a fat. (In Chapter 4, we saw that a hydrocarbon containing all single C—C bonds was referred to as "saturated.")

In Figure 7.21a, you can see that, in the structure of the fat molecule, the hydrocarbon tails of the three fatty acids are physically closer to each other than those of the oil in Figure 7.21b. This proximity creates disturbances in the electron distribution around each of the tails as they interact with the other tails. These disturbances result in London forces—the tails are attracted to one another. The saturated tails allow for many surface contacts, which increases the attractions between them and slows down (restricts) the molecular motions in the molecules, allowing the molecules to form a solid.

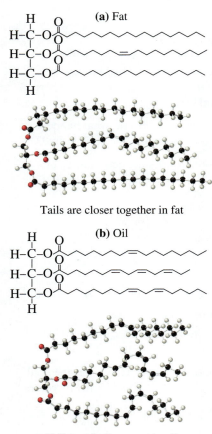

(a) Fat

Tails are closer together in fat

(b) Oil

Tails are farther apart in oil

FIGURE 7.21 (a) A typical fat molecule and (b) an oil molecule. Fats are solids and oils are liquids at room temperature. Notice that the hydrocarbon tails "pack" closer together in the fat than in the oil. This is due to the multiple cis double bonds in the hydrocarbon tails of the oil.

So fats are solids at room temperature because the motion of their hydrocarbon tails is restricted by the London forces holding them together. Instead of the unrestricted movement found in a single hydrocarbon tail like a soap, the tails of a fat are attracted to one another and their movement is more restricted. But why is the melting point of a fat low enough to melt at body temperature?

When a substance melts, it changes from a solid to a liquid. As we saw with boiling, when heat energy is added to a substance, attractive forces are disrupted, and molecular motion increases. For a fat to melt, the forces that have to be disrupted are weak London forces. Fats are typically solid substances but with low melting points. Even the heat of your hand (37 °C) is enough to melt butter. Because the melting points are low, fats are often said to be semisolid.

Oils Are Liquids

Dietary **oils** are derived from plants like corn, soybean, or peanuts and are also triglycerides—three fatty acids bonded to a glycerol. The difference in the name "fat" versus "oil" indicates the physical state at room temperature—oils are liquids and fats are solids. How can two molecules that are so similar in structure exist in different states?

The answer lies in the structure of the hydrocarbon tails of the oil molecules. Compare the tails in the oil molecule to those in the fat molecule in Figure 7.21. The hydrocarbon tails in the oil molecules contain more cis double bonds than those in the fat. (Recall that naturally occurring fatty acids contain *only* cis double bonds—Section 4.5.) Because they contain several carbon–carbon double bonds, the tails of the oil are called polyunsaturated. So how does this simple difference translate to a difference in the physical state (solid versus liquid) of these molecules?

Notice that the introduction of cis double bonds in the tails of the oil creates kinks in the normally straight hydrocarbon chain (see Figure 7.21b). The tails of the unsaturated oil cannot stack together as closely as those in the fat can. The kinked tails interact less with each other through London forces because they are not physically as close to one another. Less interaction means weaker forces of attraction, which in turn means that the hydrocarbon chains in the oil are less attracted to one another and, therefore, move more freely. The greater molecular freedom of motion among the hydrocarbon tails in the oil does not allow enough stacking of the tails for a solid to form, so oils remain liquid.

Partial Hydrogenation and Trans Fats

Oils derived from plants have long been considered a healthier alternative for human dietary fat requirements. However, liquids are not always the most convenient form of triglycerides—imagine coating your morning toast with corn oil instead of a soft solid spread. Not only does it sound unappetizing, it would not be particularly easy to do. To overcome this problem, chemists have developed a method to convert the liquid, plant-derived triglycerides (oils) into solids to produce what we know today as margarine.

The production of margarine—a solid from a plant-derived oil such as corn oil—involves an understanding of attractive forces. The only difference between fats and oils is the number of double bonds in the fatty acid chains. If double bonds could be removed and single bonds left in their place, the attractions between the chains would increase, and the liquids would be solids. This is how margarine is formed from vegetable oil.

In Chapter 5, we explored **hydrogenation,** an addition reaction to an alkene. Recall that, in this reaction, one H_2 molecule reacts with the double bond in the compound. The second bond between the two carbons is removed (leaving a single bond) and one hydrogen atom is added to each of the original carbon atoms of the double bond. A platinum or palladium catalyst is used to speed up the reaction. In a hydrogenation reaction, the carbons of the unsaturated compound become saturated with hydrogen.

The hydrogenation of plant-derived oils is controlled so that some double bonds become saturated while others remain intact. This process, known as *partial hydrogenation*, allows producers to create margarines that are solids that, although more saturated than oils, are less saturated than butter, making them easier to spread even when cold. One of the consequences of partial hydrogenation is that some of the double bonds are incompletely hydrogenated. Because hydrogenation is a difficult reaction to control, some of the double bonds in these compounds re-form, but when they do, they switch from the less stable cis form to the more stable trans form, resulting in the compounds known as *trans fats* (see **Figure 7.22**).

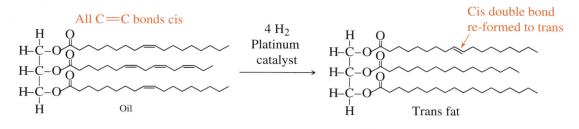

FIGURE 7.22 Incomplete hydrogenation of an oil forming a trans fat. The cis C=C bonds in the oil become hydrogenated to form alkanes or re-form as trans C=C bonds.

Partial hydrogenation enhances desirable qualities for manufacturers. It increases shelf life and flavor stability and allows the control of spreadability in butter substitutes. While the trans double bonds do allow the fatty acid chains to stack more easily, enhancing the solidity of the fat, there is public concern because some studies show that trans fats have deleterious health effects. This has forced many manufacturers of partially hydrogenated oils to consider alternatives. One alternative is to use more saturated plant oils in foods such as coconut and palm oils, which are naturally solid at room temperature. Today all nutritional labeling must give the amounts of trans fat present in food.

sample problem 7.8 Hydrogenation of Oils

How many hydrogen molecules (H_2) are required to convert the oil shown in the figure into a completely saturated fat?

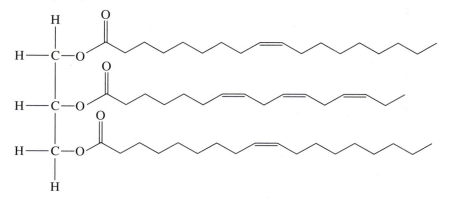

Solution

One molecule of H_2 is required to hydrogenate each carbon–carbon double bond in an unsaturated compound. Therefore, count the number of carbon–carbon double bonds in the oil molecule (the carbon–oxygen double bonds will not react under these conditions). This number will be the same as the number of H_2 molecules necessary to convert the oil molecule into a saturated fat molecule.

The oil molecule shown has five carbon–carbon double bonds, so five H_2 molecules will be required for this reaction. The saturated product is shown.

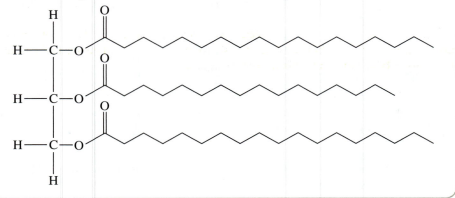

practice
problems

7.31 What is a triglyceride? Draw the skeletal structure for a saturated triglyceride whose hydrocarbon tails each contain six carbons.

7.32 Draw the structural formula for a simple triglyceride whose hydrocarbon tails each contain eight carbons, two of which are saturated and one of which is unsaturated between carbons 3 and 4 from the tail end.

7.33 Explain why fats are solids, but oils are liquids under normal conditions, despite the fact that both are triglycerides.

7.34 A certain triglyceride required three molecules of H_2 for the hydrogenation reaction to be completed. How many double bonds were present in the original triglyceride?

7.35 Balance the following reaction equation by providing the missing number of hydrogen molecules:

7.36 Balance the following reaction equation by providing the missing number of hydrogen molecules:

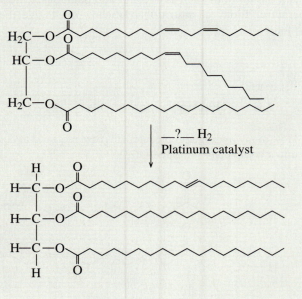

7.6 Attractive Forces and the Cell Membrane

The cells of our bodies are encased in an amazing membrane. The structure of the cell membrane, which is similar in cells throughout the body, has the ability to be flexible yet firm. As we will see in Chapter 8, the cell membrane serves to allow some molecules to move into and out of the cell while inhibiting passage of other molecules. What is it about the structure and composition of the cell membrane that enables it to be selective?

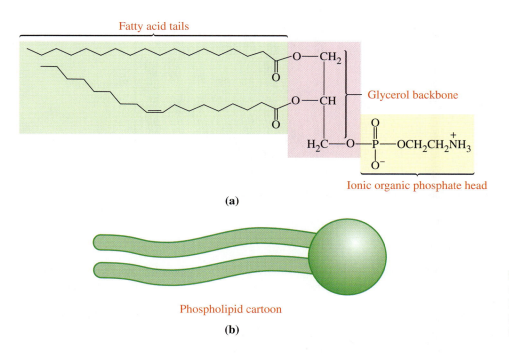

7.6 Inquiry Question:
How do attractive forces apply to the structure of a cell membrane?

A Look at Phospholipids

The main structural components of cell membranes are lipid molecules called **phospholipids.** Structurally similar to triglycerides, phospholipids have a glycerol backbone with fatty acids linked to it through an ester bond. Unlike triglycerides, phospholipids have only *two* fatty acids on their glycerol backbone, arranged as shown in Figure 7.23a. The third OH group of the glycerol is bonded to a phosphate-containing group. The phosphate-containing group is ionic (polar), which is in contrast to the fatty acid tails (nonpolar). There are many different phospholipids, but they all share this similar structure: a glycerol backbone with two nonpolar fatty acid tails and a polar phosphate-containing head. Because these molecules have both polar and nonpolar parts, they are amphipathic.

Fatty acid tails

Glycerol backbone

Ionic organic phosphate head

(a)

Phospholipid cartoon

(b)

FIGURE 7.23 Phospholipid in (a) line structure and (b) cartoon. Phospholipids have a polar phosphate-containing head group and two nonpolar hydrocarbon tails.

Phospholipids are often represented by a cartoonlike drawing (see Figure 7.23b) highlighting the dual character of these molecules (strongly polar head and two nonpolar tails). Earlier, we saw that the fatty acid molecules of soap have a similar structure with a polar head but just one nonpolar tail. Having two nonpolar tails and a larger ionic head enhances both the nonpolar and polar characteristics of phospholipids as compared to soap. The two tails of the phospholipid also affect the overall shape of the molecule. While soap molecules tend to be more conical in their overall shape and form micelles, phospholipids tend to have a more cylindrical shape with a much larger polar head group. This cylindrical shape of the phospholipids hinders their ability to arrange themselves into micelles. So how do phospholipids behave in water?

The Cell Membrane Is a Bilayer

Consider the cell and its environment: Its surroundings are watery *and* its contents function in an aqueous environment. With water environments inside and outside the cell, how can phospholipids, molecules that have nonpolar hydrophobic tails, serve to surround and protect cells? The cylindrical shape of a phospholipid constrains them to lining up in a straight line instead of a sphere like a soap. No matter how the tails orient themselves—toward the inside or outside of the cell—the hydrocarbon portion would be forced into an aqueous environment.

A cell membrane composed of phospholipids cannot exist as a single layer. Instead, the phospholipids form a double layer called a bilayer as shown in **Figure 7.24**. The phospholipids of the "outer" layer of the membrane are oriented such that the polar heads are directed out into the surrounding aqueous environment. In turn, the phospholipids of the "inner" layer of the membrane are oriented with the polar heads directed into the aqueous environment of the interior of the cell. This arrangement leaves the nonpolar tails of both layers directed toward each other, creating a nonpolar interior region. The phospholipid bilayer, as it is known, is the foundation for a cell's membrane. Next let's examine the other molecules found in the membrane.

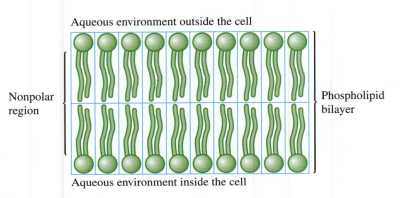

Aqueous environment outside the cell

Nonpolar region

Phospholipid bilayer

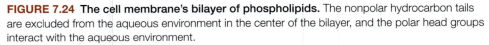

Aqueous environment inside the cell

FIGURE 7.24 The cell membrane's bilayer of phospholipids. The nonpolar hydrocarbon tails are excluded from the aqueous environment in the center of the bilayer, and the polar head groups interact with the aqueous environment.

Among the most important components in the phospholipid bilayer of the cell membrane are large biomolecules called proteins. Protein molecules can span the lipid bilayer extending outward on either side (integral membrane proteins) or they can associate with one polar surface of the bilayer (peripheral membrane proteins). While the phospholipids provide the basis for the cell membrane's structure, the proteins are the membrane's functional components, allowing selected molecules to move into and out of the cell. (Read more about this in Chapter 8.) The exterior surface of the cell membrane also contains carbohydrates (like the ABO blood groups mentioned in Chapter 6) that act as cell signals.

The current model for the cell membrane, called the **fluid mosaic model** shown in **Figure 7.25**, puts all these pieces together. This model creates what can be thought of as "icebergs" of protein floating in a "sea" of lipids, implying that the membrane itself is fluidlike because the phospholipids are able to move freely within their bilayer.

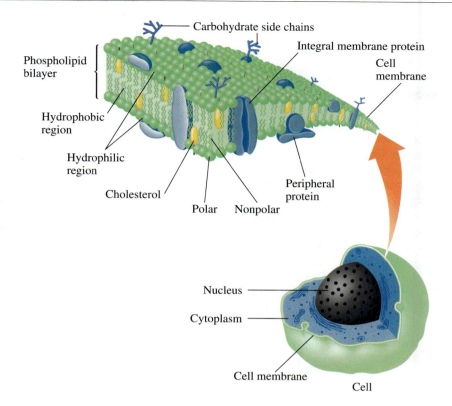

Carbohydrate side chains

Integral membrane protein

Cell membrane

Phospholipid bilayer

Hydrophobic region

Hydrophilic region

Cholesterol

Peripheral protein

Polar   Nonpolar

Nucleus

Cytoplasm

Cell membrane

Cell

FIGURE 7.25 The fluid mosaic model of the cell membrane. Proteins and cholesterol are embedded in a lipid bilayer of phospholipids. The bilayer forms a hydrophobic barrier around the cell contents. Polar heads line the membrane surface and the nonpolar tails face the center away from water.

Steroids in Membranes: Cholesterol

Take a close look at the membrane piece shown in Figure 7.25. Notice that there are cholesterol molecules positioned in the phospholipid bilayer. What is cholesterol and why is it included in a cell membrane?

Let's take a closer look at the structure of cholesterol shown. Cholesterol belongs to a class of lipids called steroids. **Steroids** are lipids with a structure that contains a four-membered fused ring called a *steroid nucleus*. Even though the steroid structure differs greatly from fatty acids and triglycerides, because they are mainly nonpolar, steroids are classified as lipids. These molecules have a variety of functions in the body including regulating sexual development (testosterone and estrogens) and emulsifying dietary fats (bile acids). One of the vitamins, vitamin D, also has a steroid form, ergosterol.

Cholesterol contains the steroid nucleus. Cholesterol, though, has a polar end (the OH group) and a nonpolar portion (the rest of the molecule), so like a phospholipid, it is amphipathic. Cholesterol situates itself in the phospholipid bilayer, so that the OH group protrudes out into the surrounding aqueous environment, while the rest of the rigid rings of the molecule are nestled in the nonpolar interior of the membrane.

Cholesterol is an important component of the cell membrane. It modulates the fluidity or flexibility of a cell membrane depending on the amount present. Cholesterol can slip in between the hydrocarbon tails of the phospholipids, disrupting the London forces and increasing the membrane's fluidity. Cholesterol can also interact with the unsaturated hydrocarbon tails via London forces, decreasing their motion and increasing the rigidity of the membrane.

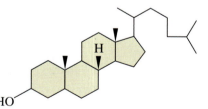

A steroid nucleus

Lipids that contain a four-membered fused ring like this one are classified as steroids.

HO

Cholesterol contains the steroid ring structure and is amphipathic.

7.37 How will phospholipids (mainly nonpolar) arrange themselves when they are put in water?

7.38 Draw a possible hydrogen bond between a molecule of cholesterol and a molecule of water.

practice **problems**

SUMMARY

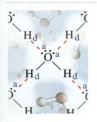

7.1 Types of Attractive Forces

7.1 Inquiry Question: How does the chemical structure of a compound dictate the attractive forces present?

The attractive forces present in a substance can be determined by the polarity of the compound. The types of attractive forces present between compounds (from weakest to strongest) are London forces, dipole–dipole, hydrogen bonding, ion–dipole, and ionic attractions. Each of these forces involves the attraction of an area of negative charge (either partial or full) on one species (molecule or ion) to an area of positive charge on a second species. The ionic attraction occurs between a full + and – charge and occurs between any two oppositely charged ions.

As heating begins, bubbles of vapor are formed in the liquid.

When boiling, vapor bubbles rise to the surface and escape. (36 °C is boiling point for _____)

7.2 Liquids and Solids: Attractive Forces Are Everywhere

7.2 Inquiry Question: How are the boiling points and melting points of compounds predicted?

The boiling point is the temperature where the change of state from liquid to gas occurs. For a compound to boil, the molecules of the compound must overcome the attractive forces between molecules. The boiling points and melting points of compounds can be predicted by examining the attractive forces present. Compounds with stronger attractive forces require more heat energy to disrupt those attractions and have higher boiling points. Water has a comparatively high boiling point. In contrast, for alkanes that have only London forces between molecules, the boiling point depends largely on the surface area of the molecule—the greater the surface area, the higher the boiling point.

Oil

Water

7.3 Attractive Forces and Solubility

7.3 Inquiry Question: How is the solubility of a compound in water determined?

The solubility of one substance in another is described by the golden rule of solubility, *like dissolves like,* which means that polar compounds dissolve polar compounds and nonpolar compounds dissolve other nonpolar compounds. In other words, the more attractive forces that two compounds have in common, the more soluble they will be in each other. Soap is an amphipathic compound. In water, soap molecules form spherical structures called micelles with the hydrophobic tails gathered together in the nonpolar interior of the micelle while the hydrophobic heads cover the surface and interact with the aqueous external environment. Most organic compounds containing functional groups are soluble to various extents in water based on the number and distribution of polar or ionic groups.

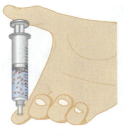

7.4 Gases: Attractive Forces Are Limited

7.4 Inquiry Question: How are Boyle's law and Charles's law applied to gases?

Gases have few attractive forces because particles in a gas are often too far apart for an attraction to occur. Pressure is force exerted on a given area and is typically measured in units of psi or mmHg. Changing the pressure or temperature of a gas causes a change in the volume of the gas. Boyle's law compares volume and pressure. Pressure and volume are inversely related at constant temperature. As volume decreases, pressure increases. Charles's law compares volume and temperature. Temperature (in Kelvins) and volume are directly related at constant pressure. As temperature increases, volume increases.

7.5 Dietary Lipids and Trans Fats

7.5 Inquiry Question: How are fats and oils converted into one another?

Melting is the change of state from solid to liquid. This change involves a decrease in attractive forces and an increase in the motion of the molecules in the substance. Fats are solid triglyceride compounds that have low melting points. Oils are also triglycerides, but they are liquids at room temperature. The fatty acid tails of oils are highly unsaturated, and they have fewer London forces between their fatty acid tails than fats. Oils can be converted into fats by hydrogenation of the double bonds to increase the saturation of the fatty acid tails. This is the process used to create margarine from oils derived from plants. Hydrogenation often results in the conversion of naturally occurring cis double bonds to trans isomers. The products of this process are less healthful trans fats.

7.6 Attractive Forces and the Cell Membrane

7.6 Inquiry Question: How do attractive forces apply to the structure of a cell membrane?

Cell membranes are composed of phospholipids, which are amphipathic. Phospholipids have a strongly polar head and two nonpolar tails. Because of their shape, phospholipids of the cell membrane arrange themselves in a bilayer around the cellular contents with the polar heads of one layer oriented out into the surrounding aqueous environment and the polar heads of the second layer directed inward toward the aqueous interior of the cell. This results in a hydrophobic interior layer formed by the nonpolar tails. The fluid mosaic model describes the complete structure of the cell membrane. Cell membranes also contain cholesterol, a steroid lipid, which helps modulate the fluidity of the membrane.

The study guide will help you check your understanding of the main concepts in Chapter 7. You should be able to

7.1 Types of Attractive Forces

- Describe five types of attractive forces present in compounds.
- Determine the attractive forces present in a compound from its chemical structure.

7.2 Liquids and Solids: Attractive Forces Are Everywhere

- Describe the process of boiling.
- Predict relative boiling points for liquids based on the attractive forces present.
- Predict relative melting points for solids based on the attractive forces present.

7.3 Attractive Forces and Solubility

- State the golden rule of solubility.
- Predict the solubility of a molecule in water.
- Recognize an amphipathic molecule.
- Define the role of an emulsifier.
- Draw a fatty acid micelle.

7.4 Gases: Attractive Forces Are Limited

- Contrast the attractive forces present in a gas with those in a solid or liquid.
- Define pressure.
- Apply Boyle's law.
- Apply Charles's law.

7.5 Dietary Lipids and Trans Fats

- Distinguish a fat from an oil.
- Describe the differences in melting points of fats and oils based on their attractive forces.
- Predict the products of the complete hydrogenation of a triglyceride.

7.6 Attractive Forces and the Cell Membrane

- Draw a phospholipid bilayer.
- Describe the structure of a cell membrane.
- Locate the polar and nonpolar regions of a phospholipid and cholesterol.
- Classify molecules as steroids based on their structure.

Key Terms

amphipathic—A compound that has both polar and nonpolar parts.

attractive force—An attraction of opposite charges between compounds.

boiling—The formation of gas bubbles of a liquid.

boiling point—The temperature at which the molecules of a substance change from a liquid to a gas (boil).

change of state—The change that occurs when a substance that exists in one state of matter moves to another state of matter; for example, a solid becomes a liquid or a liquid becomes a gas.

condensation—The change in state from a gas to a liquid.

deposition—The change in state from a gas directly to a solid.

dipole–dipole attraction—The attraction of the partially positive end of one polar molecule for the partially negative end of a second polar molecule.

emulsifier—An amphipathic compound that allows polar and nonpolar substances to mix.

esterification—A condensation reaction of a carboxylic acid and an alcohol to form an ester.

fat—A lipid molecule composed of three fatty acids joined to a glycerol backbone which exists as a solid or semisolid at room temperature; also known as a triglyceride.

fluid mosaic model—A model used to describe the nature of the cell membrane. The model incorporates the structural components (phospholipids and cholesterol) and the functional components (proteins) of the membrane.

freezing—The change in state from a liquid to a solid.

golden rule of solubility—Molecules having similar polarities and participate in similar attractive forces will dissolve in each other.

hydration—For an ionic compound, the dissociation of the compound into its component ions by the action of many water molecules interacting with the ions and then surrounding each of the newly separated ions.

hydrogen bonding—The strong dipole–dipole attraction of a hydrogen, covalently bonded to an O, N, or F (*the hydrogen-bond donor*), to a nonbonding pair of electrons on an O, N, or F (*the hydrogen-bond acceptor*).

hydrogenation—The addition of hydrogen (H_2) to a carbon–carbon double bond to convert it to a single bond.

hydrophilic—Literally, "water-loving"; polar compounds or parts of compounds that readily dissolve in water.

hydrophobic—Literally, "water-fearing"; nonpolar compounds or parts of compound that do not dissolve in water.

induced dipole—A temporary, uneven distribution of the electrons over the surface of a compound creating a brief separation of charge in the compound.

intermolecular forces—Attractive forces between two or more molecules that are caused by uneven distribution of electrons within the molecules.

ionic attraction—An attraction between a + and a − charge in or between compounds; also called an ionic bond or a salt bridge.

ion–dipole attraction—An electrical attraction between an ion and a polar molecule.

London forces—The weakest attractive force; an attraction formed from temporary (or induced) dipoles on molecules.

melting—The change in state from solid to liquid.

melting point—The temperature at which the molecules of a substance change from a solid to a liquid (melt).

micelle—A three-dimensional spherical arrangement of soap molecules in water with the hydrophobic tails pointing inside, away from the water, and the hydrophilic heads pointing out into the water.

oil—A lipid molecule composed of three fatty acids joined to a glycerol backbone that exists as a liquid at room temperature; also known as a triglyceride.

permanent dipole—An uneven distribution of the electrons in a compound caused by differences in the electronegativity of the atoms.

phospholipid—The primary structural component of cell membranes consisting of a glycerol backbone esterified with two fatty acids and a phosphate-containing group.

pounds per square inch (psi)—A measure of pressure expressed as force (pounds) on a defined area (a square inch).

pressure—The amount of force exerted against a given area.

solubility—The maximum amount of a substance that can dissolve in a specified amount of water at a given temperature.

salt—An ionic compound.

salt bridge—See ionic attraction.

steroid—A lipid containing a four-membered fused ring structure called a steroid nucleus.

sublimation—The change in state from a solid directly to a gas.

triglyceride—The product of a condensation reaction of three fatty acids and glycerol; also known as a fat or an oil.

vaporization—The change in state from a liquid to a gas.

Additional Problems

7.39 Compare and contrast a dipole–dipole attraction and an ion–dipole attraction.

7.40 Compare and contrast ion-dipole attraction and an ionic attraction.

7.41 Show a possible hydrogen bond between acetone (CH₃COCH₃) and water. Label the donor and the acceptor. Is there more than one possible hydrogen bond between these two molecules? Explain.

7.42 Show a possible hydrogen bond between methanol (CH₃OH) and dimethyl ether (CH₃OCH₃). Label the donor and the acceptor. Is there more than one possible hydrogen bond between these two molecules? Explain.

7.43 List all attractive forces found in each of the pure substances:

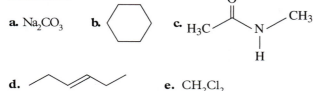

a. Na₂CO₃   **b.**   **c.** H_3C — N — CH_3 / H

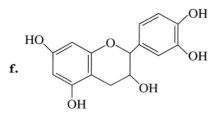

d.   **e.** CH₂Cl₂

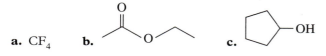

f.

7.44 List all attractive forces found in each of the pure substances:

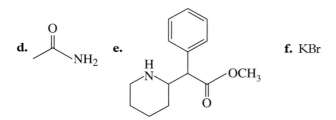

a. CF₄   **b.**   **c.** —OH

d.   **e.**   **f.** KBr

7.45 Explain how an emulsifier can combine a nonpolar substance with a polar substance.

7.46 In terms of attractive forces, explain why oil and water do not mix.

7.47 Predict which member of each of the following pairs of compounds will be more soluble in vegetable oil:
  a. fatty acid or fatty acid salt
  b. fat or water
  c. CH₃NH₂ or CS₂
  d. KCl or CH₃CH₂CH₂CH₂CH₂CH₃

7.48 Predict which member of each of the following pairs of compounds will be more soluble in vegetable oil:
  a. NH₃ or CH₃CH₂CH₂ CH₂CH₃
  b. NaCl or CH₃CH₂OCH₂CH₃
  c. cyclopentane or CH₃CH₂NH₂
  d. CH₃OH or CH₃OCH₃

7.49 A stain on your shirt will not come out when you wipe it with a wet cloth. What clue does this give you to the nature of the stain?

7.50 In order to remove stains from delicate fabrics without using soap and water, dry cleaners once used perchloro-ethylene, which was banned in 1998 for environmental reasons. Its structure is shown in the figure. Explain why perchloroethylene makes a good stain remover.

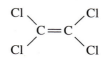

7.51 Rank the four compounds in the figure in decreasing order of boiling point. Explain the reasoning behind your ranking.

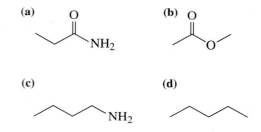

7.52 Rank the four compounds in the figure in decreasing order of boiling point. Explain the reasoning behind your ranking.

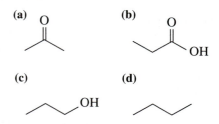

7.53 The boiling point of octane is much greater than that of hexane (see Table 7.3). Explain the difference in their boiling points.

7.54 The boiling point of heptane is over 60 °C higher than the boiling point of pentane. Draw the structure of both compounds and explain the difference in their boiling points.

7.55 Explain why a straight-chain alkane has a higher boiling point than a branched-chain alkane with the same molecular formula.

7.56 Based on your knowledge of boiling point trends in alkanes, predict which compound has the higher boiling point, butane or 2-methylpropane (isobutane).

7.57 Predict which of the following compounds is likely to be more soluble in water.

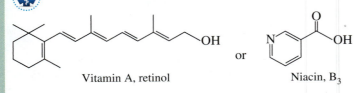

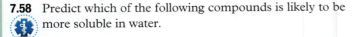

Vitamin A, retinol   or   Niacin, B$_3$

7.58 Predict which of the following compounds is likely to be more soluble in water.

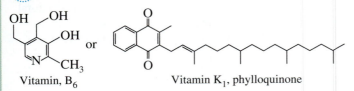

Vitamin, B$_6$ Vitamin K$_1$, phylloquinone

7.59 A sealed syringe contains 5 cc of oxygen gas under 14.7 psi of pressure. If the pressure on the syringe is increased to 36.8 psi, but the temperature remains constant, what is the volume of the oxygen gas in the syringe?

7.60 Underwater pressure increases by 14.7 psi for every 10 m of depth. Standing on a boat in scuba gear, you fill a balloon with helium.

  a. If you take the balloon with you as you dive to a depth of 40 m, what, if anything, will happen to the size of the balloon (assume temperature stays the same)?

  b. If the balloon slips out of your hand and floats upward before you dive, what, if anything, will happen to the size of the balloon (assume temperature stays the same)?

  c. If the balloon had a volume of 2 L at a depth of 40 m, what was the original volume of the balloon if we assume the pressure at the surface of the water is 14.7 psi?

7.61 The lowest recorded atmospheric pressure occurred during a typhoon in the Pacific Ocean in 1979 and measured 652 mmHg. The highest atmospheric pressure on record occurred in Mongolia in 2001 and measured 814 mmHg. If a balloon were inflated with air to a volume of 10. L under the conditions in Mongolia and then rapidly transported to the conditions described for the typhoon in the Pacific, what would the volume of the balloon be under these new conditions, assuming no change in temperature?

7.62 Scuba divers wear tanks of compressed air to help them breathe underwater. Based on your knowledge of

Boyle's law, explain why the air in a scuba tank must be pressurized for a deep-sea diver to breathe while underwater.

7.63 A beach ball is filled with 10.0 L of air in a hotel room where the temperature is 65 °F. What is the volume of the air in the beach ball out on the beach where the temperature is 38 °C?

7.64 The night before a big January "polar bear" bike ride, you take your bike inside to clean it up and pump up the tires. The next morning, after spending the night in your cold garage, the tires on your bike feel "squishy." Assuming they did not leak overnight, why do the tires feel less firm this morning than they did last night?

7.65 If a hot air balloon has an initial volume of 1000 L at 50 °C, what is the temperature (in °C) of the air inside the balloon if the volume expands to 2500 L?

7.66 Liquid nitrogen is an extremely cold liquid (-196 °C) used in chemistry "magic" shows to "shrink" balloons (and do other fun things). If a balloon is inflated to 2 L at 26 °C and then cooled in liquid nitrogen, what is the final volume of the balloon?

7.67 Draw the structure of the product of the following hydrogenation reaction:

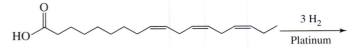

7.68 Draw the structure of the product of the following hydrogenation reaction:

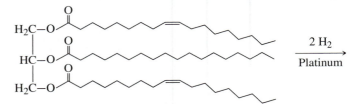

7.69 Sometimes olive oil will solidify at refrigerator temperatures whereas soybean oils do not. Predict which oil has more unsaturated fatty acids present.

7.70 Sketch a cartoon model of the phospholipids bilayer of a cell membrane and label the polar and nonpolar parts.

7.71 Compare the structure of a soap molecule to a phospholipid and explain why a soap's polar head is smaller than that of a phospholipid.

7.72 Soap, phospholipids, and cholesterol are all amphipathic molecules. Provide a drawing of each and label the polar and nonpolar portion.

Challenge Problems

7.73 If very small amounts of soap are dissolved in water, the soap molecules can form a structure known as a mono-layer on the surface of the water with part of the soap molecule dissolved in the water and part sticking into the air. Using the cartoon figures for soap molecules, draw a small part of a monolayer showing the water, the air, and the soap molecules as they would be arranged.

7.74 Getem Clean, Inc. has hired you as chief chemist and put you in charge of a corporate project to build a bet-ter soap. Given your understanding of how soap works, what would you do to improve a fatty acid soap mole-cule to make it more effective in dissolving greasy dirt and in being dissolved by water?

7.75 Mayonnaise is a thick mixture containing oil, vinegar (water-based), and a phospholipid molecule called leci-thin found in the yolks of eggs. Mayonnaise cannot be made without lecithin. Explain why lecithin is a critical ingredient.

7.76 The structures and boiling points of cyclopentane and tetrahydrofuran are shown in the following figure.

Explain why the boiling point of tetrahydrofuran is higher than that of cyclopentane.

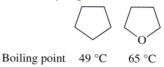

Boiling point   49 °C   65 °C

7.77 Some other physical properties of liquids include vis-cosity, the resistance to flow, which can be thought of as the "thickness of a liquid," and surface tension, which measures the forces at a liquid surface. Consider the following liquids: honey (mainly glucose and fructose) versus water. (You may need to think back to Chapter 6 to consider the structures of glucose and fructose.)

  a. In your experience, which liquid has a higher viscosity?

  b. What attractive forces are present? Which one can form more hydrogen bonds?

  c. Explain viscosity in terms of attractive forces.

  d. It is possible to float a small paper clip on the surface of a cup of water. Try it. What happens if you add soap to the water? Can you float the same paper clip?

Answers to Odd-Numbered Problems

Practice Problems

7.1 Covalent bonds involve the sharing of electrons within a molecule creating a chemical bond. Attractive forces involve attractions between positive and negative areas of different molecules. In attractive forces, electrons are not shared between atoms as in a chemical bond.

7.3

Molecule	Hydrogen Bond	Dipole–Dipole	London
CH_3OH	Yes	Yes	Yes
NH_3	Yes	Yes	Yes
CH_3CH_3	No	No	Yes

7.5 The dipole in London forces is temporary; the dipole in dipole–dipole forces is permanent.

7.7 A hydrogen bond acceptor is a pair of electrons on an O, N, or F.

7.9 Polar molecules like water are not ionic and pure ionic compounds are attracted more strongly through ionic attractions.

7.11 **a.** ionic

  b. London forces

  c. hydrogen bonding

  d. dipole–dipole

7.13 Three hydrogen bonds can be formed.

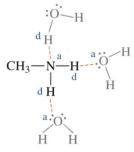

7.15 **a.** CH_4, London force; H_2O, H-bonding; H_2O has the higher boiling point.

  b. NH_3, H-bonding; CO_2, London force; NH_3 has the higher boiling point.

  c. Acetic acid, hydrogen bonding; acetone, dipole-diople; acetic acid has the higher boiling point.

  d. KCl, ionic; CH_3OCH_3, dipole–dipole; KCl has the higher boiling point.

7.17 The highest boiling point is found in the compound with the strongest attractive forces (hydrogen bonding), CH_3OH. The lowest boiling point is found in the compound with the weakest attractive forces (London force), CH_4.

7.19 Of the two alkanes, the one with the more branching has the lower boiling point because of its decreased surface area. The alcohol molecule on the right has the highest boiling point because it has the strongest attractive forces present. It is an alcohol and has hydrogen bonding and dipole–dipole attractions.

7.21   **a.** insoluble
   b. soluble
   c. forms a micelle

7.23   **a.** The fatty acid salt is more soluble in water. The salt can form more attractions to water through ion–dipole interactions not available to the fatty acid.
   b. KCl. The ions in KCl readily dissolve in water and form ion–dipole interactions, which are the strongest attraction that a molecule can have with water.

7.25   **a.** It would look like balloon C.
   b. It would look like balloon B.
   c. It would look like balloon A.

7.27   17 L

7.29   **a.** It would look most like balloon C.
   b. It would look most like balloon A.
   c. It would look most like balloon B.

7.31 A triglyceride is a dietary lipid containing three fatty acids bonded to a glycerol molecule.

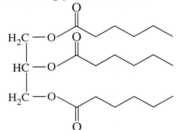

7.33 Even though fats and oils are both triglycerides, the large number of cis double bonds in the hydrocarbon tails of oils prevents them from interacting with each other as much as the saturated tails of the fats. More interactions mean that the tails move more slowly, becoming more like a solid, forming a fat.

7.35   2

7.37 Phospholipids will form two layers called a bilayer (see Figure 7.24) with the nonpolar tails of the two layers facing each other and their polar heads facing the water.

Additional Problems

7.39 Both are attractive forces that occur between positive and negative areas. A dipole–dipole attraction occurs between two opposite partial charges. An ion–dipole attraction occurs between opposite ionic charges and partial charges.

7.41

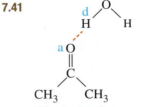

The molecules can form more than one hydrogen bond. Acetone can form two hydrogen bonds acting as an acceptor twice (lone pairs on oxygen).

7.43   **a.** ionic attraction
   b. London forces
   c. London forces, dipole–dipole, hydrogen bonding
   d. London forces
   e. London forces, dipole–dipole
   f. London forces, dipole–dipole, hydrogen bonding

7.45 An emulsifier can attract both a nonpolar and a polar substance, allowing both substances to be suspended in a mixture.

7.47   **a.** fatty acid
   b. fat
   c. CS_2
   d. $CH_3CH_2CH_2CH_2CH_2CH_3$

7.49 The stain must be hydrophobic because it is not soluble in water.

7.51 Highest to lowest boiling point:
   a. five possible H-bonds per molecule (strongest)
   c. three possible H-bonds per molecule
   b. polar molecule
   d. nonpolar molecule (weakest)

7.53 Because the hydrocarbon chain is longer in octane (meaning the molecule has a greater surface area), the London forces between molecules are stronger which makes the boiling point higher for octane.

7.55 A branched alkane has less surface contact with neighboring molecules than does a straight-chain alkane. The attractive forces are therefore stronger between the straight-chain alkane molecules, raising the boiling point.

7.57 Niacin. The niacin overall contains more polar groups versus nonpolar areas than the vitamin A.

7.59 2 cc

7.61 12 L

7.63 10.7 L

7.65 535 °C

7.67

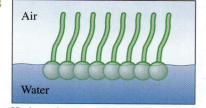

7.69 The soybean oil has more unsaturated fatty acids because it has a lower melting point.

7.71 The phospholipid head group contains many more atoms than the three atoms in the carboxylate head group of a soap.

7.73

Hydrocarbons would interact with nonpolar air, polar heads with water at surface.

7.75 Phospholipids are amphipathic molecules. They act to emulsify the oil and the water in the mayonnaise, forming a thick mixture.

7.77 **a.** Honey

   b. Hydrogen bonding, dipole–dipole attractions, and London forces. A glucose in ring form can make many more hydrogen bonds to water.

   c. Because there are so many more hydrogen bonds possible between sugar molecules in honey, this makes the honey more viscous than water.

   d. By adding soap to the water, the hydrogen bonding network between the water molecules is disrupted, and the paper clip can no longer float on the surface.

Dissolving a teaspoon of sugar in a glass of iced tea may seem simple enough, but why do some substances dissolve in water while others don't? How do the properties of solutions affect the movement of molecules into a cell? Read Chapter 8 to find out.

8 Solution Chemistry—How Sweet Is Your Tea?

IF YOU DRINK ICED TEA, you might sweeten it with sugar or a sugar substitute. Even before you add sweetener, a glass of iced tea has many different substances in it—water, caffeine, and other chemicals extracted from tea leaves—so iced tea represents a mixture. Once the sweetener has completely disappeared, all of the substances in a glass of iced tea are dissolved in water to make a refreshing drink. The liquid in the iced tea is now a homogeneous mixture called a *solution*. (See Section 1.1 for a review of homogeneous and heterogeneous mixtures.) If the tea is not sweet enough for you, you can change the taste by dissolving more sugar in it, or we could say you are increasing the sugar concentration in your tea. If you accidentally make your tea too sweet, you can always add more unsweetened tea and in so doing decrease the concentration of the sugar in the tea solution.

Understanding more about solutions and concentrations of solutions is important for many reasons. Body fluids, for example, are complex watery solutions containing dissolved substances like glucose and ions like K^+, Na^+, Cl^-, and HCO_3^-. Our bodies have some pretty sophisticated ways of keeping the concentration of dissolved molecules and ions constant.

8.1 Solutions Are Mixtures

8.1 Inquiry Question: How is a solution identified?

A glass of iced tea represents a type of homogeneous mixture called a **solution.** A solution consists of at least one substance—the **solute**—evenly dispersed throughout a second substance—the **solvent.** The components in a solution do not react with each other, so the sugar mixed into the tea is still sugar and still tastes sweet. The solute is the substance in a solution present in the smaller amount, and the solvent is the substance present in the larger amount. In the case of iced tea, the sugar is one of the solutes and the water is the solvent. A glass of iced tea, although brownish in color, is transparent; if held up to a light, you can see through the liquid. Once the sugar is dissolved in the water, it will not undissolve over time. These properties of solutions provide a quick way for you to determine whether a substance is a solution. See **Table 8.1** for some examples of other properties of solutions.

States of Solutes and Solvents

We tend to think of solutions as being solids, like sugar, dissolved in liquids like water. However, this does not have to be the case. Solutions can be homogeneous mixtures of gases. The air we breathe, for example, is a homogeneous mixture of gases, so it is also a solution in which nitrogen is the solvent, and oxygen and other gases are the solutes. Brass metal is a solution of solids in solids because it is the solute metal zinc in the solvent metal

Solute: The substance present in lesser amount

Salt

Water

Solvent: The substance present in greater amount

Solutions are homogeneous mixtures.

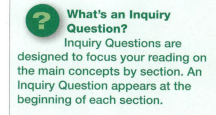

What's an Inquiry Question?
Inquiry Questions are designed to focus your reading on the main concepts by section. An Inquiry Question appears at the beginning of each section.

TABLE 8.1 Some Properties of Solutions

Particles are evenly distributed.
Components do not chemically react with each other.
Aqueous solutions are transparent.
Components do not separate upon standing.
Concentration can be changed.

a.

b.

c.

FIGURE 8.1 Solutions can be solid, liquid, or gas. (a) Brass is a solution of copper and zinc solids, (b) air is a solution of gases, and (c) carbonated beverages represent a solution of a gas in a liquid.

copper. The solute and solvent can either be a solid, liquid, or gas—the three states of matter—in a solution (see **Figure 8.1**).

Our study of solutions will focus on dilute solutions where water is the solvent. These solutions are referred to as **aqueous solutions.**

sample problem 8.1 Identifying Solute and Solvent

Identify the solute and solvent in solutions composed of the following:

a. 20.0 mL of ethanol and 250 mL of water

b. a pinch of salt in a pot of boiling water

c. a 5-gallon tank filled with methane (natural gas) and a small amount of methyl mercaptan that gives an odor if any gas should leak

Solution

Remember the solute is the substance in the smaller amount and the solvent is the substance in the larger amount.

a. Water is the solvent, and ethanol is the solute.

b. Water is the solvent, and salt is the solute.

c. Methane is the solvent, and methyl mercaptan is the solute.

Colloids and Suspensions

Is homogenized milk a solution? If we look back to the properties of solutions in Table 8.1, we would have to say that it is not, because milk is not a transparent liquid. Why is milk not a transparent liquid and therefore not a solution? Homogenized milk is uniform throughout and is therefore a homogeneous mixture, yet milk contains proteins and fat molecules that do not dissolve in water, the solvent for milk. Instead, these molecules aggregate to form larger particles in solution much as the soap molecules, discussed in Chapter 7, formed micelles. The nonpolar areas collect in the center and the polar areas of the milk proteins face the exterior and interact with the solvent.

Homogenized milk is a **colloid** (or colloidal mixture) because of the proteins and fat molecules that do not dissolve. By definition, the particles in a colloid must be between 1 nanometer (1×10^{-9} meter) and 1000 nanometers in diameter. Particles of this size remain suspended in solution, so a colloid does not separate over time.

What kind of mixture is muddy water? The dirt in this mixture will separate from the water upon standing. If the diameter of the particles in a mixture is greater than 1000 nanometers (1 micrometer), the mixture is considered a **suspension.**

Blood is another example of a suspension. Blood contains blood cells that are larger than 1 micrometer, and these cells will settle to the bottom of a test tube upon standing. Blood cells are routinely separated from plasma through centrifugation, a spinning process that accelerates settling (see **Figure 8.2**).

colloid

Blood after
centrifugation

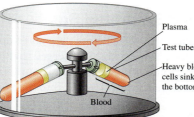

Plasma

Test tube

Heavy blood
cells sink to
the bottom

Blood

a.

b.

FIGURE 8.2 Blood is a suspension.
The cells in blood can be isolated
using centrifugation (a) into the soluble
components (plasma) and cells (b).

sample problem
8.2   Type of Mixture

Are the following solutions or colloids?

a.   apple juice   b. mayonnaise   c. fog

Solution

a.   Apple juice is transparent and uniform throughout; therefore it is a solution.
b.   Mayonnaise is opaque and uniform throughout; therefore, it is a colloid, not a solution.
c.   Fog is uniform throughout and contains small water droplets, but it is not transparent; therefore, it is a colloid, not a solution.

practice problems

8.1   Identify the solute and solvent in solutions composed of the following:
a. 80% nitrogen and about 20% oxygen in the air
b. 100.0 g of copper and 20.0 g of zinc
c. 500.0 mL of ethanol and a few drops of blue food coloring

8.2   Identify the solute and solvent in solutions composed of the following:
a. 0.80 L of N_2 and 0.01 L of CO_2
b. a spoonful of sugar and a pint of ice water
c. 3 tablespoons of vinegar (acetic acid) and 1 cup of water used in a salad dressing

8.3   Identify the following as solutions or colloids:
a. yogurt
b. cola drink
c. gasoline

8.4   Identify the following as solutions or colloids:
a. Kool-Aid®
b. salt water
c. cheddar cheese

8.2 Formation of Solutions

In Chapter 7, the golden rule of solubility, *like dissolves like*, was introduced to explain why some compounds dissolve in water and others do not. Polar covalent compounds, like glucose, and ionic compounds, like NaCl, will dissolve in water because strong attractive forces exist between them. Nonpolar covalent compounds, like hydrocarbons and oils, are not soluble in the polar molecules of water. Amphipathic molecules, like fatty acids that contain polar and nonpolar parts, can interact with water through their polar part, but these molecules typically do not form solutions.

?   **8.2 Inquiry Question:** How do temperature and pressure affect the formation of a solution?

The dissolving process requires that individual solvent particles surround the solute molecules and interact through attractive forces. This process is referred to as **solvation** for all solutions and, more specifically, as **hydration** in the case of an aqueous solution. Because the particles are just redistributing themselves, dissolving is a physical change, not a chemical change. For a review of this topic, see Section 7.4.

Factors Affecting Solubility and Saturated Solutions

How much sugar can you dissolve in your glass of iced tea before the sugar stops dissolving and starts to settle to the bottom of your glass? You may have observed that it is easier to dissolve sugar in hot tea than in cold tea. Solubility depends on a number of factors including the polarity of the solute and the solvent (the attractive forces present), as well as the temperature of the solvent.

If a solution does *not* contain the maximum amount of the solute that the solvent can hold, it is referred to as an **unsaturated solution.** If a solution contains all the solute that can possibly dissolve, the solution is referred to as a **saturated solution.** If more solute is added to a saturated solution, the additional solute will remain undissolved. A solution that is saturated reaches an **equilibrium** state between the dissolved solute and undissolved solute. At equilibrium, the speed or rate of dissolving solute and the rate of dissolved solute re-forming crystals (called recrystallization) are the same. This can be represented in an equation where a double arrow or *equilibrium arrow* is used between the products and reactants in the chemical equation.

$$\text{Undissolved solute} \underset{\text{crystallization}}{\overset{\text{solvation}}{\rightleftharpoons}} \text{dissolved solute}$$

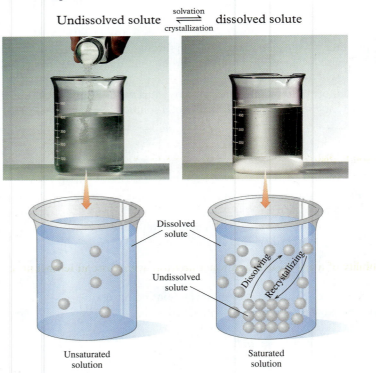

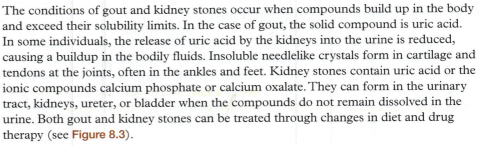

Gout, Kidney Stones, and Solubility

integrating chemistry

The conditions of gout and kidney stones occur when compounds build up in the body and exceed their solubility limits. In the case of gout, the solid compound is uric acid. In some individuals, the release of uric acid by the kidneys into the urine is reduced, causing a buildup in the bodily fluids. Insoluble needlelike crystals form in cartilage and tendons at the joints, often in the ankles and feet. Kidney stones contain uric acid or the ionic compounds calcium phosphate or calcium oxalate. They can form in the urinary tract, kidneys, ureter, or bladder when the compounds do not remain dissolved in the urine. Both gout and kidney stones can be treated through changes in diet and drug therapy (see **Figure 8.3**).

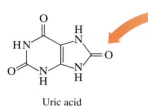

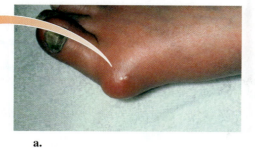

Uric acid

a.

b.

FIGURE 8.3 Exceeding solubility limits. (a) When uric acid levels in the blood exceed the solubility limit, the excess forms the needlelike crystals responsible for gout. (b) Kidney stones, like the ones shown here, are typically composed of the ionic compounds calcium phosphate ($Ca_3(PO_4)_2$) or calcium oxalate (CaC_2O_4). Even small crystals are painful when passing through the urethra.

Solubility and Temperature

It is easier to dissolve sugar in hot tea than in cold tea. In fact, more sugar will dissolve in hot tea than in cold tea. This demonstrates the effect of temperature on solubility. The solubility of most solids dissolved in water *increases* with temperature. This property of solutions is important because solubility can be manipulated by changing the temperature of a solution.

Now, think about what might happen to a bottle of club soda left in the trunk of a car on a hot summer day. You may return to your car after several hours to find that the container exploded and the trunk of your car is soaking wet! Soft drinks are aqueous solutions containing dissolved CO_2 gas as one of the solutes. The reason the container burst is related to gas solubility, pressure, and temperature. (For a review of gas pressure, refer to Section 7.4.) The solubility of a gas dissolved in water *decreases* with a rise in temperature. The warmer soda could not hold as much dissolved gas. As the CO_2 escaped from the solution, it moved into the space above the liquid in the container. Because there are now more molecules in the volume, the pressure is increased, ultimately causing the container to rupture.

Solubility and Pressure—Henry's Law

If you open a can of soda that is at room temperature and pour it into a glass, the CO_2 gas will still fizz and escape from the liquid even though the temperature is not changing. Why is the soda fizzing? When you first open the can of soda or untwist the cap on a bottle, you hear a hissing sound. The hiss is because the little space above the beverage in the container is filled with CO_2 at a higher pressure than atmospheric pressure, and the CO_2 escapes quickly once the seal is broken. The CO_2 that was dissolved in the drink at the higher pressure (before opening) will not all stay dissolved at the lower pressure of the atmosphere, and the CO_2 begins to bubble out of the solution.

This relationship between gas solubility and pressure was summarized by the English chemist William Henry (1775–1836) in **Henry's law,** which states that the solubility of a gas in a liquid is directly related to the pressure of that gas over the liquid. That is, the amount of a gas that can dissolve in a liquid increases as the pressure of the gas in the space above the liquid increases.

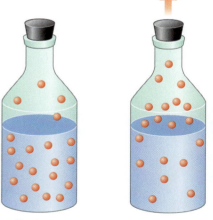

T = lower   T = higher

At higher temperature, the solubility of a gas in a liquid decreases.

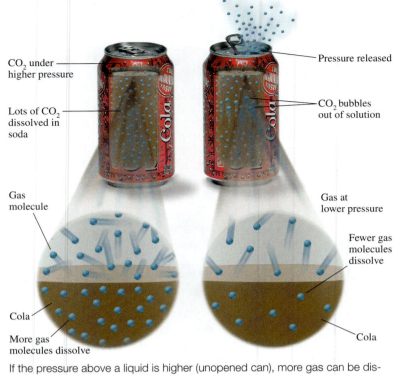

CO$_2$ under higher pressure

Lots of CO$_2$ dissolved in soda

Gas molecule

Cola

More gas molecules dissolve

Pressure released

CO$_2$ bubbles out of solution

Gas at lower pressure

Fewer gas molecules dissolve

Cola

If the pressure above a liquid is higher (unopened can), more gas can be dissolved in the liquid. Once the can is opened, the pressure above the can is lowered and the gas bubbles out of the liquid.

In Section 7.4 we discussed the relationship between pressure and volume (Boyle's law) and its application to breathing. Here the relationship between gas solubility and pressure in Henry's law explains gas exchange while breathing. The pressure exerted by carbon dioxide (P_{CO_2}) produced in the tissues or the pressure of oxygen (P_{O_2}) inhaled at the lungs results in an exchange of gases. Just as dissolved carbon dioxide bubbles out of an open can of soda, if the pressure of CO$_2$ is higher in the blood delivered back to the lungs (coming from the tissues) than the pressure of CO$_2$ found at the lungs, the gaseous CO$_2$ will pass out of the bloodstream into the lungs (see **Figure 8.4**). Similarly, oxygen dissolves into the blood at the lungs because, like the unopened soda can, the pressure of oxygen in the air is higher, allowing it to dissolve in the bloodstream. This oxygenated blood then circulates throughout the body.

FIGURE 8.4 Henry's law in the lungs. The CO$_2$ levels in blood returning from the tissues are higher than the level in the atmosphere, which allows it to be released in the expired air. The O$_2$ levels in the atmosphere are higher than in the returning blood, allowing more O$_2$ to be dissolved in the blood during inhalation.

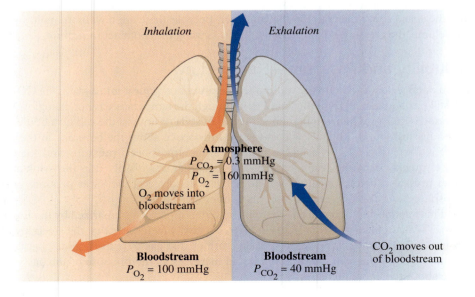

Inhalation

Exhalation

Atmosphere
P_{CO_2} = 0.3 mmHg
P_{O_2} = 160 mmHg

O$_2$ moves into bloodstream

Bloodstream
P_{O_2} = 100 mmHg

Bloodstream
P_{CO_2} = 40 mmHg

CO$_2$ moves out of bloodstream

sample problem 8.3 — Effect of Temperature on Solubility

Will the solubility of the solute increase or decrease in each of the following situations?

a. Sucrose is dissolved in water at 95 °C instead of 20 °C.
b. A can of soda containing dissolved CO_2 is stored at room temperature instead of in the refrigerator.

Solution

a. The solubility of sucrose, a solid, will increase if dissolved at 95 °C.
b. The solubility of the CO_2, a gas, will decrease at room temperature.

sample problem 8.4 — Effect of Pressure on Gas Solubility

Will the pressure of oxygen (P_{O_2}) in the bloodstream increase or decrease in each of the following situations?

a. A person climbs a mountain 10,000 ft high where the pressure of oxygen in the air is lower.
b. A person in shock receives mouth-to-mouth resuscitation and the pressure of air in the lungs increases.

Solution

a. The pressure of oxygen (P_{O_2}) in the lungs will be lower than normal, and thus the solubility of oxygen in the bloodstream will decrease.
b. The pressure of oxygen (P_{O_2}) increases because it is being forced into the lungs, and therefore the amount of oxygen in the bloodstream will increase.

practice problems

8.5 Do the following represent a saturated solution?
  a. throwing a pinch of salt into a pot of boiling water
  b. a layer of sugar forms at the bottom of a glass of tea as ice is added

8.6 Do the following represent a saturated solution?
  a. a piece of rock salt does not change in size when added to a salt solution
  b. a teaspoon of honey dissolves in hot tea

8.7 Fill in the blank in the statements below with the words "increase," "decrease," or "stay the same."
  a. If the temperature of a solution increases, the solubility of most solid solutes will _____.
  b. If the temperature of a solution increases, the solubility of a gaseous solute will _____.
  c. If the pressure above a solution increases, the solubility of a gaseous solute will _____.

8.8 Fill in the blank in the statements below with the words "increase," "decrease," or "stay the same."
  a. If the temperature of a solution decreases, the solubility of a gaseous solute will _____.

  b. If the temperature of a solution decreases, the solubility of a solid solute will _____.
  c. If the pressure above a solution decreases, the solubility of a gaseous solute will _____.

8.9 Explain what is happening in the following situations in terms of solubility:
  a. An opened bottle of soda goes flat more slowly when recapped and stored in the refrigerator.
  b. A bottle of club soda keeps its fizz longer if the lid is screwed on tightly.
  c. Less sugar dissolves in cold iced tea than in hot tea.

8.10 Explain what is happening in the following situations in terms of solubility:
  a. Fish die in a lake that gets heated with water from a power plant cooling tower.
  b. Crystals form in a jar of honey that has been sitting for several months.
  c. A person receives oxygen through a breathing device.

Discovering the Concepts

? Inquiry Activity—Electrolytes

Part 1. Information

nonelectrolyte—A substance that does not conduct electricity because it does not ionize (dissociate forming ions) in aqueous solution. A polar covalent compound like glucose is a nonelectrolyte.

strong electrolyte—A substance that conducts electricity because it completely dissociates into ions in aqueous solution. Ionic compounds that dissolve in water are strong electrolytes.

weak electrolyte—A substance that weakly conducts electricity because it partially dissociates in aqueous solution. A weak acid like a carboxylic acid is a weak electrolyte.

The warning in the figure appears on some electrical appliances.

QUESTIONS

1. Electrical current is carried by a flow of charged particles; electrons in metal or ions in solution. This flow can only happen in an aqueous solution if whole (not partial) charges are present. Based on this information,
    a. does tap water conduct electricity? Consider whether there are enough charges for ions to flow in a solution of tap water.
    b. does pure water conduct electricity? Consider whether there are enough charges for ions to flow in a solution of pure water.
2. Are the following compounds likely to dissociate, partially dissociate, or not dissociate when dissolved in solution? Classify each as a nonelectrolyte, strong electrolyte, or weak electrolyte.
    a. $NaNO_3(s)$
    b. Sucrose, $C_{12}H_{22}O_{11}(s)$
    c. Formic acid, $HCOOH(l)$

8.3 Chemical Equations for Solution Formation

8.3 Inquiry Question:
How are hydration equations written for electrolytes, nonelectrolytes, and weak electrolytes?

A warning symbol on hair dryers and other small electrical appliances used in the bathroom cautions us that we should not use them near the bathtub. Why? Because tap water conducts electricity, and if a plugged-in electrical appliance falls into a tub, a person in the tub could be electrocuted. Pure water does not conduct electricity, but tap water does. What is making the difference? Tap water contains dissolved ions, and those dissolved ions can conduct electricity. Solutes that produce ions in solution are called **electrolytes.**

We can place all substances that dissolve in water (hydrate) into one of three categories based on their ability to conduct electricity. Any aqueous solution can be tested to see whether it will conduct electricity with a simple circuit consisting of a battery, wires to two electrodes, and a light bulb (see **Figure 8.5**). If a solution conducts electricity, the light bulb glows when the electrodes are immersed in the solution. An ionic compound completely dissociates into its charged ions when dissolved in water. The large number of charged ions present makes the light bulb in the simple circuit burn brightly.

Ionic compounds that dissolve in water are considered **strong electrolytes.** Covalent compounds like sugar do not **ionize** (dissociate to form ions) when they dissolve in solution. They exist as molecules, and no ions are formed upon dissolving. Soluble covalent compounds do not conduct electricity and are therefore termed **nonelectrolytes.** Some covalent compounds *can* partially ionize in water. These compounds contain a highly polar bond that can dissociate, forming some ions in water. These covalent compounds weakly conduct electricity and are termed **weak electrolytes.** We can write balanced chemical equations to describe the hydration of each of these substances.

In Chapters 1 and 5, we balanced chemical equations. Here we balance chemical equations for hydration as solutes dissolve into solution. Remember that reactants (represented by their chemical formula) always appear on the left side and the products (also represented by their chemical formula) always appear on the right side. These two are separated by an arrow meaning *react to form* or *yield*.

$$\text{Reactants} \xrightarrow[\text{(react to form)}]{} \text{products}$$

Strong Electrolytes

Let's look at the parts of a balanced chemical equation for the hydration of the ionic compound magnesium chloride ($MgCl_2$), a strong electrolyte. The following equation can be verbally stated as "solid magnesium chloride dissolves in water to yield magnesium ions and chloride ions."

$$MgCl_2(s) \xrightarrow{H_2O} Mg^{2+}(aq) + 2\,Cl^-(aq)$$

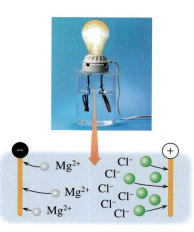

Strong electrolyte

First, notice that the number of magnesium and chloride ions formed as products is the same as the number of each found in the reactant, regardless of whether the ions are together in solid form or dissolved in solution. This was first observed and stated by the French chemist Antoine Lavoisier (1743–1794) as the **law of conservation of matter:** Matter can neither be created nor destroyed. Matter merely changes forms. In other words, atoms cannot just appear or disappear.

As outlined in previous chapters, it is important to check an equation after balancing to make sure the same number of each element appears on each side of the equation. In the preceding sample equation, the coefficient 2 in front of the chloride ion indicates that there are two chloride ions produced for every one $MgCl_2$ that dissociates. Besides balancing the number of each element, the charges in a hydration equation must be the same on both sides of the chemical equation. In this example, $MgCl_2$ has no net charge (reactant) and one Mg^{2+} and two Cl^- sum to a total charge of zero (products), so the charges are also balanced.

$$MgCl_2(s) \xrightarrow{H_2O} Mg^{2+}(aq) + 2\,Cl^-(aq)$$

Charge in   Charge in products
reactants = 0   2 + 2(−1) = 0

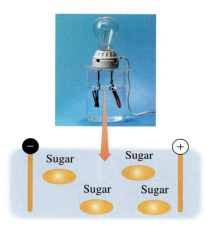

Nonelectrolyte

Second, note the reaction arrow points in one direction, implying that the process occurs in only one direction, that is, it is irreversible.

Third, note the phases of matter. For ionic compounds, the reactant will usually be a solid that dissolves. In the products, the phases will always be aqueous, because the substance is hydrated, that is, it is dissolved in water.

Finally, notice the water (H_2O) appearing below the reaction arrow. Substances that are not involved in the balanced equation, like the solvent in this case, are often placed at the arrow to give the reader more information regarding the conditions of the reaction. For the hydration equations in this chapter, water will appear here.

Nonelectrolytes

Nonelectrolytes are polar compounds that dissolve in water but do not ionize in water. As an example, we can examine the following chemical equation for the carbohydrate glucose ($C_6H_{12}O_6$) dissolving in water:

$$C_6H_{12}O_6(s) \xrightarrow{H_2O} C_6H_{12}O_6(aq)$$

Covalent compounds like glucose do not dissociate. The only difference between the reactant and the product is the phase. Because glucose dissolves in water, we indicate this on the products side of the chemical equation by changing the phase of the glucose molecules to aqueous.

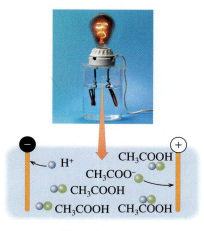

Weak electrolyte

FIGURE 8.5 Simple experiments for determining whether a solution contains strong, weak, or nonelectrolytes. A pair of electrodes connected to a light bulb and inserted in the solution will either conduct electricity strongly, weakly, or not at all, based on dissociation into ions by the solute. Notice that the strong electrolyte $MgCl_2$ dissociates into three particles.

Weak Electrolytes

Which covalent compounds act as weak electrolytes? If we re-examine the functional group list, we can see that there are two functional groups on the list that contain a form with an ionic charge, the carboxylic acid/carboxylate group and the amine/protonated amine.

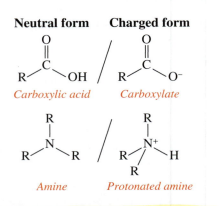

Neutral form   **Charged form**

Carboxylic acid   *Carboxylate*

Amine   *Protonated amine*

The functional groups carboxylic acid and amine are weak electrolytes that have ionic forms differing from the neutral form by a hydrogen. Recall from Table 4.3 that amines can be primary, secondary, or tertiary and have protonated forms. The tertiary amine and its protonated form is shown.

A carboxylic acid group contains a very polar O—H bond that can dissociate to form carboxylate ions and H^+ ions when dissolved. Carboxylic acids and amines are used as examples of weak electrolytes in this chapter. Consider the partial ionization of acetic acid, the compound in vinegar:

$$CH_3-\overset{\overset{\text{O}}{\|}}{C}-OH(l) \underset{H_2O}{\rightleftharpoons} CH_3-\overset{\overset{\text{O}}{\|}}{C}-O^-(aq) + H^+(aq)$$

As the number of H^+ and CH_3COO^- ions builds up in the solution, some of them will recombine to re-form CH_3COOH. Eventually, the rates of the forward and reverse reactions equalize, and the amount of the acetic acid and the ions no longer changes—thus, an equilibrium exists. An equilibrium arrow, $\rightleftharpoons$ is used in this chemical equation to indicate this. When inspecting the equation, we see that the number of atoms and the total charge on each side is balanced, so the equation is balanced. The phase of the weak electrolyte before hydration may be solid, liquid, or gas.

---

sample problem 8.5   **Characterizing Strong, Weak, and Nonelectrolyte Solutions**

Predict whether the following will fully dissociate, partially dissociate, or not dissociate when dissolved in solution:

a. $NaNO_3(s)$

b.   Sucrose, $C_{12}H_{22}O_{12}(s)$

c.   Formic acid, $HCOOH(l)$

Solution

a.   The ionic compound $NaNO_3$ will fully dissociate, forming Na^+ and NO_3^- ions when dissolved in solution.

b.   The covalent compound $C_{12}H_{22}O_{12}$ will not dissociate when dissolved in solution.

c.   The carboxylic acid $HCOOH$ will *partially* dissociate, forming the ions H^+ and $HCOO^-$, as well as $HCOOH$ molecules.

Solving a Problem

Balancing Hydration Equations

Write a balanced equation for the hydration of the following:

a. $CaCl_2(s)$, a strong electrolyte

b. $CH_3OH(l)$, a nonelectrolyte

c. $NH_4OH(aq)$, a weak electrolyte

Solution

STEP 1: Write the reactants and products with their state of matter. The arrow is irreversible for strong and nonelectrolytes and reversible for weak electrolytes. Because they dissolve, the products will all be aqueous. For our examples,

a. $CaCl_2(s) \xrightarrow{\;H_2O\;} Ca^{2+}(aq) + Cl^-(aq)$

b. $CH_3OH(l) \xrightarrow{\;H_2O\;} CH_3OH(aq)$

c. $NH_4OH(aq) \underset{H_2O}{\rightleftharpoons} NH_4^+(aq) + OH^-(aq)$

STEP 2: Examine and balance the equations. Be sure to balance the number of each element on either side. Upon examination, only part (a) is unbalanced. If there are two chlorides in the solid, $CaCl_2$, there must be two Cl^- present on the product side. By adding the coefficient 2 in front of the Cl^-, we have balanced both the elements and the charge.

a. $CaCl_2(s) \xrightarrow{\;H_2O\;} Ca^{2+}(aq) + $ 2 $Cl^-(aq)$

b. $CH_3OH(l) \xrightarrow{\;H_2O\;} CH_3OH(aq)$

c. $NH_4OH(aq) \underset{H_2O}{\rightleftharpoons} NH_4^+(aq) + OH^-(aq)$

STEP 3: Check the equations by confirming equal charge on each side. Checking the charges of the reactants and products in each of the equations confirms that the charges are equal on either side.

a. $CaCl_2(s) \xrightarrow{\;H_2O\;} Ca^{2+}(aq) + $ 2 $Cl^-(aq)$
    Charge in reactants = 0 · · · · Charge in products $(2^+) + 2(1^-) = 0$

b. $CH_3OH(l) \xrightarrow{\;H_2O\;} CH_3OH(aq)$
    Charge in reactants = 0 · · · · Charge in products = 0

c. $NH_4OH(aq) \underset{H_2O}{\rightleftharpoons} NH_4^+(aq) + OH^-(aq)$
    Charge in reactants = 0 · · · · Charge in products $(1^+) + (1^-) = 0$

sample problem 8.6   Balancing Hydration Equations

Write a balanced equation for the hydration of the following:

a. $KCl(s)$, a strong electrolyte
b. $NH_2CONH_2(l)$, a nonelectrolyte
c. $HF(l)$, a weak electrolyte

Solution

a.   **STEP 1: Write the reactants and products with their state of matter.**

$$KCl(s) \xrightarrow{\text{H}_2\text{O}} K^+(aq) + Cl^-(aq)$$

STEP 2: Examine and balance the equations. Examination indicates that the equation is balanced.

STEP 3: Check the equations by confirming equal charge on each side. The charges are balanced to zero on both sides.

b.   **STEP 1: Write the reactants and products with their state of matter.** A nonelectrolyte will dissolve but not ionize, merely changing states. Since the formulas did not change, the equation will be balanced after STEP 1.

$$NH_2CONH_2(l) \xrightarrow{\text{H}_2\text{O}} NH_2CONH_2(aq)$$

c.   **STEP 1: Write the reactants and products with their state of matter.** HF is a common weak acid. (Other properties of weak acids are discussed in detail in Chapter 9.)

$$HF(l) \underset{\text{H}_2\text{O}}{\rightleftharpoons} H^+(aq) + F^-(aq)$$

STEP 2: Examine and balance the equations. Examination indicates that the equation is balanced.

STEP 3: Check the equations by confirming equal charge on each side. The charges are balanced to zero on both sides.

TABLE 8.2 The Relationship between Equivalents and Moles Expressed as Conversion Factors

Ion	Conversion Factor
Na^+	$\dfrac{1 \text{ Eq Na}^+}{1 \text{ mole Na}^+}$
Ca^{2+}	$\dfrac{2 \text{ Eq Ca}^{2+}}{1 \text{ mole Ca}^{2+}}$
Cl^-	$\dfrac{1 \text{ Eq Cl}^-}{1 \text{ mole Cl}^-}$
SO_4^{2-}	$\dfrac{2 \text{ Eq SO}_4^{2-}}{1 \text{ mole SO}_4^{2-}}$

Ionic Solutions and Equivalents

Blood and other bodily fluids contain many electrolytes as dissolved ions like Na^+, K^+, Cl^-, and HCO_3^-. The amount of a dissolved ion found in fluids can be expressed by the unit **equivalent (Eq)**. An equivalent relates the charge in a solution to the number of ions or the moles of ions present (see Table 8.2). (Recall from Chapter 3 that moles can be related to the measurable quantity grams through the atomic mass.)

For example, 1 mole of Na^+ has one equivalent of charge because the charge on a sodium ion is plus 1. One mole of Ca^{2+} has two equivalents of charge because one mole of calcium contains two charges (or equivalents) per mole. *The number of equivalents present per mole of an ion equals the charge on that ion.*

sample problem 8.7   Electrolytes in Solution

How many equivalents of Cl^- are present in a solution that contains 2.50 moles of Cl^-?

Solution

Use the conversion factor that relates the number of moles of Cl^- to the number of equivalents.

$$2.50 \text{ moles Cl}^- \times \frac{1 \text{ Eq Cl}^-}{1 \text{ mole Cl}^-} = 2.50 \text{ EqCl}^-$$

sample problem 8.8 Electrolytes in Solution

How many equivalents of Ca^{2+} are present in a solution that contains 2.50 moles of Ca^{2+}?

Solution

Use the conversion factor that relates the number of moles of Ca^{2+} to the number of equivalents.

$$2.50 \; \text{Moles } Ca^{2+} \times \frac{2 \text{ Eq } Ca^{2+}}{1 \text{ mole } Ca^{2+}} = 5.00 \text{ Eq } Ca^{2+}$$

Electrolytes in Blood Plasma

integrating chemistry

The amount of electrolytes present in bodily fluids and typical intravenous fluid replacements are often represented as the number of milliequivalents per liter of solution (mEq/L).

Ionic solutions have a balance in the number of positive and negative charges present because they are formed by dissolving ionic compounds that have no net charge. Typical blood plasma has a total electrolyte concentration of 150 mEq/L, meaning that the total concentration of both positive and negative ions is 150 mEq/L (see **Table 8.3**). Examples of the concentrations of other ionic solutions used as intravenous fluids are shown in **Table 8.4**. These also show a balance of total charge.

TABLE 8.3 Some Typical Concentrations of Electrolytes in Blood Plasma

Electrolyte	Concentration (mEq/L)
Cations	
Na^+	138
K^+	5
Mg^{2+}	3
Ca^{2+}	4
Total	150
Anions	
Cl^-	110
HCO_3^-	30
HPO_4^{2-}	4
Proteins	6
Total	150

TABLE 8.4 Electrolyte Concentrations in Intravenous Replacement Solutions

Solution	Electrolytes (mEq/L)	Use
Sodium chloride (0.90%)	Na^+ 154, Cl^- 154	Replacement of fluid loss
Potassium chloride with 5.0% dextrose	K^+ 40, Cl^- 40	Treatment of malnutrition (low potassium levels)
Ringer's solution	Na^+ 147, K^+ 4, Ca^{2+} 4, Cl^- 155	Replacement of fluids and electrolytes lost through dehydration
Maintenance solution with 5.0% dextrose	Na^+ 40, K^+ 35, Cl^- 40, lactate$^-$ 20, HPO_4^{2-} 15	Maintenance of fluid and electrolyte levels
Replacement solution (extracellular)	Na^+ 140, K^+ 10, Ca^{2+} 5, Mg^{2+} 3, Cl^- 103, acetate$^-$ 47, citrate^{3-} 8	Replacement of electrolytes in extracellular fluids

sample problem 8.9 Using Equivalents

A Ringer's solution for intravenous fluid replacement has a concentration of 155 mEq Cl^- per liter of solution. If a patient receives 1250 mL of Ringer's solution, how many equivalents of Cl^- were given?

Solution

This problem asks for the number of equivalents of Cl^- present, which can be determined from the number of mEq given through the conversion factor $\dfrac{1 \text{ Eq}}{1000 \text{ mEq}}$.

Using conversion factors, the information given (concentration and volume), and canceling appropriate units, the problem is solved as

$$\frac{155 \; \text{mEq } Cl^-}{1 \; \text{L Solution}} \times \frac{1 \text{ Eq}}{1000 \; \text{mEq}} \times \frac{1 \; \text{L}}{1000 \; \text{mL}} \times 1250 \; \text{mL Solution} = 0.194 \text{ Eq } Cl^-$$

8.11 Predict whether the following will fully dissociate, partially dissociate, or not dissociate when dissolved in solution:
a. potassium fluoride, $KF(s)$, a strong electrolyte
b. hydrogen cyanide, $HCN(g)$, a weak electrolyte
c. fructose, $C_6H_{12}O_6(s)$, a nonelectrolyte

8.12 Predict whether the following will fully dissociate, partially dissociate, or not dissociate when dissolved in solution:
a. sodium sulfate, $Na_2SO_4(s)$, a strong electrolyte
b. ethyl alcohol, $CH_3CH_2OH(l)$, a nonelectrolyte
c. hypochlorous acid, $HClO(aq)$, a weak electrolyte

8.13 Provide a balanced equation for the hydration of each compound in Problem 8.11.

8.14 Provide a balanced equation for the hydration of each compound in Problem 8.12.

8.15 For the following ionic compounds, write a balanced equation for their hydration in water:
a. $CaCl_2$ b. $NaOH$ c. KBr d. $Fe(NO_3)_3$

8.16 For the following ionic compounds, write a balanced equation for their hydration in water:
a. $FeCl_3$ b. $NaHCO_3$ c. K_2SO_4 d. $Ca(NO_3)_2$

8.17 How many equivalents of K^+ are present in a solution that contains 4.25 moles of K^+?

8.18 How many equivalents of Mg^{2+} are present in a solution that contains 2.50 moles of Mg^{2+}?

8.19 A sodium chloride intravenous (IV) fluid contains 154 mEq of Na^+ per liter of solution. If a patient receives exactly 500 mL of the IV solution, how many moles of Na^+ were given?

8.20 A 5% dextrose IV fluid contains 35 mEq of K^+ per liter of solution. If a patient receives 235 mL of the IV solution, how many moles of K^+ were delivered?

8.4 Concentrations

8.4 Inquiry Question:
How is the concentration of a solution expressed?

Whether you drink a can of soda or a glass of the same soda from a 2-liter bottle, the soda tastes the same. Beverage manufacturers go to great lengths to maintain a consistent product. Each beverage must have exactly the same amount of ingredients (see **Figure 8.6**).

The amount of each ingredient dissolved in the soda determines the **concentration.** The concentration of a solution can be expressed in many different units, but the units we will encounter in this chapter relate the amount of each solute dissolved to the total amount of solution present.

$$\text{Concentration} = \frac{\text{amount of solute}}{\text{amount of solution}}$$

Note that the denominator is the *total* amount of solution (includes the solute), not just the amount of solvent.

Table 8.5 shows the normal concentration ranges for some substances typically tested in a blood profile. Notice that the units vary depending on the type of solute. Why all the different units? Different units are used so that all the numbers fall in ranges that are readable without using numbers after a decimal. For example, if the units for cholesterol were g/dL instead of mg/dL, the numbers in the normal range would be 0.120–0.240 g/dL, which is not as easy to read and discuss with a patient as a number ranging in the hundreds.

FIGURE 8.6 Concentration and volume. The concentration of the sugar and the taste of the beverage are the same in all containers, even though the volume is different.

TABLE 8.5 Normal Concentration Ranges for Five Common Substances Measured in a Blood Test*

Substance	Normal Range
Na^+	136–146 mEq/L
Cl^-	97–110 mEq/L
Glucose	65–110 mg/dL
Cholesterol	120–240 mg/dL
Hemoglobin (a protein)	12–16 g/dL

In this section, the common concentration units of molarity, percent, parts per million, and parts per billion are introduced.

Millimoles per Liter (mmol/L) and Molarity (M)

Sometimes the units for electrolytes like Na^+ are given in mmoles/L instead of the previously mentioned mEq/L. The charge on an ion is the number of equivalents present in 1 mole (Section 8.3). So, for an ion with a charge like sodium, the number of equivalents is equal to the number of moles, and the units mEq/L and mmole/L are the same. We can show this mathematically by converting mEq/L to mmole/L using the following conversion factor:

$$\frac{1 \text{ mmole } Na^+}{1 \text{ mEq } Na^+}$$

Using this conversion factor, we can determine the concentration of a 135 mEq/L Na^+ solution in mmole/L:

$$\frac{135 \text{ mEq } Na^+}{1 \text{ L}} \times \frac{1 \text{ mmole } Na^+}{1 \text{ mEq } Na^+} = \frac{135 \text{ mmole } Na^+}{1 \text{ L}}$$

Chemists use a unit related to mmole/L to describe the concentrations of solutions prepared in the laboratory. This unit is called **molarity, M,** and it is defined as

$$M = \frac{\text{mole solute}}{\text{L solution}}$$

The *mole* is directly related to the number of *molecules* present (through Avogadro's number). A 5.0 M (read "five molar") solution of ethanol (CH_3CH_2OH) has the same number of molecules present as a 5.0 M glucose ($C_6H_{12}O_6$) solution, even though the *mass* of ethanol and *mass* of glucose used to make the two solutions are different. The unit molarity is useful for chemists because they want to know how many particles of a solute are available to react in a chemical reaction.

Solving a Problem

Calculating Molarity

What is the molarity of a solution prepared by dissolving 2.0 g of NaCl in water giving a total volume of 250 mL?

Solution

STEP 1: Examine the problem. Decide what information is given and what information is being sought. In this problem you are solving for molarity (M), so your final units should be moles/L. To arrive at moles/L, you must use some conversion factors because you are given grams of NaCl and the volume in mL.

STEP 2: Find appropriate conversion factors. To get NaCl in moles, you must find the molar mass of NaCl, which is the sum of the atomic masses from the periodic table:

Na:	22.99
Cl:	35.45
NaCl:	58.44 g/mole

The conversion factor for mL to L is

$$\frac{1000 \text{ mL}}{1 \text{L}}$$

STEP 3: Solve the problem. Be sure that the units you don't want cancel and you are left with the units of moles/L, which is molarity.

$$\frac{2.0 \text{ g NaCl}}{250 \text{ mL solution}} \times \frac{1000 \text{ mL}}{1 \text{ L}} \times \frac{1 \text{ mole}}{58.44 \text{ g NaCl}} = \frac{0.14 \text{ mole}}{1 \text{ L}} = 0.14 \text{ M}$$

sample problem 8.10 Molarity

What is the molarity of a solution prepared by dissolving 5.00 g of $MgCl_2$ in water to give a total volume of 500. mL?

Solution

STEP 1: Examine the problem. Decide what information is given and what information is being sought. In this problem you are solving for molarity (M), so your final units should be moles/L. To arrive at moles/L, you must use some conversion factors because you are given grams of $MgCl_2$ and the volume in mL.

STEP 2: Find appropriate conversion factors. To get $MgCl_2$ in moles, you must find the molar mass of $MgCl_2$, which is the sum of the atomic masses from the periodic table:

$$
\begin{array}{ll}
\text{Mg:} & 24.31 \\
\text{2 Cl:} & 35.45 \times 2 = 70.90 \\
\hline
MgCl_2\text{:} & 95.21 \text{ g/mole}
\end{array}
$$

The conversion factor for mL to L is

$$\frac{1000 \text{ mL}}{1 \text{L}}$$

STEP 3: Solve the problem. Be sure that the units that you don't want cancel and that you are left with the units of moles/L, which is molarity.

$$\frac{5.00 \text{ g } MgCl_2}{500. \text{ mL solution}} \times \frac{1000 \text{ mL}}{1 \text{ L}} \times \frac{1 \text{ mole}}{95.21 \text{ g } MgCl_2} = \frac{0.0105 \text{ mole}}{1 \text{ L}} = 0.105 \text{ M}$$

sample problem 8.11 Molarity

a. How many moles of KNO_3 are present in 125 mL of a 2.0-M KNO_3 solution?
b. How many grams of KNO_3 are present?

Solution

a. In this problem, the concentration is known, but the actual number of moles is not. The units on the final answer are moles. By definition, 2.0 M is 2.0 moles/L. Set up an equation to solve for moles with the given information:

$$\frac{2.0 \text{ moles } KNO_3}{1 \text{ L}} \times \frac{1 \text{ L}}{1000 \text{ mL}} \times 125 \text{ mL} = 0.25 \text{ mole } KNO_3$$

b. The number of grams can be determined from the number of moles found in part a by multiplying by the molar mass of KNO_3. The molar mass is the sum of a compound's atomic masses.

$$
\begin{array}{ll}
\text{K:} & 39.10 \\
\text{N:} & 14.01 \\
\text{3 O:} & 16.00 \times 3 = 48.00 \\
\hline
KNO_3\text{:} & 101.11 \text{ g/mole}
\end{array}
$$

$$0.25 \text{ mole } KNO_3 \times \frac{101.11 \text{ g } KNO_3}{1 \text{ mole } KNO_3} = 25 \text{ g } KNO_3$$

Percent (%) Concentration

There are three common concentration units that use percent: mass/volume percent, mass/mass percent, and volume/volume percent. The concentration unit percent is no different in definition from the everyday definition of percent discussed in Chapter 1, which indicates the number of *parts per one hundred parts*. As an example, a dime (10 pennies) is 10% of a dollar (100 pennies), or 10 parts out of a total 100 parts.

100 cents (1 dollar)

10 cents = 10% of a dollar

A dime is 10% of a dollar.

$$\% \text{ Concentration} = \frac{\text{parts of solute}}{100 \text{ parts of solution}}$$

Percent Mass/Mass, % (m/m)

A solution can easily be prepared in a hospital pharmacy or laboratory with the unit % (m/m) by measuring both the solute and the solvent on a balance and mixing, realizing that mass of solute + mass of solvent = mass of solution. The concentration unit percent mass/mass, % (m/m), can be determined by

$$\% \text{ (m/m)} = \frac{\text{g solute}}{\text{g solution}} \times 100\%$$

This unit is also known as percent weight/weight (% wt/wt). Suppose we weighed out 8.0 g of glucose (the solute) on a scale and mixed 42.0 g of water (the solvent) with it. The total mass of the solution is the mass of the solute and the solvent (8.0 g $+ 42.0$ g $= 50.0$ g). The concentration of the resulting solution would be solved as follows:

$$\frac{8.0 \text{ g glucose}}{50.0 \text{ g solution}} \times 100 = 16\%$$

8.0 g glucose + 42.0 g water
(Solute)   (Solvent)

Percent Volume/Volume, % (v/v)

The unit % (v/v) is typically used when liquids or gases are the solute and solvent, for example, ethanol and water. A bottle of wine that is 14% (v/v) alcohol means that 14 mL of alcohol is present in 100 mL of the wine. The concentration unit percent volume/volume, % (v/v), can be determined by

$$\% \text{ (v/v)} = \frac{\text{mL solute}}{\text{mL solution}} \times 100\%$$

The percent of alcohol in wine is a % (v/v) concentration.

Percent Mass/Volume % (m/v)

The unit % (m/v) is often used in the preparation of intravenous fluids. Normal saline (NS) solution used to replace body fluids is 0.90% (m/v) NaCl, which means the solution contains 0.90 g of NaCl in 100 mL of solution. The concentration unit percent mass/volume, % (m/v), can be determined by

$$\% \text{ (m/v)} = \frac{\text{g solute}}{\text{mL solution}} \times 100\%$$

This unit is also known as percent weight/volume (% w/v). Many times in labeling, the type of percent concentration (m/m, v/v, or m/v) is not indicated. The state of matter of the solute indicates the percent concentration unit. For example, one type of milk is 2%, which indicates the amount of milk fat present in the milk. Because fat is semisolid, the percent unit is % (m/v).

In a hospital pharmacy, a solution can be prepared by weighing the solute on a balance and mixing it with a solvent to form a final total volume of solution. For this reason, it is important to determine the amount of solute you might add to a solution in addition to calculating the concentration of a solution. Both types of problems are illustrated on the next page.

2%

---

sample problem 8.12   **Calculating Percent Concentration**

What is the % (m/m) concentration of a NaCl solution prepared with 60.0 g of NaCl and 500.0 g of water?

Solution

The mass of the solute is 60.0 g of NaCl. The mass of the solution is the sum of the solute and the solvent:

$$60.0 \text{ g NaCl} + 500.0 \text{ g water} = 560.0 \text{ g solution}$$

$$\% \text{ (m/m)} = \frac{60.0 \text{ g NaCl}}{560.0 \text{ g solution}} \times 100\% = 10.7\%$$

---

sample problem 8.13   **Solute in Percent Concentration**

How many grams of fat are present in 250 mL of 2% milk?

Solution

Remember the definition of this concentration unit. The percent here for milk fat is grams/100 mL milk. A simple ratio can be set up, and the number of grams in 250 mL of milk determined:

$$\frac{2 \text{ g milk fat}}{100 \text{ mL milk}} = \frac{? \text{ g milk fat}}{250 \text{ mL milk}}$$

$$\frac{2 \text{ g milk fat}}{100 \text{ mL milk}} \times 250 \text{ mL milk} = ? \text{ g milk fat}$$

$$? = 5 \text{ g milk fat}$$

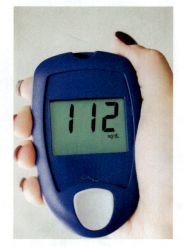

Glucose monitor
measuring mg/dL

Relationship to Other Common Units

The unit typically used for measuring the oxygen carrying protein hemoglobin in the blood is g/dL, which is the same unit as % (m/v). A deciliter is equal to 100 mL, so the unit g/dL is the same as g/100 mL. Hemoglobin levels between 13–18 g/dL (13–18 g per 100 mL of blood) for men and 12–16 g/dL for women are considered in the normal range.

A common unit used when measuring molecules like glucose and cholesterol levels in the blood is milligrams per deciliter (mg/dL). This unit is also referred to as mg% (milligram percent). The mg in front of the % symbol indicates that the definition is mg per 100 mL instead of the usual g per 100 mL definition of % (m/v). Glucose levels over 110 mg% (110 mg per 100 mL blood) are considered higher than normal. Total cholesterol levels below 200 mg/dL (200 mg per 100 mL blood) are considered desirable.

Parts per Million (ppm) and Parts per Billion (ppb)

Parts per million (ppm) and parts per billion (ppb) are convenient concentration units used to describe very dilute solutions. To get an idea of how small a part per million is, consider that there are 1 million pennies in $10,000, or a penny is a ppm of $10,000. In terms of volume, about 5 drops of food coloring in a bathtub of water represents a part per million, and about 1 drop of food coloring in an Olympic-sized swimming pool of water is about a part per billion—a very dilute solution.

Fluoride is often added to tap water at a level of less than 4 parts per million to promote strong teeth. The maximum contaminant level of lead in drinking water is 15 parts per billion. In the case of the fluoride, 4 ppm means 4 g of fluoride in every million mL of tap water. In the case of the lead and ppb, 15 ppb means 15 g of lead in every billion mL of tap water. It is easier to think in terms of liters of water rather than millions or billions of mL of water, so the unit ppm is sometimes referred to as 1 mg/L and ppb as 1 μg/L.

$$1\text{ ppm} = \frac{1\text{ g solute}}{1{,}000{,}000\text{ mL solution}} \times \frac{1000\text{ mg}}{1\text{ g}} \times \frac{1000\text{ mL}}{1\text{ L}} = \frac{1\text{ mg solute}}{1\text{ L solution}}$$

Alternatively, recall that a percent mass/volume (% m/v) is *parts per hundred* so, similarly, parts per million and parts per billion can be determined by multiplying by a million or a billion, respectively:

$$\text{ppm} = \frac{\text{g solute}}{\text{mL solution}} \times 1{,}000{,}000 \longleftarrow 1 \times 10^6$$

$$\text{ppb} = \frac{\text{g solute}}{\text{mL solution}} \times 1{,}000{,}000{,}000 \longleftarrow 1 \times 10^9$$

sample problem 8.14 — Calculating ppm and ppb

What is the concentration in (a) parts per million (ppm) and (b) parts per billion (ppb) of a solution that contains 15 mg of lead in 1250 mL of solution?

Solution

a. A part per million is equivalent to a mg per L, so convert the units given to mg/L:

$$\frac{15\text{ mg lead}}{1250\text{ mL solution}} \times \frac{1000\text{ mL}}{1\text{ L}} = 12\text{ mg/L} = 12\text{ ppm}$$

b. A part per billion is equivalent to a μg per L (it is 1000 times smaller than a ppm), so convert the units given to μg/L. There are 1000 μg per 1 mg.

$$\frac{15\text{ mg lead}}{1250\text{ mL solution}} \times \frac{1000\text{ mL}}{1\text{ L}} \times \frac{1000\text{ }\mu\text{g}}{1\text{ mg}} = 12{,}000\text{ }\mu\text{g/L} = 12{,}000\text{ ppb}$$

---

practice problems

8.21 What is the concentration in mmole/L of a Ca^{2+} solution that is 2.50 mEq/L?

8.22 What is the concentration in mEq/L of a Mg^{2+} solution that is 100 mmole/L?

8.23 What is the molarity of 250 mL of solution containing 4.20 moles of $ZnCl_2$?

8.24 What is the molarity of 750 mL of solution containing 2.10 moles of Na_2SO_4?

8.25 What is the molarity of 1.0 L of solution containing 25 g of NaCl?

8.26 What is the molarity of 500 mL of solution containing 45 g of $NaHCO_3$?

8.27 How many grams of KBr are present in 300.0 mL of a 1.00 M solution?

8.28 How many grams of $CaCl_2$ are present in 25.0 mL of a 0.500 M solution?

8.29 What is the concentration in % (m/v) of a solution containing 0.30 g of glucose and 60 mL of solution?

8.30 Calculate the percent mass/volume (% m/v) for the solute in each of the following solutions:
   a. 75 g of Na_2SO_4 in 250 mL of Na_2SO_4 solution
   b. 39 g of sucrose in 355 mL of a carbonated drink

8.31 Calculate the percent mass/volume (% m/v) for the solute in each of the following solutions:
   a. 2.50 g of KCl in 50.0 mL of solution
   b. 7.5 g of casein in 120 mL of low-fat milk

8.32 What is the concentration in % (m/m) of a solution prepared by mixing 10.0 g of KCl with 100.0 g of distilled water?

8.33 How many grams of insulin are present in 26.0 g of a 0.450% (m/m) solution?

8.34 How many grams of NaCl must be combined with water to prepare 500 mL of a 0.90% (m/v) physiological saline solution?

8.35 What is the concentration in ppm of a solution containing 0.30 mg of fluoride and 60 mL of tap water? What is the concentration in ppb?

8.36 What is the concentration in ppm of a solution containing 1.50 mg of copper and 100.0 mL of tap water?

8.5 Dilution

8.5 Inquiry Question:
How is the concentration
of a solution determined if
a dilution is made?

If you have ever prepared orange juice from a frozen can of concentrate by adding 3 cans of water, you have made a dilute solution from a more concentrated one. One way to prepare solutions of lower concentration is to dilute a solution of higher concentration by adding more solvent.

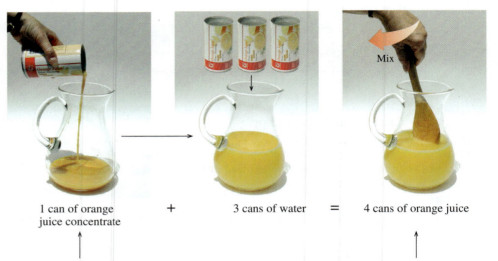

1 can of orange   +   3 cans of water   =   4 cans of orange juice
juice concentrate

More concentrated juice solution   Less concentrated juice solution

Amount of juice in can = Amount of juice in pitcher

FIGURE 8.7 Dilution. The amount of
orange juice (solute) is the same both
before and after adding water (solvent)
to the concentrated can of orange juice.

When you add water to the can of orange juice concentrate, the amount of orange juice present does not change even though you have a lot more solution present. The *amount of solute* stayed the same, but the *volume* of solution increased, so the *concentration* of the solution decreased (see **Figure 8.7**).

The following dilution equation represents this mathematically, where $C_{initial}$ represents the initial concentration, C_{final} represents the final concentration, $V_{initial}$ represents the initial volume, and V_{final} represents the final volume. If three of the four variables are known, the fourth one can be determined:

$$C_{initial} \times V_{initial} = C_{final} \times V_{final}$$

For example, if you diluted 150 mL of a normal saline solution (0.90% m/v) to a final volume of 450 mL with water, what would the concentration of the final, diluted solution be?

$C_{initial} = 0.90\%$ m/v $\quad V_{final} = 450$ mL dilute solution
$V_{initial} = 150$ mL $\quad\quad C_{final} = ?$

$$C_{initial} \times V_{initial} = C_{final} \times V_{final}$$

$$\frac{C_{initial} \times V_{initial}}{V_{final}} = C_{final}$$

$$\frac{(0.90\%) \times (150 \,\text{mL})}{(450 \,\text{mL})} = C_{final}$$

$$0.30\% = C_{final}$$

The dilution equation works with any concentration unit where the amount of solution (the denominator) is expressed in volume units. (The equation will not work with % m/m.) The dilution equation is useful in the health fields because many pharmaceuticals are often prepared as concentrated aqueous solutions and must be diluted before being administered to the patient.

Solving a Problem

Using the Dilution Equation

Hydrocortisone is used as an anti-inflammatory for localized pain. You must prepare a 50-mg/mL solution of hydrocortisone for an injection. You have 5 mL of a 200-mg/mL stock solution of hydrocortisone available. How many milliliters of the 50-mg/mL solution can you prepare?

Solution

STEP 1: Establish the given information. In this problem, two concentrations and one volume are given. The volume of the initial stock solution is given. You will solve for the final volume.

$C_{initial} = 200$ mg/mL $C_{final} = 50$ mg/mL
$V_{initial} = 5$ mL $V_{final} = ?$

STEP 2: Arrange the dilution equation to solve for the unknown quantity.

$$C_{initial} \times V_{initial} = C_{final} \times V_{final}$$

$$\frac{C_{initial} \times V_{initial}}{C_{final}} = V_{final}$$

STEP 3: Solve for the unknown quantity.

$$\frac{(200 \text{ mg/mL}) \times (5 \text{ mL})}{(50 \text{ mg/mL})} = V_{final}$$

$$20 \text{ ml} = V_{final}$$

Does your answer make sense? Because you are diluting, the final volume determined should be more than the initial volume.

---

sample problem 8.15   Preparing a Solution

How would you prepare 500 mL of a 5% (m/v) glucose solution from a 25% (m/v) stock solution?

Solution

STEP 1: Establish the given information. In this problem, two concentrations and one volume—the volume of the final solution—are given. You will solve for the initial volume.

$C_{initial} = 25\%$ $C_{final} = 5\%$
$V_{initial} = ?$ $V_{final} = 500$ mL

STEP 2: Arrange the dilution equation to solve for the unknown quantity.

$$C_{initial} \times V_{initial} = C_{final} \times V_{final}$$

$$V_{initial} = \frac{C_{final} \times V_{final}}{C_{initial}}$$

STEP 3: Solve for the unknown quantity.

$$V_{initial} = \frac{(5\%) \times (500 \text{ mL})}{(25\%)}$$

$$V_{initial} = 100 \text{ mL}$$

The answer to the equation tells us that 100 mL of the stock solution should be used. The question asks how we would prepare the solution, so the final answer requires that we interpret our answer to the dilution equation by stating that to prepare the 5% solution we should combine 100 mL of the 25% solution and enough water for a total solution volume of 500 mL.

8.37 How many liters of a 0.90% (m/v) NaCl solution can be prepared from 600 mL of a 9.0% (m/v) stock solution?

8.38 How many mL of a 0.45% (m/v) KCl solution can be prepared from 500 mL of a 9.0% (m/v) stock solution?

8.39 What would the concentration of the resulting glucose solution be if 250 mL of a 12% (m/v) glucose solution were diluted to a final volume of 1 L with distilled water?

8.40 If 40 mL of a 6 M NaOH solution is diluted to a final volume of 200 mL, what is the resulting concentration of the solution?

8.41 How would you prepare 250 mL of a 0.225% (m/v) NaCl solution from a 0.90% (m/v) NaCl stock solution?

8.42 How would you prepare 2 L of 1 M $MgCl_2$ from a 5 M $MgCl_2$ stock solution?

practice
problems

Discovering the Concepts

? Inquiry Activity—Osmosis

Part 1. Information

Cells and the solutions surrounding them are separated by the cell's membrane. The cell membrane is semipermeable. This means that some molecules are able to pass through while others are not. Water can spontaneously flow through the cell membrane in a direction that will equalize the concentrations inside and outside the cell. Three possibilities exist for a cell and its surrounding solution: the concentrations of both are the same, the exterior solution is of higher concentration than the solution inside the cell, or the interior solution is higher concentration than the solution outside the cell. The solution types are outlined in the following table and figure:

Solution Type	Amount of Solute (amount of water) Outside the Cell	Amount of Solute (amount of water) Inside the Cell	Movement of Water	Effect on the Cells
Isotonic	The same as inside	The same as outside	No net movement	No effect
Hypotonic	Lower than inside (higher than inside)	Higher than outside (lower than outside)	Water moves into the cells	Cells swell and can burst (lyse)
Hypertonic	Higher than inside (lower than inside)	Lower than outside (higher than outside)	Water moves out of the cells	Cells shrivel (crenate)

Isotonic solution   Hypotonic solution   Hypertonic solution

Red blood cells are isotonic with a NaCl solution that is 0.90% (m/v), also known as normal saline and 5% glucose, also known as D5W (dextrose 5% in water). Intravenous fluids must be isotonic to avoid cell lysis (by bursting) or the opposite effect, shriveling, also called *crenation*.

Demonstration—Common Osmosis

You will need two carrots and two lemon slices. Dissolve as much table salt as possible into two glasses of tap water to make a saturated salt solution. Place one carrot and one lemon slice into separate glasses of salt solution and one carrot and one lemon into separate glasses of tap water. Allow them to sit overnight. Then observe each and answer the following questions.

Questions

1. Describe the appearance of the foods in (a) the tap water and (b) the salt water.
2. How did the foods change when they were placed in (a) the tap water and (b) the salt water?
3. In which glass did the water in the solution move into the food cells?
4. In which glass did the solution remove water from the food cells?
5. Which of the solutions (tap water or salt water) is (a) hypotonic or (b) hypertonic to the vegetables?
6. A cell is placed in a 10% (m/v) glucose solution.
   a. Which direction will water flow to equalize the concentrations: into the cell or out of the cell?
   b. Will the cell swell or crenate?
7. If a person pours a concentrated salt water solution on an earthworm, the worm will wriggle and die. What type of solution (isotonic, hypertonic, or hypotonic) was the salt solution? What are the earthworm's cells doing?
8. If a person drinks too much water too quickly, a condition known as hyponatremia will occur. In this condition, there is not enough sodium in the body fluids outside the cell. What type of solution (isotonic, hypertonic, or hypotonic) would a person's cells be in? What are the cells likely to do?

8.6 Osmosis and Diffusion

Osmosis

Because our bodies are mostly water, we could consider ourselves to be composed of a set of specialized aqueous solutions. These solutions are found both inside and outside cells. The concentration of these solutions is highly controlled by the cells. The solutions are separated by a semipermeable barrier called the *cell membrane* whose structure was described in Chapter 7. A **semipermeable membrane** allows some molecules to pass through the barrier but not others. Under normal physiological conditions, these solutions are considered to be **isotonic solutions,** meaning that the concentration of dissolved solutes is the same on both sides of the membrane.

Consider this: Why is it that there is a limit to the amount of water that we can drink? And why is it that we cannot survive by drinking sea water? The short answer is that tap water is not as concentrated as the body's solutions, and sea water is more concentrated. Let's look more closely at these two questions.

A person can drink too much fresh water. It is recommended that we drink eight to twelve 8-ounce glasses of water a day. The kidneys can process up to 15 liters of water a day, so the average person would really have to overdo it to drink too much water.

People engaging in vigorous exercise and infants whose formula is overly diluted can actually drink too much water. What is happening in the body? In this instance, the concentration of dissolved solutes in the body's internal solutions is higher than the concentration of dissolved solutes in tap water or dilute formula. When a person drinks large quantities of tap water, it dilutes the blood. The concentration of solutes in the blood goes down, resulting in an imbalance between the concentration of bodily solution outside the cells (lower concentrated solution, less dissolved solutes) and bodily solution inside the cells (higher concentrated solution, more dissolved solutes). In this case, the solution outside of the cells is said to be a **hypotonic solution** (*hypo* is a prefix meaning "lower than").

8.6 Inquiry Question:
How is the direction of osmosis predicted?

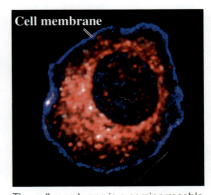

Cell membrane

The cell membrane is a semipermeable membrane.

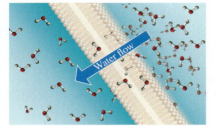

Water will pass through a cell membrane in an attempt to equalize the concentration on either side.

When the solution concentrations inside and outside the cells are different, water will travel across the cell membrane in an attempt to equalize the concentrations. This passage of *water* through a semipermeable membrane such as a cell membrane is called **osmosis.** So, if we drink too much water, a condition known as hyponatremia (low sodium concentration here caused by excess water) can result from the bloodstream becoming diluted by all the ingested water. Through osmosis, too much water enters the cells in an attempt to equalize the concentrations, and as a result, the cells swell up and could even burst (a phenomenon called *lysing*).

As water flows through a semipermeable membrane during osmosis, the water molecules in the more concentrated solution exert a certain amount of pressure (recall that pressure is force over an area) on the membrane as they attract the water molecules from the less concentrated solution to the more concentrated solution. This pressure is termed **osmotic pressure.** The greater the relative number of solute molecules present in a solution (that is, the more concentrated the solution), the greater the number of water molecules that will pass into the more concentrated solution to equalize the concentrations, and, as a result, the higher the osmotic pressure. Pure water has an osmotic pressure of zero. Applying pressure in opposition to the osmotic pressure will stop osmosis.

If we were marooned on a desert island without fresh water, why could we not drink sea water to quench our thirst? The concentration of dissolved ions in salt water is about three times that of the blood, so when sea water is consumed, it actually draws water out of the cells through osmosis in an effort to equalize the concentrations. This dehydrates the cells. If a person were to drink sea water, the concentration of solutes in the bloodstream would go up, resulting in an imbalance between the concentration outside the cells (higher concentrated solution, more dissolved solutes) and inside the cells (lower concentrated solution, less dissolved solutes). The solution outside the cells is said to be a **hypertonic solution** (*hyper* is a prefix meaning "higher than"). During dehydration, the cells shrivel in a process known as **crenation.** A summary of how cells behave in different types of solutions is shown in Figure 8.8.

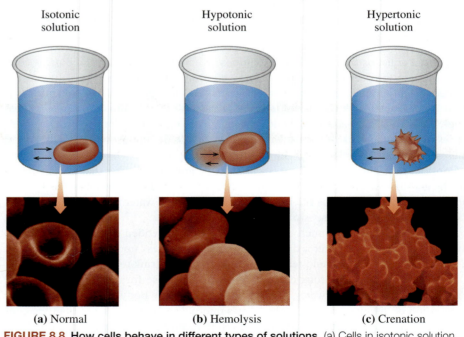

FIGURE 8.8 **How cells behave in different types of solutions.** (a) Cells in isotonic solution. (b) Cells in hypotonic solution will swell. (c) Cells in hypertonic solution will shrink (crenate).

Let's look more closely at how osmosis occurs across a cell membrane. Given that cell membranes separate solutions of different concentration, consider the situation in the following illustration. In which direction would osmosis occur (water flow) to equalize the concentration?

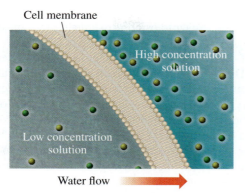

Cell membrane

High concentration solution

Low concentration solution

Water flow

During osmosis, water flows through the cell membrane from an area where the solution is less concentrated to that where it is more concentrated to even out the concentrations.

The net flow of water will be from the solution with the lower solute concentration into the solution of higher solute concentration to dilute the solute. Water moves from a solution containing more water molecules (dilute solution) into a solution containing fewer water molecules (more concentrated solution) to equalize the concentrations.

Because cell membranes are semipermeable, osmosis is an ongoing process and is used by the cells to maintain the concentrations inside and outside the cells at about the same level. Body fluids like blood, plasma, and lymph all exert some osmotic pressure. Intravenous (IV) solutions that are delivered into patients' bloodstreams are isotonic. They have solute concentrations equal to the solute concentrations inside of cells. Isotonic solutions minimize osmosis, which is desirable when introducing IV solutions into the blood and eventually to cells. Common isotonic IV solutions used in hospitals include 0.90% (m/v) NaCl (normal saline, NS) and a 5% (m/v) D-glucose (dextrose) solution commonly referred to as D5W ("Dextrose 5% in Water"). These solutions are called **physiological solutions** because they exert the same osmotic pressure as the cells and are isotonic with cells.

Diffusion and Dialysis

Suppose we put a drop of green food coloring into a large beaker of water. Over time, the green dye molecules (solute) will mix with the water (solvent) and the resulting solution will have a uniform light green tinge to it. The two solutions (one with a high concentration of green molecules and one with no green molecules) spontaneously mix, and the green solute molecules *diffuse* into the water to form one dilute solution with a final green color intermediate between green food coloring from the dropper bottle and water.

Time

Over time, a drop of concentrated dye will diffuse through water, making the whole container a uniform color.

integrating chemistry

Kidney Dialysis

Small molecules and ions can also pass through the cell membrane. If a membrane is permeable to other molecules in addition to water, those molecules will also move across the membrane by a process similar to osmosis to equalize the concentrations of the two solutions.

The movement of *solute* molecules across a semipermeable membrane is a diffusion process, not osmosis. **Diffusion** is the movement of molecules in a direction that equalizes the concentration. One notable place where diffusion occurs in the body is in the kidneys. The kidneys act to remove small waste molecules out of the blood through diffusion across membranes in the kidneys. Cells and larger molecules, such as proteins found in the blood, are too large to pass through the membrane. These larger molecules are reabsorbed into the bloodstream after passing through the kidney. At the same time, small molecules like urea diffuse out of the blood (higher concentration) and move into urine (lower concentration) in a process called **dialysis.**

If the kidneys cannot dialyze waste products out of the bloodstream, increased levels of urea and other wastes in the bloodstream can become life-threatening. A person whose kidneys are failing can undergo artificial dialysis—called *hemodialysis*—to cleanse the blood (see **Figure 8.9**). In this process, blood is removed from the patient and passes through one side of a semipermeable membrane in contact on the opposite side with a dialyzing solution that is isotonic with normal blood solute concentrations. Urea and small waste molecules are present in greater concentration in the patient's blood and diffuse out of the passing blood and into the dialyzing solution, and the dialyzed blood returns to the patient.

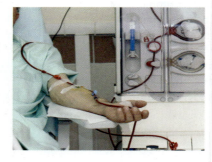
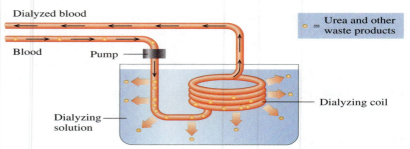

FIGURE 8.9 Hemodialysis. Hemodialysis removes waste products from the blood while outside the body, much as a healthy kidney removes waste products from the blood inside the body.

sample problem

8.16 Osmosis and Osmotic Pressure

A cell with concentration equal to 5% (m/v) glucose is placed in a 10% (m/v) glucose solution.

a. In which direction will water flow to equalize the concentrations (into the cell or out of the cell)?

b. Which solution is exerting the greater osmotic pressure?

c. Will the cell swell or crenate?

Solution

a. Water will flow from the area where there is more water—the 5% (m/v) solution—to the area where there is less water—the 10% (m/v) solution—so water will flow out of the cell.

b. The 10% (m/v) glucose solution has the higher solute concentration, so it has the greater osmotic pressure.

c. Because water is moving out of the cell, the cells will crenate.

---

sample problem

8.17 Isotonic, Hypotonic, and Hypertonic Solutions

Are each of the following solutions considered isotonic, hypotonic, or hypertonic with respect to body fluids?

a. 3% (m/v) NaCl   b. 0.90% (m/v) NaCl   c. 0.09% (m/v) NaCl

Solution

Compare each solution to physiological saline, which is 0.90% (m/v) NaCl.

a.   hypertonic   b. isotonic   c. hypotonic

---

practice problems

8.43 A cucumber placed in a briny salt water solution makes a pickle.
  a. Does water leave or enter the cucumber's cells?
  b. Does the cucumber swell or crenate?
  c. Is the salt water solution hypertonic or hypotonic to the cucumber?

8.44 A lettuce leaf in the produce aisle at the grocery is accidentally submerged in water.
  a. Will water exit or enter the lettuce cells?
  b. Will the cells swell or crenate?
  c. Is tap water hypertonic or hypotonic to the lettuce leaf cells?

8.45 A patient is undergoing hemodialysis. As the blood leaves the patient, its solute concentrations are _____ (hypertonic, hypotonic, isotonic) to the dialyzing solution. As the blood reenters the patient, its solute concentrations are _____ (hypertonic, hypotonic, isotonic) to the dialyzing solution.

8.46 A patient is undergoing hemodialysis. As the blood leaves the patient, it has _____ (higher, lower, the same) osmotic pressure than the dialyzing solution. As the blood reenters the patient, its osmotic pressure is _____ (higher, lower, the same as) the dialyzing solution.

8.47 Are the following solutions isotonic, hypotonic, or hypertonic with physiological solutions?
  a. 10% (m/v) NaCl   b. 0.90% (m/v) glucose
  c. 5% (m/v) NaCl   d. 5% (m/v) glucose

8.48 Will a red blood cell undergo crenation, lysis, or no change in each of the following solutions?
  a. 10% (m/v) NaCl   b. distilled water
  c. 5% (m/v) NaCl   d. 5% (m/v) glucose

8.7 Transport across Cell Membranes

We saw in Section 8.6 that water can cross the cell membrane through a process called osmosis, but how do polar molecules like glucose or ions like Na^+ move across? As we discussed in Section 7.6, the main structural components of a cell membrane are the phospholipids, which have a polar (hydrophilic) "head" containing a phosphate and a

? 8.7 Inquiry Question:
How does the cell membrane control diffusion in and out of the cell?

nonpolar (hydrophobic) part containing long hydrocarbon "tails." **Figure 8.10** shows the arrangement of the phospholipids in a cell membrane creating a nonpolar barrier enclosing its contents.

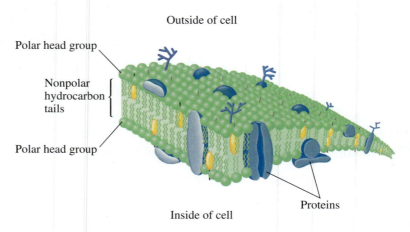

Outside of cell

Polar head group

Nonpolar hydrocarbon tails

Polar head group

Proteins

Inside of cell

FIGURE 8.10 The cell membrane. Phospholipids create a nonpolar barrier to the passage of polar molecules.

Ions, nonpolar molecules, and polar molecules move across cell membranes in different ways. We will focus on three main forms of transport across cell membranes: passive diffusion, facilitated transport, and active transport.

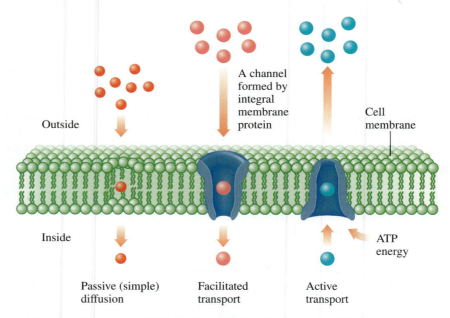

Outside

A channel formed by integral membrane protein

Cell membrane

Inside

ATP energy

Passive (simple) diffusion

Facilitated transport

Active transport

Transport through a cell membrane can occur with diffusion (no energy required) or against diffusion (energy required).

Small molecules that are in high concentration like water and the nonpolar molecules O_2, N_2, and CO_2 can diffuse directly through the cell membrane. Diffusion moves solutes in a direction that attempts to equalize the concentrations on either side of a membrane. This process does not require any additional energy, so this simple diffusion process is also referred to as **passive diffusion.** Other nonpolar molecules like steroids can also passively diffuse through cell membranes.

Because of their polar character, ions and small polar molecules diffuse very slowly across the nonpolar barrier (refer to Figure 8.10) found in the center of the cell membrane, water being one exception. To enable small molecules and ions to pass through the cell membrane, some proteins found in the cell membrane have polar channels that open and close, allowing small polar molecules and ions to be transported across the cell membrane, to equalize concentrations. These proteins are often integral membrane proteins, spanning the phospholipid bilayer (Chapter 7). This type of transport is called **facilitated transport** and such transport does not require energy. Glucose transporter proteins are found in virtually all cell membranes and facilitate the transport of glucose into the cell during times when glucose concentrations in the bloodstream are high (for example, after a meal).

Sometimes ions or small molecules must be moved across the cell membrane in the opposite direction of diffusion (making the concentrations more different, less equal). This can also occur, but not without the use of energy. Transporting ions or small polar molecules across the cell membrane in a direction opposite to equalizing concentrations is possible and also requires the assistance of a protein channel or pump. This pumping in the opposite direction is called **active transport** and it requires energy usually in the form of the energy molecule adenosine triphosphate (ATP) (Chapter 12).

One of these active transport pumps known as the $K^+/H^+ATPase$ controls the concentration of potassium and hydrogen ions in the stomach and is responsible for keeping the acid concentration in your stomach constant. Medications like Tagamet®, Zantac®, and Pepcid® prevent the effects of heartburn (high levels of stomach acid) by blocking the normal production of stomach acid through these pumps.

sample problem 8.18   Transport across Membranes

In the preceding figure, in which direction would passive diffusion occur across the cell membrane (to the right or to the left)?

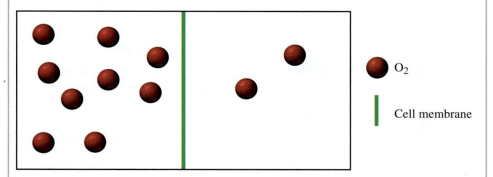

O_2

Cell membrane

Solution

Because the O_2 will diffuse to equalize the concentrations, the O_2 molecules will move to the right side in an attempt to increase the concentration on that side.

practice problems

8.49 Identify the type of transport (passive diffusion, facilitated transport, or active transport) that will occur for the following molecules:

   a. oxygen

   b. glucose, no energy required

   c. Na^+, energy required

   d. K^+, no energy required

8.50 Identify the type of transport (passive diffusion, facilitated transport, or active transport) that will occur for the following molecules:

   a. cholesterol, no energy required

   b. nitrogen

   c. Cl^-, no energy required

   d. H^+, energy required

SUMMARY

8.1 Solutions Are Mixtures

8.1 Inquiry Question: How is a solution identified?

A solution forms when a solute dissolves in a solvent. In a solution, the particles of a solute are evenly distributed in the solvent. The solute and solvent may be solid, liquid, or gas. Solutions are transparent. Mixtures with particles suspended in a solution are colloids and are usually not transparent. Mixtures that contain particles that settle upon standing are suspensions.

8.2 Formation of Solutions

8.2 Inquiry Question: How do temperature and pressure affect the formation of solutions?

An increase in temperature increases the solubility of most solids in water but decreases the solubility of gases in water. Henry's law describes the relationship between pressure and gas solubility. Increasing the pressure above a solution with a dissolved gas in it increases the solubility of the gas. A solution that contains the maximum amount of dissolved solute is a saturated solution. A solution that is saturated reaches an equilibrium state between the dissolved solute and undissolved solid solute where the rate of dissolving and re-forming solid is the same.

8.3 Chemical Equations for Solution Formation

8.3 Inquiry Question: How are hydration equations written for electrolytes, nonelectrolytes, and weak electrolytes?

Hydration equations can be written for solutes dissolving in solvents. The form of this equation depends on the ability of the solute to dissociate in solution. Substances that release ions when they dissolve in water are called electrolytes because the solution will conduct an electrical current. Strong electrolytes are ionic compounds that completely dissociate in water. Weak electrolytes only partially dissociate into ions. Nonelectrolytes are substances (usually covalent compounds) that dissolve in water but do not dissociate. The unit known as an equivalent expresses the amount of dissolved ions in fluids. The number of equivalents per mole of an ion equals the charge on that ion.

8.4 Concentration

8.4 Inquiry Question: How is the concentration of a solution expressed?

The concentration of a solution is the amount of solute dissolved in a certain amount of solution. Fluid replacement solutions are often expressed in units of mEq/L or in some cases mmole/L. Molarity is the moles of solute per liter of solution. Percent mass/ volume expresses the ratio of the mass of solute (in g) to the volume of solution (in mL) multiplied by 100. This percent mass/volume is equivalent to the unit g/dL. Percent concentration is also expressed as mass/mass and volume/volume ratios. Parts per million and parts per billion describe very dilute solutions.

8.5 Dilution

8.5 Inquiry Question: How is the concentration of a solution determined if a dilution is made?

When a solution is diluted, the amount of solute stays the same while the volume of solution increases. The concentration of the solution decreases.

8.6 Osmosis and Diffusion

8.6 Inquiry Question: How is the direction of osmosis predicted?

In osmosis, solvent (water) passes through a semipermeable membrane from a solution of lower solute concentration to a solution of higher solute concentration. The osmotic pressure exerted on the membrane is directly related to the number of water molecules pushing against that membrane. Isotonic solutions have osmotic pressures equal to those of bodily fluids. Cells maintain their volume in an isotonic solution, but they swell and may burst in a hypotonic solution and shrivel in a hypertonic solution. In dialysis, water and small solute particles pass through a dialyzing membrane in a related process called *diffusion* while large particles like proteins are retained.

8.7 Transport across Cell Membranes

8.7 Inquiry Question: How does the cell membrane control diffusion in and out of the cell?

The semipermeable membrane surrounding cells separates the cellular contents from the external fluids. Molecules can be transported across the cell membrane by passive diffusion, facilitated transport, or active transport, depending on their concentration inside and outside the cell and their polarity.

The study guide will help you check your understanding of the main concepts in Chapter 8. You should be able to

8.1 Solutions Are Mixtures

- Distinguish solute and solvent.
- Identify solutions, colloids, and suspensions.

8.2 Formation of Solutions

- Define saturated and dilute solutions.
- Predict the effect of temperature on the solubility of a solute.
- Predict the effect of pressure on the solubility of a gas in a liquid.

8.3 Chemical Equations for Solution Formation

- Write chemical equations for hydration of electrolytes, nonelectrolytes, and weak electrolytes.
- Calculate the number of milliequivalents present for an ionic compound that fully dissociates in solution.
- Convert from mEq to moles.

8.4 Concentrations

- Express concentration in molarity units.
- Express concentration in percent units.
- Express concentration in parts per million and parts per billion.

8.5 Dilution

- Calculate concentrations or determine volumes using the dilution equation.

8.6 Osmosis and Diffusion

- Predict the direction of osmosis or diffusion given the concentration on both sides of a semipermeable membrane.

8.7 Transport across Cell Membranes

- Characterize three forms of transport across a cell membrane.

Key Terms

active transport—The transport of ions or small molecules across the cell membrane in a direction opposite to diffusion. This process involves an integral membrane protein and requires energy.

aqueous solution—A solution in which water is the solvent.

colloid—A homogeneous mixture containing particles ranging from 1 to 1000 nm in diameter. Colloidal mixtures do not separate upon standing.

concentration—The amount of solute dissolved in a certain amount of solution.

crenation—The shriveling of a cell due to water loss when the cell is placed in a hypertonic solution.

dialysis—A process in which water and small molecules pass through a selectively permeable membrane.

diffusion—The movement of solutes that lowers the concentration of a solution. This movement does not require energy (it is a passive process).

electrolyte—A substance that conducts electricity because it dissociates into ions in aqueous solution.

equilibrium—A state where two processes are occurring at the same rate in opposite directions. In a saturated solution, the rate of a solid dissolving and the rate of a dissolved solid re-forming are the same.

equivalent—A unit of concentration that relates the charge in a solution to the number of ions or the moles of ions present in the solution.

facilitated transport—The diffusion of ions or small polar molecules across the cell membrane with the assistance of a protein.

hydration—The solvation process when water is the solvent.

Henry's law—The solubility of a gas in a liquid is directly related to the pressure of that gas over the liquid.

hypertonic solution—A solution outside of a cell having a higher concentration of solutes than the solution inside the cell.

hypotonic solution—A solution outside of a cell having a lower concentration of solutes than the solution inside the cell.

ionize—To dissociate into ions when dissolved.

isotonic solution—A solution outside of a cell having the same concentration of solutes as the solution on the inside of a cell.

law of conservation of matter—Matter can neither be created nor destroyed; it merely changes forms.

molarity—A unit of concentration defined as the number of moles of solute dissolved per liter of solution.

nonelectrolyte—A substance that does not conduct electricity because it does not ionize in aqueous solution.

osmosis—The passage of water across a semipermeable membrane in an effort to equalize the solution concentrations on either side. This passage does not require energy (it is a passive process).

osmotic pressure—The pressure that water exerts during osmosis. This amount of pressure applied to the more concentrated solution of the two separated solutions would stop osmosis.

passive diffusion—Movement of solutes across the cell membrane to equalize their concentration.

physiological solution—A solution that is isotonic with normal body fluids.

saturated solution—A solution containing the maximum amount of a solute capable of dissolving in the solution at a given temperature.

semipermeable membrane—A membrane that allows passage of some solutes and blocks the passage of others.

solubility—The maximum amount of a solute capable of dissolving in solvent at a given temperature.

solute—The substance in a solution present in the smaller amount. Solutions can have more than one solute.

solution—A homogeneous mixture where particles (ions or molecules) are dispersed individually and evenly throughout.

solvation—The process involved when a solute dissolves in a solvent.

solvent—The substance in a solution present in the larger amount. There can be only one solvent in a solution.

strong electrolyte—A substance that conducts electricity because it completely ionizes in aqueous solution.

suspension—A mixture containing particles greater than 1000 nm (1 micrometer) in diameter. Suspensions separate upon standing.

unsaturated solution—A solution containing less than the maximum amount of solute capable of dissolving.

weak electrolyte—A substance that weakly conducts electricity because it partially ionizes in aqueous solution.

Important Equations

$$\text{Concentration} = \frac{\text{amount of solute}}{\text{amount of solution}}$$

$$\% \ (m/v) = \frac{\text{g solute}}{\text{mL solution}} \times 100\%$$

$$C_{\text{initial}} \times V_{\text{initial}} = C_{\text{final}} \times V_{\text{final}}$$

$$\text{Molarity, (M)} \ \ M = \frac{\text{mole solute}}{\text{L solution}}$$

$$\% \ (m/m) = \frac{\text{g solute}}{\text{g solution}} \times 100\%$$

$$\text{ppm} = \frac{\text{g solute}}{\text{mL solution}} \times 1,000,000$$

$$\% \ \text{concentration} = \frac{\text{parts of solute}}{100 \ \text{parts of solution}}$$

$$\% \ (v/v) = \frac{\text{mL solute}}{\text{mL solution}} \times 100\%$$

$$\text{ppb} = \frac{\text{g solute}}{\text{mL solution}} \times 1,000,000,000$$

Additional Problems

8.51 Identify the following as solutions, colloids, or suspensions.

a. whipped cream

b. wine

c. coffee with cream and sugar

8.52 Identify the following as solutions, colloids, or suspensions.

a. hair spray

b. gasoline

c. hand lotion

8.53 Does the solubility of the solute increase or decrease in each of the following situations?

a. Sugar is dissolved in iced tea instead of hot tea.

b. A bottle of soda (solute is CO_2 gas) is placed in the refrigerator instead of in a pantry at room temperature.

c. An opened bottle of champagne (solute is CO_2 gas) is allowed to sit open to the air for hours.

8.54 Does the solubility of the solute increase or decrease in each of the following situations?

a. Ice is placed in a glass of cola (solute is CO_2 gas).

b. Honey is dissolved in hot water instead of cold water.

c. The pressure of O_2 over a solution is increased (solute is O_2 gas).

8.55 Predict whether the following will fully dissociate, partially dissociate, or not dissociate when dissolved in solution:

a. sodium iodide, $NaI(s)$, a strong electrolyte

b. pyruvic acid, a weak electrolyte (as shown to the right)

c. glucose, $C_6H_{12}O_6(s)$, a nonelectrolyte

8.56 Predict whether the following will fully dissociate, partially dissociate, or not dissociate when dissolved in solution:

a. potassium nitrate, $KNO_3(s)$, a strong electrolyte

b. isopropyl alcohol, $CH_3CH(OH)CH_3(l)$, a nonelectrolyte

c. boric acid, $H_3BO_3(s)$, a weak electrolyte

8.57 Provide a balanced equation for the hydration of each compound in Problem 8.55.

8.58 Provide a balanced equation for the hydration of each compound in Problem 8.56.

8.59 How many equivalents of Ca^{2+} are present in a solution that contains 1.34 moles of Ca^{2+}?

8.60 How many equivalents of Cl^- are present in a solution that contains 6.00 moles of Cl^-?

8.61 A Ringer's solution for intravenous fluid replacement contains 4 mEq Ca^{2+} per liter of solution. If a patient receives 1750 mL of Ringer's solution, how many milliequivalents of Ca^{2+} were given?

8.62 An extracellular replacement fluid contains 3 mEq/L citrate. The conversion factor for citrate is 3 Eq citrate/ 1 mole of citrate. If a patient receives 550 mL of this solution, how many millimoles of citrate were delivered?

8.63 What is the concentration in mmole/L of a $SO_4{}^{2-}$ solution that is 25.0 mEq/L?

8.64 What is the concentration in mEq/L of a Na^+ solution that is 115 mmole/L?

8.65 What is the molarity of 1.50 L of solution containing 2.80 moles KCl?

8.66 What is the molarity of 750 mL of solution containing 6.2 moles $CaCl_2$?

8.67 What is the molarity of 0.50 L of solution containing 40.0 g of $CaCO_3$?

8.68 What is the molarity of 450 mL of solution containing 15 g NaCl?

8.69 A 750 mL bottle of wine contains 12% (v/v) ethanol. How many milliliters of ethanol are in the bottle of wine?

8.70 What is the concentration in % (v/v) of a methanol solution prepared by mixing 30.0 mL of methanol with 650 mL of distilled water?

8.71 How many grams of dextrose must be combined with water to prepare 500 mL of a 5.0% (m/v) dextrose maintenance solution?

8.72 The normal range for blood hemoglobin in females is 12–16 g/dL. What is this value in % (m/v)?

8.73 The normal level of urea nitrogen in adults is 7–18 mg/dL. What is this value in ppm?

8.74 The normal creatinine levels in adults are 0.6–1.2 mg/dL. What is this value in ppm?

8.75 Calculate the final concentration of each of the following diluted solutions:

   a. 2.0 L of a 3.0 M HNO_3 solution is added to water so that the final volume is 6.0 L.

   b. Water is added to 0.50 L of a 6.0 M KOH solution to make 4.0 L of a diluted KOH solution.

8.76 Calculate the final concentration of each of the following diluted solutions:

   a. A 50.0-mL sample of 4.0% (m/v) NaOH is diluted with water so that the final volume is 100.0 mL.

   b. A 15.0-mL sample of 35% (m/v) acetic acid (CH_3COOH) solution is added to water to give a final volume of 25 mL.

8.77 What is the final volume in liters that can be prepared from each of the following concentrated solutions?

   a. a 1.00 M HCl solution prepared from 150.0 mL of a 6.0 M HCl solution

   b. a 4.00% (m/v) $NaHCO_3$ solution prepared from 250 mL of a 20.0% (m/v) $NaHCO_3$

8.78 What is the initial volume in liters necessary to prepare each of the following diluted solutions?

   a. 2.0 L of a 0.90% NaCl using a 18.0% (m/v) NaCl stock solution

   b. 1.5 L of a 5.0% glucose solution using a 15.0% glucose solution

8.79 How would you prepare 1.0 L of a normal saline solution (0.9% (m/v) from a stock solution that is 18% (m/v)?

8.80 How would you prepare 500 mL of a 5% D5W (dextrose in water) solution from a 25% stock solution?

8.81 Consider a cell placed in solution as shown in the following figure:

If the inside of the cell has a concentration equivalent to 0.90% (m/v) NaCl, is the cell in a hypertonic, hypotonic, or isotonic solution if the solution outside of the cell is the following:

   a. 5% NaCl

   b. 0.090% NaCl

   c. 0.90% NaCl

8.82 Under the conditions in Problem **8.81**a–c, in which direction will water move by osmosis: into the cell, out of the cell, or will no net movement occur?

8.83 Edema, commonly referred to as water retention, is characterized by a swelling of the tissues. Individuals with kidney disease will not excrete normal amounts of Na^+, leading to higher levels of Na^+ in the tissues. In terms of osmosis, explain how high levels of Na^+ in the tissues can lead to edema.

8.84 Many people gain relief from swollen feet at the end of the day by soaking their feet in Epsom salts, which creates a hypertonic solution of $MgSO_4$. In terms of osmosis, explain how soaking in Epsom salts can reduce swelling in the feet.

8.85 A process called reverse osmosis purifies tap water by pushing water through a set of semipermeable membrane filters. Explain why this process is called *reverse* osmosis.

8.86 Consider the cell in the following figure.

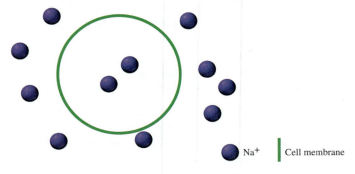

In which direction (into the cell or out of the cell) would the Na^+ be transported if the transport protein in the cell membrane was

   a. an ATPase pump?

   b. an ion transport protein operating under facilitated transport?

8.87 Do the following processes require energy when transporting molecules across the cell membrane?

   a. passive diffusion   **b.** facilitated transport

   c. active transport

8.88 Describe the concentration of the solution outside the cell as hypertonic or hypotonic if that solute is being transported across the cell membrane by

   a. passive diffusion.   **b.** facilitated transport.

   c. active transport.

Challenge Problems

8.89 A scuba diver diving 100 m down in the ocean experiences 10 times more pressure on her body than a person swimming at sea level experiences from the atmosphere. The high pressure of the air in a scuba diver's air tank affects the solubility of gases in the bloodstream. A condition called "the bends," which is related to nitrogen gas solubility, can occur in a scuba diver who ascends too quickly from a deep dive to the surface. Can you give an explanation for this condition?

8.90 How would you prepare 500 mL of a 6.0 M NaOH solution using solid NaOH pellets? What is the % (m/v) concentration of a 6.0 M NaOH solution?

8.91 Two containers of equal volume are separated by a membrane that allows free passage of water but completely restricts passage of solute molecules. Solution A has 3 molecules of the protein albumin (molecular weight 66,000) and solution B contains 16 molecules of glucose (molecular weight 180). Into which compartment would water flow, or will there be no net movement of water? Which solution is at the higher concentration? Which chamber is exerting the higher osmotic pressure?

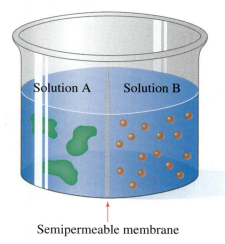

Semipermeable membrane

8.92 Proteinuria is a condition in which excessive protein is found in the urine. What must be happening to the kidney's membrane filters if such large molecules are found in the urine?

Answers to Odd-Numbered Problems

Practice Problems

8.1 **a.** solute, oxygen, solvent, nitrogen
b. solute, zinc, solvent, copper
c. solute, blue food coloring, solvent, ethanol

8.3 **a.** colloid
b. solution
c. solution

8.5 **a.** No, not saturated
b. Yes, saturated

8.7 **a.** increase
b. decrease
c. increase

8.9 **a.** Decreasing the temperature increases the solubility of the gas in the soda, so more of the gas stays in the solution and capping the bottle increases the CO_2 pressure over the solution and increases the solubility of CO_2.
b. If the lid is on tight, no gas can escape and gas that escapes from the solution will build up a pressure above the solution. At some point an equilibrium is reached in which no more gas will escape from solution.
c. The solubilities of most solid solutes decrease with lower temperature, so less sugar will dissolve in the iced tea.

8.11 **a.** fully dissociate
b. partially dissociate
c. not dissociate

8.13 **a.** $KF(s) \xrightarrow[H_2O]{} K^+(aq) + F^-(aq)$

b. $HCN(g) \underset{H_2O}{\rightleftharpoons} H^+(aq) + CN^-(aq)$

c. $C_6H_{12}O_6(s) \xrightarrow[H_2O]{} C_6H_{12}O_6(aq)$

8.15 **a.** $CaCl_2(s) \xrightarrow[H_2O]{} Ca^{2+}(aq) + 2Cl^-(aq)$

b. $NaOH(s) \xrightarrow[H_2O]{} Na^+(aq) + OH^-(aq)$

c. $KBr(s) \xrightarrow[H_2O]{} K^+(aq) + Br^-(aq)$

d. $Fe(NO_3)_3(s) \xrightarrow[H_2O]{} Fe^{3+}(aq) + 3NO_3^-(aq)$

8.17 4.25 Eq

8.19 0.0770 mole Na^+

8.21 1.25 mmoles Ca^{2+}/L

8.23 17 M

8.25 0.43 M

8.27 35.7 g KBr

8.29 0.5% (m/v)

8.31 **a.** 5.00%

b. 6.3%

8.33 0.117 g insulin

8.35 5 ppm, 5000 ppb

8.37 6 L

8.39 3% (m/v)

8.41 Add 62.5 mL of the 0.90% NaCl stock solution to enough distilled water for a total volume of 250 mL solution.

8.43 **a.** leave

b. crenate

c. hypertonic

8.45 hypertonic, isotonic

8.47 **a.** hypertonic

b. hypotonic

c. hypertonic

d. isotonic

8.49 **a.** passive diffusion

b. facilitated transport

c. active transport

d. facilitated transport

Additional Problems

8.51 **a.** colloid

b. solution

c. colloid

8.53 **a.** decrease

b. increase

c. decrease

8.55 **a.** fully dissociate

b. partially dissociate

c. not dissociate

8.57 **a.** $NaI(s) \xrightarrow[H_2O]{} Na^+(aq) + I^-(aq)$

b.

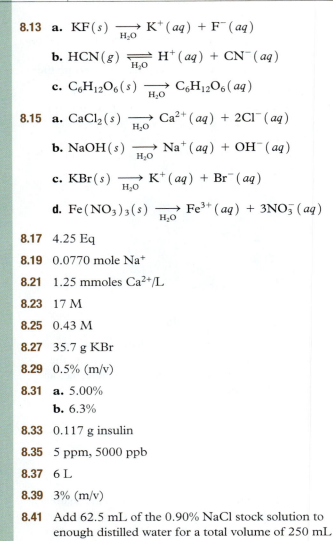

c. $C_6H_{12}O_6(s) \xrightarrow[H_2O]{} C_6H_{12}O_6(aq)$

8.59 2.68 Eq Ca^{2+}

8.61 7 mEq Ca^{2+}

8.63 12.5 mmole/L

8.65 1.87 M

8.67 0.80 M

8.69 90. mL ethanol

8.71 25 g dextrose

8.73 70–180 ppm urea nitrogen

8.75 **a.** 1.0 M HNO_3

b. 0.75 M KOH

8.77 **a.** 0.90 L

b. 1.25 L

8.79 Combine 50mL of the stock solution (18% (m/v)) with enough water to make 1 L of solution.

8.81 **a.** hypertonic

b. hypotonic

c. isotonic

8.83 The cells will attempt to dilute the higher than normal concentration of Na^+ in the tissues by moving more water into the tissues causing fluid retention.

8.85 If osmosis is water moving from a lower concentrated solution to a higher concentrated solution, balancing out the concentrations, the reverse would require water moving in the opposite direction. Less pure water (higher concentration of solutes) is forced (requires energy) through a filter (semipermeable membrane), removing more impurities, making drinkable water.

8.87 **a.** No
b. No
c. Yes

8.89 According to Henry's law, more nitrogen would dissolve in the bloodstream at lower depths (higher pressure) than at the surface (lower pressure). When a diver ascends to the surface quickly, the pressure of the air in the tank (and therefore the air in the lungs) lessens more rapidly than a person can expel the nitrogen.

Unable to stay dissolved in the bloodstream, nitrogen gas bubbles begin forming in the bloodstream causing the condition.

8.91 Water will move from solution A to solution B because solution B is a higher concentrated solution (more solute/solution). It does not matter that the albumin particles are huge; osmotic flow depends on the concentration, not the size of the particles. Since solution B has a higher concentration, it exerts a higher osmotic pressure.

The water in swimming pools where swimmers train is maintained at a constant temperature and pH, a measure of acidity. This swimmer's body regulates temperature and blood pH through biochemical reactions. Read Chapter 9 to understand more about pH and what happens if your blood pH deviates from normal.

9 Acids, Bases, and Buffers in the Body

WHEN YOU EAT a dill pickle, how does it taste? Sour, right? Citrus fruits like lemons and grapefruit also have a sour taste. Pickles, citrus fruits, and even some candies taste sour because they all contain acid. Our stomachs produce acid to help digest the food we eat, and our muscles produce lactic acid when we exercise. An acid can be neutralized by substances called bases. Soaps are mild bases, and, like all other bases, they feel slippery to the touch.

Maintaining a swimming pool requires a pool test kit that measures the concentration of dissolved ions and pH. The pH refers to the level of acidity of a solution. Life in general operates under very strict pH conditions. For example, amino acids, the building blocks of proteins, will change form if the acidity of their environment changes. Proteins change their shape and their ability to function if the pH of their surroundings changes. Our body fluids, including blood and urine, contain compounds called buffers that maintain pH. In this chapter, we discuss the bicarbonate buffer system in the blood and the physiological conditions of acidosis and alkalosis, but first we consider acids, bases, and equilibrium.

9.1 Acids and Bases—Definitions

9.1 Inquiry Question: How do acids and bases differ?

Acids

The first person to chemically define **acids** was the Swedish chemist Svante Arrhenius (pronounced *Ar-RAY-nee-us*). He described acids as substances that dissociate, producing hydrogen ions (H^+) when dissolved in water. The presence of hydrogen ions (H^+) gives acids their sour taste and allows acids to corrode some metals. In the early twentieth century, Johannes Brønsted and Thomas Lowry, working independently, expanded the definition of an acid to include the concept that an acid is a compound that *donates* a proton.

$$HCl(g) \xrightarrow{\text{H}_2\text{O}} H^+(aq) + Cl^-(aq)$$

Dissociates into ions
An acid is a substance that dissociates in water producing a hydrogen ion, H^+.

How is the hydrogen ion (H^+) described by Arrhenius related to the proton used in the Brønsted–Lowry definition of an acid? We know that a proton is a subatomic particle and that H^+ is an ion. Because most hydrogen atoms contain one proton and one electron, a hydrogen ion (H^+), which is a hydrogen atom that has lost its electron, and a proton are one and the same. In an aqueous solution, a free proton (H^+) rarely exists. The partial negative charge on the oxygen atom in water is strongly attracted to the positive charge of a proton. This attraction is so strong that the proton and the oxygen atom in water form a covalent bond, creating a **hydronium ion, H_3O^+.**

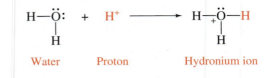

Bases

According to Arrhenius, **bases** are ionic compounds that, when dissolved in water, dissociate to form a metal ion and a hydroxide ion (OH^-). Most Arrhenius bases are formed from Group 1A and 2A metals, such as NaOH, KOH, LiOH, and $Ca(OH)_2$. These hydroxide bases are characterized by a bitter taste and a slippery feel. The Brønsted–Lowry definition of a base mirrors the acid definition in that a Brønsted–Lowry base is a compound that *accepts* a proton.

Acids and Bases Are Both Present in Aqueous Solution

The Brønsted–Lowry definition states that acids *donate* protons and bases *accept* protons, implying that a proton is usually transferred in an acidic or basic solution. Often water can act as an acid or a base by donating or accepting a proton. We can write the formation of a hydrochloric acid solution as a transfer of a proton from hydrogen chloride to water. By accepting a proton in the reaction, water is acting as a base here according to the Brønsted–Lowry theory.

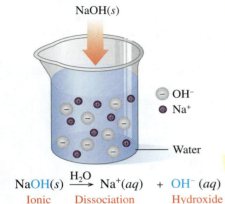

$$NaOH(s) \xrightarrow{H_2O} Na^+(aq) + OH^-(aq)$$

Ionic compound — Dissociation — Hydroxide ion

An Arrhenius base is an ionic compound that dissociates producing hydroxide, OH^-.

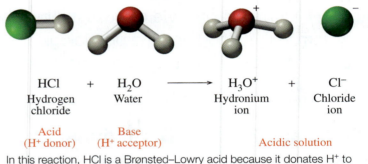

HCl	+	H_2O	$\longrightarrow$	H_3O^+	+	Cl^-
Hydrogen chloride		Water		Hydronium ion		Chloride ion
Acid (H^+ donor)		Base (H^+ acceptor)		Acidic solution		

In this reaction, HCl is a Brønsted–Lowry acid because it donates H^+ to water forming hydronium ion.

In another reaction, ammonia (NH_3) reacts with water. Because the nitrogen atom of NH_3 has a stronger attraction for a proton than the oxygen of water, water acts as an acid in this case by donating a proton.

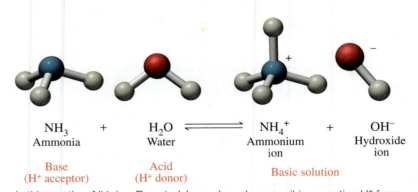

NH_3	+	H_2O	$\rightleftharpoons$	NH_4^+	+	OH^-
Ammonia		Water		Ammonium ion		Hydroxide ion
Base (H^+ acceptor)		Acid (H^+ donor)		Basic solution		

In this reaction, NH_3 is a Brønsted–Lowry base because it is accepting H^+ from water producing hydroxide.

Acids and Bases

In each of the equations, identify the reactant that is an acid (H^+ donor) and the reactant that is a base (H^+ acceptor):

a. $HBr(aq) + H_2O(l) \longrightarrow H_3O^+(aq) + Br^-(aq)$

b. $H_2O(l) + CN^-(aq) \rightleftharpoons HCN(aq) + OH^-(aq)$

Solution

a.   Examine what is happening to each of the reactants as it changes to product. When HBr forms Br^-, a proton is donated (H^+), so HBr is acting as the acid in this equation. When H_2O forms H_3O^+, it is accepting an H^+, which makes water a base in this equation.

b.   Examine what is happening to each of the reactants as it changes to product. When H_2O forms OH^-, a proton is donated (H^+) forming OH^-, so H_2O is acting as the acid in this equation. When CN^- forms HCN, it is accepting an H^+ (+ and − charges make HCN a neutral compound), making CN^- a base in this equation.

practice problems

9.1   Indicate if each of the following statements is characteristic of an acid or a base:
a. has a sour taste
b. accepts a proton
c. produces H^+ ions in water
d. is named potassium hydroxide

9.2   Indicate if each of the following statements is characteristic of an acid or a base:
a. neutralized by a base
b. produces OH^- in water
c. has a slippery feel
d. donates a proton

9.3   In your own words, explain how a proton and the hydrogen ion represent the same thing.

9.4   In your own words, explain how in aqueous solution H^+ and H_3O^+ represent similar things.

9.5   In each of the following equations, identify the reactant that is an acid (H^+ donor) and the reactant that is a base (H^+ acceptor):
a. $HI(aq) + H_2O(l) \longrightarrow H_3O^+(aq) + I^-(aq)$
b. $F^-(aq) + H_2O(l) \rightleftharpoons HF(aq) + OH^-(aq)$

9.6   In each of the following equations, identify the reactant that is an acid (H^+ donor) and the reactant that is a base (H^+ acceptor):
a. $CO_3^{2-}(aq) + H_2O(l) \longrightarrow HCO_3^-(aq) + OH^-(aq)$
b. $H_2SO_4(aq) + H_2O(l) \longrightarrow H_3O^+(aq) + HSO_4^-(aq)$

9.2 Strong Acids and Bases

In Section 9.1, we saw that acids and bases are classified by their ability to donate or accept protons, respectively. There are six common **strong acids** (see Table 9.1) that completely (~100%) dissociate, meaning they break up completely into ions when placed in water, forming hydronium ions and anions.

9.2 Inquiry Question:
How does an acid react with a base?

$$HCl(g) + H_2O(l) \rightarrow H_3O^+(aq) + Cl^-(aq)$$

Acids that only partially dissociate (~5%) are considered **weak acids** and will be examined more closely in Section 9.4. One such example is acetic acid (CH_3COOH), which is the main component of vinegar (see Figure 9.1).

TABLE 9.1 Six Common Strong Acids

Acid Name	Formula
Perchloric acid	$HClO_4$
Sulfuric acid*	H_2SO_4
Hydroiodic acid	HI
Hydrobromic acid	HBr
Hydrochloric acid	HCl
Nitric acid	HNO_3

*Only the first proton is 100% dissociated. The product, HSO_4^-, is a weak acid.

FIGURE 9.1 Strong versus weak acid.
A strong acid such as HCl is completely dissociated (100%), whereas a weak acid such as CH_3COOH contains mostly molecules and few ions in solution.

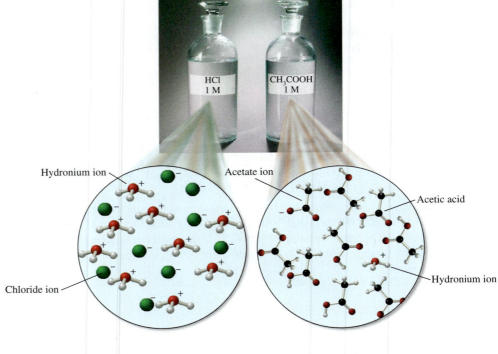

$$CH_3COOH(l) + H_2O(l) \rightleftharpoons CH_3COO^-(aq) + H_3O^+(aq)$$

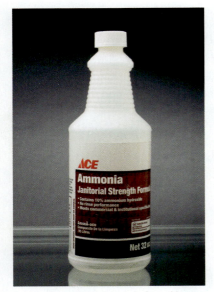
Ammonia, a weak base, is found in many cleaning supplies.

Strong bases, like NaOH (also known as lye), are used in household products such as oven cleaners or drain openers. The Arrhenius bases such as LiOH, KOH, NaOH, and $Ca(OH)_2$ are **strong bases** that dissociate completely (100%) in water. Because they are ionic compounds, they dissociate in water to give an aqueous solution of a metal ion and a hydroxide ion.

Bases that only partially dissociate (~5%) are considered **weak bases.** Many common weak bases, such as those found in household cleaners, contain ammonia (NH_3).

Neutralization

What happens when a strong acid and strong base are mixed? Consider the HCl and NaOH used as previous examples. Because both completely dissociate to form ions in water, the water contains sodium ions and chloride ions as well as hydronium ions and hydroxide ions. The protons in the hydronium ions are strongly attracted to the hydroxide ions and combine to form water molecules. This chemical reaction produces a lot of energy as heat and is therefore an exothermic reaction.

$$H_3O^+(aq) + OH^-(aq) \longrightarrow 2H_2O(l) + heat$$

The sodium ions and the chloride ions remain in solution. Suppose the water were removed by boiling. The ionic compound, sodium chloride (NaCl), would remain. The reaction of a strong acid and strong base therefore always produces water and an ionic compound called a **salt.** This reaction is called a **neutralization** reaction because the acid and base neutralize each other when they react to produce water. We can write the neutralization reaction for HCl and NaOH as follows:

$$HCl(aq) + NaOH(aq) \longrightarrow NaCl(aq) + H_2O(l)$$

Strong   Strong   Salt   Water
acid   base

Note that the chemical equation is balanced as written; that is, the number of atoms on the reactant side is equal to the number of atoms on the product side.

Completing a Neutralization Reaction

Complete and balance the following neutralization reaction.

$$HNO_3(aq) + KOH(s) \longrightarrow$$

Solution

STEP 1: Form the products. The products will always be (a) a salt and (b) H_2O. The salt produced must be a neutral ionic compound (refer to Section 3.3 for a refresher on forming ionic compounds like salts). The potassium ion has a 1+ charge (potassium is a Group 1A element) and the nitrate anion has a 1− charge (see Table 3.2), so the salt formed has a 1:1 ratio of cations to anions with a formula of KNO_3.

$$HNO_3(aq) + KOH(s) \longrightarrow KNO_3(aq) + H_2O(l)$$

Strong   Strong   Salt   Water
acid   base

STEP 2: Balance the chemical equation. Balancing is done by adding coefficients in front of the product or reactant compounds where appropriate. Remember that the same number of atoms must appear in both the reactants and products. If we inspect this reaction after forming the salt, we see that the same number of each atom is present (two H, four O, one N, and one K) so no further balancing is needed.

---

sample problem
9.2   Completing a Neutralization Reaction

Complete and balance the following neutralization reaction.

$$HBr(aq) + Ca(OH)_2(s) \longrightarrow$$

STEP 1: Form the products. The calcium ion has a 2+ charge (a Group 2A element), and the bromide anion has a 1− charge (a Group 7A element), so the salt formed combines in a 1:2 ratio of cations to anions with a formula of $CaBr_2$.

$$HBr(aq) + Ca(OH)_2(s) \longrightarrow CaBr_2(aq) + H_2O(l)$$

Strong   Strong   Salt   Water
acid   base

STEP 2: Balance the chemical equation. Two atoms of Br appear on the product side, and therefore, a coefficient of 2 is needed on the reactant side in front of the HBr. Adding the coefficient produces a total of 4 H atoms and 2 O atoms on the reactant side, so a coefficient of 2 is also need in front of the H_2O on the product side. Now the equation is balanced.

$$2HBr(aq) + Ca(OH)_2(s) \longrightarrow CaBr_2(aq) + 2H_2O(l)$$

Antacids

integrating chemistry

Antacids are substances that are used to neutralize excess stomach acid (HCl). Some antacids are mixtures of aluminum hydroxide and magnesium hydroxide. These hydroxides are not very soluble in water, so the levels of available OH^- are not damaging to the intestinal tract. However, aluminum hydroxide has the side effects of producing constipation and binding phosphate in the intestinal tract, which may cause weakness and loss of appetite. Magnesium hydroxide has a laxative effect. These side effects are less likely when a combination is used.

$$Mg(OH)_2(s) + 2HCl(aq) \longrightarrow MgCl_2(aq) + 2H_2O(l)$$

Some antacids use carbonates to neutralize excess stomach acid. When carbonates are used to neutralize acid, the reaction produces a salt, water, *and* carbon dioxide gas. When calcium carbonate is used, about 10% of the calcium is absorbed into the bloodstream, where it elevates the levels of serum calcium. Calcium carbonate is not recommended for people who have peptic ulcers or a tendency to form kidney stones.

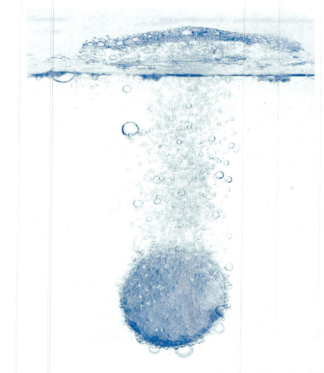

Sodium bicarbonate can affect the acidity level of the blood and elevate sodium levels in the body fluids. It is also not recommended in the treatment of peptic ulcers.

The neutralizing substances in some antacid preparations are shown in **Table 9.2**.

TABLE 9.2 Basic Compounds in Some Antacids

Antacid	Base(s)
Amphojel	$Al(OH)_3$
Milk of magnesia	$Mg(OH)_2$
Mylanta, Maalox, Di-Gel, Gelusil, Riopan	$Mg(OH)_2$, $Al(OH)_3$
Bisodol	$CaCO_3$, $Mg(OH)_2$
Titralac, Tums, Pepto-Bismol	$CaCO_3$
Alka-Seltzer	$NaHCO_3$, $KHCO_3$

practice problems

9.7 Which of the following are strong acids?
a. H_2SO_4 b. HCl
c. HF   d. HNO_3
e. H_3PO_4

9.8 Which of the following compounds completely ionize in water?
a. HBr   b. CH_3COOH
c. HCN   d. HI
e. NH_3

9.9 Complete and balance the following neutralization reactions:
a. $HNO_3(aq) + LiOH(s) \longrightarrow$
b. $H_2SO_4(aq) + Ca(OH)_2(s) \longrightarrow$

9.10 Complete and balance the following neutralization reactions:
a. $HNO_3(aq) + Mg(OH)_2(s) \longrightarrow$
b. $HCl(aq) + NaHCO_3(s) \longrightarrow$

9.11 Complete and balance the following neutralization reactions:
a. $HBr(aq) + Al(OH)_3(s) \longrightarrow$
b. $HI(aq) + CaCO_3(s) \longrightarrow$

9.12 Complete and balance the following neutralization reactions:
a. $H_2SO_4(aq) + NaOH(s) \longrightarrow$
b. $HBr(aq) + LiOH(s) \longrightarrow$

9.3 Chemical Equilibrium

Before we continue into weak acids and bases, let's consider the general principles of chemical equilibrium.

Imagine a concert with general admission seating. Everyone tries to arrive in plenty of time to get a seat. When the doors open, people rush in and fill up all the seats in the concert hall. When the hall is full, event security stops people from entering the hall even though some people who waited in line did not get a seat. During the concert, the security personnel allow people to leave and others to enter to keep the hall full but not overfilled. This steady flow of people in and out of the concert results in a state of dynamic equilibrium in which the rate of people entering the hall and the rate of people leaving is the same. Note that the number of people in the concert hall stays the same once the hall fills, even though the individuals in the concert hall differ.

? **9.3 Inquiry Question:** How do the properties of equilibrium apply to Le Châtelier's principle?

People entering the concert   Security manages flow   People leaving the concert

When the flow of people entering (like reactants forming products) and leaving (like products forming reactants) a concert is the same, a state of equilibrium exists.

Like a concert, some chemical reactions will, after forming product (people entering the concert), reverse and re-form reactants (people leaving the concert). These are reversible reactions. To explore these reversible reactions more, let's start with a chemical reaction for the generation of ammonia,

$$N_2(g) + 3H_2(g) \rightleftharpoons 2NH_3(g)$$

Reactants   Equilibrium arrow   Products

The generation of ammonia is a reversible reaction. Once ammonia is formed, the reaction will reverse, re-forming nitrogen and hydrogen. Eventually, the rate of the formation of ammonia (the forward reaction) and the rate of re-formation of nitrogen and hydrogen gases (the reverse reaction) become equal. This balance of the rates of the reactions is called **chemical equilibrium.** A special type of reaction arrow, called an equilibrium arrow (shown in the equation), is used in this chemical equation to indicate that both the forward and reverse reactions can take place simultaneously. Equilibrium was introduced previously in Section 8.2. This does not mean that the *amounts* of ammonia and of nitrogen and hydrogen are equal, but because the rates of the forward and reverse reactions are equal, there is no net change in amounts. In other words, the amounts of products and reactants stay the same.

The Equilibrium Constant *K*

Going back to the concert, suppose the concert hall remains completely full for the duration of the concert and the number of people waiting to get in remains the same (although the actual individuals waiting may have changed). The concert attendance has reached equilibrium, and we could observe that the ratio of people inside the concert versus people outside the concert would be constant.

Similarly, the ammonia reaction shown reaches equilibrium. If we measure the concentrations of ammonia, nitrogen, and hydrogen present, the ratio of products to reactants would be a constant. This value is called the **equilibrium constant, K,** and it is a characteristic of equilibrium reactions at a given temperature. *K* is defined as

$$K = \frac{[\text{Products}]}{[\text{Reactants}]}$$

The brackets, [ ], mean "molar concentration of." So, the equation states that the equilibrium constant, *K*, is equal to the molar concentration of the products divided by that of the reactants. If there is more than one reactant or product, the concentrations are multiplied together. For the generation of ammonia, the expression for *K* is shown. The superscripts in the expression come from the coefficients (number of moles of each) found in the balanced chemical equation.

$$K = \frac{[\text{NH}_3]^2}{[\text{N}_2][\text{H}_2]^3}$$

In general, for an equilibrium reaction of the form

$$aA + bB \rightleftharpoons cC + dD$$

the general equilibrium expression is given as

$$K = \frac{[\text{C}]^c[\text{D}]^d}{[\text{A}]^a[\text{B}]^b}$$

Only substances with concentrations that change appear in an equilibrium expression. Solids (*s*) and pure liquids (*l*) have constant concentrations and do not appear in the equilibrium expression.

Inside
—————————— = Constant

Outside

The ratio of people leaving (products) versus entering (reactants) is constant for a particular concert (reaction).

---

sample problem 9.3   **Equilibrium Constant Expressions**

Write an equilibrium constant expression for the following reactions:

a. $CH_4(g) + H_2O(g) \rightleftharpoons CO(g) + 3H_2(g)$

b. $CaCO_3(s) \rightleftharpoons CO_2(g) + CaO(s)$

Solution

K is defined as the ratio of product concentrations to reactant concentrations. Any coefficients in front of molecules appear as superscripts in the expression.

a. $K_{eq} = \dfrac{[\text{CO}][\text{H}_2]^3}{[\text{CH}_4][\text{H}_2\text{O}]}$

H_2O appears in this expression because it appears as a gas in the equilibrium equation.

b. $K_{eq} = [\text{CO}_2]$

Because $CaCO_3$ and CaO are solids, they have constant concentrations and do not appear in the equilibrium expression.

The value of K for the formation of ammonia is 9.6 at 300 °C. A value of 9.6 indicates a ratio of 9.6:1 for products:reactants. Values for K vary greatly depending on temperature and reaction. If K has a value equal to 1, the ratio of products:reactants is 1:1 or the [products] = [reactants]. A value of K greater than (>) 1 indicates that the amount of products (numerator) is larger than the amount of reactants (denominator) or [products] > [reactants]. A value of K less than (<) 1 indicates that the amount of reactants (denominator) is larger than the amount of products (numerator) or [products] < [reactants] (see Table 9.3).

TABLE 9.3 Interpreting Values of K

Value of K	Predominant Species at Equilibrium
$K = 1$	Equal amounts of products and reactants
$K > 1$	Products
$K < 1$	Reactants

sample problem 9.4 — Interpreting Values of K

K for the following reaction at 25 °C is 1×10^2.

$$CO(g) + H_2O(g) \rightleftharpoons H_2(g) + CO_2(g)$$

Based on the value for K, at equilibrium are the products or reactants greater?

Solution

Because the value of $K > 1$, the products are present in greater amounts.

Effect of Concentration on Equilibrium— Le Châtelier's Principle

Suppose that we had our reaction for the generation of ammonia sitting in a reaction vessel and that it was at equilibrium. What would happen if we decided to inject more nitrogen into the vessel?

$$N_2(g) + 3H_2(g) \rightleftharpoons 2NH_3(g)$$

According to **Le Châtelier's principle** (pronounced *leh-shat-lee-AYS*), if we disturb an equilibrium—chemists refer to this as applying stress to the equilibrium—the rate of the forward or reverse reaction will change to offset the stress and regain equilibrium. Applying this principle, if we add more N_2 to our system (it appears on the reactant side of our equation), the rate of the forward reaction will increase, shifting the equilibrium to produce more products. This rate increase occurs because adding more N_2 molecules increases the chance that N_2 will collide with H_2 molecules, forming NH_3 more rapidly.

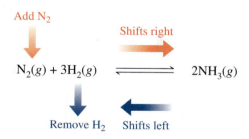

As a contrast, suppose that we remove one of the reactants, H_2. To regain equilibrium, the reverse reaction must be faster than the forward reaction, allowing the H_2 to be replenished. The equilibrium shifts to the left, forming more of the reactants and reestablishing the equilibrium between the two rates.

In general, we can think of equilibrium as a balancing act between forward and reverse reactions. If one side of the reaction gains a substance, the reaction shifts to the other side to regain its equilibrium. If one side of the reaction loses a substance, the reaction will shift toward that side to regain its equilibrium.

Effect of Temperature on Equilibrium

What would happen to the equilibrium of our ammonia reaction if we change the temperature of the reaction? First, we have to know whether the reaction itself is one that produces heat, an exothermic reaction, or one that absorbs heat from its surroundings, an endothermic reaction. The generation of ammonia is known to be an exothermic reaction. The chemical equation can be written as follows to signify that heat is produced:

$$N_2(g) + 3H_2(g) \rightleftharpoons 2NH_3(g) + \text{heat}$$

Because heat is a product of the reaction, if the temperature of the reaction is raised (heat added), the rate of the reverse reaction increases to offset the stress of adding heat. This causes the equilibrium to shift to the left. If the reaction were cooled down (heat removed), the rate of the forward reaction would increase to replenish the heat produced, shifting the equilibrium to the right.

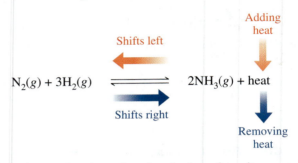

In an endothermic reaction (one that absorbs heat from its surroundings), like the reaction for production of NO gas shown, heat appears as a reactant, so the opposite shifts occur. **Table 9.4** summarizes the effects of stress on a chemical equilibrium.

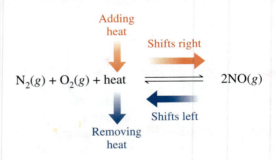

TABLE 9.4 Effects of Changes on Equilibrium

Factor	Change (stress)	Reaction Shifts toward
Concentration	Add reactant	Right
	Remove reactant	Left
	Add product	Left
	Remove product	Right
Temperature	Raise temperature of endothermic reaction	Right
	Lower temperature of endothermic reaction	Left
	Raise temperature of exothermic reaction	Left
	Lower temperature of exothermic reaction	Right

sample problem 9.5   Factors Affecting Equilibrium

How does each of the following actions affect the equilibrium of the reaction shown?

$$2SO_2(g) + O_2(g) \rightleftharpoons 2SO_3(g) + heat$$

a. add O_2

b. lower the temperature

c. add SO_3

Solution

a. The addition of O_2 increases the forward reaction; the equilibrium shifts to the right.

b. Lowering the temperature removes heat, which favors the exothermic direction, so the equilibrium shifts to the right.

c. The addition of SO_3 favors the reverse reaction toward reactants; the equilibrium shifts to the left.

practice problems

9.13 What is meant by the term *reversible reaction*?

9.14 When does a reversible reaction reach equilibrium?

9.15 Write an equilibrium constant expression for the following reactions:

a. $CO(g) + H_2O(g) \rightleftharpoons H_2(g) + CO_2(g)$

b. $CH_3COOH(aq) + H_2O(l) \rightleftharpoons H_3O^+(aq) + CH_3COO^-(aq)$

9.16 Write an equilibrium constant expression for the following reactions:

a. $2N_2(g) + 3Br_2(g) \rightleftharpoons 2NBr_3(g)$

b. $C(s) + O_2(g) \rightleftharpoons CO_2(g)$

9.17 For the following values of K, indicate whether the products or reactants are present in larger amounts:

a. 1×10^{-5} b. 156   c. 1

9.18 For the following values of K, indicate whether the products or reactants are present in larger amounts:

a. 1×10^7

b. 0.0045

c. 0.00000079

9.19 Hydrogen chloride can be made by reacting hydrogen gas and chlorine gas.

$$H_2(g) + Cl_2(g) \rightleftharpoons 2HCl(g) + heat$$

What effect does each of the following changes have on the equilibrium?

a. add H_2

b. add heat (raise temperature)

c. remove HCl

d. remove Cl_2

9.20 Sulfur trioxide is produced by reacting sulfur dioxide with oxygen.

$$2SO_2(g) + O_2(g) \rightleftharpoons 2SO_3(g) + heat$$

What effect does each of the following changes have on the equilibrium?

a. add O_2

b. lower temperature

c. remove SO_2

d. remove O_2

9.21 In the lower atmosphere, oxygen is converted to ozone (O_3) by the energy provided from lightning.

$$3O_2(g) + heat \rightleftharpoons 2O_3(g)$$

What effect does each of the following changes have on the equilibrium?

a. add O_3

b. add O_2

c. raise temperature

d. lower temperature

9.22 When heated, carbon reacts with water to produce carbon monoxide and hydrogen.

$$C(s) + H_2O(g) + heat \rightleftharpoons CO(g) + H_2(g)$$

What effect does each of the following changes have on the equilibrium?

a. add heat

b. lower temperature

c. remove CO

d. add H_2O

9.23 After you open a bottle of soda, the drink eventually goes flat. How can Le Châtelier's principle explain this using the following reaction?

$$H^+(aq) + HCO_3^-(aq) \rightleftharpoons CO_2(g) + H_2O(l)$$

9.24 When you exercise, energy is produced by increasing the rate of the following reaction involving glucose. Why do you breathe faster when you exercise?

$$C_6H_{12}O_6(aq) + 6O_2(g) \rightleftharpoons 6CO_2(g) + 6H_2O(l) + Energy$$

Discovering the Concepts

Inquiry Activity—Weak Acids

Information

Weak acids and bases establish an equilibrium in solution. In contrast to strong acids, only a few molecules are dissociated (<5%). Some weak acids can donate more than one proton. In this case, two separate equilibria are established. These acids are called *polyprotic* (literally "many proton") acids. The Brønsted–Lowry definition is useful to describe the behavior of weak acids and bases.

 Acid—Any substance that can donate a proton (H^+) to another substance.
 When an acid donates a proton, a conjugate base is produced.
 Base—Any substance that can accept a proton (H^+) from another substance.
 When a base accepts a proton, a conjugate acid is produced.

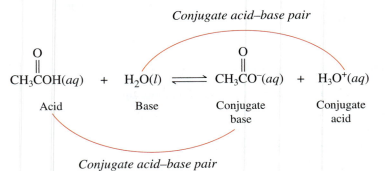

Conjugate acid–base pair

$$CH_3COH(aq) \; + \; H_2O(l) \; \rightleftharpoons \; CH_3CO^-(aq) \; + \; H_3O^+(aq)$$

Acid   Base   Conjugate   Conjugate
           base   acid

Conjugate acid–base pair

Questions

1. Does the acid in the example donate or accept a proton?
2. Does the conjugate base in the example donate or accept a proton?
3. Label the acid, base, conjugate acid, and conjugate base in the reaction of ammonia with water shown.

$$NH_3(aq) \; + \; H_2O(l) \; \rightleftharpoons \; NH_4^+(aq) \; + \; OH^-(aq)$$

4. Provide a definition for *conjugate acid–base pair*.
5. Provide the acid–base reaction for lactic acid ($CH_3CH(OH)COOH$), a weak acid, when it reacts with water. Label the acid, base, conjugate acid, and conjugate base.

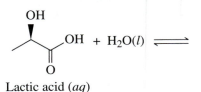

Lactic acid (*aq*)

6. Provide the
  a. conjugate base of H_2S. _____
  b. conjugate acid of OH^-. _____
  c. conjugate acid of HPO_4^{2-}. _____
  d. conjugate base of HCO_3^-. _____

9.4 Weak Acids and Bases

All the principles of equilibrium from Section 9.3 apply to weak acids and bases because weak acids and bases only partially dissociate into ions, establishing an equilibrium in aqueous solution. For example, the dissociation of the weak acid acetic acid (CH_3COOH) into acetate anions (CH_3COO^-) and hydronium ions is as follows:

$$CH_3\overset{\overset{O}{\|}}{C}OH(aq) + H_2O(l) \rightleftharpoons CH_3\overset{\overset{O}{\|}}{C}O^-(aq) + H_3O^+(aq)$$

The equilibrium constant expression representing this reaction would be

$$K = \frac{[CH_3COO^-][H_3O^+]}{[CH_3COOH]}$$

Remember that pure liquids like $H_2O(l)$ are present in large amounts that do not change significantly as a reaction approaches equilibrium. Therefore, they are considered constant and not included in the equilibrium expression.

The Equilibrium Constant K_a

All weak acids dissociate the same way in water by donating a proton to form hydronium ion. However, because they dissociate much less than 100%, each weak acid has an equilibrium constant called an **acid dissociation constant, K_a**. The K_a value for acetic acid is 1.75×10^{-5}. Notice that this number is less than 1, which indicates that more acetic acid molecules are present at equilibrium than acetate anions. Because all weak acids can attain equilibrium in solution, they all have a K_a value at a given temperature. Each weak acid dissociates to a different extent so the K_a values are different from acid to acid.

The strength of a weak acid can be determined from the K_a value. The larger the K_a value, the stronger the acid (the more protons dissociated). You may recall the two weak acids mentioned in Section 8.3 that were weak electrolytes: carboxylic acids and protonated amines. Their dissociation reactions with water are shown in **Figure 9.2**.

9.4 Inquiry Question: How are weak acid–base equations written?

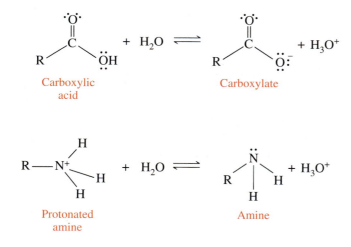

FIGURE 9.2 Functional groups as acids and bases. Two common organic functional groups act as weak acids in aqueous solution: carboxylic acids and protonated amines.

TABLE 9.5 K_a Values for Substances Acting as Weak Acids (25 °C)

Name	Formula	K_a
Hydrogen sulfate ion	HSO_4^-	1.0×10^{-2}
Phosphoric acid	H_3PO_4	7.5×10^{-3}
Hydrofluoric acid	HF	6.5×10^{-4}
Nitrous acid	HNO_2	4.5×10^{-4}
Formic acid	$HCOOH$	1.8×10^{-4}
Acetic acid	CH_3COOH	1.75×10^{-5}
Carbonic acid	H_2CO_3	4.5×10^{-7}
Water	H_2O	1.0×10^{-7}
Dihydrogen phosphate ion	$H_2PO_4^-$	6.6×10^{-8}
Ammonium ion	NH_4^+	6.3×10^{-10}
Hydrocyanic acid	HCN	6.2×10^{-10}
Bicarbonate ion	HCO_3^-	4.8×10^{-11}
Hydrogen phosphate ion	HPO_4^{2-}	1.0×10^{-12}

Increasing
acid
strength

sample problem 9.6   Strength of Weak Acids

Of the following acids—H_2CO_3, HF, and HCOOH—which is (a) the strongest? (b) the weakest?

Solution

Use the K_a values in **Table 9.5**. The larger values (numbers closer to 1) are stronger acids.

a.   The strongest acid of the group is HF, with a K_a value of 6.5×10^{-4}.

b.   The weakest acid of the group is H_2CO_3, with a K_a value of 4.5×10^{-7}.

Conjugate Acids and Bases

According to the Brønsted–Lowry theory, the reaction between an acid and base involves a proton transfer, so if a weak acid is mixed with water, water will act as a base. Consider the dissociation of acetic acid, CH_3COOH:

Conjugate acid–base pairs differ by one H^+.

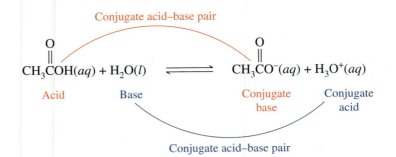

The acid CH_3COOH donates a proton to a molecule of water that accepts the proton, forming a hydronium ion, H_3O^+. What remains of the acid after the donation, CH_3COO^-—a carboxylate called acetate anion—is the **conjugate base** of CH_3COOH.

The term *conjugate base* comes from the fact that in the reverse reaction, the CH_3COO^- acts as a base and accepts the proton from the hydronium ion to form CH_3COOH. Likewise, when the water acts as a base, accepting a proton from acetic acid, a hydronium ion is formed. Hydronium ion is called the **conjugate acid** in this reaction because, during the reverse reaction, hydronium ion acts as an acid donating its proton to the acetate anion. Molecules or ions related by the loss or gain of one H^+ are referred to as **conjugate acid–base pairs.** The functional groups carboxylic acid and carboxylate are conjugate acid–base pairs. Weak acids are generically designated as HA and their conjugate base as A^-.

A weak base such as an amine will accept a proton to form a protonated amine, in which case the water acts as an acid. The protonated amine and amine are conjugate acid–base pairs of the same functional group.

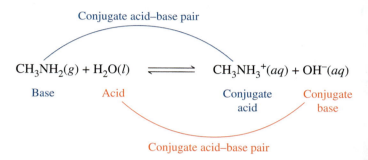

When two weak acids are mixed, the stronger weak acid will provide a H^+ to the stronger base making a weaker conjugate acid. The conjugate base formed is the weaker of the two bases. In other words, the stronger acid and base will form a weaker base and acid when two weak acids are mixed.

sample problem 9.7   Identifying Conjugate Acid–Base Pairs

Label the acid, base, conjugate acid, and conjugate base in the following reaction:

$$HF(aq) + H_2O(l) \rightleftharpoons F^-(aq) + H_3O^+(aq)$$

Solution

Compare HF on the reactant side to F^- on the product side. A proton is donated from the HF, leaving F^- as a product. So HF is acting as an acid. The product (F^-) is the conjugate base. Similarly, H_2O is accepting a proton to form the hydronium ion, so water is acting as a base, producing a conjugate acid, the hydronium ion.

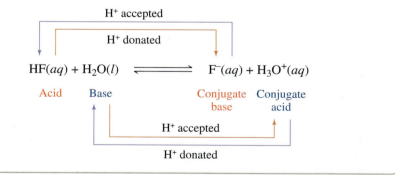

sample problem 9.8   Determining Formulas of Conjugates

Provide the following:

a.   conjugate acid of OH^-

b.   conjugate acid of HPO_4^{2-}

c.   conjugate base of H_2S

d.   conjugate base of HCO_3^-

Solution

Keep in mind the definitions of Brønsted–Lowry acids and bases when trying to answer this type of problem.

a. To find the conjugate acid of OH^-, OH^- would be acting as a base by accepting a proton. The addition of H^+ to OH^- produces the neutral molecule H_2O as its conjugate acid because the 1+ and 1− charges cancel out.

b. If HPO_4^{2-} acts as a base, it would be accepting a proton (H^+), thereby forming $H_2PO_4^-$ as its conjugate acid. (Note that when H^+ and HPO_4^{2-} combine, the charge changes from 2− to 1−.)

c. To find the conjugate base of H_2S, H_2S would be acting as an acid by donating a proton, H^+. When a proton is donated, what is left over from the H_2S is HS^-. The 1− charge is left behind when H^+ is donated.

d. If HCO_3^- acts as an acid, it would be donating a proton, forming CO_3^{2-} as its conjugate base. The donation of H^+ leaves behind a negative charge, making the total charge on conjugate base 2−.

Solving a Problem

Writing Weak Acid–Base Equations

Complete the following reaction if $H_2PO_4^-$ acts as a base. Label the acid, conjugate acid, and conjugate base for the following reaction:

$$H_2PO_4^-\,(aq) + H_2O(l) \rightleftharpoons$$

Base

Solution

Keep in mind the following rules to correctly write and balance a weak acid–base equilibrium equation.

Rules for writing products of a weak acid–base equation.

• The conjugates will always appear on the product side of a chemical equilibrium.
• When a proton is donated, the conjugate base formed will have one more negative charge than the acid from which it was formed.
• When a proton is accepted, the conjugate acid formed will have one more positive charge than the base from which it was formed.
• The total charge of the reactants equals the total charge of the products.

STEP 1: Write the products. In this problem, $H_2PO_4^-$ acts as a base, so its conjugate acid will be H_3PO_4. The water acts as an acid, donating a proton to form the conjugate base OH^-. Using the rules outlined, the final equation is written

$$H_2PO_4^-\,(aq) \;+\; H_2O(l) \;\rightleftharpoons\; H_3PO_4\,(aq) \;+\; OH^-\,(aq)$$

Base   Acid   Conjugate acid   Conjugate base

STEP 2: Check your work. If you have assigned charges correctly, the total charge on either side of the equation should be the same. Here, the reactants have a total charge of 1−, and the products have a total charge of 1−.

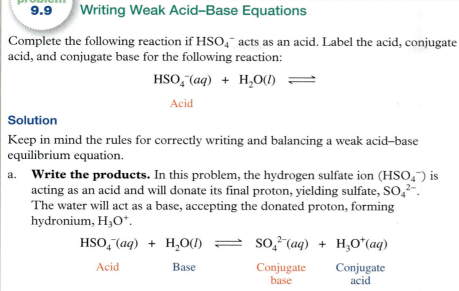

sample problem 9.9   **Writing Weak Acid–Base Equations**

Complete the following reaction if HSO_4^- acts as an acid. Label the acid, conjugate acid, and conjugate base for the following reaction:

$$HSO_4^-(aq) + H_2O(l) \rightleftharpoons$$

Acid

Solution

Keep in mind the rules for correctly writing and balancing a weak acid–base equilibrium equation.

a. **Write the products.** In this problem, the hydrogen sulfate ion (HSO_4^-) is acting as an acid and will donate its final proton, yielding sulfate, SO_4^{2-}. The water will act as a base, accepting the donated proton, forming hydronium, H_3O^+.

$$HSO_4^-(aq) + H_2O(l) \rightleftharpoons SO_4^{2-}(aq) + H_3O^+(aq)$$

Acid   Base   Conjugate   Conjugate
                                       base   acid

b. **Check your work.** The total charge on either side of the equation should be the same. In this case, both the reactants and products have a charge of $1-$.

Weak Acids, Oxygen Transport, and Le Châtelier's Principle

integrating chemistry

The binding and release of oxygen in the body is actually controlled by weak acid–base equilibria. The protein hemoglobin (abbreviated Hb) carries oxygen from the lungs to the tissues via the bloodstream. Both protons (H^+) and oxygen (O_2) bind to hemoglobin, but with opposite affinity. That is, one or the other is transported. This reaction can be represented by the equilibrium equation

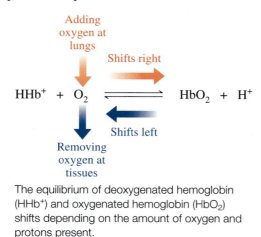

Adding
oxygen at
lungs

Shifts right

$$HHb^+ + O_2 \rightleftharpoons HbO_2 + H^+$$

Shifts left

Removing
oxygen at
tissues

The equilibrium of deoxygenated hemoglobin (HHb^+) and oxygenated hemoglobin (HbO_2) shifts depending on the amount of oxygen and protons present.

If the concentrations of either acid or oxygen change, the equilibrium will shift. For example, at the lungs, the oxygen concentration is higher, so the equilibrium will shift to the right of the equation shown, binding more oxygen and releasing H^+. As the oxygenated hemoglobin (HbO_2) reaches the tissues where oxygen concentrations are low, oxygen is released, shifting the equilibrium to the left and forming more of the deoxygenated hemoglobin (HHb^+) for transport back to the lungs where the process repeats.

9.25 Using Tables 9.1 and 9.5, identify the stronger acid in each pair:
a. $H_2PO_4^-$ or HPO_4^{2-}
b. HF or HCOOH
c. HBr or HNO_2

9.26 Using Tables 9.1 and 9.5, identify the stronger acid in each pair:
a. HCN or H_2CO_3
b. H_3PO_4 or $H_2PO_4^-$
c. HCl or HSO_4^-

9.27 Identify the acid and base on the reactant side of the following equations and identify their conjugate species on the product side:
a. $HSO_4^-(aq) + H_2O(l) \rightleftharpoons H_3O^+(aq) + SO_4^{2-}(aq)$
b. $NH_4^+(aq) + H_2O(l) \rightleftharpoons H_3O^+(aq) + NH_3(aq)$
c. $HCN(aq) + NO_2^-(aq) \rightleftharpoons CN^-(aq) + HNO_2(aq)$

9.28 Identify the acid and base on the reactant side of the following equations and identify their conjugate species on the product side:
a. $H_3PO_4(aq) + H_2O(l) \rightleftharpoons H_3O^+(aq) + H_2PO_4^-(aq)$
b. $CO_3^{2-}(aq) + H_2O(l) \rightleftharpoons OH^-(aq) + HCO_3^-(aq)$
c. $H_3PO_4(aq) + NH_3(aq) \rightleftharpoons NH_4^+(aq) + H_2PO_4^-(aq)$

9.29 Write the formula and name of the conjugate base formed from each of the following acids:
a. HF   b. H_2O c. H_2CO_3 d. HSO_4^-

9.30 Write the formula and name of the conjugate base formed from each of the following acids:
a. HCO_3^- b. H_3O^+ c. HPO_4^{2-} d. HNO_2

9.31 Write the formula and name of the conjugate acid formed from each of the following bases:
a. CO_3^{2-} b. H_2O c. $H_2PO_4^-$ d. Br^-

9.32 Write the formula and name of the conjugate acid formed from each of the following bases:
a. SO_4^{2-}
b. CN^-
c. OH^-
d. ClO_2^-, chlorite ion

9.33 Complete the following reactions and label the conjugate acid–base pairs.
a. $\underset{\text{Acid}}{HA(aq)} + \underset{\text{Base}}{H_2O(l)} \rightleftharpoons$

b. $\underset{\text{Acid}}{H_2PO_4^-(aq)} + \underset{\text{Base}}{H_2O(l)} \rightleftharpoons$

c. $\underset{\text{Base}}{NH_3(aq)} + \underset{\text{Acid}}{H_2O(l)} \rightleftharpoons$

9.34 Complete the following reactions and label the conjugate acid–base pairs.
a. $\underset{\text{Base}}{A^-(aq)} + \underset{\text{Acid}}{H_2O(l)} \rightleftharpoons$

b. $\underset{\text{Base}}{HCO_3^-(aq)} + \underset{\text{Acid}}{H_2O(l)} \rightleftharpoons$

c. $\underset{\text{Acid}}{HCO_3^-(aq)} + \underset{\text{Base}}{H_2O(l)} \rightleftharpoons$

9.5 pH and the pH Scale

The Autoionization of Water, K_w

? 9.5 Inquiry Question: How is the acidity of a solution determined?

We have seen that water can act as a weak acid or base depending upon whether a base or acid is present in the solution. In pure water, the molecules spontaneously react with each other, donating and accepting protons.

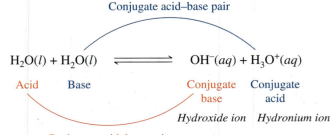

Water can act as an acid or base and spontaneously dissociates to a small extent.

This reaction, which is always present in water, is called the **autoionization of water.** If we were to write the equilibrium constant expression for water, K_w, (keeping in mind that pure liquid water will not appear), we would write

$$K_w = [OH^-][H_3O^+]$$

Wait, didn't we say in Chapter 8 that pure water does not conduct electricity because there are no dissolved ions present? Now we are saying there *are* ions found in pure water? Let's consider the measured value for K_w:

$$K_w = 1 \times 10^{-14} \text{ at } 25\ °C$$

From the small value of K_w and remembering the definition of K ([products]/[reactants]), we can see that the ionized products are found in small amounts in pure water. In fact, K_w is 100 times smaller than the weakest acid found in Table 9.5. Pure water has some H_3O^+ and OH^- ions present, but the amounts are so small that there are not enough ions present in pure water to conduct electricity.

Because the autoionization of water always occurs in aqueous solution, all aqueous solutions have small amounts of H_3O^+ and OH^- present.

$[H_3O^+]$, $[OH^-]$, and pH

In pure water, both the hydroxide and hydronium ion are being formed equally from the transfer of protons between water molecules, so in pure water $[H_3O^+] = [OH^-]$. At 25 °C both these values are 1×10^{-7} M. (Remember that "M" is the concentration unit molarity, or moles per liter.) When these concentrations are equal, the solution is neutral. Keeping in mind K_w is constant, if an acid is added to water, there is an increase in $[H_3O^+]$ and a decrease in $[OH^-]$, which makes the solution acidic. If base is added, $[OH^-]$ increases and $[H_3O^+]$ decreases, making a basic solution (see **Figure 9.3**). Most aqueous solutions are *not* neutral and have unequal concentrations of H_3O^+ and OH^-.

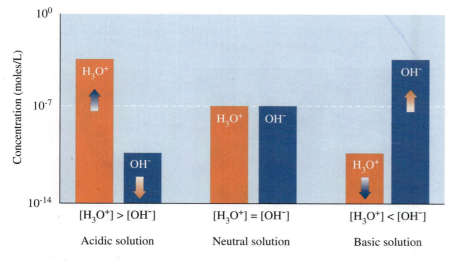
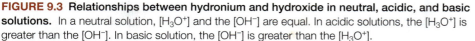

FIGURE 9.3 Relationships between hydronium and hydroxide in neutral, acidic, and basic solutions. In a neutral solution, $[H_3O^+]$ and the $[OH^-]$ are equal. In acidic solutions, the $[H_3O^+]$ is greater than the $[OH^-]$. In basic solution, the $[OH^-]$ is greater than the $[H_3O^+]$.

The amount of H_3O^+ in an aqueous solution defines the acidity of the solution. Concentrations of H_3O^+ usually range from about 1 M to 1×10^{-14} M (really close to zero). This range of numbers is extremely wide. For this reason, the mathematical function "log" is used to compare $[H_3O^+]$ because it gives a set of numbers that usually falls between 0 and 14. This set of numbers describes the **pH** scale (see **Figure 9.4**). This scale was developed for the simple comparison of $[H_3O^+]$ values. The letter "p" literally means the mathematical function, negative log (-log), so a pH can be determined from the $[H_3O^+]$ as

$$pH = -\log[H_3O^+]$$

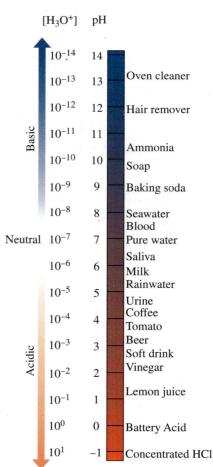

FIGURE 9.4 The relationship between $[H_3O^+]$ and pH of some common substances. Notice that the higher the pH, the smaller the $[H_3O^+]$ and vice versa.

TABLE 9.6 Relationship between pH and $[H_3O^+]$

Solution Acidity	pH	$[H_3O^+]$
Basic	> 7	$< 1.0 \times 10^{-7}$ M
Neutral	$= 7$	$= 1.0 \times 10^{-7}$ M
Acidic	< 7	$> 1.0 \times 10^{-7}$ M

Table 9.6 shows the relationship between pH and $[H_3O^+]$.

sample problem 9.10 **The Relationship among Acidity, pH, and $[H_3O^+]$**

State whether solutions with the following conditions would be considered acidic, basic, or neutral.

a. pH = 5.0
b. $[H_3O^+] = 2.3 \times 10^{-9}$ M
c. pH = 12.0
d. $[H_3O^+] = 4.3 \times 10^{-4}$ M
e. $[H_3O^+] = 1 \times 10^{-7}$ M

Solution

Refer to Table 9.6 or Figure 9.4.

a. acidic
b. basic
c. basic
d. acidic
e. neutral

Measuring pH

Living things prefer environments that are kept at a constant pH. For example, normal blood pH is strictly regulated between 7.35–7.45. The pH of a solution is commonly measured either electronically by using an instrument called a pH meter or by using pH indicator paper. The pH paper is embedded with indicators that change color depending on the pH of the test solution (see **Figure 9.5**).

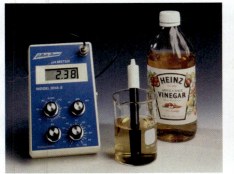

a.   **b.**

FIGURE 9.5 pH measurement.
The pH of a solution can be determined using (a) a pH meter or (b) pH indicator paper.

Calculating pH

Suppose we have a 0.050 M HCl solution, and we want to calculate its pH. Because strong acids fully ionize in solution, $[HCl] = [H_3O^+]$. This means that we can calculate the pH directly for a strong acid as

$$pH = -\log[H_3O^+]$$
$$pH = -\log(0.050)$$
$$pH = -(-1.30) = 1.30$$

Most scientific calculators have a $\boxed{\log}$ button on them. Be sure that you can locate this button on your calculator and can perform the example calculation correctly. (There is another logarithm function used in mathematics called the natural log, or *ln*, which is a completely different function and will not be discussed in this chapter.)

Let's take a look at the significant figures in the preceding calculation. In pH calculations with logarithms, the number of significant figures in the $[H_3O^+]$ will be the number of decimal places in the pH value.

$$[H_3O^+] = 0.050 \qquad\qquad pH = 1.30$$

Two significant figures   *Two decimal places*

---

sample problem 9.11   **Calculating pH from [H₃O⁺]**

Determine the pH for the following solutions:

a. $[H_3O^+] = 1.0 \times 10^{-5}\,M$
b. $[H_3O^+] = 5 \times 10^{-8}\,M$

Solution

To do these problems, you will be using the log function, $\boxed{\text{log}}$, on your calculator. Depending on the type of scientific calculator that you have, you may be inputting the numbers differently. Check with your instructor if you are having difficulty producing the correct answers using your calculator.

a. $pH = -\log[H_3O^+] = -\log(1.0 \times 10^{-5}) = -(-5.00) = 5.00$
b. $pH = -\log[H_3O^+] = -\log(5 \times 10^{-8}) = -(-7.3) = 7.3$

Calculating [H₃O⁺]

Because we can easily measure the pH of most solutions, we are often more interested in finding the $[H_3O^+]$ from a measured pH value. Suppose we measured the pH of a solution to be 3.00 and we want to find the corresponding $[H_3O^+]$:

$$pH = -\log[H_3O^+]$$

To solve the pH equation for $[H_3O^+]$, we will have to multiply both sides of the equation by negative 1 and find the inverse log function, INV log. In case you don't have an $\boxed{\text{INV}}$ button on your calculator, keep in mind that the inverse log function is 10^x so to solve the equation for $[H_3O^+]$,

$$\boxed{\text{INV}}\ \log\,(-pH) = [H_3O^+]$$

or alternatively,

$$10^{-pH} = [H_3O^+]$$

From this equation we can solve for the $[H_3O^+]$ of a pH 3.00 solution as

$$\text{INV}\ \log\,(-3.00)\ \text{or}\ 10^{-3.00} = [H_3O^+]$$
$$1.0 \times 10^{-3}\,M = [H_3O^+]$$

Many programmable calculators have an $\boxed{\text{INV}}$ button that can be used to solve the preceding equation. Most other scientific calculators have a $\boxed{10^x}$ button typically found as a second function above the log button. The $\boxed{y^x}$ button can also be used if 10 is entered as y. Be sure that you can perform the example calculation correctly before continuing.

Regarding significant figures, the number of decimal places given in the pH measurement tells us the number of significant figures in the $[H_3O^+]$:

$$pH = 3.00 \qquad\qquad [H_3O^+] = 1.0 \times 10^{-3}$$

Two decimal places   *Two significant figures*

sample problem 9.12 **Calculating [H_3O^+] from a Measured pH**

Determine the [H_3O^+] for solutions having the following measured pH values:

a. pH = 7.35

b. pH = 11.0

Solution

To do these problems, you will be using the inverse log function, either [INV] [log] or [10^x], on your calculator. Depending on the type of scientific calculator that you have, you will be inputting the numbers differently. Check with your instructor if you are having difficulty producing the correct answers using your calculator.

a. [H_3O^-] = INV log (−pH) or 10^{-pH} = INV log (−7.35) or $10^{-7.35}$ = 4.5 × 10^{-8} M

b. [H_3O^+] = INV log (−pH) or 10^{-pH} = INV log (−11.0) or $10^{-11.0}$ = 1 × 10^{-11} M

practice problems

9.35 State if each of the following solutions is acidic, basic, or neutral:

a. blood, pH 7.38

b. vinegar, pH 2.8

c. drain cleaner, pH 11.2

d. coffee, pH 5.5

9.36 State if each of the following solutions is acidic, basic, or neutral:

a. soda, pH 3.2

b. shampoo, pH 5.7

c. laundry detergent, pH 9.4

d. rain, pH 5.8

9.37 State if each of these following solutions is acidic, basic, or neutral:

a. [H_3O^+] = 1.2 × 10^{-8} M

b. [H_3O^+] = 7.0 × 10^{-3} M

c. [H_3O^+] = 4.7 × 10^{-11} M

d. [H_3O^+] = 1.0 × 10^{-7} M

9.38 State if each of the following solutions is acidic, basic, or neutral:

a. [H_3O^+] = 5.6 × 10^{-10} M

b. [H_3O^+] = 6.2 × 10^{-8} M

c. [H_3O^+] = 5 × 10^{-2} M

d. [H_3O^+] = 1.8 × 10^{-6} M

9.39 Calculate the pH of each of the solutions in Problem 9.37.

9.40 Calculate the pH of each of the solutions in Problem 9.38.

9.41 Calculate the [H_3O^+] for each of the following measurements of pH:

a. drain cleaner, pH 12.10

b. coffee, pH 5.5

c. gastric juice, pH 2.00

d. toothpaste, pH 8.3

9.42 Calculate the [H_3O^+] for each of the following measurements of pH:

a. hand soap, pH 9.5

b. seawater, pH 8.2

c. apple juice, pH 3.5

d. beer, pH 4.5

9.6 pK_a

? 9.6 Inquiry Question: How is the pK_a used to determine the strength of a weak acid?

Acetic acid and the ammonium ion are both weak acids, but do they have the same strength? That is, do they dissociate by the same amount? We can determine dissociation by comparing their K_a values as we did in Section 9.4. Notice from **Table 9.7** that the K_a value for acetic acid, 1.75 × 10^{-5}, is a number closer to 1 than the K_a for ammonium ion, 6.3 × 10^{-10}, which tells us that acetic acid is the stronger of these two weak acids.

In the previous section we learned that the mathematical function "p" means take the *negative log* of a number. As seen with the pH scale, it is easier to compare whole numbers than those in scientific notation. To make this comparison for acid strength, we can use pK_a values, which range from 0 to about 60.

TABLE 9.7 pK_a and K_a Values for Substances Acting as Weak Acids (25 °C)

Name	Formula	pK_a	K_a
Hydrogen sulfate ion	HSO_4^-	2.00	1.0×10^{-2}
Phosphoric acid	H_3PO_4	2.12	7.5×10^{-3}
Hydrofluoric acid	HF	3.19	6.5×10^{-4}
Nitrous acid	HNO_2	3.35	4.5×10^{-4}
Formic acid	$HCOOH$	3.74	1.8×10^{-4}
Acetic acid	CH_3COOH	4.76	1.75×10^{-5}
Carbonic acid	H_2CO_3	6.35	4.5×10^{-7}
Water	H_2O	7.00	1.0×10^{-7}
Dihydrogen phosphate ion	$H_2PO_4^-$	7.18	6.6×10^{-8}
Ammonium ion	NH_4^+	9.20	6.3×10^{-10}
Hydrocyanic acid	HCN	9.21	6.2×10^{-10}
Bicarbonate ion	HCO_3^-	10.32	4.8×10^{-11}
Hydrogen phosphate ion	HPO_4^{2-}	12.00	1.0×10^{-12}

Increasing acid strength

Notice that the pK_a value for acetic acid is a smaller number, 4.76, than that of ammonium ion, 9.20. This comparison illustrates the following rule for determining the strength of a weak acid using pK_a values: *The smaller the pK_a value, the stronger the acid.* Table 9.7 compares the pK_a and K_a values for some common weak acids. The reactions for these two weak acids in water are shown in **Figures 9.6** and **9.7**.

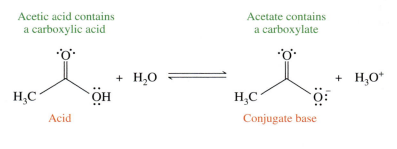

$$K_a = 1.75 \times 10^{-5}$$
$$pK_a = -\log K_a = 4.76$$

FIGURE 9.6 Equilibrium reaction for the weak acid, acetic acid. When acetic acid donates a proton to water, the carboxylate named acetate is formed. Recall that carboxylic acid and carboxylate are different forms of the same functional group.

$$K_a = 6.3 \times 10^{-10}$$
$$pK_a = -\log K_a = 9.20$$

FIGURE 9.7 Equilibrium reaction for the weak acid, ammonium ion. When ammonium ion donates a proton to water, ammonia is formed. Recall that a protonated amine and an amine are different forms of the same functional group.

sample problem 9.13 Determining Strength of Weak Acids Using pK_a Values

Use Table 9.7 to determine the stronger acid of the following pairs of weak acids:

a.   ammonium ion or nitrous acid

b.   phosphoric acid or dihydrogen phosphate ion

Solution

Using the rule, the smaller the pK_a value the stronger the weak acid, and Table 9.7,

a.   Nitrous acid is the stronger acid.

b.   Phosphoric acid is the stronger acid.

practice problems

9.43 Using Table 9.7, determine the stronger acid from the pairs of following acids:
a. $H_2PO_4^-$ or HPO_4^{2-}
b. H_2SO_4 or acetic acid
c. formic acid or carbonic acid
d. ammonium ion or hydrocyanic acid

9.44 Using Table 9.7, determine the stronger acid from the following pairs of weak acids:
a. HCl or HCN
b. acetic acid or ammonium ion
c. carbonic acid or bicarbonate ion
d. ammonium ion or hydrogen phosphate ion

9.7 Amino Acids: Common Biological Weak Acids

9.7 Inquiry Question: How does the charge of an amino acid change with pH?

Consider the circled functional groups on the following molecule. These groups are identified as a protonated amine and a carboxylate.

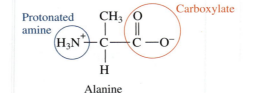

Alanine

This molecule is alanine and it belongs to a class of molecules called **amino** (from the amine functional group) **acids** (from the carboxylate functional group). Amino acids are the building blocks of proteins. We will explore proteins more in Chapter 10. Remember that these two functional groups have an acidic form and a basic form. The amino acid shown is in the form that predominates in water at a physiological pH of 7.4. This ionic form containing no net charge (+ and − cancel each other out) is called a **zwitterion.**

Amino acids and many other biological molecules contain more than one weak acid group and have more than one pK_a value, so they exist in different acid/base forms depending on the pH of the solution. Each molecule has a unique pH value at which only the zwitterion is present. This point is called the **isoelectric point (pI).** At the pI of alanine, the negative charge on the carboxylate is balanced by the positive charge on the ammonium ion, and the net charge of the amino acid is zero. The pI for the amino acid alanine is 6.0. Its pI is halfway between the pK_a values for the protonated amine and carboxylic acid as shown.

In **Figure 9.8** the amino acid alanine is shown with the functional groups in various forms. At pH values below the pI, the acidic forms of the functional groups are the predominant form. At pH values above the pI, the basic forms of the functional groups predominate.

Alanine ion
pH more acidic than 6
(net charge = 1+)

Alanine zwitterion
pH = 6.0
(net charge = 0)

Alanine ion
pH more basic than 6
(net charge = 1−)

$pK_{a-COOH} = 2.3$

$pK_{a-\overset{+}{N}H_3} = 9.7$

FIGURE 9.8 At acidic pH, both amine and carboxylic acid functional groups are in their acidic forms. At the isoelectric point, the amine group is in its acidic form and the carboxylic acid is in its conjugate base form (carboxylate). At basic pH, both functional groups are in their basic forms.

sample problem 9.14   Amino Acids in Acid or Base

Write the zwitterionic form of the amino acid cysteine shown in the following figure. If the pI for cysteine is 5.1, what will the net charge of the amino acid be at the following pH values?

a.   2.0

b.   5.1

c.   10.5

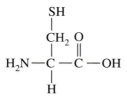

Solution

The zwitterionic form contains a protonated amine and carboxylate group.

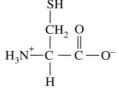

Zwitterion

a.   At a pH of 2.0, the amine and the carboxylic acid will be in their acidic forms. The net charge will be 1+.

b.   When pH = pI, the zwitterion form is present and the net charge is 0.

c.   At a pH of 10.5, the amine and carboxylic acid will be in their basic forms. The net charge will be 1−.

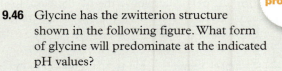

9.45 Valine has the zwitterion structure shown in the following figure. What form of valine will predominate at the indicated pH values?

$$CH(CH_3)_2$$

$$H_3N^+\!\!-\!C\!-\!C\!-\!O^-$$

pI = 6.0

a. pH 1.0   b. pH 6.0   c. pH 11.0

9.46 Glycine has the zwitterion structure shown in the following figure. What form of glycine will predominate at the indicated pH values?

$$H_3N^+\!\!-\!C\!-\!C\!-\!O^-$$

pI = 6.0

a. pH 1.5   b. pH 12.0   c. pH 6.0

Discovering the Concepts

Inquiry Activity—The Bicarbonate Buffer System

Information
The bicarbonate buffer system is an effective physiological buffer in the blood, helping maintain a normal physiological pH of 7.4.

$$H_2O(l) + CO_2(g) \underset{equilibrates}{\overset{Rapidly}{\rightleftharpoons}} H_2CO_3(aq) + H_2O(l) \rightleftharpoons H_3O^+(aq) + HCO_3^-(aq)$$

Carbonic acid (Acid)   Bicarbonate ion (Conjugate base)

CO$_2$ produced at the tissues rapidly equilibrates through carbonic acid to bicarbonate ion in the blood.

The ventilation rate (rate of breathing) controls the amount of CO$_2$ present and, therefore, the amount of acid (H$_3$O$^+$) in the blood. When normal ventilation rates are altered, the normal pH balance becomes disrupted, causing distress. An increase in ventilation rate would expel more CO$_2$ than normal.

Questions
1. If the lungs fail to expel normal amounts of CO$_2$ due to shallow exhaling that can occur in lung disease, a condition called hypoventilation occurs. According to Le Châtelier's principle, if more CO$_2$ is added to the system illustrated in the equilibrium (not enough expelled), what happens to the acidity of the blood? Which of the four substances in the preceding equilibrium is a measure of acidity? How might a patient with this condition be treated?

2. If the lungs expel CO$_2$ faster than normally (ventilation rate becomes high—hyperventilation), which can occur during heavy exercise, and more CO$_2$ is removed from the system than normally, what will happen to the pH of the blood? How might a patient with this condition be treated?

3. Diabetics may produce excess acid in their tissues when they metabolize fats for energy (metabolic acidosis). How would a diabetic's ventilation rate adjust to correct this condition?

4. Suppose a person drank too much of a baking soda solution (sodium bicarbonate) to combat heartburn (which is excess stomach acid). What would happen if too much acid were removed from the bloodstream?

9.8 Buffers and Blood: The Bicarbonate Buffer System

How do we keep the pH levels constant in our blood when we eat a variety of foods at varying pH levels? In our bodies, solutions of weak acids containing both acid and conjugate base help neutralize incoming bases and acids, thereby maintaining pH levels. A solution that contains relatively equal amounts of both a weak acid and its conjugate base or a weak base and its conjugate acid is called a **buffer.** A buffer solution will resist a change in its pH if small amounts of acid or base are added. Strong acids and bases are not components of buffers.

Our blood is buffered mainly by the bicarbonate buffer system. Dissolved CO_2 produced during cellular respiration travels through the bloodstream for exhalation at the lungs. Although dissolved, this CO_2 rapidly equilibrates with carbonic acid and bicarbonate ions. The intermediates (H_2CO_3 and H_2O) are often omitted in this reaction (see **Figure 9.9**).

9.8 Inquiry Question: How do properties of buffers apply to the bicarbonate buffer in the blood?

Reaction pK is 6.1

$$H_2O(l) + CO_2(g) \underset{\text{equilibrates}}{\overset{\text{Rapidly}}{\rightleftharpoons}} H_2CO_3(aq) + H_2O(l) \rightleftharpoons H_3O^+(aq) + HCO_3^-(aq)$$

Carbonic
acid
(Acid)

Bicarbonate
ion
(Conjugate base)

FIGURE 9.9 The bicarbonate buffer equilibrium. CO_2 produced at the tissues rapidly returns to equilibrium through carbonic acid to bicarbonate ion.

A buffer system like the bicarbonate buffer can be denoted H_2CO_3/HCO_3^-, which shows the acid and its conjugate base. Sometimes, the conjugate base is represented as an ionic compound, a salt, like this: $H_2CO_3/NaHCO_3$.

Maintaining Physiological pH with Bicarbonate Buffer: Homeostasis

The bicarbonate buffer system helps to maintain optimal physiological pH in our bodies. The ability of an organism to regulate its internal environment by adjusting factors such as pH, temperature, and solute concentration is called **homeostasis.** Let's examine how the bicarbonate buffer system is able to regulate blood pH under adverse conditions.

Changes in Ventilation Rate

During normal breathing, appropriate amounts of CO_2 gas are removed from the bloodstream upon exhalation in the lungs, and blood pH is maintained. A person who **hypoventilates** may fail to exhale enough CO_2 from the lungs due to shallow breathing, causing CO_2 gas to build up in the bloodstream. Hypoventilation can occur, for example, when a person is suffering from emphysema, a condition that blocks gas diffusion in the lungs.

According to Le Châtelier's principle (Section 9.3), a buildup of CO_2 ultimately produces more H_3O^+ in the bicarbonate equilibrium, making the blood more acidic. This condition is known as **respiratory acidosis** and can occur in a number of health conditions (**Table 9.8**). Persons suffering from this condition must be treated to raise their blood pH back into the normal range. Treatment is done by administering a bicarbonate solution intravenously, which will have the effect of shifting the equilibrium to the left.

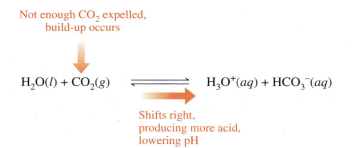

Not enough CO_2 expelled, build-up occurs

$$H_2O(l) + CO_2(g) \rightleftharpoons H_3O^+(aq) + HCO_3^-(aq)$$

Shifts right, producing more acid, lowering pH

TABLE 9.8 Acidosis and Alkalosis: Symptoms, Causes, and Treatments

Respiratory Acidosis: $CO_2 \uparrow pH \downarrow$		Metabolic Acidosis: $H_3O^+ \uparrow pH \downarrow$	
Symptoms:	Failure to ventilate, suppression of breathing, disorientation, weakness, coma	Symptoms:	Increased ventilation, fatigue, confusion
Causes:	Lung disease blocking gas diffusion (e.g., emphysema, pneumonia, bronchitis, and asthma); depression of respiratory center by drugs, cardiopulmonary arrest, stroke, poliomyelitis, or nervous system disorders	Causes:	Renal disease, including hepatitis and cirrhosis; increased acid production in diabetes mellitus, hyperthyroidism, alcoholism, and starvation; loss of alkali in diarrhea; acid retention in renal failure
Treatment:	Correction of disorder, infusion of bicarbonate	Treatment:	Sodium bicarbonate given orally, dialysis for renal failure, insulin treatment for diabetic ketosis
Respiratory Alkalosis: $CO_2 \downarrow pH \uparrow$		Metabolic Alkalosis: $H_3O^+ \downarrow pH \uparrow$	
Symptoms:	Increased rate and depth of breathing, numbness, light-headedness, tetany	Symptoms:	Depressed breathing, apathy, confusion
Causes:	Hyperventilation because of anxiety, hysteria, fever, exercise; reaction to drugs such as salicylate, quinine, and antihistamines; conditions causing hypoxia (e.g., pneumonia, pulmonary edema, and heart disease)	Causes:	Vomiting, diseases of the adrenal glands, ingestion of excess alkali
Treatment:	Elimination of anxiety-producing state, rebreathing into a paper bag	Treatment:	Infusion of saline solution, treatment of underlying diseases, administer NH_4Cl

Breathing into a paper bag can help a person hyperventilating inhale more CO_2.

Conversely, a person who **hyperventilates** (exhaling excessively versus inhaling), which can occur under conditions of anxiety, is exhaling too much CO_2 from the lungs. Hyperventilation draws H_3O^+ from the bloodstream, making the blood more basic. This condition is known as **respiratory alkalosis.** A person suffering from this condition needs to get more CO_2 back into the bloodstream by breathing a CO_2-rich atmosphere. This can easily be done by having the person breathe into a paper bag, which keeps some of the previously exhaled CO_2 available for the next breath. This increases the CO_2 present in the bloodstream, shifting the equilibrium back to the right, thereby producing more H_3O^+ and re-establishing the equilibrium.

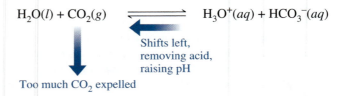

$$H_2O(l) + CO_2(g) \rightleftharpoons H_3O^+(aq) + HCO_3^-(aq)$$

Shifts left, removing acid, raising pH

Too much CO_2 expelled

Changes in Metabolic Acid Production

The chemical reactions that occur in our bodies can also directly change the pH of our blood by producing too much or too little H_3O^+. Diabetics use much less glucose for energy than a normal person because glucose is unable to get inside their cells for use. Because of this, diabetics often use fatty acids as a carbon source for energy production. A byproduct of fatty acid chemical breakdown is acid production. In this case the imbalance is not caused by breathing, but by chemical reactions in the body, so it is termed **metabolic acidosis.** As in the case of respiratory acidosis, administering bicarbonate neutralizes the excess acid, forming more CO_2 that can then be exhaled.

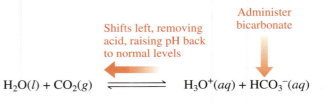

Administer bicarbonate

Shifts left, removing acid, raising pH back to normal levels

$$H_2O(l) + CO_2(g) \rightleftharpoons H_3O^+(aq) + HCO_3^-(aq)$$

The body can also lose too much acid, thereby making the blood basic. This condition, known as **metabolic alkalosis,** can occur under conditions of excessive vomiting, when the body is trying to replace the acid lost from the stomach, thereby making the blood basic. To lower the pH back to normal, ammonium chloride (NH_4Cl, a weak acid) can be administered to neutralize the excess base buildup.

practice problems

9.47 During stress or trauma, a person can start to hyperventilate. The person may then be instructed to breathe into a paper bag to avoid fainting.

$$CO_2(g) + H_2O(l) \rightleftharpoons H_3O^+(aq) + HCO_3^-(aq)$$

a. How does blood pH change during hyperventilation?

b. In which direction will the bicarbonate equilibrium shift during hyperventilation?

c. What is this condition called?

d. How does breathing into a paper bag help return the blood pH to normal?

e. In which direction will the bicarbonate equilibrium shift as the pH returns to normal?

9.48 A person who overdoses on antacids may neutralize too much stomach acid, which causes an imbalance in the bicarbonate equilibrium in their blood.

$$CO_2(g) + H_2O(l) \rightleftharpoons H_3O^+(aq) + HCO_3^-(aq)$$

a. How does blood pH change when this occurs?

b. In which direction will the bicarbonate equilibrium shift during an Alka-Seltzer® overdose?

c. What is this condition called?

d. To treat severe forms of this condition, patients are administered NH_4Cl, which acts as a weak acid. How does this restore the bicarbonate equilibrium?

e. In which direction will the bicarbonate equilibrium shift as the pH returns to normal?

NaOH(s)

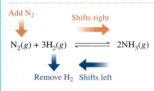

9.1 Acids and Bases—Definitions

9.1 Inquiry Question: How do acids and bases differ?

The Arrhenius definition states that acids produce H^+ and bases produce OH^- when dissolved in water. The Brønsted–Lowry definition builds on the Arrhenius definition by stating that an acid donates an H^+ and a base accepts an H^+. An H^+ or proton in solution exists in contact with a water molecule, thereby forming the hydronium ion, H_3O^+.

$$HCl(aq) + NaOH(aq) \longrightarrow NaCl(aq) + H_2O(l)$$

Strong acid · Strong base · Salt · Water

9.2 Strong Acids and Bases

9.2 Inquiry Question: How does an acid react with a base?

When a strong acid combines with a strong base, a neutralization reaction occurs. The products of neutralization are a salt (ionic compound) and water. Neutralization reactions are typically exothermic. The use of an antacid provides a common example of a neutralization reaction. Antacids are basic compounds that neutralize excess stomach acid. Strong acids and bases completely ionize or dissociate in solution. There are six common strong acids: $HClO_4$, H_2SO_4, HI, HBr, HCl, and HNO_3. Water can act as either an acid or a base in a chemical reaction.

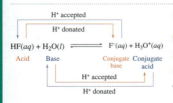

Add N_2 · Shifts right

$$N_2(g) + 3H_2(g) \rightleftharpoons 2NH_3(g)$$

Remove H_2 · Shifts left

9.3 Chemical Equilibrium

9.3 Inquiry Question: How do the properties of equilibrium apply to Le Châtelier's principle?

Many chemical reactions are reversible, operating in the forward direction up to a point when some of the products revert back to reactants. A chemical equilibrium is established when forward and reverse reactions occur at the same rate. The equilibrium constant, K, defines the extent of a chemical reaction as a ratio of the concentration of the products to the concentration of the reactants. If a chemical reaction at equilibrium is disturbed, the reaction can regain its equilibrium according to Le Châtelier's principle by shifting to offset the disturbance. For example, if more reactants are added to a chemical reaction that is currently at equilibrium, the equilibrium will shift to the right, forming more products to offset the addition. Equilibrium is also affected by changes in temperature.

H^+ accepted

H^+ donated

$$HF(aq) + H_2O(l) \rightleftharpoons F^-(aq) + H_3O^+(aq)$$

Acid · Base · Conjugate base · Conjugate acid

H^+ accepted

H^+ donated

9.4 Weak Acids and Bases

9.4 Inquiry Question: How are weak acid–base equations written?

Weak acids can establish equilibrium because they only partially dissociate (<5%) in solution. The equilibrium constant for a weak acid is called the acid dissociation constant, K_a. When a weak acid dissociates, it produces a conjugate base. When a weak base reacts with water, it produces a conjugate acid. These are referred to as conjugate acid–base pairs and differ by one H^+.

The more a weak acid dissociates, the higher its K_a value, and the stronger the acid.

9.5 pH and the pH Scale

9.5 Inquiry Question: How is the acidity of a solution determined?

The amount of H_3O^+ in an aqueous solution determines its acidity. Water ionizes slightly, producing hydronium (H_3O^+) and hydroxide (OH^-) ions. An excess of H_3O^+ in a solution makes a solution acidic. An excess of OH^- makes a solution basic. When the concentrations of hydronium and hydroxide ions are equal, a solution is neutral. The pH scale is a measure of acidity with values typically falling between 0–14. Neutral solutions have a pH value of 7, acidic solutions have values less than 7, and basic solutions have pH values higher than 7. The pH is mathematically related to the concentration of H_3O^+ by the following equation: $pH = -\log[H_3O^+]$.

9.6 pK_a

9.6 Inquiry Question: How is the pK_a used to determine the strength of a weak acid?

$K_a = 1.75 \times 10^{-5}$

$pK_a = -\log K_a = 4.76$

The pH value changes with $[H_3O^+]$, yet the pK_a value is constant for a specific weak acid at a certain temperature. The smaller the pK_a value for a weak acid, the stronger the acid.

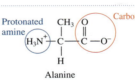

Protonated amine · CH_3 · Carboxylate

$$H_3N^+-C-C-O^-$$

H

Alanine

9.7 Amino Acids: Common Biological Weak Acids

9.7 Inquiry Question: How does the charge of an amino acid change with pH?

An amino acid is a biological molecule that contains the acid–base functional groups amine and carboxylic acid. The charge of the functional group changes as the pH changes. At acidic pH, the acid forms of these functional groups is the main form present (protonated amine and carboxylic acid, net charge = 1+). At basic pH, the conjugate base form predominates (amine and carboxylate, net charge = 1−). The isoelectric point (pI) exists when the solution contains no net charge.

9.8 Buffers and Blood: The Bicarbonate Buffer System

9.8 Inquiry Question: How do the properties of buffers apply to the bicarbonate buffer in the blood?

Buffer solutions consist of approximately equal amounts of a weak acid and its conjugate base. Buffers resist changes in pH when acid or base is added to a solution. The bicarbonate buffer is an important buffer system in the blood. Blood pH is maintained in a narrow range of 7.35–7.45. If the blood pH drops below this range, a condition called acidosis occurs. If the blood pH becomes elevated, a condition called alkalosis exists.

The study guide will help you check your understanding of the main concepts in Chapter 9. You should be able to

9.1   Acids and Bases—Definitions

- Describe an acid and a base using the Arrhenius or Brønsted–Lowry definition.
- Describe the physical characteristics of an acid and a base.

9.2   Strong Acids and Bases

- Name the six strong acids.
- Characterize strong bases.
- Compare a strong acid to a weak acid.
- Write and balance a neutralization reaction.

9.3   Chemical Equilibrium

- Define chemical equilibrium.
- Write an equilibrium expression for K.
- Apply Le Châtelier's principle to chemical equilibrium.

9.4   Weak Acids and Bases

- Apply the principles of chemical equilibrium to weak acids and bases.
- Determine strengths of weak acids based on their pK_a values.
- Identify conjugate acid–base pairs.
- Complete a chemical equation for a conjugate acid–base in water.

9.5   pH and the pH Scale

- Determine if a solution is acidic, basic, or neutral if given its pH.
- Calculate the pH if given the $[H_3O^+]$.
- Calculate the $[H_3O^+]$ if given the pH.

9.6 pK_a

- Predict the relative strength of a weak acid from its pK_a value.

9.7   Amino Acids: Common Biological Weak Acids

- Define isoelectric point.
- Predict the charge of an amino acid below, at, and above the pI value.

9.8   Buffers and Blood: The Bicarbonate Buffer System

- Describe the properties of a buffer.
- Predict the direction the bicarbonate buffer equilibrium will shift with changes in ventilation rate.

Key Terms

acid—A substance that dissolves in water and produces hydrogen ions (H^+), according to the Arrhenius theory. All acids are proton donors, according to the Brønsted–Lowry theory.

acid dissociation constant, K_a—A constant that establishes the ratio of products to reactants for a weak acid at equilibrium.

acidic—A term that describes an aqueous solution where the concentration of hydronium ions (H_3O^+) is greater than the concentration of hydroxide ions (OH^-).

amino acid—The building blocks of protein containing the functional groups protonated amine and carboxylate, which have weak acid–base forms.

autoionization of water—Spontaneous reaction of two water molecules to form a hydronium ion (H_3O^+) and a hydroxide ion (OH^-).

base—A substance that dissolves in water and produces hydroxide ions (OH^-), according to the Arrhenius theory. All bases are proton acceptors according to the Brønsted–Lowry theory.

basic—A term that describes an aqueous solution in which the concentration of hydroxide ions (OH^-) is greater than the concentration of hydronium ions (H_3O^+).

buffer—A solution consisting of a weak acid and its conjugate base or a weak base and its conjugate acid that resists a change in pH when small amounts of acid or base are added.

chemical equilibrium—A state where the rate of formation of products and the rate of formation of reactants are equal.

conjugate acid—The product after a base accepts a proton. It is capable of donating a proton in the reverse reaction.

conjugate acid–base pair—An acid and base that differ by one H^+.

conjugate base—The product after an acid donates a proton. It is capable of accepting a proton in the reverse reaction.

equilibrium constant, K—A constant that is equal to the ratio of concentrations of products to reactants at equilibrium.

homeostasis—The ability of an organism to regulate its internal environment by adjusting its physiological processes.

hydronium ion, H_3O^+—The ion formed by the attraction of a proton (H^+) to an H_2O molecule.

hyperventilation—A condition of deep, rapid breathing where too much CO_2 is exhaled from the lungs, thereby upsetting the bicarbonate equilibrium.

hypoventilation—A condition of shallow breathing where too little CO_2 is exhaled from the lungs, thereby upsetting the bicarbonate equilibrium.

isoelectric point (pI)—The pH at which an amino acid has a net charge of zero.

Le Châtelier's principle—A chemical reaction that has been disturbed from equilibrium (concentration change, temperature change) will shift its equilibrium to offset the disturbance.

metabolic acidosis—A physiological condition caused by cellular metabolism where the blood pH is lower than 7.35.

metabolic alkalosis—A physiological condition caused by cellular metabolism where the blood pH is higher than 7.45.

neutral—The term that describes a solution with equal concentrations of H_3O^+ and OH^-.

neutralization—The reaction between an acid and a base to form a salt and water.

pH—A measure of the H_3O^+ concentration in a solution. pH = -log $[H_3O^+]$.

pK_a—A measure of the acidity of a weak acid. pK_a = -log K_a.

respiratory acidosis—A physiological condition in which the blood pH is lower than 7.35, caused by changes in breathing.

respiratory alkalosis—A physiological condition in which the blood pH is higher than 7.45, caused by changes in breathing.

salt—An ionic compound containing a metal or a polyatomic ion (like NH_4^+) as the cation and a nonmetal or polyatomic ion as the anion (exception: OH^-).

strong acid—An acid that completely ionizes in water.

strong base—A base that completely ionizes in water.

weak acid—An acid that ionizes only slightly in solution.

weak base—A base that ionizes only slightly in solution.

zwitterion—The ionic form of an amino acid existing at physiological pH containing a $-NH_3^+$ and $-COO^-$ group and with no net charge.

Additional Problems

9.49 Identify the following as strong or weak acids:
- **a.** H_2SO_4
- **b.** H_2CO_3
- **c.** HCl
- **d.** HF

9.50 Identify the following as strong or weak acids:
- **a.** HNO_3
- **b.** HNO_2
- **c.** H_3PO_4
- **d.** HBr

9.51 What are some similarities and differences between strong and weak acids?

9.52 What are some ingredients found in antacids? What do they do?

9.53 For the following reaction,

$$2HI(g) \rightleftharpoons H_2(g) + I_2(g)$$

a. Write an equilibrium constant expression.
b. If the value for $K = 2.1 \times 10^{-2}$ at 720 K, which substance predominates, HI or H_2 and I_2?

9.54 For the following reaction,

$$N_2O_4(g) \rightleftharpoons 2 NO_2(g)$$

a. Write an equilibrium constant expression.
b. If the value for $K = 46$ at 500 K, which gas predominates, N_2O_4 or NO_2?

9.55 Consider the following reaction:

$$CO(g) + 2 H_2(g) \rightleftharpoons CH_3OH(g) + heat$$

a. Is the reaction exothermic or endothermic as written?
b. Provide an equilibrium expression (K) for the reaction.
c. How will the equilibrium shift if hydrogen gas is added to the reaction?
d. How will the equilibrium shift if carbon monoxide is removed from the reaction?
e. How will the equilibrium shift if the temperature is lowered?

9.56 Consider the following reaction:

Glycogen(aq)(stored glucose) + $H_2O(l)$

⇅

blood glucose(aq) + heat

a. Is the reaction exothermic or endothermic as written?
b. Provide an equilibrium expression (K) for the reaction.
c. How will the equilibrium shift if blood glucose is removed from the reaction?
d. How will the equilibrium shift if the temperature is raised?

9.57 Write the formula of the conjugate base formed from each of the following weak acids.
a. nitrous acid, HNO_2
b. methylammonium, $CH_3NH_3^+$
c. formic acid, $HCOOH$

9.58 Write the formula of the conjugate base formed from each of the following weak acids.
a. phosphoric acid, H_3PO_4
b. pyruvic acid,

c. chlorous acid, $HClO_2$

9.59 Write the formula of the conjugate acid formed from each of the following weak bases.
a. lactate, $CH_3CH(OH)COO^-$
b. aminomethane, CH_3NH_2
c. phosphate, PO_4^{3-}

9.60 Write the formula of the conjugate acid formed from each of the following weak bases.
a. HS^-
b. $H_2BO_3^-$
c. acetate, CH_3COO^-

9.61 Determine the pH for the following solutions. State whether each solution is acidic, basic, or neutral.
a. $[H_3O^+] = 2.5 \times 10^{-8}$ M
b. $[H_3O^+] = 7.0 \times 10^{-2}$ M
c. $[H_3O^+] = 4 \times 10^{-11}$ M

9.62 Determine the pH for the following solutions. State whether each solution is acidic, basic, or neutral.
a. $[H_3O^+] = 1 \times 10^{-4}$ M
b. $[H_3O^+] = 6 \times 10^{-9}$ M
c. $[H_3O^+] = 5.2 \times 10^{-2}$ M

9.63 Calculate the H_3O^+ concentration for a solution with the following pH:
a. 3.52
b. 5.0
c. 9.25

9.64 Calculate the H_3O^+ concentration for a solution with the following pH:
a. 8.2
b. 6.75
c. 12.1

9.65 Consider the acetic acid buffer system with acetic acid, CH_3COOH, and its salt sodium acetate, $NaCH_3COO$:

$$CH_3COOH(aq) + H_2O(l) \rightleftharpoons H_3O^+(aq) + CH_3COO^-(aq)$$

a. What is the purpose of a buffer system?
b. What is the purpose of $NaCH_3COO$ in the buffer?
c. How does the buffer react when acid (H_3O^+) is added?
d. How does the buffer react when base (OH^-) is added?

9.66 Consider the lactic acid buffer with lactic acid, $CH_3CH(OH)COOH$ and its salt sodium lactate, $NaCH_3CH(OH)COO$:

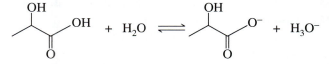

a. What is the purpose of a buffer system?
b. Why is a salt of the acid needed?
c. How does the buffer react when acid (H_3O^+) is added?
d. How does the buffer react when base (OH^-) is added?

9.67 In blood plasma, pH is maintained by the carbonic acid–bicarbonate buffer equilibrium as follows:

$$CO_2(g) + H_2O(l) \rightleftharpoons H_3O^+(aq) + HCO_3^-(aq)$$

  a. If excess acid is added to the bloodstream by a physiological process other than breathing, what is this condition called?
  b. Is the pH of the blood below or above normal?
  c. How could such a condition be treated?

9.68 Adding a few drops of a strong acid to water will lower the pH appreciably. However, adding the same number of drops to a buffer does not appreciably alter the pH. Why?

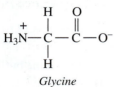

| pH = 7.0 | pH = 3.0 | pH = 7.0 | pH = 6.9 |

Water   *Buffer*

9.69 Consider the amino acid valine shown in its zwitterion form. The pI of valine is 6.0.

  a. Draw the structure of the predominant form of valine at pH = 10.0.
  b. Draw the structure of the predominant form of valine at pH = 3.0.

Challenge Problems

9.70 Phenol is a very weak acid. Write the conjugate acid–base reaction for phenol in water.

OH

9.71 Write the conjugate base for the amino acid glycine shown here.

Glycine

9.72 Explain why the following amino acid cannot exist in the form shown in aqueous solution:

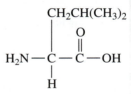

9.73 To determine the concentration of an unknown weak acid solution, a known amount of strong base can be titrated (added) to it until complete neutralization occurs (an endpoint is reached). If a 10.0-mL sample of vinegar (aqueous acetic acid, CH_3COOH) requires 16.5 mL of 0.500 M NaOH to reach the endpoint, what is the molarity of the acetic acid solution? [HINT: *1 mole of NaOH will neutralize 1 mole of acetic acid.*]

Answers to Odd-Numbered Problems

Practice Problems

9.1   **a.** acid   **b.** base   **c.** acid   **d.** base
9.3   Most hydrogen atoms contain one proton and one electron and no neutrons. Therefore, a positively charged hydrogen ion (H^+) is one proton.
9.5   **a.** HI is the acid (proton donor) and H_2O is the base (proton acceptor).
  b. H_2O is the acid (proton donor) and F^- is the base (proton acceptor).
9.7   The strong acids are (a.) H_2SO_4, (b.) HCl, and (d.) HNO_3.
9.9   **a.** $HNO_3(aq) + LiOH(s) \rightarrow H_2O(l) + LiNO_3(aq)$
  b. $H_2SO_4(aq) + Ca(OH)_2(s) \rightarrow 2\ H_2O(l) + CaSO_4(aq)$

9.11   **a.** $3\ HBr(aq) + Al(OH)_3(s) \rightarrow 3\ H_2O(l) + AlBr_3(aq)$
  b. $2\ HI(aq) + CaCO_3(s) \rightarrow CaI_2(aq) + CO_2(g) + H_2O(l)$
9.13   Reversible reactions are reactions that can proceed in both the forward and reverse directions.
9.15   **a.** $K = \dfrac{[CO_2][H_2]}{[CO][H_2O]}$
  b. $K = \dfrac{[CH_3COO^-][H_3O^+]}{[CH_3COOH]}$
9.17   **a.** reactants   **b.** products
  c. both reactants and products present in equal amounts
9.19   **a.** Products are favored.   **b.** Reactants are favored.
  c. Products are favored.   **d.** Reactants are favored.

9.21 **a.** Reactants are favored. **b.** Products are favored.
c. Products are favored. **d.** Reactants are favored.

9.23 The escape of CO_2 gas removes CO_2; equilibrium favors the formation of the products. The soda no longer bubbles.

9.25 **a.** $H_2PO_4^-$ **b.** HF **c.** HBr

9.27 **a.** acid HSO_4^-; conjugate base SO_4^{2-}; base H_2O; conjugate acid H_3O^+

b. acid NH_4^+; conjugate base NH_3; base H_2O; conjugate acid H_3O^+

c. acid HCN; conjugate base CN^-; base NO_2^-; conjugate acid HNO_2

9.29 **a.** F^-, fluoride ion **b.** OH^-, hydroxide ion
c. HCO_3^-, bicarbonate ion **d.** SO_4^{2-}, sulfate ion

9.31 **a.** HCO_3^-, bicarbonate ion **b.** H_3O^+, hydronium ion
c. H_3PO_4, phosphoric acid **d.** HBr, hydrobromic acid

9.33 **a.** $HA(aq) + H_2O(l) \rightleftharpoons A^-(aq) + H_3O^+(aq)$

    Acid   Base   Conjugate Conjugate
                                    base   acid

b. $H_2PO_4^-(aq) + H_2O(l) \rightleftharpoons HPO_4^{2-}(aq) + H_3O^+(aq)$

    Acid   Base   Conjugate   Conjugate
                                    base   acid

c. $NH_3(aq) + H_2O(l) \rightleftharpoons NH_4^+(aq) + OH^-(aq)$

    Base   Acid   Conjugate   Conjugate
                                    acid   base

9.35 **a.** basic **b.** acidic **c.** basic **d.** acidic

9.37 **a.** basic **b.** acidic **c.** basic **d.** neutral

9.39 **a.** 7.92 **b.** 2.15 **c.** 10.33 **d.** 7.00

9.41 **a.** 7.9×10^{-13} M **b.** 3×10^{-6} M
c. 1.0×10^{-2} M **d.** 5×10^{-9} M

9.43 **a.** $H_2PO_4^-$ **b.** H_2SO_4
c. formic acid **d.** ammonium ion

9.45 **a.**

$CH(CH_3)_2$

$H_3N^+ - C - C - OH$ (with O double bond, H below)

pH = 1.0

b.

$CH(CH_3)_2$

$H_3N^+ - C - C - O^-$ (with O double bond, H below)

pH = 6.0

c.

$CH(CH_3)_2$

$H_2N - C - C - O^-$ (with O double bond, H below)

pH = 11.0

9.47 **a.** Hyperventilation will lower the CO_2 level in the blood, which decreases the H_3O^+ and increases the blood pH.

b. The equilibrium will shift to the left.

c. This condition is called respiratory alkalosis.

d. Breathing into a bag will increase the CO_2 level, which increases H_3O^+ and lowers the blood pH.

e. The equilibrium will shift to the right.

Additional Problems

9.49 **a.** strong **b.** weak **c.** strong **d.** weak

9.51 Both strong and weak acids produce H_3O^+ in water. Weak acids are only slightly ionized, whereas a strong acid exists as ions in solution (fully ionizes).

9.53 **a.** $K = \dfrac{[H_2][I_2]}{[HI]^2}$

b. The HI predominates.

9.55 **a.** exothermic

b. $K = \dfrac{[CH_3OH]}{[CO][H_2]^2}$

c. It will shift to the right.

d. It will shift to the left.

e. It will shift to the right.

9.57 **a.** NO_2^- **b.** CH_3NH_2 **c.** $HCOO^-$

9.59 **a.** $CH_3CH(OH)COOH$ **b.** $CH_3NH_3^+$
c. HPO_4^{2-}

9.61 **a.** 7.60, basic **b.** 1.15, acidic **c.** 10.4, basic

9.63 **a.** 3.0×10^{-4} M **b.** 1×10^{-5} M **c.** 5.6×10^{-10} M

9.65 **a.** A buffer system keeps the pH constant.

b. The conjugate base, CH_3COO^-, from the salt $NaCH_3COO$ neutralizes any acid added.

c. Added H_3O^+ reacts with CH_3COO^-

d. Added OH^- reacts with CH_3COOH.

9.67 **a.** metabolic acidosis **b.** below normal
c. administer bicarbonate (HCO_3^-)

9.69 **a.**

$CH(CH_3)_2$

$H_2N - C - C - O^-$ (with O double bond, H below)

pH = 10

b.

$CH(CH_3)_2$

$H_3N^+ - C - C - OH$ (with O double bond, H below)

pH = 3

9.71

$H_2N - C - C - O^-$ (with H top, O double bond, H below)

9.73

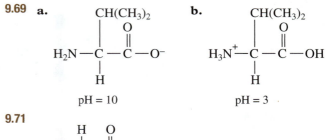

$\dfrac{0.500 \text{ mole NaOH}}{1 L} \times \dfrac{1 L}{1000 mL} \times 16.5 mL$

$= 0.00825$ mole NaOH

$\dfrac{0.00825 \text{ mole acetic acid}}{10.0 mL} \times \dfrac{1000 mL}{1 L}$

$= 0.825$ M acetic acid

Fruits like pears and apples contain an enzyme that will cause the fruit to brown if the skin is broken. Lemon juice can slow the browning reaction. Enzymes are proteins. To learn more about the browning reaction and the structure and function of proteins, read Chapter 10.

10   Proteins—Workers of the Cell

WHAT DO INSULIN, hemoglobin, lactase, ion channels, and collagen have in common? They are all proteins. Much of the structure and function of our bodies builds on these large biomolecules called proteins. They transport oxygen in blood, serve as the main components of skin and muscle, defend our bodies against infection, direct cellular chemistry as enzymes, and control metabolism as hormones.

Proteins are polymers (*poly*—"many," *meros*—"part") composed entirely of amino acids covalently bonded in specific sequences. There are 20 different commonly occurring amino acids, and proteins can contain as few as 50 or as many as several hundreds or even thousands of these amino acids. The ordering of the amino acids in a protein determines both its structure and biological function.

In this chapter, we discover that each amino acid has its own characteristics and chemical behavior. When the amino acids are linked together in a sequence like colored beads on a string, this string can twist and fold as a result of attractive forces to form a unique three-dimensional structure that has a specific function.

Further, we look more closely at a specific type of protein called an *enzyme*. Enzymes are nature's catalysts. We study the action and structure of enzymes, making each uniquely suited to function as a catalyst in the chemistry of a cell.

Discovering the Concepts

? Inquiry Activity—Characteristics of Amino Acids

Information

In plants and animals, a set of 20 amino acids is required to make proteins. Human beings can manufacture only 10 of them. The other 10 must be obtained through the diet. The latter 10 are termed *essential* amino acids. Dietary proteins are complete or incomplete depending on whether they contain all essential amino acids. Plant sources are more likely to contain incomplete proteins, but when combined, they can provide complete protein sources. For example, protein from rice is deficient in the amino acid lysine, but rich in the amino acid methionine. Beans are deficient in methionine and rich in lysine. So a meal of rice and beans can provide complete protein.

All 20 amino acids have some features in common:

$$H_3N^+-\overset{\overset{\displaystyle R}{|}}{C}-\overset{\overset{\displaystyle O}{\|}}{C}-O^-$$
$$\underset{\displaystyle H}{|}$$

The general formula of an amino acid.

? **What's an Inquiry Question?**
Inquiry Questions are designed to focus your reading on the main concepts by section. An Inquiry Question appears at the beginning of each section.

The R is the part of the 20 amino acids that is different. In amino acids, R is often referred to as the "side chain." All the naturally occurring amino acids are *L-enantiomers*. The 20 different amino acids can be classified by the polarity of their side chain: polar neutral, polar charged (acidic or basic), or nonpolar.

At physiological pH, the amine group is protonated, and the carboxyl group is a conjugate base (carboxylate), as discussed in Section 9.7. This neutral ionic form is called a *zwitterion*.

Questions

1. What is an *essential* amino acid?
2. Gelatin is produced from an animal protein called *collagen* that does not contain all 20 amino acids and lacks the essential amino acid tryptophan. Is gelatin considered a complete protein or an incomplete protein?
3. Identify the chiral center (using an asterisk) in the general formula of an amino acid shown in the preceding figure.
4. Name the two functional groups common to all 20 amino acids.
5. Locate the side chain (R) on each amino acid shown. Identify the functional group(s) found in the side chain (R). Is the R group polar neutral, polar charged, or nonpolar?

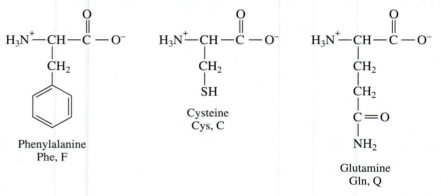

Phenylalanine
Phe, F

Cysteine
Cys, C

Glutamine
Gln, Q

10.1 Amino Acids—A Second Look

10.1 Inquiry Question:
What are the characteristics of an amino acid?

As mentioned in Section 9.7, the term *amino acid* gives us a way to remember its structure. "Amino" reminds us that the structure contains a protonated amine ($-NH_3^+$), and "acid" reminds us that it contains a carboxylic acid in the form of carboxylate ($-COO^-$). These two functional groups are bonded to a central carbon atom called the **alpha (α) carbon.** The protonated amine bonded to this carbon is often referred to as the **alpha (α) amino group** and the carboxylate ion as the **alpha (α) carboxylate group.** The (α) carbon is also bonded to a hydrogen atom and a larger side chain designated by R, which is responsible for the unique identity and characteristics of each amino acid.

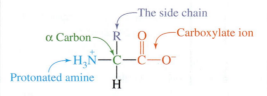

In all but one amino acid (glycine), the α carbon is a chiral center because it has four different groups bonded to it. As we saw in Chapter 4, compounds that contain a single chiral center can have two forms called *enantiomers*, which are nonsuperimposable mirror images.

We can draw Fischer projections for amino acids like those drawn for carbohydrates. In the following figure, the carbon chain is drawn vertically with the carboxylate group at the top and the R group at the bottom. An amino acid has an enantiomer with the protonated amine on the left-hand side of the structure and one with the protonated amine on the right-hand side of the structure. The enantiomer with the protonated amine on the left is called the L-*stereoisomer* and its mirror image is the D-*stereoisomer*.

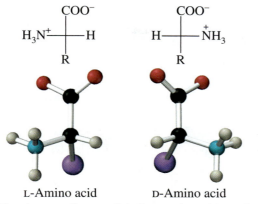

L-Amino acid   D-Amino acid

Each amino acid has two enantiomers. Only the L-amino acids are found in proteins.

The L-amino acids are the building blocks of proteins. Some D-amino acids do occur in nature but rarely in proteins.

The side chain or R group gives each amino acid its unique identity and characteristics. Twenty amino acids are found in most proteins. Nine different families of organic compounds are represented in the structures of the different amino acid side chains—alkanes (hydrocarbon), aromatics, thioethers, alcohols, phenols, thiols, amides, carboxylic acids (present as carboxylate ions), and amines (present as protonated amines). You may want to refer back to Table 4.3 to get reacquainted with the structures of these functional groups.

The chemical properties and behaviors of the functional groups are present in the amino acids and are used to classify the amino acids into a few simple categories (nonpolar, polar, acidic, and basic). The structures of the amino acids, including their side chains, names, functional groups, and abbreviations are shown in **Table 10.1**.

TABLE 10.1 The 20 Common Amino Acids in Proteins (Common Functional Groups Noted in Side Chains)

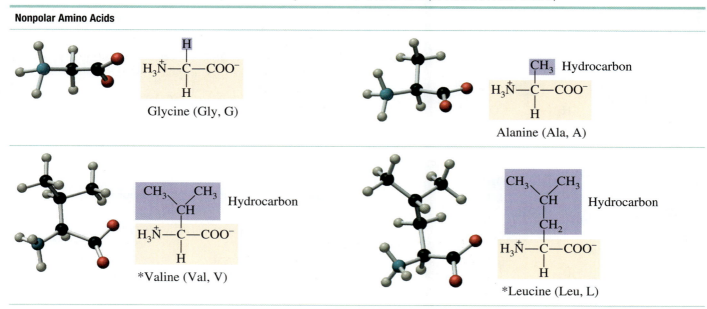

Nonpolar Amino Acids

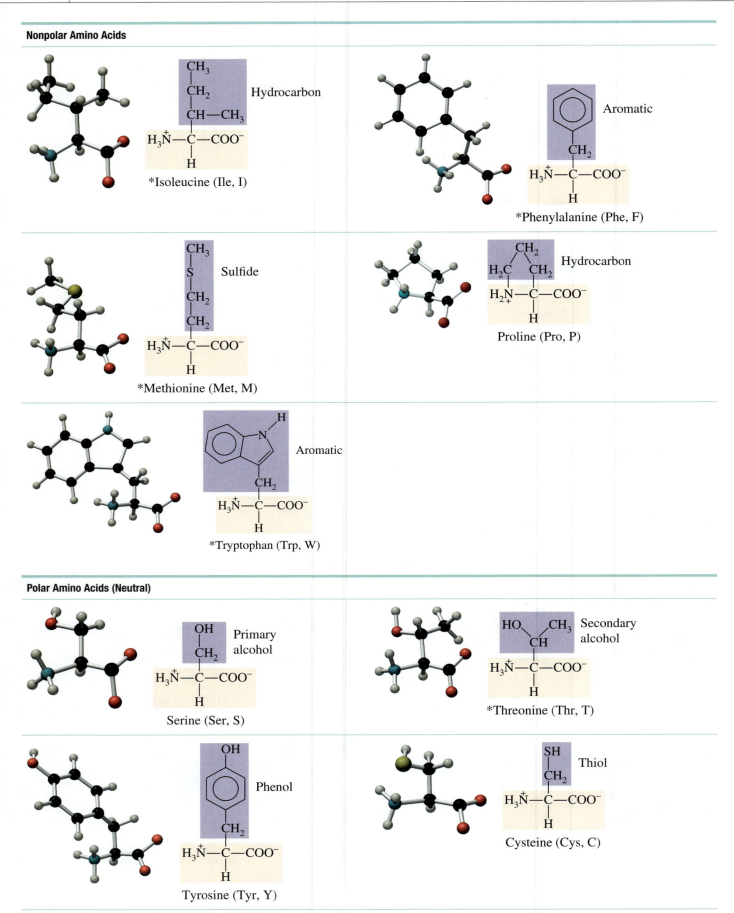

CH₃
|
CH₂
|
CH—CH₃   Hydrocarbon

$H_3\overset{+}{N}$—C—COO⁻
|
H

*Isoleucine (Ile, I)

Aromatic

CH₂
|
$H_3\overset{+}{N}$—C—COO⁻
|
H

*Phenylalanine (Phe, F)

CH₃
|
S   Sulfide
|
CH₂
|
CH₂
|
$H_3\overset{+}{N}$—C—COO⁻
|
H

*Methionine (Met, M)

CH₂
/   \
H₂C   CH₂   Hydrocarbon
|
$H_2\overset{+}{N}$—C—COO⁻
|
H

Proline (Pro, P)

H
|
N

Aromatic

CH₂
|
$H_3\overset{+}{N}$—C—COO⁻
|
H

*Tryptophan (Trp, W)

Polar Amino Acids (Neutral)

OH   Primary
| alcohol
CH₂
|
$H_3\overset{+}{N}$—C—COO⁻
|
H

Serine (Ser, S)

HO   CH₃   Secondary
\   /   alcohol
CH
|
$H_3\overset{+}{N}$—C—COO⁻
|
H

*Threonine (Thr, T)

OH

Phenol

CH₂
|
$H_3\overset{+}{N}$—C—COO⁻
|
H

Tyrosine (Tyr, Y)

SH   Thiol
|
CH₂
|
$H_3\overset{+}{N}$—C—COO⁻
|
H

Cysteine (Cys, C)

Polar Amino Acids (Neutral)

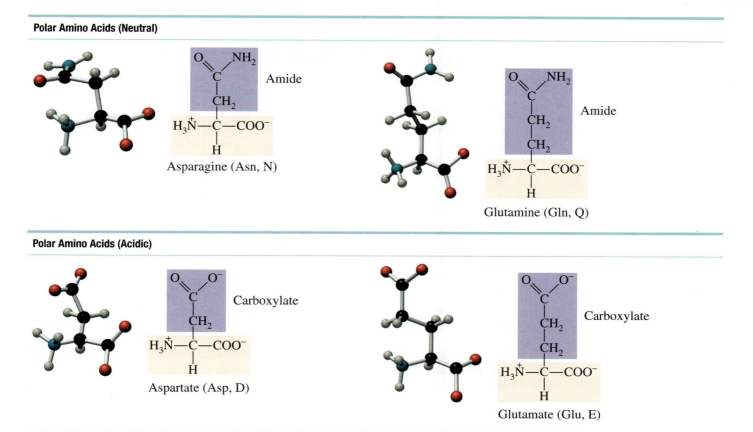

Asparagine (Asn, N)

Amide

Glutamine (Gln, Q)

Amide

Polar Amino Acids (Acidic)

Carboxylate

Aspartate (Asp, D)

Carboxylate

Glutamate (Glu, E)

Polar Amino Acids (Basic)

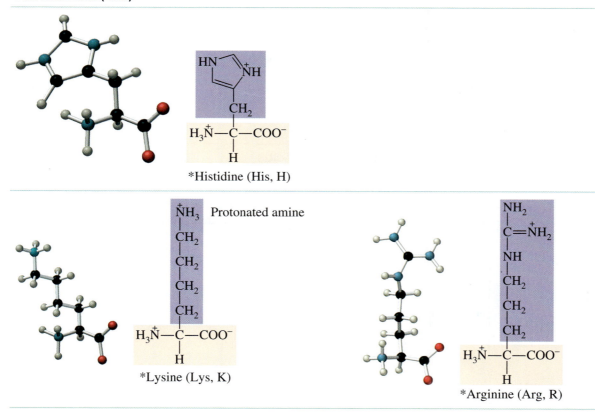

*Histidine (His, H)

Protonated amine

*Lysine (Lys, K)

*Arginine (Arg, R)

*Essential amino acids.

The 10 amino acids designated with an asterisk (*) in the table are called essential amino acids because they cannot be synthesized in the human body and must be obtained in the diet. Two of these amino acids, arginine and histidine, are essential in the diets of children but not adults. The nonessential amino acids can be synthesized from the essential amino acids as needed by the body.

Proteins that contain all of the essential amino acids are known as *complete proteins*. Soybeans and most proteins found in animal products such as eggs, milk, fish, poultry, and meat are complete proteins. Some proteins from plant sources are *incomplete proteins* because they lack one or more of the essential amino acids. A complete protein meal can be obtained by combining foods like rice and beans or peanut butter on whole-grain bread.

Classification of Amino Acids

The amino acids in Table 10.1 are separated into two broad classifications based on the characteristics of the side chains: **nonpolar amino acids** and **polar amino acids.** The polar amino acids are further divided into neutral, acidic, and basic.

Now let's examine the side chains of the amino acids that are classified as "nonpolar." We see that with few exceptions, the side chains of these amino acids are composed entirely of carbon and hydrogen. Recall from Chapter 7 that the carbon–hydrogen bond is nonpolar and compounds composed of only carbon and hydrogen are nonpolar and hydrophobic (water-fearing). So, the side chains of this group of amino acids are nonpolar and hydrophobic. (In the case of tryptophan, the nitrogen atom makes up such a small part of the side chain that it contributes very little to its polarity. In the case of methionine, the C—S bonds present are nonpolar because C and S have the same electronegativity— see Chapter 3, Figure 3.14.)

Continue by examining the side chains of each of the amino acids classified as "polar." These side chains contain functional groups such as hydroxyl (—OH) and amide (—CONH$_2$) that create an uneven distribution of electrons in the side chain, making them polar. All but one of the polar side chains can form hydrogen bonds with water and, therefore, are hydrophilic (water-loving). The exception is cysteine, which has a polar thiol (—SH) group but does not form hydrogen bonds. Notice that the polar acidic and polar basic amino acids have charged side chains, allowing them to form ion–dipole interactions with water, making them even more polar and hydrophilic than are the amino acids classified as polar neutral.

sample
problem
10.1   **Identifying Amino Acids**

Name the following amino acids, provide the three-letter and one-letter abbreviations, identify the functional group in the side chain, identify any chiral centers with an asterisk, and indicate whether the side chain is polar neutral, polar charged, or nonpolar.

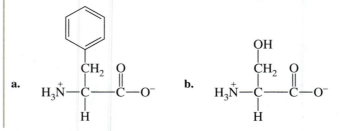

Solution

a.   Referring to Table 10.1, this amino acid is phenylalanine, abbreviated Phe or F.
     It contains one chiral center and an aromatic functional group. The side chain
     contains only C and H, so this amino acid is nonpolar (hydrophobic).

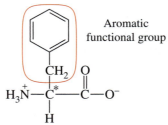

b.   Referring to Table 10.1, this amino acid is serine, abbreviated Ser or S. It con-
     tains one chiral center. Because the side chain contains an OH, this is a polar
     (hydrophilic) neutral amino acid.

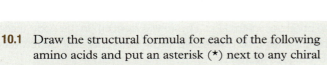

10.1   Draw the structural formula for each of the following
           amino acids and put an asterisk (★) next to any chiral
           carbon centers in your structure:
           a. alanine
           b. lysine
           c. tryptophan
           d. aspartate

10.2   Classify each of the amino acids in Problem 10.1 as
           polar neutral, polar charged, or nonpolar.

10.3   Give the three-letter and one-letter abbreviations and
           identify the functional group in the side chain for
           each amino acid in Problem 10.1.

10.4   Draw the structural formula for each of the following
           amino acids and put an asterisk (★) next to any chiral
           carbon centers in your structure:
           a. leucine
           b. glutamate
           c. methionine
           d. threonine

10.5   Classify each amino acid in Problem 10.4 as
           polar neutral, polar charged, or nonpolar.

10.6   Give the three-letter and one-letter
           abbreviations and identify the functional
           group in the side chain for each amino acid
           in Problem 10.4.

10.7   Draw the structure for the amino acid represented
           by each of the following abbreviations:
           a. G
           b. His
           c. Q
           d. Ile

10.8   Draw the structure for the amino acid represented
           by each of the following abbreviations:
           a. Pro
           b. N
           c. Val
           d. Y

10.9   Give the names of each amino acid you drew in
           Problem 10.7.

10.10 Give the names of each amino acid you drew in
           Problem 10.8.

Discovering the Concepts

? Inquiry Activity—Condensation and the Peptide Bond

Information

Amino acids combine to form peptides and, ultimately, proteins through a *condensation* (also known as *dehydration*) reaction. One water molecule is removed when a carboxylate and protonated amine react. The condensation of two amino acids is shown in the following figure.

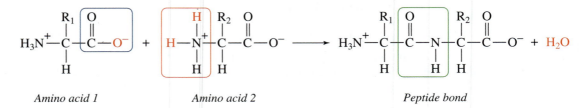

Amino acid 1   Amino acid 2   Peptide bond

When two amino acids combine as shown, a *dipeptide* is formed. A string of many amino acids bonded through peptide bonds is a *polypeptide*. A *protein* is a polypeptide that has a biological function.

Questions

1. Name the organic functional group in a peptide bond.
2. In the preceding condensation reaction, a dipeptide is formed. How many amino acids would be present in a tetrapeptide? _____ How many peptide bonds would be formed in a tetrapeptide? _____
3. The breaking of a peptide bond occurs through a hydrolysis reaction. In the body, these reactions are catalyzed by enzymes called *proteases*. Draw the products of the following reaction if the dipeptide undergoes hydrolysis.

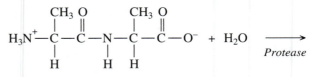

The dipeptide, Ala—Ala

10.2 Protein Formation

 10.2 Inquiry Question:
How is a peptide bond formed?

In Chapter 6, we showed how two monosaccharides bonded together through a condensation reaction to form a disaccharide. In Chapter 7, we showed that a glycerol and three fatty acids react to form a triglyceride through a condensation reaction.

Condensation reactions can also occur between amino acids, and the product formed is a dipeptide. In this case, the carboxylate ion ($-COO^-$) of one amino acid molecule reacts with the protonated amine ($-NH_3^+$) of a second amino acid. A water molecule is removed, and an **amide** functional group is formed.

Amides are a family of organic compounds characterized by a functional group containing a carbonyl ($C=O$) bonded to a nitrogen atom. The nitrogen of the amide is bonded to two other atoms, which may be two hydrogen atoms, two carbon atoms, or one carbon atom and one hydrogen atom.

Biological Condensation Reactions

What do most condensation reactions have in common? The reacting functional groups have an —H and an —OH that can be removed during the condensation, resulting in different functional groups as outlined in **Table 10.2**. Remember from Chapter 5 that a condensation reaction operating in reverse is a hydrolysis reaction and the —H and —OH are added to the atoms whose bond was hydrolyzed.

TABLE 10.2 Common Condensation Reactions between Functional Groups

Condensation Reaction	Representative Reaction in Biomolecules
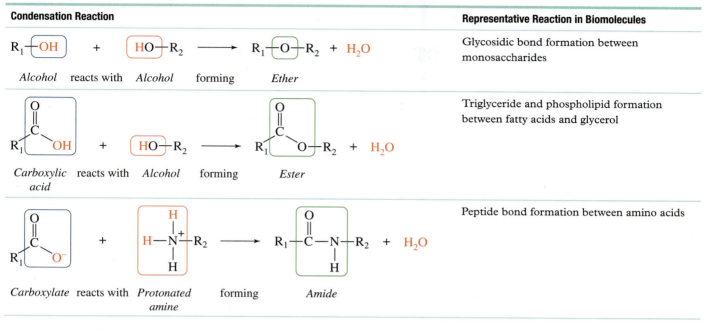	Glycosidic bond formation between monosaccharides
	Triglyceride and phospholipid formation between fatty acids and glycerol
	Peptide bond formation between amino acids

Before we continue discussing protein formation, let's review condensation and hydrolysis reactions as they apply to the biomolecules we have seen.

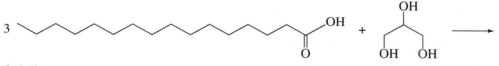

Solving a Problem

Writing Condensation and Hydrolysis Products

Write the products for the reaction shown.

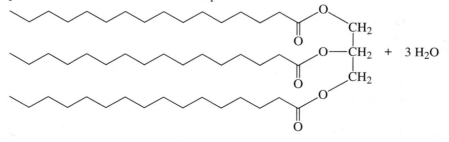

Solution

STEP 1: Decide whether a condensation or hydrolysis is occurring. Two molecules are combining, so this is a condensation.

STEP 2: Identify the functional groups present. In this problem, three carboxylic acids and three alcohol functional groups are present. These will combine to form esters.

STEP 3: Write the products. In a condensation, an —H and —OH will be removed from the functional groups combining to form a bond. The molecule produced will have three ester bonds present.

sample problem 10.2 — Writing Condensation and Hydrolysis Products

Write the products for the reaction shown.

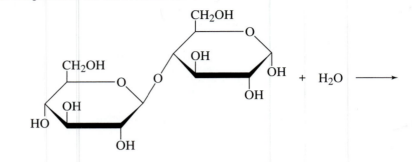

Solution

STEP 1: Decide whether a condensation or hydrolysis is occurring. One molecule containing two monosaccharides is reacting with water, so a hydrolysis is occurring.

STEP 2: Identify the functional groups present. In this problem, there are several alcohol groups and some ether groups.

STEP 3: Write the products. A hydrolysis is going to break the glycosidic bond between the two monosaccharides. The atoms from water are added as an —H and —OH (shown in red). The positioning of the —OH groups formed at the glycosidic bond remains as it was prior to hydrolysis.

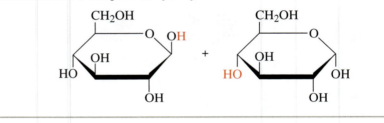

Amides and the Peptide Bond

When amino acids combine in a condensation reaction, the amide bond that is formed is called a **peptide bond.**

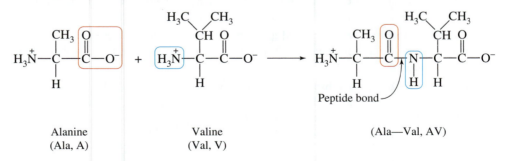

| Alanine (Ala, A) | Valine (Val, V) | (Ala—Val, AV) |

The product of the condensation reaction of alanine and valine, as shown in the preceding figure, is a **dipeptide,** which is represented as Ala—Val, or AV. In this dipeptide, alanine is called the **N-terminus** (or amino terminus) because it has an unreacted (free) α-amino group. Similarly, valine is called the **C-terminus** (or carboxy terminus) because it has an unreacted carboxylate group. By convention, peptides (and larger peptide-containing structures) are always written from the N-terminus to the C-terminus (left to right) whether using full structures or the one- or three-letter abbreviations.

Dipeptides can be formed from any two amino acids. Each pair of different amino acids can combine in two ways forming two different dipeptides. The two dipeptides, Ala—Val and Val—Ala, are structural isomers, different compounds, and have different properties. Even if the order of the amino acids may not seem important, it is critical to both the structure and the function of the compound.

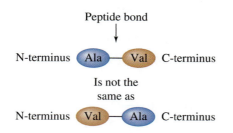

Dipeptides are the smallest members of the peptide class. Any compound containing amino acids joined by a peptide bond can be called a peptide. A compound with three amino acids is a tripeptide, one with four amino acids is a tetrapeptide, and so on. As the number of amino acids increases, the compound is referred to as a **polypeptide.** A biologically active polypeptide typically containing 50 or more amino acids joined together by peptide bonds is a **protein.**

---

sample problem 10.3   **Identifying the Components of a Peptide**

Consider the following dipeptide:

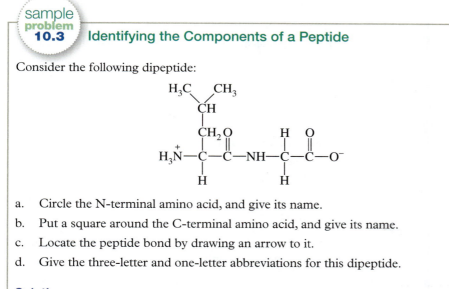

a.   Circle the N-terminal amino acid, and give its name.

b.   Put a square around the C-terminal amino acid, and give its name.

c.   Locate the peptide bond by drawing an arrow to it.

d.   Give the three-letter and one-letter abbreviations for this dipeptide.

Solution

a., b., and c.

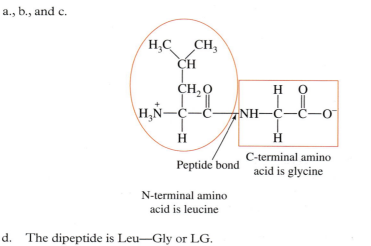

d.   The dipeptide is Leu—Gly or LG.

10.11 Write the products for the following condensation or hydrolysis reactions:

a.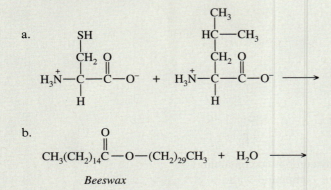

b.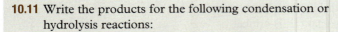

Beeswax

10.12 Write the products for the following condensation or hydrolysis reactions:

a.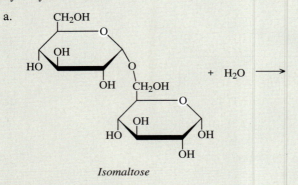

Isomaltose

b.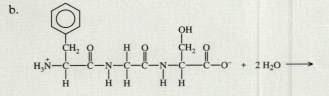

10.13 Consider the following tripeptide:

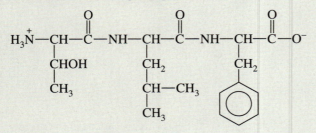

a. Draw a circle around the N-terminal amino acid, and give its name. Draw a square around the C-terminal amino acid, and give its name.

b. Give the one-letter and three-letter abbreviations of this tripeptide.

10.14 Consider the following tripeptide:

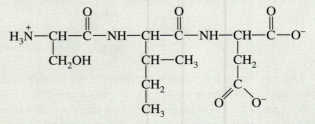

a. Draw a circle around the N-terminal amino acid, and give its name. Draw a square around the C-terminal amino acid, and give its name.

b. Give the one-letter and three-letter abbreviations of this tripeptide.

10.15 Draw the structural formula for each of the following peptides:
a. Phe—Glu
b. KCG
c. His—Met—Gln
d. WY

10.16 Draw the structural formula for each of the following peptides:
a. Ala—Asn—Thr
b. DS
c. Val—Arg
d. IYP

10.17 Label the N-terminus and the C-terminus for each of the peptides in Problem 10.15.

10.18 Label the N-terminus and the C-terminus for each of the peptides in Problem 10.16.

10.3 The Three-Dimensional Structure of Proteins

In Chapter 3, we saw that molecules are three-dimensional and that the shapes of simple molecules are determined by the charge clouds around a given atom using the VSEPR model. Peptides and proteins also have three-dimensional shapes or structures, but even the simplest dipeptide composed of two glycines has 17 atoms. It would be nearly impossible to determine its overall three-dimensional shape using the VSEPR model.

Because proteins are more complex than simple organic molecules, a hierarchy was developed to describe the structure of proteins. This system organizes the structure of a protein into four levels: primary (1°), secondary (2°), tertiary (3°), and quaternary (4°). Each of these levels is a result of interactions between amino acids in the protein. Taken together, these four levels provide a complete description of the three-dimensional structure of a protein.

10.3 Inquiry Question: How is a protein's structure held together?

Primary Structure

Like different colored beads on a string, the **primary (1°) structure** of a protein is the order in which the amino acids (beads) are joined together by peptide bonds to form the **protein backbone** that lines up from the N-terminus to the C-terminus. The side chains of the amino acids are substituents dangling from this backbone.

The actual sequence of amino acids is important. To illustrate, consider the word *protein*. Rearranging the letters in the word, we can make the word *pointer*, which has no meaning in common with the word *protein*. The same seven letters are used to make each word, but the meaning of each word is different. So it is with the primary structure of proteins. Arranging the amino acids in a different order creates a different peptide that no longer has the same function.

Figure 10.1 shows the primary structure of the eight amino acid peptide *angiotensin II*. This small peptide is involved in blood-pressure regulation in humans. Its amino acid sequence is Asp—Arg—Val—Tyr—Ile—His—Pro—Phe (DRVYIHPF). Any other ordering of these same eight amino acids would not function as angiotensin II. The order of the amino acids determines the function of a protein.

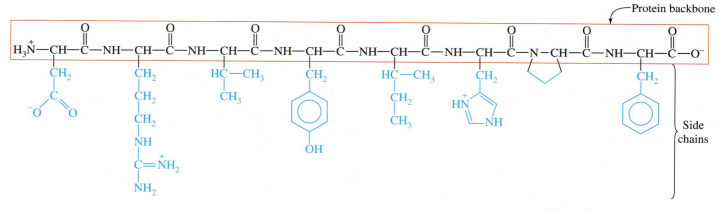

FIGURE 10.1 Primary structure of angiotensin II. Angiotensin II is a peptide involved in the regulation of blood pressure in humans. Notice that the side chains (blue) are not part of the backbone. Can you locate the N-terminus and C-terminus?

Secondary Structure

Have you ever stretched a Slinky® toy or a spring and noticed the spiral shape of its coils? The **secondary (2°) structure** of a protein can form coils or other repeating patterns of structure within the three-dimensional shape of a protein.

The two most common secondary structures are the **alpha helix (α helix)** and the **beta-pleated sheet (β-pleated sheet).** The secondary structures in a protein are stabilized by hydrogen bonding along the protein backbone and involve amino acids that are near each other in the primary structure. (Recall from Chapter 7 that hydrogen bonding is not a covalent bond, but a strong attractive force.)

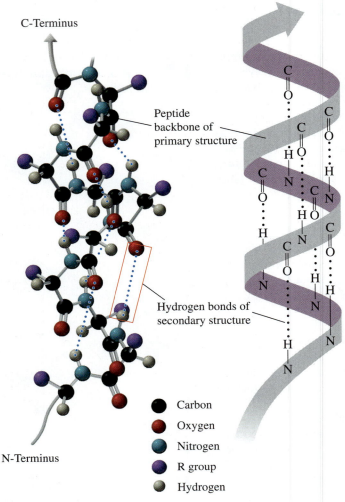

C-Terminus

Peptide backbone of primary structure

Hydrogen bonds of secondary structure

● Carbon
● Oxygen
● Nitrogen
● R group
● Hydrogen

N-Terminus

FIGURE 10.2 The α helix. The α helix has the hydrogen bonds positioned along the axis of the helix. There are three amino acids between the carbonyl and amine hydrogen that form the hydrogen-bonding interaction. Often the backbone is represented by a ribbon as shown on the right.

FIGURE 10.3 β-pleated sheet. Adjacent strands of the protein backbone are held together by hydrogen bonds in the β-pleated sheet. Alternating R groups (front and back) give the backbones a zigzag or pleated shape.

The α helix is a coiled structure in which the protein backbone adopts the form of a right-handed coiled spring. As shown in **Figure 10.2**, the helical structure is stabilized by hydrogen bonds formed between the carbonyl (C=O) oxygen atom (δ^-) of one amino acid and the N—H hydrogen atom (δ^+) of the amino acid located on the fourth amino acid away from it in the primary structure. The positioning of the hydrogen bonds along the axis of the helix allows it to stretch and recoil. Multiple hydrogen-bonding interactions along the backbone make the helix a strong structure. In the α helix, the side chains of the amino acids project outward away from the axis of the helix.

The β-pleated sheet is an extended structure in which segments of the protein chain align to form a zigzag structure much like a folded paper fan. As shown in **Figure 10.3**, strands called beta strands are held together side by side by hydrogen-bonding interactions between their backbones. In the β-pleated sheet, the side chains of the amino acids project above and below the sheet.

In summary, secondary structures involve hydrogen bonding along the backbone of the protein chain. Interactions of the amino acid side chains within the structure of a protein lead to the next level of structure.

Tertiary Structure

Imagine a coiled phone-charger cord or an old telephone cord lying on a desk with short lengths of the coils arranged next to or piled on top of each other. Like the coils of a cord, the α helices and β-pleated sheets of the polypeptide chain interact with each other and their environment to create a three-dimensional shape that is described by the **tertiary structure (3°).**

The folding and twisting of the polypeptide chain is caused by both hydrophobic and hydrophilic interactions between the side chains of the amino acids and the surrounding aqueous

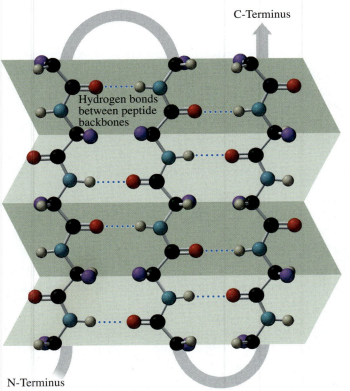

C-Terminus

Hydrogen bonds between peptide backbones

N-Terminus

environment. Like the soap micelle that we saw in Chapter 7, the nonpolar side chains in proteins are repelled by an aqueous environment and turn toward the interior of the protein even as the polar side chains are attracted to the aqueous surroundings and appear on the surface. The tertiary structure is stabilized by the attractive forces between the side chains and the aqueous environment of the protein as well as by the attractive forces between the side chains themselves. In an attempt to satisfy all the competing interactions, the protein folds into a specific three-dimensional shape.

The interactions involved when a protein folds into its tertiary structure (see **Figure 10.4**) include the following:

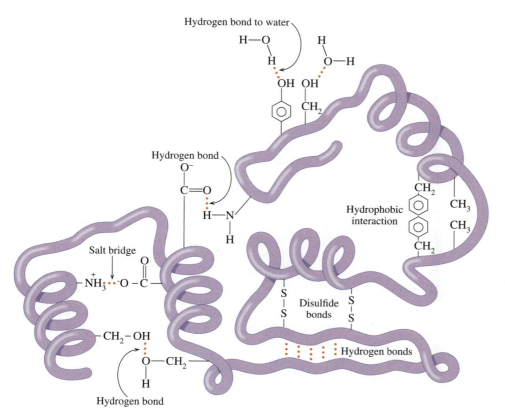

FIGURE 10.4 Protein tertiary structure. The interactions of amino acid side chains with their environment and each other stabilize the tertiary structure of a protein.

1. **Nonpolar interactions** The nonpolar amino acid side chains are *unlike* the aqueous (polar) environment. They are repelled by the aqueous environment and aggregate in the interior of the protein. The ability of water to exclude the nonpolar side chains is referred to as the **hydrophobic effect** and drives the three-dimensional folding of proteins. The nonpolar side chains, gathered together in the interior of the protein, interact with each other through London forces, helping to stabilize the tertiary structure. Figure 10.4 shows an example of two aromatic side chains that stack and two methyl groups in hydrophobic interactions.

2. **Polar interactions** Polar amino acid side chains are *like* the surrounding aqueous environment and easily interact with water and each other through the following hydrophilic interactions: dipole–dipole, ion–dipole, and hydrogen bonding interactions. Because they are charged, polar acidic and polar basic side chains form ion–dipole interactions with water. The side chains of most of the polar amino acids (cysteine being an exception) can interact with each other and with water in the surrounding environment through hydrogen bonding (see Figure 10.4).

3. **Salt bridges (ionic interactions)** Acidic and basic amino acid side chains exist in their ionized form in an aqueous environment. In other words, the acids are carboxylate ions ($-COO^-$), and the bases exist as the protonated amines ($-NH_3^+$). When a protein folds and a carboxylate ion on one amino acid side chain folds up near a protonated amine on a second amino acid side chain, the opposite charges attract, thereby forming a stabilizing ionic interaction also called a *salt bridge* (see Figure 10.4 for an example).

4. **Disulfide bonds** If two cysteine side chains are brought close together during the folding of a protein, the two —SH groups (thiols) can react with each other through an oxidation reaction (losing hydrogens) to form a disulfide bond (—S—S—). An example is shown in Figure 10.4. Note that the disulfide bond is a covalent bond (electrons being shared between two atoms), in contrast to the other stabilizing attractive forces.

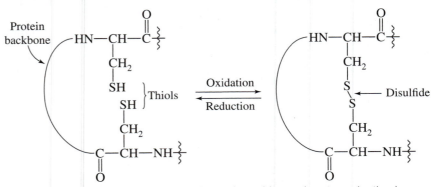

A disulfide bond can form when two cysteine amino acids are close to each other in space.

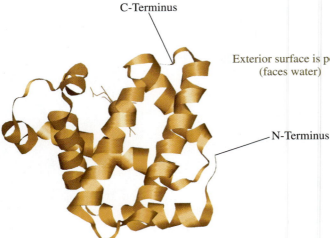

C-Terminus

Exterior surface is polar (faces water)

N-Terminus

FIGURE 10.5 Globular protein. Globular proteins like myoglobin fold into spherical structures with polar amino acid side chains on their surface and nonpolar amino acid side chains in the interior. The ribbon shown represents the protein backbone.

Proteins can be loosely classified into groups based on how they fold into their resulting three-dimensional shape. **Globular proteins** are proteins that fold into a compact, spherical shape. Such proteins have the tertiary structure just described; that is, they have polar amino acid side chains on their surface, enhancing their solubility in water while the nonpolar amino acid side chains gather in the center of the spherical structure forming a nonpolar core. Myoglobin is an example of a globular protein (see Figure 10.5). It stores oxygen in skeletal muscle. Although myoglobin is present in human muscle, it exists in much higher concentration in the muscles of sea mammals such as dolphins and whales and is the reason these animals can stay underwater for long periods of time without surfacing to breathe. Enzymes and many cellular proteins are globular proteins (see Section 10.5).

Fibrous proteins, on the other hand, have long, thread-like structures. Together, the aligned helices form strong, durable structures. Two familiar examples of fibrous proteins are keratins and collagen (see Figure 10.6). Keratins are found in hair, nails, and the scales of reptiles. Collagen is found in connective tissue, skin, tendons, ligaments, and cartilage. Fibrous proteins tend to be insoluble in water.

FIGURE 10.6 Fibrous protein. (a) The wool of sheep contains the fibrous protein alpha keratin. (b) Tendons and connective tissue contain collagen, which has three alpha helices twisted around each other in a triple helix providing rope-like strength.

a.

α-Keratin

α Helix

b.

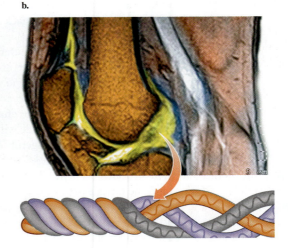

Collagen and Vitamin C

Scurvy, a disease caused by a deficiency of vitamin C in the diet, affects collagen formation. Collagen contains mainly the amino acids glycine, proline, and alanine along with a modified amino acid called hydroxyproline. The additional hydroxyl group on hydroxyproline allows the polypeptide chains to form hydrogen bonds between the chains, adding extra strength to the triple helix formed in collagen. Vitamin C is critical to the conversion of proline to hydroxyproline. When hydroxyproline is absent, collagen becomes weakened and results in the common symptoms of spongy and bleeding gums, opening of healed scars, and nail loss. Scurvy can be reversed by resuming a diet containing normal amounts of vitamin C. In the United States, several groups are at risk for vitamin C deficiency. These include smokers and people who do not get enough fresh produce or do not take vitamin supplements.

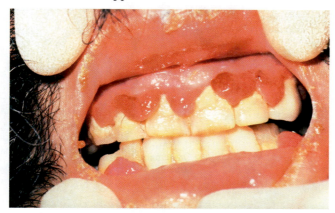

Individuals suffering from scurvy (vitamin C deficiency) can have bleeding gums due to poor collagen formation.

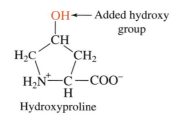

Hydroxyproline

Quaternary Structure

Imagine two or more of the piled-up cords mentioned in the description of tertiary structure neatly arranged next to each other. The **quaternary (4°) structure** describes the interactions of two or more polypeptide chains to form a single biologically active protein. The individual polypeptide chains or subunits are held together by the same interactions that stabilize the tertiary structure of a single protein.

The best-studied example of a protein with a quaternary structure is hemoglobin, which transports oxygen in blood. Hemoglobin consists of four polypeptide chains or subunits—two identical alpha subunits and two identical beta subunits (as in Figure 10.7d). In adult hemoglobin, all four subunits must be present for the protein to properly function as an oxygen carrier.

It is important to note that not all biologically active proteins have a quaternary (4°) structure. Some proteins, myoglobin for example, are made up of a single polypeptide chain.

Figure 10.7 and Table 10.3 summarize the structural levels of proteins.

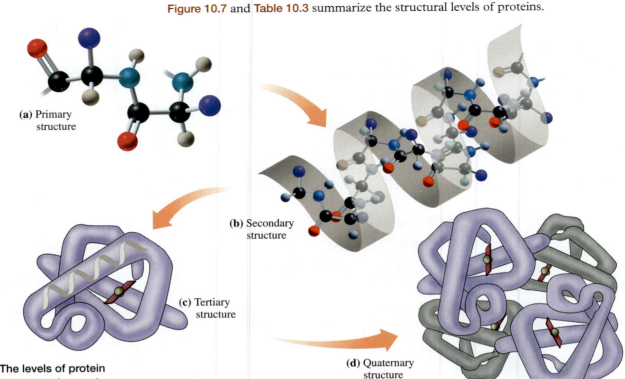

FIGURE 10.7 The levels of protein structure. Primary, secondary, and tertiary protein structures build upon each other as shown. Some proteins also have a quaternary structure.

(a) Primary structure

(b) Secondary structure

(c) Tertiary structure

(d) Quaternary structure

TABLE 10.3 Levels of Protein Structure and Stabilizing Forces Present

Level of Structure	Forces Stabilizing Structure
Primary (1°)	Peptide bonds
Secondary (2°)	Hydrogen bonding along the protein backbone atoms between amino acids close together in sequence
Tertiary (3°)	London forces, hydrogen bonding, dipole–dipole and ion–dipole interactions, ionic salt bridges, and disulfide bonds between the R groups of amino acids far away from each other in sequence
Quaternary (4°)	Same as tertiary structure but between subunits

sample problem 10.4   Side Chains in Tertiary Structure

Predict whether each of the following amino acid side chains would more likely be on the surface or in the interior of a globular protein after it has folded into its tertiary structure. Assume the protein is in an aqueous environment.

a.   tyrosine   b.   leucine

Solution

a.   In Table 10.1, tyrosine is classified as a polar amino acid. Therefore, it would be on the surface of the protein. The —OH group of the tyrosine side chain would form hydrogen bonds with water.

b.   In Table 10.1, leucine is classified as a nonpolar amino acid. This means that it would be repelled by the aqueous environment and would be found along with other nonpolar amino acids in the interior of the protein.

10.19 What type of bonding is present in the primary structure of a protein?

10.20 How many different tripeptides that contain one leucine, one glutamate, and one tryptophan are possible?

10.21 Name the stabilizing intermolecular force in the secondary structure of a protein.

10.22 When a protein folds into its tertiary structure, how does the primary structure change?

10.23 Describe the differences in the shape of an α helix and a β-pleated sheet.

10.24 What type of interaction would you expect between the side chains of each of the following pairs of amino acids in the tertiary structure of a protein?
a. histidine and aspartate
b. alanine and valine
c. two cysteines
d. serine and asparagine

10.25 What type of interaction would you expect between the side chains of each of the following pairs of amino acids in the tertiary structure of a protein?
a. lysine and glutamate
b. leucine and isoleucine
c. threonine and tyrosine
d. glutamine and arginine

10.26 Determine whether each of the following statements describes the primary, secondary, tertiary, or quaternary structure of a protein.
a. Side chains interact to form disulfide bonds.
b. Peptide bonds join amino acids in a polypeptide chain.

c. Two polypeptide chains are held together by hydrogen bonds.
d. Hydrogen bonding between amino acids in the same polypeptide gives a coiled shape to the protein.

10.27 Determine whether each of the following statements describes the primary, secondary, tertiary, or quaternary structure of a protein.
a. Three polypeptide chains interact to form a biologically active protein.
b. Hydrogen bonds form between adjacent segments of the backbone of the same protein to form a "folded-fan" structure.
c. Nonpolar side chains are repelled by water and move to the interior of the protein.
d. Amino acids react in a condensation reaction to form a peptide bond.

10.28 Myoglobin is a protein containing 153 amino acids. Approximately half of the amino acids in myoglobin have polar side chains.
a. Where would you expect these amino acid side chains to be located in the tertiary structure of the protein?
b. Where would you expect the nonpolar side chains to be?
c. Would you expect myoglobin to be more or less soluble in water than a protein composed mainly of nonpolar amino acids? Why?

10.4 Denaturation of Proteins

Have you ever cracked open an egg and dropped it into a hot pan? When you do so, the clear part of the egg, called the egg white, quickly turns from clear to white. You are observing the **denaturation** of the proteins in the egg white.

Denaturation of a protein is a process that disrupts the stabilizing attractive forces in the secondary, tertiary, or quaternary structure. When a protein is denatured, its primary structure is not changed.

In the example of the egg white, the denaturing agent is heat. An increase in temperature disrupts attractive forces such as hydrogen bonding, London forces, and other interactions. A change in pH can also denature a protein. When the pH of a protein's environment changes, the side chains of the acidic and basic amino acids alter their charges and, therefore, their ability to form salt bridges. Certain organic compounds or heavy-metal ions can also denature a protein by disrupting disulfide bridges. Table 10.4 gives several examples of denaturing agents and their effects on the various stabilizing forces of protein structure.

Without the stabilizing interactions necessary to maintain its three-dimensional structure, a protein will unfold into a shapeless string of amino acids. When a protein loses its three-dimensional shape, it also loses its biological activity.

? **10.4 Inquiry Question:**
What causes proteins to denature?

TABLE 10.4 Examples of Protein Denaturation

Denaturing Agent	Forces or Bonds Disrupted	Examples
Heat above 50 °C	Hydrogen bonds, hydrophobic interactions	Cooking food
Acids, bases, ionic compounds	Salt bridges, hydrogen bonds	Lactic acid from bacteria, which denatures milk proteins in the preparation of yogurt and cheese
Reducing agents	Disulfide bonds	Thiols, which are used in hairstyling for hair straightening or permanent waves
Detergents	Hydrophobic interactions	Membrane proteins
Heavy metal ions Ag^+, Pb^{2+}, Hg^{2+}	Disulfide bonds, salt bridges	Mercury and lead poisoning
Mechanical agitation	Hydrogen bonds, London forces	Whipped cream and meringue made from egg whites

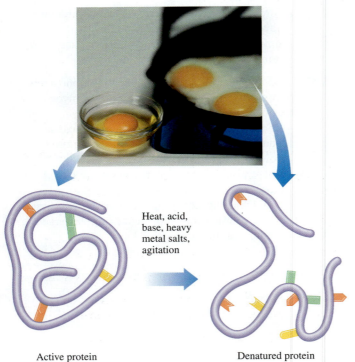

Heat, acid, base, heavy metal salts, agitation

Active protein   Denatured protein

When the attractive forces in a protein are disrupted, the protein is denatured.

Hair relaxing and perming involve protein denaturation. If you have curly hair and want it to be straight, you get it relaxed or straightened. If you have straight hair and you want it to be curly, you get a permanent wave. Both of these processes involve denaturing the proteins in your hair by disrupting the disulfide bonds found in the hair protein keratin, reshaping the hair, and forcing the disulfide bonds to reform at different points. Ammonium thioglycolate is one of the chemicals used to reduce the disulfides, and hydrogen peroxide is often used to oxidize the resulting thiols back to disulfide bonds.

Proteins and their denaturation can be used to treat lead or mercury poisoning. A person who has accidentally ingested a heavy metal like lead or mercury is given egg whites to drink. The proteins in the egg whites are denatured by the mercury or lead and the combination forms a precipitate. Typically, an emetic is then administered to induce vomiting to discharge the metal–protein precipitate from the body.

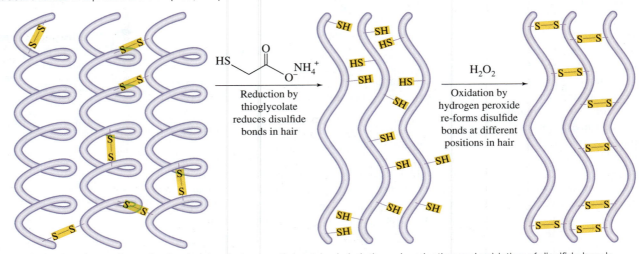

Hair relaxation and perming involve denaturing keratin proteins in hair through reduction and oxidation of disulfide bonds.

10.29 List the type of attractive force disrupted and the level of protein structure changed by the following denaturing treatments:
   a. adding salt to soy milk to make tofu
   b. whipping egg whites into stiff peaks to fold into an angel food cake

10.30 List the type of attractive force disrupted and the level of protein structure changed by the following denaturing treatments:
   a. adding a reducing agent to hair to relax (straighten) it
   b. baking the proteins in dough to make bread

practice problems

10.5 Protein Functions

Next, we consider some important roles of proteins and examples. Even though all proteins are made of the same twenty amino acids, the sequence of amino acids dictates the structure that the protein adopts, which, in turn, dictates protein function. Proteins have many functions in the body. Table 10.5 shows a variety of protein functions and gives some common examples.

? 10.5 Inquiry Question: What functions do proteins play in the body?

TABLE 10.5 Protein Functions and Examples

Function	Description	Protein Examples
Messenger	Control activities between cells	*Insulin* released from pancreas binds to receptors on other cell types.
Receptor	Control activities in the cell	*Insulin receptor* binds insulin on the cell surface, stimulating glucose uptake into cell.
Transport	Transport nutrients	*Glucose transporter* transports glucose from the bloodstream into the cell. *Hemoglobin* transports O_2 through the bloodstream to the tissues.
Storage	Store nutrients	*Casein* stores protein in milk. *Ferritin* stores iron in cells.
Contraction	Contract muscles	*Myosin* and *actin* make up the thick and thin filaments in muscle tissue.
Protection	Protect cells	*Antibodies* bind to foreign substances in the body, identifying them for elimination.
Structure	Maintain shape	*Collagen* supports the structure in ligaments, tendons, skin, nails, and horns. *Actin* is also found in the supportive structure of the cell (cytoskeleton).
Catalysis	Catalyze biochemical reactions	*Pepsin* breaks peptide bonds in proteins in the stomach during digestion.

Proteins as Messengers, Receptors, and Transporters

Insulin is a protein released from the pancreas as we digest sugar. Insulin's purpose is to signal the cells to take in glucose. Insulin is a hormone that acts as a messenger between cells. A **hormone** is a chemical, sometimes a peptide or protein, created in one part of the body that affects another part of the body.

How does insulin get its message to a cell? Sending a message leads us to another type of protein, a receptor. **Receptors** are proteins facing the outer surface of a cell that bind to a hormone or other messenger, triggering a signal inside the cell. In the case of insulin, it signals the synthesis of another protein, the glucose transporter (GLUT), a protein facilitating transport of glucose across the cell membrane. (For a discussion of transport across the cell membrane, refer to Section 8.7.) The glucose transporter is an **integral membrane protein** spanning a phospholipid bilayer. Figure 10.8 shows how insulin binds to the insulin receptor and affects glucose concentrations in the cell.

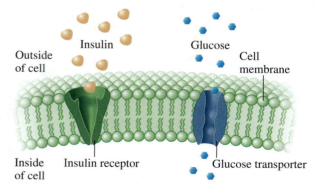

FIGURE 10.8 Insulin receptor and glucose transporter. When the glucose level increases in the bloodstream, insulin is released and travels through the bloodstream binding to insulin receptor proteins on cells. Insulin binding signals the cell to activate glucose transporter proteins at the cell membrane allowing glucose to flow into the cell.

Hemoglobin—Your Body's Oxygen Transporter

Another well-studied transport protein that we could not live without is hemoglobin. As mentioned in Section 10.3, the globular protein hemoglobin transports oxygen in the blood. Figure 10.9 shows hemoglobin's four polypeptide subunits: two α subunits in red and two β subunits in blue. Note that each of the subunits in hemoglobin contains many α helices in its secondary structure (represented by the curly ribbons) folded and packed together. These four folded subunits of hemoglobin are attracted to each other through attractive forces including hydrogen bonds, London forces, and salt bridges to form a protein quaternary structure that is biologically active.

Each subunit contains a nonprotein part called a **prosthetic group** that is vital to the protein's function. Hemoglobin's prosthetic group is called a heme (shown in green in Figure 10.9). Each heme group binds Fe^{2+}, which, in turn, binds oxygen (O_2). Each Fe^{2+} can bind one oxygen molecule for transport. Therefore, one molecule of hemoglobin can transport up to four molecules of oxygen.

The binding of O_2 to the Fe^{2+} of hemoglobin in the oxygen-rich environment of the lungs induces a change in the shape of the hemoglobin molecule. Biochemists call this reshaping where bond angles are only rotated and not broken a **conformational change.** The conformational change allows hemoglobin to hold the oxygen molecules long enough to deliver the oxygen to the tissues where oxygen levels are low. At the tissues, the oxygen dissociates from the hemoglobin, and the shape of the hemoglobin changes back to its deoxygenated form. Because of hemoglobin's unique structure, it is able to interact with its environment and function as an oxygen transport and delivery system.

Antibodies—Your Body's Defense Protein

When foreign substances like bacteria enter your body, your immune system produces proteins called antibodies (also called *immunoglobulins)* to recognize and destroy the foreign substances. The foreign substance recognized by an antibody is called an **antigen.**

Figure 10.10 shows the structure of an antibody. It consists of four polypeptide subunits, two heavy chains (higher mass, shown in blue and red), and two light chains (lighter mass, shown in yellow and green). The secondary structure contains β-pleated sheets (represented by the flat ribbons) that are stacked tightly together. As shown in the figure, the quaternary structure is held together through disulfide bridges between the polypeptide chains forming a unique "Y" shape, which is the biologically active form of the protein.

The stem of the Y is similar in all antibodies and can bind to receptors on a variety of cells in the body. Antibodies bind antigens at the top of each arm of the Y. Each antibody has a unique primary sequence at the top of the Y that recognizes a single antigen. This ability to recognize a single antigen and the distinctive Y structure of an antibody are well suited to bind antigens and present them for future destruction by the immune system.

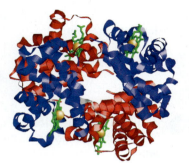

Oxygen unbound
(Deoxygenated state)

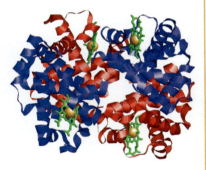

Oxygen bound
(Oxygenated state)

FIGURE 10.9 Hemoglobin. Hemoglobin consists of four polypeptide subunits (two α chains shown in red and two β chains shown in blue) that are held together mainly by salt bridges to form the biologically active protein. The heme is shown in green. When oxygen binds to Fe^{2+} (orange ball), a conformational change occurs as salt bridges between the subunits collapse, closing the hole formed in the center of the protein. Can you see the difference?

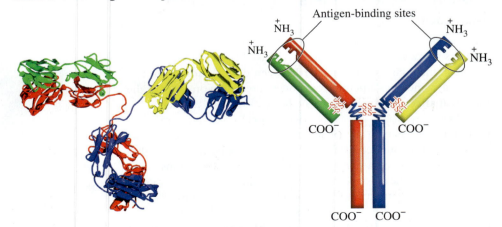

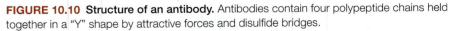

FIGURE 10.10 Structure of an antibody. Antibodies contain four polypeptide chains held together in a "Y" shape by attractive forces and disulfide bridges.

integrating chemistry

10.31 Match each protein in Column A with its function in Column B:

Column A	Column B
Collagen	Storage protein
Hemoglobin	Structural connector
Antibody	Oxygen transporter
Casein	Bind foreign substances in body

10.32 Match each protein in Column A with its shape in Column B:

Column A	Column B
Collagen	Y-shaped
Hemoglobin	Ropelike
Antibody	Integral membrane protein
Glucose transporter	Roughly spherical

10.33 The glucose transporter provides facilitated transport across the cell membrane. Do you think this transporter is active (a) after you eat a meal? (b) after you wake up in the morning? Support your answers.

10.34 Hemoglobin is found in blood, which is composed mostly of water. Would you expect the amino acid side chains on the surface of hemoglobin to be polar or nonpolar? Support your answer.

10.6 Enzymes—Life's Catalysts

Have you ever wondered how the food you eat is transformed into energy to move muscles and maintain your body? The process involves remarkable proteins called enzymes. Enzymes are typically large globular proteins and are present in every cell of your body. Enzymes act as *catalysts,* compounds that accelerate the reactions of *metabolism* (chemical reactions in the body) but are not consumed or changed by those reactions. Like all other catalysts, an enzyme cannot force a reaction to occur that would not normally occur. Instead, an enzyme simply makes a reaction occur faster.

Like most other proteins, enzymes are large molecules with complex three-dimensional shapes. Because many enzymes in your body function in an aqueous environment, they fold so that polar amino acids are on their surface. The final folded (tertiary) structure of an enzyme plays an important role in its function.

Let's look at the features of an actual enzyme, hexokinase. This enzyme's job is to catalyze the transfer of a phosphate group from the energy molecule adenosine triphosphate (ATP) to a six-carbon sugar (hexose) like D-glucose. It functions in the first step of glycolysis (the process by which glucose is broken down in the body) by adding a phosphate to the sixth carbon of glucose, forming the product glucose-6-phosphate.

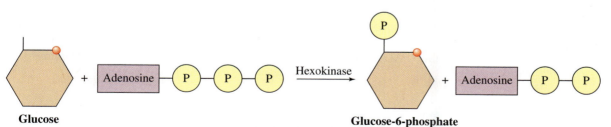

Glucose   **Glucose-6-phosphate**

The enzyme hexokinase catalyzes the transfer of a phosphate group from ATP to D-glucose.

In the chemical equation, the enzyme name usually appears above or below the reaction arrow. The phosphate group is represented by a P in a circle. It contains a phosphorus surrounded by four oxygen atoms. In section 3.2, we saw phosphate as one of the polyatomic ions, and in Chapter 4, phosphate was reintroduced as a functional group. In the body, phosphate is the anion hydrogen phosphate, HPO_4^{2-}, also called inorganic phosphate and abbreviated P_i.

? 10.6 Inquiry Question:
How does an enzyme catalyze a biochemical reaction?

The Active Site

Figure 10.11 shows the folded structure of the enzyme hexokinase. The protein backbone of hexokinase is represented as a ribbon. As hexokinase folds into its tertiary structure, an indentation forms on one part of its surface. This pocket is known as the **active site,** and it is lined with amino acid side chains. As its name implies, the active site is the functional part of the enzyme where catalysis occurs. The active site of hexokinase is labeled in Figure 10.11.

FIGURE 10.11 The active site of hexokinase. Hexokinase's protein structure is shown as a ribbon. During catalysis, glucose fits in the pocket of the active site between the upper and lower lobes of the protein. It is held in place by many hydrogen-bonding interactions (red dashes) with amino acid side chains.

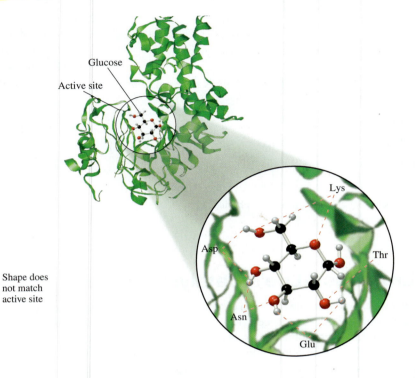

FIGURE 10.12 Enzymes and enantiomers. The pocket of an enzyme's active site (yellow) has a specific shape that can distinguish between the mirror image shapes of enantiomers.

Like a left foot fitting into a left shoe, glucose, the reactant for hexokinase, fits snugly into its active site. In an enzyme-catalyzed reaction, the reactant is called the **substrate.** The active site of hexokinase fits D-glucose only. Not even L-glucose, its enantiomer, or even D-galactose, its epimer (one chiral center difference), will fit into the active site.

Because the shape of an enzyme's active site is complementary to the shape of its substrate, most enzymes have only a few substrates, and many only one. This property of enzymes is called **substrate specificity.** Just as we know which shoe fits on which foot, many enzymes are specific to one enantiomer of the same compound (see **Figure 10.12**). The opposite enantiomer will not fit perfectly in the active site and cannot undergo catalysis.

Hexokinase also has a nonprotein "helper" in the form of a magnesium ion (Mg^{2+}) in its active site. Mg^{2+} assists with catalysis. There are two categories of nonprotein helpers—**cofactors** and **coenzymes.** Cofactors are inorganic substances like the magnesium ion that can be obtained in a normal diet from minerals. Coenzymes are small organic molecules, many of them derived from vitamins. For example, the coenzyme flavin adenine dinucleotide (FAD) is synthesized in the body from the vitamin riboflavin (vitamin B_2).

---

sample problem
10.5 **Enzymes and Substrates**

Describe the function of the active site of an enzyme.

Solution

The active site of an enzyme is a pocket uniquely fitted to the substrate and contains the amino acid side chains that catalyze the reaction.

Enzyme–Substrate Models

How does a substrate like glucose find the active site of an enzyme like hexokinase and situate itself for catalysis? Glucose is drawn to the active site of hexokinase by attractive forces like hydrogen bonding. Five polar amino acid side chains—asparagine (Asn), aspartate (Asp), glutamate (Glu), lysine (Lys), and threonine (Thr)—form hydrogen bonds to glucose, orienting it correctly within hexokinase's active site for catalysis (see Figure 10.11).

This initial interaction of an enzyme with a substrate is called the **enzyme–substrate complex (ES).** The formation of this complex occurs before catalysis can begin. This complex is held together by attractive forces.

Several models describe how the enzyme interacts with the substrate to form the ES. Two will be described here. The first model to be introduced was the **lock-and-key model.** In this model, the active site is thought of as a rigid, inflexible shape that is an exact complement to the shape of its substrate. According to the lock-and-key model, each enzyme accommodates one and only one substrate, much as a lock works with only one key (see Figure 10.13a).

Now we know that many enzymes can react with two or more similar substrates, so a new model called the **induced-fit model** was developed. In this model, the enzyme's active site is flexible and has a shape that is roughly complementary to the shape of its substrate. As the substrate interacts with the enzyme, the active site undergoes a conformational change, adjusting to fit the shape of the substrate (see Figure 10.13b). The shape of the substrate may change slightly as well.

Hexokinase and glucose form an enzyme–substrate complex that is a good example of the induced-fit model. In Figure 10.14, the shape of hexokinase is shown in a ball-and-stick representation. Notice how the shape changes when the enzyme interacts with glucose. The two lobes of the enzyme close snugly around the substrate to form the ES.

a. Lock-and-key Model

Substrate
+

Enzyme

Enzyme–substrate complex

b. Induced-fit Model

Substrate
+

Enzyme

Enzyme–substrate complex

FIGURE 10.13 Models for enzyme–substrate interaction. (a) In the lock-and-key model, the enzyme and substrate have rigid complementary shapes that fit together to form ES. (b) In the induced-fit model, the enzyme and substrate have flexible but similar shapes that adjust when the substrate gets closer to the enzyme, thereby forming a unique enzyme–substrate complex.

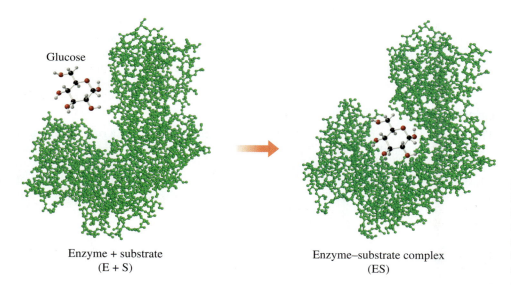

Glucose

Enzyme + substrate
(E + S)

Enzyme–substrate complex
(ES)

FIGURE 10.14 Interaction of hexokinase and glucose through induced fit. When glucose enters the active site of hexokinase (left), the enzyme's two lobes come together, wrapping glucose snugly into its active site forming the enzyme–substrate complex (right).

Rates of Reaction

Recall from Chapter 5 that a catalyst speeds up a reaction by *lowering* the activation energy. Because enzymes are catalysts, they do not affect the thermodynamics of the reaction (whether the reaction is exergonic or endergonic). Enzymes only speed up the formation of the products. An enzyme-catalyzed reaction does this by first forming ES and only then proceeding to form products. Figure 10.15 shows how a reaction energy diagram first introduced in Chapter 5 would be modified for an enzyme-catalyzed reaction.

Going back to the hexokinase reaction, glucose and ATP react in the active site of the enzyme and a phosphate is transferred from ATP to glucose. The products of the reaction are glucose-6-phosphate and adenosine diphosphate, ADP. If hexokinase is not present, the reaction between glucose and ATP occurs more slowly. What is it about the environment of the active site of hexokinase that causes these two molecules to react more quickly to form products?

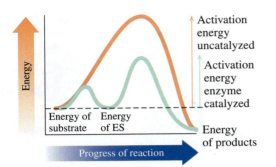

FIGURE 10.15 Effect of an enzyme on a chemical reaction. Like other catalysts, an enzyme lowers the activation energy of a reaction, thereby increasing the rate of the reaction. Enzymes accomplish this by first forming ES.

In the absence of an enzyme, it is more difficult for substrates to get close enough and orient themselves to react.

In the presence of an enzyme, it is easier for substrates to get close enough and orient themselves to react.

Thus far, we have seen that the first step in an enzyme-catalyzed reaction is the formation of ES. We have also seen that catalysts lower the activation energy of a reaction. Putting these two pieces of information together, it makes sense that the formation of ES is key to lowering the activation energy and allowing a biochemical reaction to proceed and quickly form products.

The activation energy lowering is accomplished during the formation of ES through interactions between the enzyme and the substrate. Each interaction releases a small amount of energy, stabilizing the complex. These small interactions combine to substantially lower the activation energy for an enzyme-catalyzed reaction versus its uncatalyzed reaction. Let's look at some interactions that help to speed up the reaction and lower the activation energy.

Proximity

The active site of an enzyme has a small volume. When the ES forms, the active site is "filled" with substrate. This means that the reacting molecules (for hexokinase, glucose, and ATP) are in close proximity to each other and to the catalytic side chains of the amino acids lining the active site. The closer they are, the more likely they will react. The activation energy is lowered by this effect. The substrates don't have to find each other as they would in solution. They are already close together in the active site.

The amino acid side chains in the active site of an enzyme are like tools that the enzyme uses to help facilitate the reaction. These side chains are often the functional groups of the acidic and basic amino acids.

Orientation

In the active site of an enzyme, the substrate molecules are held at the appropriate distance and in correct alignment to each other to allow the reaction to occur. The arrangement of the amino acid side chains in the active site creates interactions that orient the substrates so they will react. Without an enzyme to orient the substrates, substrates bumping into each other in solution may not react because of incorrect positioning. The enzyme guarantees that the substrates line up correctly. Correct orientation helps lower the activation energy required.

Bond Energy

Think of a bond as a rubber band. Which is easier to cut with scissors: a rubber band that is stretched or one that is slack? The stretched one is easier to cut because the rubber molecules are weakened or strained and it takes less energy to break the molecules with scissors. Likewise, when an enzyme interacts with its substrate to form ES, the bonds of the substrate molecule(s) are weakened (strained). The weakening of the bonds means that they will more readily react. In other words, weaker bonds break more easily and the activation energy is lowered by this effect.

Looking back at our hexokinase reaction, the coenzyme Mg^{2+} holds the ATP molecule in one area of the active site and the glucose interacts with another part of the site. The amino acid side chains in the active site of hexokinase form multiple hydrogen bonds with the glucose (refer to Figure 10.11). Each hydrogen-bonding interaction formed stabilizes ES and lowers the activation energy.

When glucose enters the active site, the enzyme undergoes a significant conformational change causing the two lobes of the enzyme to close around the substrate (following the induced-fit model). With the two lobes of the enzyme closer together, the ATP is moved closer to the glucose and is in proper alignment for the reaction.

When in the active site of hexokinase, glucose is positioned so that the hydroxyl group on C6, activated by an aspartate side chain, is able to remove a phosphate from ATP to form glucose-6-phosphate + ADP (see Figure 10.16). Because the enzyme is less attracted to the products than the substrates, the lobes of the enzyme move apart and the products are released.

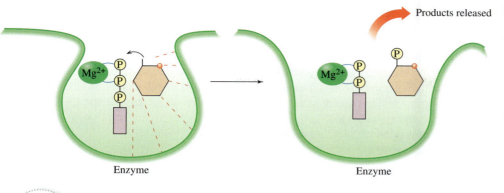

Products released

FIGURE 10.16 Catalysis by hexokinase. The coenzyme Mg^{2+} holds ATP in place (blue lines) and hydrogen bonds (red lines) hold glucose in the active site. The O on C6 (at black arrow tail) is perfectly positioned to form a bond to the end phosphate of ATP. The product do not fit the active site as snugly as the substrates and are released.

Enzyme   Enzyme

sample problem 10.6   Rate of Catalysis

List three factors that contribute to lowering the activation energy and speeding up the reaction when ES is formed.

Solution

Three factors are (1) close proximity of reactants to each other and to interactions with amino acid side chains, (2) favorable orientation of reactants, and (3) weakening of bond energy in the reactants.

practice problems

10.35 What level of protein structure is involved in the formation of an enzyme's active site?

10.36 Describe how a substrate is drawn to an enzyme to form ES.

10.37 Which model for enzyme–substrate interaction describes the action of hexokinase? Describe how the enzyme changes when the enzyme and glucose form ES.

10.38 Describe the key difference in the lock-and-key and induced-fit models.

10.39 What kind of interaction attracts the cofactor Mg^{2+} and ATP to each other? **HINT:** *Look at the structure of the phosphate group.*

10.40 In this section, we saw that glucose is held in the active site of hexokinase by hydrogen bonding. The outline of the active site of a hypothetical enzyme and its substrate glyceraldehyde is represented in the

following figure. Draw dotted lines to show the possible hydrogen bonds between glyceraldehyde and the hypothetical enzyme's active site.

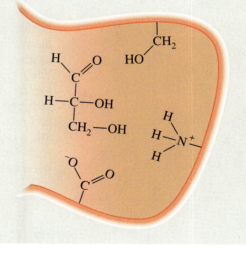

Discovering the Concepts

❓ Inquiry Activity—Factors That Affect Enzyme Activity

Information

The activity of an enzyme (a measure of how fast it catalyzes a chemical reaction) can be affected by changes in the enzyme's environment. Enzymes are most efficient at a pH and temperature optimum.

Questions

1. Pepsin and trypsin are enzymes that catalyze the digestion of proteins. Pepsin has a pH optimum of 2 and trypsin has a pH optimum of 7 to 8. Which of these two enzymes is likely to digest proteins in the stomach?
2. Lactase, the enzyme that hydrolyzes the disaccharide lactose, operates at a pH optimum of about 6.5 and a temperature optimum of 37 °C. How would the following changes affect the rate of catalysis—increase, decrease, or stay the same?
   a. lowering the pH to 2
   b. raising the temperature to 50 °C
   c. increasing the amount of lactose available

Enzyme Inhibition

An enzyme's activity can be reversibly decreased or inhibited if (a) a molecule structurally similar to the normal substrate competes for the active site or (b) if a molecule binds to a second site on the enzyme, changing the shape of the active site.

Questions

1. Which of the two scenarios, (a) or (b), would be called *competitive inhibition*?
2. Which of the two scenarios, (a) or (b), would be called *noncompetitive inhibition*?
3. Succinate is the substrate for the enzyme succinate dehydrogenase, one of the eight enzymes found in the citric acid cycle. Malonate is a reversible inhibitor of succinate dehydrogenase. What type of inhibitor is malonate likely to be? Explain your reasoning.

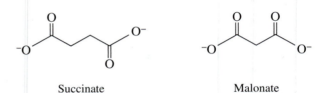

Succinate   Malonate

4. The enzyme hexokinase catalyzes the first step in glycolysis: glucose → glucose-6-phosphate. The activity of this enzyme can be inhibited by the buildup of the product glucose-6-phosphate. Binding glucose-6-phosphate to a second site on the enzyme changes the shape of the active site. Which type of inhibition is this an example of?
5. Which type of inhibition do you think could be overcome by adding more substrate? Explain your answer.
6. Which type of inhibition does not allow the enzyme to reach top speed, even if more substrate is added? Explain your answer.
7. The analgesic aspirin acts as a type of inhibitor called a "suicide" inhibitor. Aspirin forms a covalent bond to the amino acid serine, which has its side chain in the active site of the enzyme cyclooxygenase. Provide an explanation for the name of this inhibition.

10.7 Factors That Affect Enzyme Activity

? 10.7 Inquiry Question:
What factors affect enzyme activity?

If you slice a pear in half and allow it to sit untouched, the flesh of the pear will turn brown. The brown color is caused by an oxidation reaction (using oxygen from the air) catalyzed by the enzyme polyphenol oxidase. The product of this reaction is responsible for the brown color. One method often used to slow down browning is to sprinkle the pear slices with lemon juice. How does lemon juice slow down the browning of pears?

First, lemon juice contains vitamin C, which can react with the product, temporarily removing the brownish color. Second, lemon juice contains citric acid that slows down or *inhibits* the enzymatic reaction by changing the pH environment of the enzyme. Enzyme-catalyzed reactions, like most other chemical reactions, are affected by the reaction conditions. These include changes in substrate concentration, pH, temperature, and the presence of inhibitors. Changing reaction conditions affects how fast an enzyme converts its substrate to product. Measuring how fast an enzymatic reaction occurs is a measure of an enzyme's **activity.**

Substrate Concentration

How does substrate concentration affect enzyme activity? Imagine a factory that assembles bicycles. The supplier brings in enough parts to produce 10 bicycles, and the workers with their tools build 10 bicycles in one hour. The manager of the factory asks the supplier to bring in enough parts to make 15 bicycles, and the workers also complete that task in one hour.

Impressed, the manager asks the supplier to bring in enough parts to make 50 bicycles. In one hour, the workers assemble only 25 bicycles. The manager then asks the supplier to bring in enough parts to make 100 bicycles, but the workers still assemble only 25 in one hour. The workers are working at their maximum capacity by assembling 25 bicycles in one hour. Regardless of the number of extra parts the manager orders, if the resources at the factory do not increase (for example, the number of workers, number of tools) only 25 bicycles can be built in one hour. Enzymes also operate at a particular rate, just like the workers with their tools, no matter how many substrate molecules (or bicycle parts) are present.

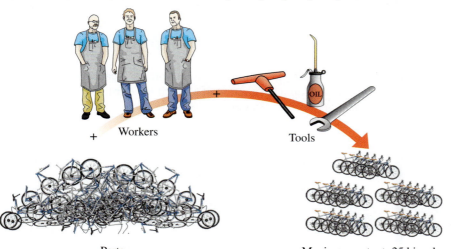

Parts   Maximum output: 25 bicycles

When workers (enzymes) are operating at a maximum rate, a steady state exists and will not increase even if an excess of bicycle parts (substrate) is present.

As we saw earlier, the first step in an enzyme-catalyzed reaction is the formation of ES. If the amount of enzyme remains unchanged, an increase in the substrate concentration increases the enzyme's activity up to the point where the enzyme becomes saturated with its substrate. That is, all of the enzyme becomes saturated with substrate, leaving no more free active sites. The enzyme is working at maximum activity. Increasing the amount of substrate will not increase the activity further. At maximum activity, the conditions under which the enzyme is operating are considered to be in a **steady state.** Under steady-state conditions, substrate is being converted to product as efficiently as possible (see Figure 10.17).

pH

Besides substrate concentration, the pH of an enzyme's environment is another condition that affects enzyme activity. Let's go back to the workers at the bicycle factory for an explanation. Suppose that it is flu season and several workers come to work with early stages of the flu, breathe on their coworkers, and infect most of the workforce. With fewer able-bodied workers at the factory, fewer bicycles can be assembled in an hour. Like the environment at the factory, when an enzyme's environment is changed (in this case the pH), its tertiary structure is disrupted, altering the active site and causing the enzyme's activity to decrease.

Enzymes are most active at a pH known as their **pH optimum.** At this pH, the enzyme maintains its tertiary structure and, therefore, its active site. Changes in the pH not only affect the structure of the enzyme but may also change the nature of an amino

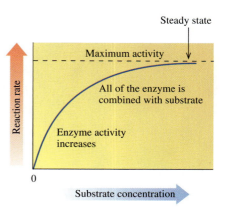

FIGURE 10.17 Effect of substrate concentration on enzyme activity. Increasing the amount of substrate increases the enzyme's activity until the enzyme becomes saturated with substrate and a steady state is reached.

acid side chain. For example, if an enzyme requires a carboxylate ion ($—COO^-$) to function properly, lowering the pH could convert the carboxylate ion to its carboxylic acid form ($—COOH$). This change would cause the enzyme's activity to decrease.

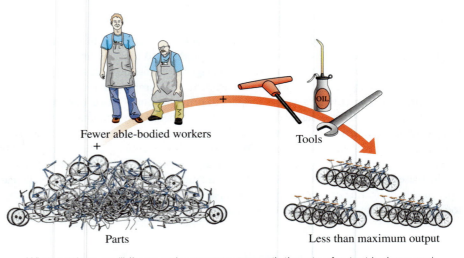

Fewer able-bodied workers
+
Parts

Tools

Less than maximum output

When workers are ill (fewer active enzymes present), the rate of output is decreased.

In the body, an enzyme's pH optimum is based on the location of the enzyme. For example, enzymes in the stomach function at a much lower pH because of the acidity. pH optimum values for several common enzymes are shown in Table 10.6.

TABLE 10.6 pH Optimum for Selected Enzymes

Enzyme	Location	Substrate	pH Optimum
Pepsin	Stomach	Peptide bonds	2
Sucrase	Small intestine	Sucrose	6.2
Urease	Liver	Urea	7.4
Hexokinase	All tissues	Glucose	7.5
Trypsin	Small intestine	Peptide bonds	8
Arginase	Liver	Arginine	9.7

Temperature

As with pH, enzymes have a **temperature optimum** at which they are most active. Let's consider the bicycle factory workers if the air conditioning broke down on a hot day in the middle of August. It is likely that the workers would not be able to work as efficiently to assemble 25 bicycles in one hour.

The temperature optimum for most human enzymes is normal body temperature, 37 °C. Above their optimum temperature, enzymes lose activity due to the disruption of the attractive forces stabilizing the tertiary structure. At high temperatures, an enzyme denatures, which in turn modifies the structure of the active site. At low temperatures, enzyme activity is reduced due to the lack of energy present for the reaction to take place at all.

Because enzymes are the major culprits in food spoilage, we store foods in a refrigerator or freezer to slow the spoilage process. The enzymes present in bacteria can also be destroyed by high temperatures, in processes like boiling contaminated drinking water or sterilizing instruments and other equipment in hospitals and laboratories.

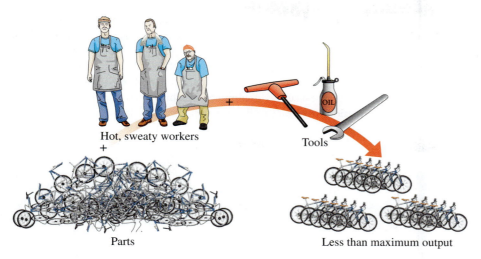

Hot, sweaty workers
+
Parts

Tools

Less than maximum output

When all the workers are present, but the temperature of the factory makes them hot (some enzymes denatured by temperature), the rate of output is decreased.

Inhibitors

Certain types of molecules known as **inhibitors** can also cause enzymes to lose activity. If we consider the bicycle factory, suppose some of the workers' tools became rusted and could not be used to assemble the bicycles. This would definitely inhibit bicycle production.

Enzyme inhibitors function in different ways, but they all prevent the active site from interacting with the substrate to form ES. Some inhibitors cause enzymes to lose catalytic activity only temporarily, while others function to cause enzymes to lose catalytic activity permanently.

In **reversible inhibition,** the inhibitor causes the enzyme to lose catalytic activity. However, if the inhibitor is removed, the enzyme becomes functional again. Reversible inhibitors can be competitive or noncompetitive.

Competitive inhibitors are molecules that compete with the substrate for the active site. A competitive inhibitor has a structure that resembles the substrate of the enzyme. The competitive inhibitor will interact with the active site to form an enzyme–inhibitor complex, but usually no reaction will take place. As long as the inhibitor remains in the active site, the enzyme cannot interact with its substrate and form product. This lowers the enzyme's activity. Figure 10.18 diagrams how a competitive inhibitor works.

A medical therapy based on competition involves the enzyme liver alcohol dehydrogenase (LAD). LAD oxidizes ethanol, the alcohol found in wine and spirits, as well as the compounds ethylene glycol and methanol, both found in antifreeze. All three are substrates of LAD and compete for the active site. Notice that the structures of the three are very similar.

$$CH_3CH_2OH \qquad HOCH_2CH_2OH \qquad CH_3OH$$
Ethanol   Ethylene glycol   Methanol

In fact, one remedy for ethylene glycol poisoning in pets and methanol poisoning in humans is the slow intravenous infusion of ethanol maintaining a controlled concentration in the bloodstream over several hours. This slows the production of the toxic metabolic products of ethylene glycol or methanol, giving the kidneys time to filter out the excess substrates in the urine.

Noncompetitive inhibitors typically do not resemble the substrate, so they do not compete for the enzyme's active site. Instead, these inhibitors bind to another site on the enzyme that is usually remote from the active site. When a noncompetitive inhibitor binds to the enzyme, it changes the shape of the enzyme. Therefore, the active site loses its shape or is distorted, so it is no longer able to interact effectively with the substrate.

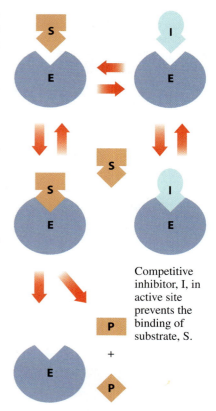

Competitive inhibitor, I, in active site prevents the binding of substrate, S.

FIGURE 10.18 Competitive inhibition. The structure of a competitive inhibitor is similar to the substrate, allowing it to compete for the active site.

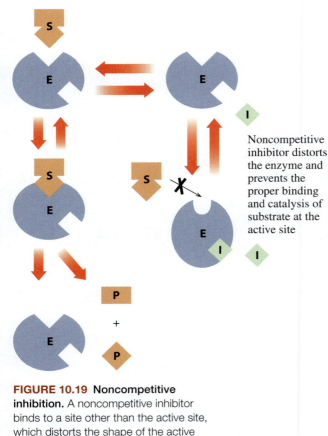

Noncompetitive inhibitor distorts the enzyme and prevents the proper binding and catalysis of substrate at the active site

FIGURE 10.19 Noncompetitive inhibition. A noncompetitive inhibitor binds to a site other than the active site, which distorts the shape of the active site, causing the enzyme to lose catalytic activity.

Again, as long as the inhibitor remains bound to the enzyme, the enzyme cannot function. Figure 10.19 diagrams how a noncompetitive inhibitor works.

As suggested in Figures 10.18 and 10.19, the inhibition caused by competitive and noncompetitive inhibitors can be reversed. Inhibition caused by a competitive inhibitor can be reversed by adding more substrate to the reaction. Because the inhibitor and the substrate compete for the active site, the higher the concentration of the substrate, the more likely that it will win the competition for the active site. In the case of the noncompetitive inhibitor, adding more substrate has no effect. Regardless of the amount of enzyme, a certain portion of the enzyme is inactivated by the inhibitor. Reversing the effect of a noncompetitive inhibitor typically requires a special chemical reagent to remove the inhibitor and restore the catalytic activity of the enzyme. Going back to the bicycle factory, if a technician removed the rust from all the workers' tools so they were again in working order, the number of bicycles produced by the factory would increase.

In **irreversible inhibition,** the inhibitor forms a covalent bond with an amino acid side chain on the enzyme. With the inhibitor covalently bonded in the active site, the substrate is excluded or the catalytic reaction is blocked. Regardless of the method, irreversible inhibitors permanently inactivate enzymes (see Figure 10.20). Likewise, if a fire destroyed a major portion of the bicycle factory, bicycle production would cease.

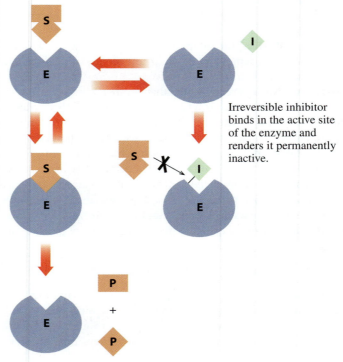

Irreversible inhibitor binds in the active site of the enzyme and renders it permanently inactive.

FIGURE 10.20 Irreversible inhibition. An irreversible inhibitor forms a covalent bond with the enzyme, rendering it permanently inactive.

Heavy transition metals like silver, mercury, and lead can inactivate enzymes irreversibly by binding to the thiol ($-SH$) functional groups of the amino acid cysteine if found in the active site. As we saw in Section 10.4, this can also denature the proteins.

Antibiotics Inhibit Bacterial Enzymes

Enzyme inhibitors have long been used in the battle against diseases. The well-known antibiotic penicillin is an irreversible inhibitor. Penicillin binds to the active site of an enzyme that bacteria use in the synthesis of their cell walls. The cell wall is a structure present in bacterial cells but not human cells. When the bacterial enzyme bonds with penicillin, the enzyme loses its catalytic activity, and the growth of the bacterial cell wall slows. Without a proper cell wall for protection, bacteria cannot survive, and the infection stops.

(R is a variable group)

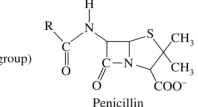

Penicillin

---

sample problem 10.7   **Factors Affecting Enzyme Activity**

Lactase catalyzes the hydrolysis of the disaccharide lactose into the monosaccharides galactose and glucose. It is active in the small intestine, where the pH is around 6.8. Supplements of lactase can help people who are lactose-intolerant to digest lactose.

$$\text{Lactose} \xrightarrow{\text{Lactase}} \text{galactose} + \text{glucose}$$

a. Would the enzyme be just as active in the stomach with a pH of 1–3? Support your answer.

b. Would a lactase supplement act just as well on lactose if it were first stirred into cold milk before drinking? Support your answer.

Solution

a. No. The enzyme would be less active in the stomach because it operates at a pH optimum of 6.8.

b. No. The enzyme has a temperature optimum of 37 °C, body temperature. Its activity would be much lower in cold milk.

---

practice problems

10.41 How would the following changes affect enzyme activity for an enzyme whose optimal conditions are normal body temperature and physiological pH?
a. raising the temperature from 37 °C to 60 °C
b. lowering the pH from 7.5 to 4.0

10.42 Chymotrypsin is an enzyme located in the small intestine that catalyzes peptide bond hydrolysis in proteins. Based on the examples given in Table 10.1, would you expect chymotrypsin to be most active at a pH of 4.5, 7.8, or 10.0?

10.43 The enzyme urease catalyzes the formation of ammonia and carbon dioxide from urea as shown:

$$\text{H}_2\text{N}-\overset{\overset{\displaystyle O}{\|}}{\text{C}}-\text{NH}_2 + \text{H}_2\text{O} \xrightarrow{\text{Urease}} 2\text{NH}_3 + \text{CO}_2$$

Describe what effect the following changes would have on the rate of this reaction assuming a steady state has been reached:

a. adding excess urea
b. lowering the temperature to 0 °C

10.44 Indicate whether each of the following describes a competitive or a noncompetitive inhibitor.
a. The structure of the inhibitor is similar to that of the substrate.
b. Adding more substrate to the reaction has no effect on the enzyme activity.
c. The inhibitor competes with the substrate for the active site.
d. The structure of the inhibitor has no resemblance to the structure of the substrate.
e. Adding more substrate to the reaction restores the enzyme activity.

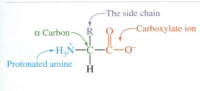

10.1 Amino Acids—A Second Look

10.1 Inquiry Question: What are the characteristics of an amino acid?

Amino acids contain a central carbon atom, called the α carbon, bonded to four different groups—a protonated amine (amino) group, a carboxylate group, a hydrogen atom, and a side chain. Because of this structure, amino acids, with the exception of glycine, are chiral compounds. The L-enantiomers of amino acids are the building blocks of proteins. The 20 different amino acids are found in most proteins. They are characterized by various side chains. The properties of the side chains determine whether the amino acids are classified as nonpolar or polar.

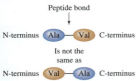

10.2 Protein Formation

10.2 Inquiry Question: How is a peptide bond formed?

Several biomolecules join through condensation reactions when functional groups combine. Amino acids join through a condensation reaction of the protonated amine group of one and the carboxylate group of the other. The bond that forms between the two amino acids is called a peptide bond, and the new structure is a dipeptide. The newly formed dipeptide has an N-terminus with a free protonated amine group and a C-terminus with a free carboxylate group. A compound containing 50 or more amino acids linked by peptide bond is a polypeptide and if it has biological activity is called a protein. Proteins are polymers of amino acids.

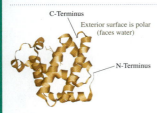

10.3 The Three-Dimensional Structure of Proteins

10.3 Inquiry Question: How is a protein's structure held together?

Four levels are used to describe protein structure. Each level is held together through covalent bonding or attractive forces. The primary structure (1°) is the sequence of the amino acids that form the protein backbone. The bonding interaction is the peptide bond. The secondary structure (2°) involves the interactions of amino acids near each other in the primary structure and describes patterns of regular or repeating structure. The most common secondary structures are the α helix and the β-pleated sheet. The secondary structure is stabilized by hydrogen bonding between atoms in the backbone. The tertiary structure (3°) is formed by folding the secondary structure onto itself and is driven by the hydrophobic interactions of amino acid side chains with their aqueous environment. This level is stabilized by the attractive forces between side chains and disulfide bonds. Proteins that fold into a roughly spherical shape are called globular proteins. Proteins that maintain elongated structures are referred to as fibrous proteins. Some proteins have a quaternary structure (4°), which involves the association of two or more peptides to form a biologically active protein. The same forces stabilize the quaternary structure as the tertiary structure.

10.4 Denaturation of Proteins

10.4 Inquiry Question: What causes proteins to denature?

Denaturation of a protein disrupts the stabilizing attractive forces in the secondary, tertiary, or quaternary structure, often unfolding the protein. When a protein is denatured, its primary structure is not changed. Proteins can be denatured by heat, a change in the pH of their environment, reaction with small organic compounds and heavy metals such as lead or mercury, or by mechanical agitation. A denatured protein is no longer biologically active.

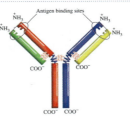

10.5 Protein Functions

10.5 Inquiry Question: What functions do proteins play in the body?

Proteins have a variety of functions in the body. They act as messengers between cells, receptors on the surface of cells, and transporters through the body or across the cell. Proteins are used to store nutrients, contract muscles, protect the cell, and support its structure. Proteins catalyze biochemical reactions as enzymes.

10.6 Enzymes—Life's Catalysts

10.6 Inquiry Question: How does an enzyme catalyze a biochemical reaction?

Enzyme–substrate complex

Enzymes are large, globular proteins that serve as catalysts in biological systems. The functional part of an enzyme is the active site, which is a small groove or cleft on the surface of the molecule where catalysis occurs. Substrates are the reactants in the reactions catalyzed by enzymes. Because of the three-dimensional shape of the active site, enzymes have few (many have only one) substrates that will bind and react. The lock-and-key and induced-fit models explain how an enzyme interacts with its substrate to form ES. The formation of ES lowers the activation energy for the catalyzed reaction in several ways. These include increasing proximity, optimizing orientation, and modifying bond energy.

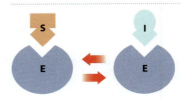

10.7 Factors That Affect Enzyme Activity

10.7 Inquiry Question: What factors affect enzyme activity?

Enzyme activity is measured by how fast an enzyme catalyzes a reaction. Environmental factors such as substrate concentration, pH, temperature, and the presence of inhibitors can affect enzyme activity. Enzymes have a pH optimum and temperature optimum. Inhibitors decrease or eliminate an enzyme's catalytic abilities. The effect of an inhibitor can be reversible or irreversible. Reversible inhibitors can be competitive inhibitors, which compete with the substrate for the enzyme's active site, or noncompetitive inhibitors, which bind to the enzyme at a site other than the active site, changing the shape of the active site.

The study guide will help you check your understanding of the main concepts in Chapter 10. You should be able to

10.1 Amino Acids—A Second Look

- Draw the general structure of an amino acid.
- Identify amino acids based on their polarity.

10.2 Protein Formation

- Predict the products of a biological condensation or hydrolysis reaction.
- Form a peptide bond between amino acids.

10.3 The Three-Dimensional Structure of Proteins

- Distinguish the levels of protein structure.
- Describe the attractive forces present as a protein folds into its three-dimensional shape.

10.4 Denaturation of Proteins

- Define protein denaturation.
- List the causes of protein denaturation and the attractive forces affected.

10.5 Protein Functions

- Identify various functions of proteins.
- Provide examples of protein structure dictating protein function.

10.6 Enzymes—Life's Catalysts

- Define active site and substrate.
- Distinguish the lock-and-key model from the induced-fit model.
- Discuss factors that lower the activation energy and speed reaction for an enzyme-catalyzed reaction.

10.7 Factors That Affect Enzyme Activity

- Describe how substrate concentration, pH, temperature, and inhibition affect enzyme activity.
- Distinguish competitive, noncompetitive, and irreversible inhibition.

Key Terms

active site—A pocket on the surface of an enzyme formed during the folding of the tertiary structure and lined with amino acid side chains; portion of the enzyme involved in the catalysis of chemical reactions.

activity—A measure of how fast an enzyme converts its substrate to product.

alpha (α) amino group—The protonated amine ($-NH_3^+$) bonded to the alpha carbon in an amino acid.

alpha (α) carbon—The carbon in an amino acid that is bonded to the carboxylate ion, the protonated amine, a hydrogen atom, and the side chain group.

alpha (α) carboxylate group—The carboxylate ion ($-COO^-$) bonded to the alpha carbon in an amino acid.

alpha (α) helix—A secondary protein structure resembling a right-handed coiled spring or a corkscrew.

amide—A family of organic compounds characterized by a functional group containing a carbonyl bonded to a nitrogen atom; amides are formed by a condensation reaction between a carboxylate ion and a protonated amine.

antigen—A molecule capable of eliciting the immune response of producing antibodies in the body.

beta (β)-pleated sheet—A secondary protein structure with a zigzag structure resembling a folded fan.

conformational change—A change in the tertiary or quaternary structure of a protein that occurs through bonds rotating, not bonds breaking.

coenzyme—A small organic molecule necessary for the function of some enzymes; many are obtained from the diet as vitamins.

cofactor—An inorganic substance necessary for the function of some enzymes; often obtained in the diet as minerals.

competitive inhibitor—A molecule structurally similar to the substrate that competes for the active site.

C-terminus (also C-terminal amino acid or carboxy terminus)—In a peptide, the amino acid with the free carboxylate ion, always written to the right of the structure.

denaturation—A process that disrupts the stabilizing attractive forces in the secondary, tertiary, or quaternary structure of a protein caused by heat, acids, bases, heavy metals, small organic compounds, or mechanical agitation.

dipeptide—Two amino acids joined by a peptide bond.

disulfide bond—Covalent bond formed between the $-SH$ groups of two cysteines in a protein; involved in stabilizing the tertiary and quaternary structure of the protein.

enzyme–substrate complex (ES)—The complex that is formed when an enzyme binds its substrate in its active site.

fibrous protein—A protein with long, thread-like structures with high helical content found in fibers such as hair, wool, silk, nails, skin, and cartilage.

globular protein—A protein with a roughly spherical tertiary structure that typically has nonpolar amino acid side chains clustered in its interior and polar amino acid side chains on its surface.

hormone—A chemical created in one part of the body that affects another part of the body.

hydrophobic effect—The movement of nonpolar amino acid side chains to the interior of a protein caused by an unfavorable interaction with the aqueous environment.

induced-fit model—A model describing the initial interaction of an enzyme to its substrate; an enzyme's active site and its substrate have flexible shapes that are roughly complementary and adjust to allow the formation of ES.

inhibitor—A molecule that causes an enzyme's catalytic activity to decrease.

integral membrane protein—A protein found within or spanning the phospholipid bilayer of a membrane; many serve as passages for polar compounds to move across the membrane.

irreversible inhibition—The permanent loss of enzyme catalytic activity.

lock-and-key model—A model describing the initial interaction of an enzyme with its substrate; each enzyme has only one substrate with a shape complementary to the active site of the enzyme.

noncompetitive inhibitor—A molecule that inhibits an enzyme's catalytic activity by interacting with a site other than the active site and distorting the shape of the active site.

nonpolar amino acids—Amino acids with side chains composed almost entirely of carbon and hydrogen, resulting in an even distribution of electrons over the side chain portion of the molecule.

nonpolar interactions—The association of nonpolar amino acid side chains with each other; London forces.

N-terminus (also N-terminal amino acid or amino terminus)—In a peptide, the amino acid with the free protonated amine, always written to the left of the structure.

peptide bond—An amide bond that joins two amino acids.

pH optimum—The pH at which an enzyme functions most efficiently.

polar amino acids—Amino acids whose side chains contain electronegative atoms, resulting in an uneven distribution of electrons over the side chain portion of the molecule.

polar interactions—The association of polar amino acid side chains with each other; includes hydrogen bonding, dipole–dipole interactions, and ion–dipole interactions.

polypeptide—A string of many amino acids, not necessarily biologically active.

primary structure—The first level of protein structure; the order of the amino acids bonded by peptide bonds forming a polypeptide chain.

prosthetic group—A non-amino acid portion of a protein required for protein function.

protein—Biologically active polymers typically containing 50 or more amino acids bonded together by peptide bonds.

protein backbone—The chain of amino acids in a protein N-terminus to C-terminus. The side chains hang from the main string.

quaternary structure—The highest level of protein structure that involves the association of two or more peptide chains to form a biologically active protein; can be stabilized by nonpolar interactions, polar interactions, ionic interactions, and disulfide bonds.

receptor—A protein facing the outer surface of a cell that binds to a hormone or other messenger triggering a signal inside the cell.

reversible inhibition—The lowering of enzyme catalytic activity caused by a competitive or noncompetitive inhibitor that can be reversed.

salt bridges (ionic interactions)—The attraction formed between a carboxylate ion on one amino acid's side chain with the protonated amine on a second amino acid's side chain.

secondary structure—The protein structural level that involves regular or repeating patterns of structure within the overall three-dimensional structure of a protein stabilized by hydrogen bonding along the protein backbone. Examples are alpha helix and beta sheet.

steady state—A reaction rate describing maximum enzyme activity when substrate is being converted to product as efficiently as possible.

substrate—The reactant in a chemical reaction catalyzed by an enzyme.

substrate specificity—The property that explains why enzymes have very few substrates and many only one substrate.

temperature optimum—The temperature at which an enzyme functions most efficiently.

tertiary structure—The protein structural level that involves the folding of the secondary structure on itself driven by the hydrophobic interactions of nonpolar amino acids with the protein's aqueous environment; stabilized by polar interactions, nonpolar interactions, ionic interactions, and disulfide bonds between amino acid side chains and by polar interactions with water molecules in the environment.

Additional Problems

10.45 What functional groups are common to all α-amino acids?

10.46 Give the name and three-letter abbreviation for the amino acid described by each of the following:
- **a.** the nonpolar amino acid with a sulfur atom in its side chain
- **b.** a polar amino acid with a single nitrogen atom in its side chain
- **c.** the nonpolar amino acid with only one carbon in its side chain

10.47 Give the name and three-letter abbreviation for the amino acid described by each of the following:
- **a.** the polar amino acid with a benzene ring in its side chain
- **b.** the nonpolar amino acid whose side chain forms a ring with its α-amino group
- **c.** the polar amino acid with a sulfur atom in its side chain

10.48 Give the name and three-letter abbreviation for the amino acid that does not have a chiral center.

10.49 Two of the amino acids have two chiral centers. Give the name and three-letter abbreviation of each.

10.50 Aspartame, which is commonly known as Nutrasweet, contains the following dipeptide:

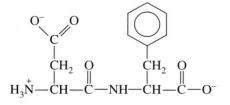

a. What are the amino acids in aspartame?
b. Give the three-letter abbreviation of the N-terminal amino acid.
c. Circle the peptide bond.
d. Draw the structure of the isomer of this dipeptide where the C-terminal and N-terminal amino acids are switched.

10.51 Vegetables, nuts, and seeds are often deficient in one or more of the essential amino acids. From the Protein Source column in the following table, choose combinations of two or more foods you could eat to get a complete protein.

Amino Acid Deficiency in Some Foods

Protein Source	Amino Acid Missing
Rice	Lys
Oatmeal	Lys
Peas	Met
Beans	Trp, Met
Almonds	Lys, Trp
Corn	Lys, Trp

10.52 Draw the structure of the possible dipeptides formed from one alanine combining with one cysteine.

10.53 Consider the amino acids glycine, proline, and lysine.
- **a.** How many tripeptides can be formed from these three amino acids if each is used only once in the structure?
- **b.** Using three-letter abbreviations for the amino acids, give the sequence for each of the possible tripeptides.
- **c.** Draw the structure of the tripeptide that has proline as its N-terminal amino acid and glycine as its C-terminal amino acid. Circle each peptide bond.

10.54 a. Draw the structure of Val—Ala—Leu.
b. Would you expect to find this segment at the center or on the surface of a globular protein? Why?

10.55 a. Draw the structure of Ser—Lys—Asp.
b. Would you expect to find this segment at the center or on the surface of a globular protein? Why?

10.56 Name the covalent bond that helps to stabilize the tertiary structure of a protein.

10.57 Identify some differences between the following pairs:
a. primary and secondary protein structures
b. complete and incomplete proteins
c. fibrous and globular proteins

10.58 Identify some differences between the following pairs:
a. α helix and β-pleated sheet
b. ionic interactions (salt bridge) and polar interactions
c. polar and nonpolar amino acids

10.59 Identify the level of protein structure associated with each of the following:
a. more than one polypeptide
b. hydrogen bonding between backbone atoms
c. the sequence of amino acids
d. intermolecular forces between R groups

10.60 Identify the level of protein structure associated with each of the following:
a. α helix
b. disulfide bridge
c. peptide bond
d. salt bridges between polypeptides

10.61 Briefly describe the structure of collagen and discuss how it is different from other helix-containing proteins.

10.62 A piece of a polypeptide is folded so that a serine side chain is located on the surface of the polypeptide. Describe the effect on the protein of changing that amino acid to isoleucine.

10.63 Indicate what type(s) of intermolecular forces are disrupted and what level of protein structure is changed by the following denaturing treatments:
a. an egg placed in water at 100 °C and boiled for 10 minutes
b. acid added to milk during the preparation of cheese
c. egg whites whipped in a mixing bowl to make meringue

10.64 Describe the changes that occur in the primary structure when a protein is denatured.

10.65 What types of covalent bonds can be disrupted when a protein is denatured? Name a denaturing agent that could accomplish this.

10.66 Describe what happens to the protein structure of hair when it is curled with a heated curling iron.

10.67 Collagen contains an amino acid that is a modified form of the naturally occurring amino acid.
a. Name the natural amino acid.
b. How is the structure of the side chain of this amino acid modified?
c. What is the result of the modification in terms of the structure and function of the collagen?

10.68 To transport oxygen, hemoglobin requires a prosthetic group. Explain what a prosthetic group is and name the one found in hemoglobin.

10.69 Transport proteins are integral membrane proteins and have nonpolar amino acids on their surface and polar amino acids in the interior. This is in contrast to proteins like hemoglobin that are found in the bloodstream. Explain why the surface of the integral membrane protein is nonpolar.

10.70 Describe the role enzymes serve for chemical reactions in the body.

10.71 What occurs at the active site of an enzyme?

10.72 Match the terms (1) ES, (2) enzyme, and (3) substrate with the following descriptions:
a. has a tertiary structure that recognizes the substrate
b. is the combination of an enzyme with the substrate
c. has a structure that fits the active site of an enzyme

10.73 Match the terms (1) active site, (2) lock-and-key model, and (3) induced-fit model with the following descriptions:
a. the portion of an enzyme where catalytic activity occurs
b. the active site adapts to the shape of a substrate
c. the active site has a rigid shape

10.74 Do the amino acids that are in the active site of an enzyme have to be near each other in the enzyme's primary structure?

10.75 The enzyme trypsin catalyzes the breakdown of many structurally diverse proteins in foods. Does the induced-fit or lock-and-key model explain the action of trypsin better? Explain.

10.76 The enzyme sucrase catalyzes the hydrolysis of the disaccharide sucrose but not the disaccharide lactose. Does the induced-fit or lock-and-key model explain the action of sucrase better? Explain.

10.77 A diet deficient in thiamine results in a condition called beriberi, which is characterized by fatigue, weight loss, and poor appetite, among other symptoms. Patients with beriberi do not properly metabolize glucose and other sugars. If thiamine is not an enzyme, what kind of compound do you suppose it is? Explain.

10.78 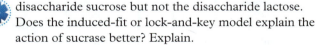 What type of interactions between an enzyme and its substrate help to stabilize ES?

10.79 Does each of the following statements describe a simple enzyme (no cofactor or coenzyme necessary), an enzyme that requires a cofactor, or an enzyme that requires a coenzyme?
  a. contains Ca^{2+} in the active site
  b. consists of one polypeptide chain in its active form
  c. contains vitamin B_6 in its active site

10.80 A substrate is held in the active site of an enzyme by attractive forces between the substrate and the amino acid side chains. For the outlined regions A, B, and C on the following substrate molecule:
  a. Name one possible attractive force it could form in the active site.
  b. Could the amino acids serine, lysine, or glutamate be present in the active site? Support your answer.

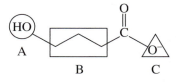

10.81 If each of the following amino acid side chains is present in the active site of an enzyme, indicate whether it would (a) serve a catalytic function, (b) serve to hold the substrate, or (c) both.
  a. aspartate

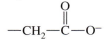

  b. phenylalanine

  —CH_2—⬡

  c. valine

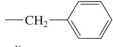

  d. lysine
  —$CH_2CH_2CH_2CH_2NH_3^+$

10.82 Chymotrypsin, an enzyme that hydrolyzes peptide bonds in proteins, functions in the small intestine at a pH optimum of 7.7 to 8.0. How is the rate of the chymotrypsin-catalyzed reaction affected by each of the following conditions?
  a. decreasing the concentration of proteins
  b. changing the pH to 3.0
  c. running the reaction at 75 °C

10.83 Pepsin, an enzyme that hydrolyzes peptide bonds in proteins, functions in the stomach at a pH optimum of 1.5 to 2.0. How is the rate of a pepsin-catalyzed reaction affected by each of the following conditions?
  a. increasing the concentration of proteins
  b. changing the pH to 5.0
  c. running the reaction at 0 °C

10.84 Problems 10.82 and 10.83 both mention enzymes that hydrolyze peptide bonds. How do you account for the fact that pepsin has a high catalytic activity at pH 1.5 but chymotrypsin has very little activity at pH 1.5?

10.85 After the reaction catalyzed by an enzyme is complete, why do the products of the reaction migrate away from the active site?

10.86 Describe the key difference between a competitive inhibitor and a noncompetitive inhibitor.

10.87 How does an irreversible inhibitor function differently than a reversible inhibitor?

10.88 When lead acts as a poison, it can do so by either replacing another ion (such as zinc) in the active site of an enzyme or it can react with cysteine side chains to form covalent bonds. Which of these is irreversible and why?

10.89 Increasing the substrate concentration of an enzyme-catalyzed reaction increases the rate of the reaction until the enzyme becomes saturated and reaches its maximum activity. If the amount of *enzyme* in the reaction is increased under saturating conditions, would the rate of the reaction increase, decrease, or remain unchanged? Explain.

10.90 A hypothetical enzyme has an optimum temperature of 40 °C. If we wish to double the rate of the reaction catalyzed by this enzyme, can we do so by increasing the temperature to 80 °C? Explain your answer.

10.91 Cadmium is a poisonous metal used in industries that produce batteries and plastics. Cadmium ions (Cd^{2+}) are inhibitors of hexokinase. Increasing the concentration of glucose or ATP, the substrates of hexokinase, or Mg^{2+}, the cofactor of hexokinase, does not change the rate of the cadmium-inhibited reaction. Is cadmium a competitive or noncompetitive inhibitor? Explain.

10.92 Meats spoil due to the action of enzymes that degrade the proteins. Fresh meats can be preserved for long periods of time by freezing them. Explain how freezing meats works to prevent spoilage.

10.93 Many drugs are competitive inhibitors of enzymes. When scientists design inhibitors to serve as drugs, why do you suppose they choose to design competitive inhibitors instead of noncompetitive inhibitors?

10.94 Fresh pineapple contains the enzyme bromelain, which degrades proteins.
  a. The directions on a package of Jell-O® (a protein-containing food product) say to add canned pineapple (which is heated to high temperatures to preserve it), not fresh pineapple. Why?
  b. Fresh pineapple is used in a marinade to tenderize tough meat. Why?

Challenge Problems

10.95 Lipoproteins transport dietary fats through the bloodstream and deliver them to the cells. Cholesterol is mainly transported as a cholesterol ester. The formation of a cholesterol ester is shown. Considering the structure of cholesterol and the fatty acid shown, draw the cholesterol ester produced.

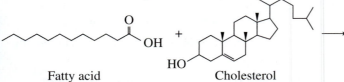

Fatty acid   Cholesterol

10.96 When lemon juice is added to milk, the milk curdles. Why does lemon juice have this effect? What is happening to the proteins in milk?

10.97 How is the structure of a soap micelle (Chapter 7) similar to the structure of a globular protein?
HINT: *Both of these are found in aqueous solution.*

10.98 Insulin is a protein hormone that functions as two polypeptide chains whose amino acid sequences are as follows:
- A chain: GIVEQCCTSICSLTQLENYCN
- B chain: FVNQHLCGDHLVEALYLVCGERGFFYTPKT

a. The highlighted region in chain B forms an α helix. In an α helix, the R groups are found on the outside of the helix, protruding out from the center, and every fourth amino acid appears on the same side of the helix. Considering the polarity of the R groups, is this helix amphipathic (having a polar part and nonpolar part)? Which side do you think faces the exterior?
b. Considering the amino acid sequences, suggest how these two polypeptide chains might be held together in an active insulin molecule.

10.99 The active site of a hypothetical enzyme contains a pocket of nonpolar amino acid side chains next to a cleft containing acidic and basic amino acid side chains.
a. Which amino acids are likely responsible for catalysis? Explain.
b. Which amino acids are present mainly to hold the substrate in the active site?
c. Would you predict that the substrate is polar or nonpolar based on your answers to part a and part b? Why?

10.100 Contact lens wearers often soak their lenses in a solution containing enzymes to remove protein deposits from the lenses. Why is it necessary to rinse the lenses thoroughly before placing them on the eyes?

Answers to Odd-Numbered Problems

Practice Problems

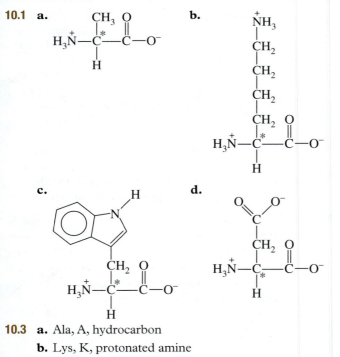

10.1 a. **b.**

c. **d.**

10.3 a. Ala, A, hydrocarbon
   b. Lys, K, protonated amine
   c. Trp, W, aromatic
   d. Asp, D, carboxylate

10.5 a. hydrophobic   **b.** hydrophilic
   c. hydrophobic   **d.** hydrophilic

10.7 a. **b.**

   c. **d.**

10.9 a. Glycine
   b. Histidine
   c. Glutamine
   d. Isoleucine

10.11 a.

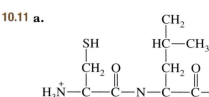

b. $CH_3(CH_2)_{14}COOH + CH_3(CH_2)_{29}OH$

10.13 a.

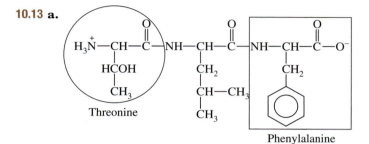

Threonine Phenylalanine

b. TLF, Thr—Leu—Phe

10.15 and 10.17

a.

b.

c.

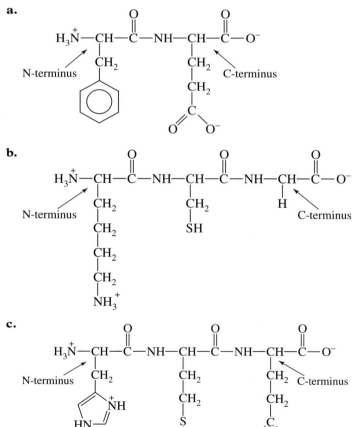

d.

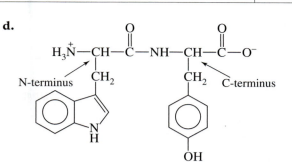

N-terminus C-terminus

10.19 covalent bonding, peptide bond

10.21 hydrogen bonding

10.23 An α helix is a coiled structure with the amino acid side chains protruding outward from the helix. The β-pleated sheet is an open, extended, zigzag structure with the side chains of the amino acids oriented above and below the sheet.

10.25 a. salt bridge

b. London force (nonpolar interaction)

c. hydrogen bonding

d. hydrogen bonding and ion–dipole

10.27 a. quaternary

b. secondary

c. tertiary

d. primary

10.29 a. ionic and hydrogen bonding; secondary, tertiary, and quaternary structure

b. hydrogen bonding and London forces; secondary, tertiary, and quaternary structure

10.31 Collagen, Structural connector

Hemoglobin, Oxygen transporter

Antibody, Bind foreign substances in body

Casein, Storage protein

10.33 After a meal, the glucose transporter is active because glucose levels are high in the bloodstream and insulin is released. Upon waking in the morning, the glucose transporter is less active because glucose and insulin levels in the bloodstream are low.

10.35 tertiary

10.37 The induced-fit model. When glucose is fit into the active site, the enzyme undergoes a conformational change, closing around the substrate.

10.39 Because the phosphates on ATP have negative charges and Mg^{2+} has positive charges, the attraction is ionic.

10.41 a. The enzyme activity will decrease if the temperature is raised above the temperature optimum.

b. The activity will be lowered if the pH is changed.

10.43 a. no effect

b. rate decreases

Additional Problems

10.45 carboxylate and protonated amine

10.47 a. Tyrosine, Tyr

    b. Proline, Pro

    c. Cysteine, Cys

10.49 Isoleucine, Ile; and Threonine, Thr

10.51 several possibilities, including rice and beans, peas, and corn

10.53 a. 6

    b. Gly—Pro—Lys, Gly—Lys—Pro, Pro—Gly—Lys, Pro—Lys—Gly, Lys—Pro—Gly, Lys—Gly—Pro

    c.

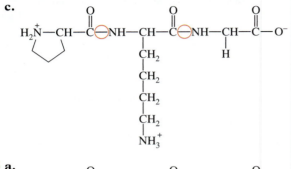

10.55 a.

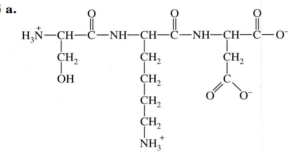

    b. This segment would be on the surface in the aqueous environment. The side chains are hydrophilic and would interact with the water in the surrounding environment.

10.57 a. Primary structures are held together by covalent bonding, and secondary structures are held together by hydrogen bonding.

    b. Complete proteins contain all essential amino acids and incomplete proteins do not.

    c. Fibrous proteins have elongated structures and globular proteins have roughly spherical structures. Globular proteins are soluble in water and fibrous proteins are not.

10.59 a. quaternary structure

    b. secondary structure

    c. primary structure

    d. tertiary and quaternary structure

10.61 Collagen's structure consists of three helices wrapped around each other to form a triple helix.

10.63 a. hydrogen bonds and nonpolar attractions; secondary, tertiary, and quaternary

    b. hydrogen bonds and salt bridges; secondary, tertiary, and quaternary

    c. hydrogen bonds and nonpolar attractions; secondary, tertiary, and quaternary

10.65 disulfide bonds, thioglycolate

10.67 a. proline

    b. An OH group is added to the side chain to form hydroxyproline.

    c. The OH on the side chain allows for the formation of more hydrogen bonds between side chains, which increases the strength of the collagen.

10.69 The middle portion of these proteins spans the nonpolar portion of the membrane. This surface interacts with the nonpolar environment and therefore must be nonpolar.

10.71 The active site is the location on an enzyme where catalysis occurs.

10.73 a. (1) active site

    b. (3) induced-fit model

    c. (2) lock-and-key model

10.75 Because it can have more than one substrate, trypsin's action is better described by the induced-fit model.

10.77 The explanation is that thiamine is involved in enzymatic reactions but is not an enzyme. It is a coenzyme.

10.79 a. requires a cofactor

    b. describes a simple enzyme

    c. requires a coenzyme

10.81 a. (c)   **b.** (b)   **c.** (b)   **d.** (c)

10.83 a. The rate would increase due to increased probability of collision until a steady state is reached.

    b. The rate would decrease due to protein denaturation.

    c. The rate would decrease due to slower movement of molecules.

10.85 The products do not fit snugly into the active site, so they are released.

10.87 An irreversible inhibitor forms covalent bonds to the enzyme and permanently inactivates it.

10.89 If the amount of enzyme were increased in a reaction occurring at a steady state, the maximum rate of the reaction would increase since more enzyme is present to convert substrate into product.

10.91 Cadmium would be a noncompetitive inhibitor because increasing the substrate and cofactor has no effect on the rate. This implies that the cadmium is binding to another site on the enzyme.

10.93 When designing an inhibitor, the substrate of an enzyme usually is known even if the structure of the enzyme is not. It is easier to design a molecule that resembles the substrate (competitive) than to find an inhibitor that binds to a second site on an enzyme (noncompetitive).

Challenge Problems

10.95

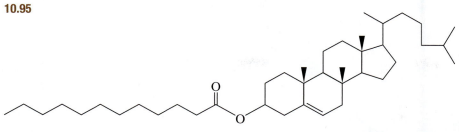

10.97 Both have nonpolar interiors and polar surfaces. On a micelle, the polar heads of the fatty acid salts face outward into the aqueous environment like the polar amino acid side chains on the globular protein. The nonpolar tails of the fatty acid salts gather together in the interior of the micelle just as the nonpolar amino acid side chains gather in the interior of the globular protein.

10.99 a. The polar amino acids are likely involved in catalysis.

b. The nonpolar amino acids are likely involved in aligning the substrate correctly.

c. The substrate is likely nonpolar with a small polar portion.

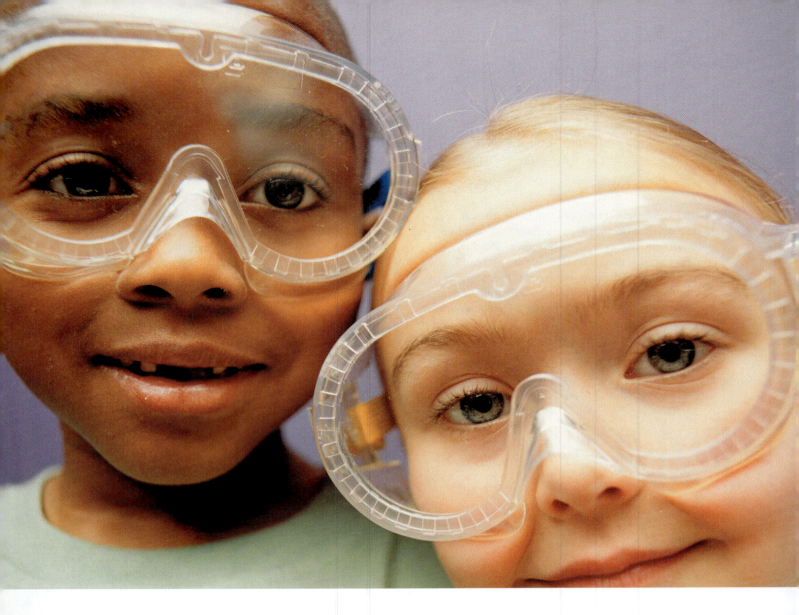

The nucleic acid called DNA defines our hereditary characteristics like eye, hair, and skin color. These large molecules are made up of smaller component parts. Chapter 11 describes how DNA is used to make proteins in the cell along with many other aspects of nucleic acids.

11

Nucleic Acids—Big Molecules with a Big Role

THE NUCLEIC ACID deoxyribonucleic acid, or DNA, appears regularly in crime shows and in courtrooms in the context of DNA testing. What is so unique and important about this molecule? What does it look like? **DNA** is the molecule in our cells that stores and directs information responsible for cell growth and reproduction. DNA, found in the nucleus of the cell, contains genetic information that is called the **genome.** One part of the genome, the **gene,** contains information to make a particular protein for the cell. The human genome contains an estimated 20,000 to 25,000 genes.

Every cell in your body contains the same DNA, so why is a skin cell different from a liver cell? Different cell types use different parts of the genome to express their characteristics. Together, the DNA in your tissues governs your physical traits like your eye and hair color, your growth rate, and your metabolic rate.

DNA molecules are so long and stringy that, when they are isolated, we can see them with the naked eye. Nucleic acids like DNA and its counterpart ribonucleic acid (RNA) are big molecules made up of repeating building blocks called *nucleotides*. Exploring the structure of the nucleotide building blocks is a starting point for understanding the more complex structure of the nucleic acids and genomics, the study of genes and their interactions.

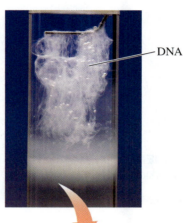
— DNA

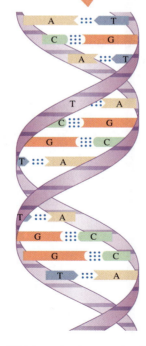

DNA is such a large molecule that it can be seen with the naked eye.

? **What's an Inquiry Question?**
Inquiry Questions are designed to focus your reading on the main concepts by section. An Inquiry Question appears at the beginning of each section.

Discovering the Concepts

? Inquiry Activity—Components of Nucleotides

Information

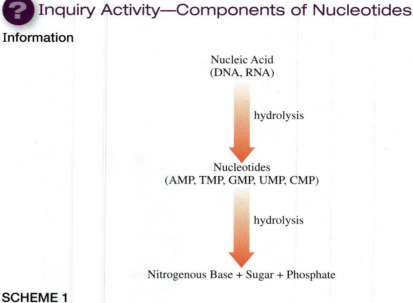

SCHEME 1
Component molecules found in nucleic acids.

Questions
1. Using Scheme 1, name the three component molecules found in a nucleotide.
2. What type of chemical reaction occurs when a nucleotide is *formed* from its component molecules?

Nitrogenous Bases
Five different aromatic rings containing nitrogen called *nitrogenous bases* fit into two families, purines and pyrimidines. The general structure of these rings is

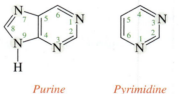

Purine   *Pyrimidine*

The five nitrogenous bases found in nucleotides and nucleic acids are

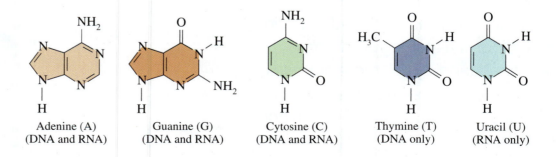

Questions
3. Which of the nitrogenous bases are purines? Which are pyrimidines?
4. Which are found in DNA? Which in RNA?
5. Compare the structures of uracil and thymine. How are they different?

Pentose Sugars

There are two aldopentose sugars found in nucleic acids, ribose and deoxyribose. They are shown here in the Haworth projection. The sugars in nucleic acids are numbered with a prime (') to distinguish their carbons from the carbons of the nitrogenous bases.

Question

6. a. Compare the structures of ribose and 2-deoxyribose. How are they different?
    b. Which one do you think is found in DNA? RNA?

Phosphate

A phosphate is a functional group that contains a phosphorus surrounded by four oxygen atoms. (See Table 4.3 for a review of the structure.)

Putting It All Together

As implied in Scheme 1, condensation reactions between phosphate, sugar, and nitrogenous base produce a nucleotide. The phosphate —OH condenses with —OH on C5′ of a sugar. A purine N9 condenses with the —OH on C1′ of a sugar. A pyrimidine N1 condenses with the —OH on C1′ of a sugar.

Questions

7. Draw the structure of the nucleotide adenosine monophosphate (AMP).

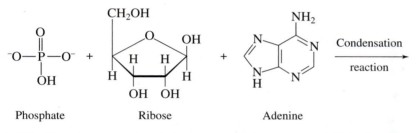

| Phosphate | Ribose | Adenine |

8. If AMP is adenosine monophosphate, how would the structure of ATP, adenosine triphosphate, differ?

Pentose sugars in RNA and DNA

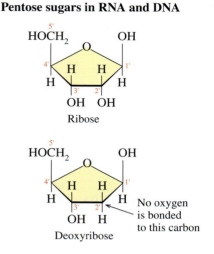

Ribose

No oxygen is bonded to this carbon

Deoxyribose

11.1 Components of Nucleic Acids

Just as proteins are strings of amino acids, nucleic acids are strings of molecules called **nucleotides.** Most proteins contain 20 different amino acids in a given sequence, while most nucleic acids are created from a set of 4 nucleotides in a given sequence. Each nucleotide has three basic components: a nitrogenous base, a five-carbon sugar (pentose), and a phosphate functional group. The differences between DNA and RNA lie in slight variations of the components found in these nucleotides.

? 11.1 Inquiry Question:
What are the components of nucleotides and nucleic acids?

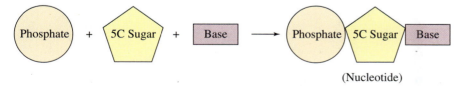

The component parts of a nucleotide are a phosphate, pentose, and a nitrogenous base.

Nitrogenous Bases

There are four different nitrogenous bases found in a nucleic acid. Each of the bases has one of two nitrogen-containing aromatic rings, either the purine ring or the pyrimidine ring.

DNA contains two purines, adenine (A) and guanine (G), as well as two pyrimidines, thymine (T) and cytosine (C). RNA contains the same bases, except that thymine (5-methyluracil) is replaced with uracil (U) (see **Figure 11.1**).

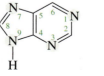

Purine   *Pyrimidine*

Purines

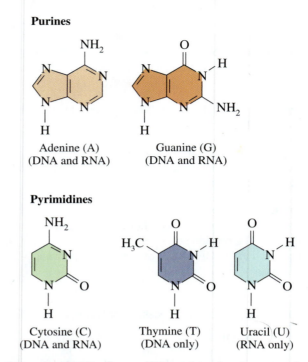

Adenine (A)
(DNA and RNA)

Guanine (G)
(DNA and RNA)

Pyrimidines

Cytosine (C)
(DNA and RNA)

Thymine (T)
(DNA only)

Uracil (U)
(RNA only)

FIGURE 11.1 **Purine and pyrimidine bases.** DNA contains the nitrogenous bases A, G, C, and T. RNA contains A, G, C, and U. Notice the structural difference between T and U is a single methyl group.

Pentose sugars in RNA and DNA

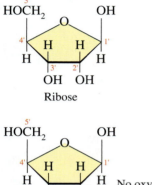

Ribose

Deoxyribose

No oxygen is bonded to this carbon

FIGURE 11.2 **Five-carbon sugars.** The five-carbon pentose sugar found in RNA is ribose and in DNA is deoxyribose. Notice that the carbon numbering includes the prime (′) symbol.

Ribose and Deoxyribose

Nucleotides also contain five-carbon pentose sugars. To distinguish the carbons in the nitrogenous bases from the carbons in the sugar rings, a prime symbol (′) is added to the carbon numbering of the sugars (1′, 2′, 3′, 4′, 5′).

RNA contains the pentose *ribose* (the "R" in RNA) and DNA contains the pentose *deoxyribose* (the "D" in DNA). Deoxyribose lacks oxygen on carbon 2′ of the pentose, but otherwise its structure is identical to ribose (see **Figure 11.2**).

sample problem 11.1   **Components of Nucleic Acids**

Identify each of the following bases as a purine or a pyrimidine:

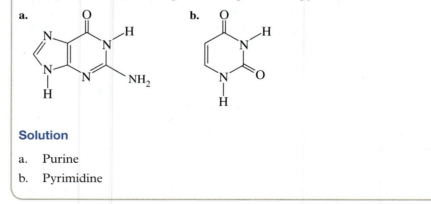

a.

b.

Solution

a.   Purine

b.   Pyrimidine

Condensation of the Components

In Chapter 6, we saw that two monosaccharides could join through a condensation reaction. In that condensation reaction, a molecule of water is removed and a glycosidic bond forms between two sugars. Similarly, a pentose and a nitrogenous base can join by this reaction when a nitrogen in the base (N1 of pyrimidines or N9 of purines) bonds to C1′ of the pentose, forming a carbon-to-nitrogen glycosidic bond and removing a water molecule. The condensation reaction between ribose and the base adenine is shown forming a molecule called *adenosine.*

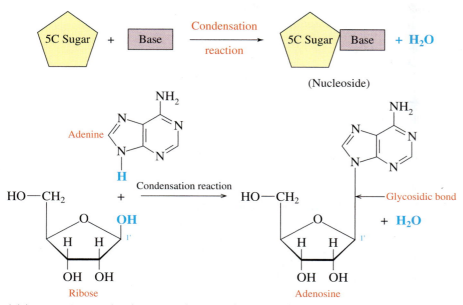

Joining a pentose and a nitrogenous base produces a nucleoside and water.

We noted earlier that a nucleotide contains a nitrogenous base, a sugar, and a phosphate, and that these nucleotides make up nucleic acids. Where does adenosine fit in? It does not contain the phosphate component. Adeno*sine* is a **nucleoside,** formed when just the pentose sugar and the nitrogenous base components combine in a condensation reaction. It is the first step in the synthesis of a nucleotide.

Suppose a hydrogen phosphate group (HPO_4^{2-}), referred to as phosphate here for simplicity, reacts with the —OH group on C5′ of an adenosine molecule. The molecule formed in this condensation reaction is a nucleo*tide*. In this case, the nucleotide *a*deno*sine* *mono*phosphate (AMP) is produced.

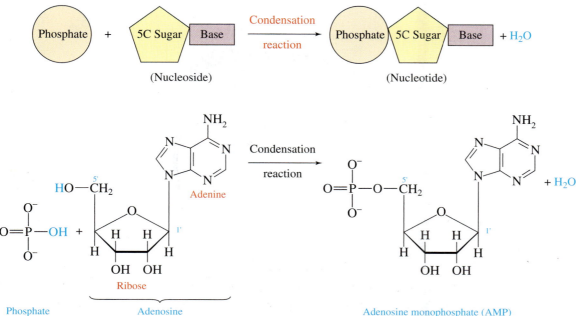

Joining a phosphate to a nucleoside produces a nucleotide and water.

Solving a Problem

Writing condensation products for nucleotide components

Write the products for the condensation reaction shown.

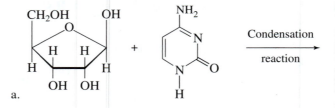

a.

Solution

STEP 1: Identify the components. The components are a pentose (ribose) and a pyrimidine (cytosine).

STEP 2: Locate the functional groups undergoing condensation. When a pentose combines with a pyrimidine, the —OH (blue) of C1′ condenses with the —H (blue) of N1, forming a glycosidic bond and H_2O.

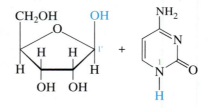

STEP 3: Write the product. The product will be the nucleoside cytidine and a molecule of water.

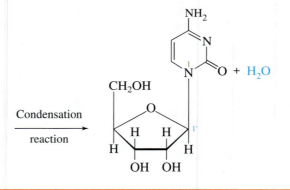

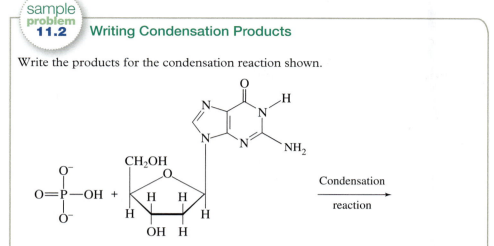

sample problem 11.2   **Writing Condensation Products**

Write the products for the condensation reaction shown.

Solution

STEP 1: Identify the components. The components are a nucleoside (deoxyguanosine) and a phosphate group.

STEP 2: Locate the functional groups undergoing condensation. When a nucleoside combines with a phosphate, the —OH (blue) on C5′ condenses with the phosphate —OH (blue), forming a nucleotide and H_2O.

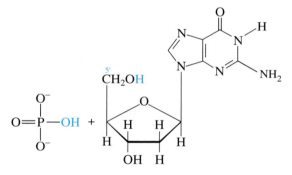

STEP 3: Write the product. The product will be the nucleotide deoxyguanosine monophosphate and a molecule of water.

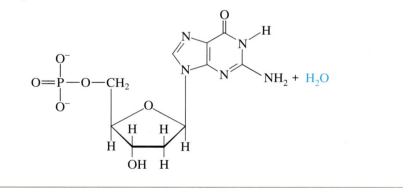

Naming Nucleotides

Nucleotides include the nucleoside name and the number of phosphates present. Because the names can become quite long, nucleotide names often are abbreviated. The abbreviation indicates the type of sugar (ribose or deoxyribose) and the nitrogenous base.

If deoxyribose is found in the nucleotide, a lowercase *d* is inserted at the beginning of the abbreviation. The names and abbreviations of the nucleotides found in nucleic acids are shown in Figure 11.3 and Table 11.1

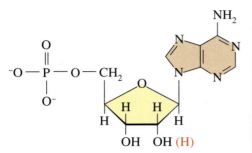

Adenosine monophosphate (AMP)
Deoxyadenosine monophosphate (dAMP)

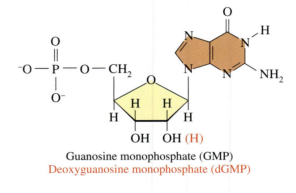

Guanosine monophosphate (GMP)
Deoxyguanosine monophosphate (dGMP)

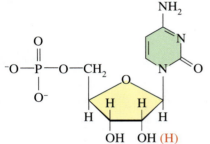

Cytidine monophosphate (CMP)
Deoxycytidine monophosphate (dCMP)

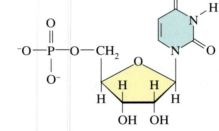

Uridine monophosphate (UMP)

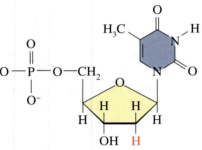

Deoxythymidine monophosphate (dTMP)

FIGURE 11.3 Naming nucleotides. The nucleotides of RNA (black names) are identical to those of DNA (red names) except in DNA the sugar is deoxyribose and deoxythymidine replaces uridine.

TABLE 11.1 Nucleoside and Nucleotide Names in DNA and RNA

Base	Nucleoside	Nucleotide
RNA		
Adenine (A)	Adenosine (A)	Adenosine monophosphate (AMP)
Guanine (G)	Guanosine (G)	Guanosine monophosphate (GMP)
Cytosine (C)	Cytidine (C)	Cytidine monophosphate (CMP)
Uracil (U)	Uridine (U)	Uridine monophosphate (UMP)
DNA		
Adenine (A)	Deoxyadenosine (A)	Deoxyadenosine monophosphate (dAMP)
Guanine (G)	Deoxyguanosine (G)	Deoxyguanosine monophosphate (dGMP)
Cytosine (C)	Deoxycytidine (C)	Deoxycytidine monophosphate (dCMP)
Thymine (T)	Deoxythymidine (T)	Deoxythymidine monophosphate (dTMP)

sample problem 11.3 Nucleotides

Identify which nucleic acid (DNA or RNA) contains each of the following nucleotides. State the components of each nucleotide:

a. deoxyguanosine monophosphate (dGMP)

b. adenosine monophosphate (AMP)

Solution

a. This is a DNA nucleotide because it contains deoxyribose. It also contains guanine and one phosphate.

b. This is an RNA nucleotide because it contains ribose (does not have deoxy prefix). It also contains adenine and one phosphate.

11.1 Identify each of the following as being a purine or a pyrimidine:
a. thymine
b.

11.2 Identify each of the following as being a purine or a pyrimidine:
a. guanine
b.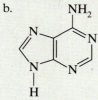

11.3 Identify the bases in Problem 11.1 as present in RNA, DNA, or both.

11.4 Identify the bases in Problem 11.2 as present in RNA, DNA, or both.

11.5 List the names and abbreviations of the four nucleotides in DNA.

11.6 List the names and abbreviations of the four nucleotides in RNA.

11.7 Provide the products for each of the following condensation reactions:

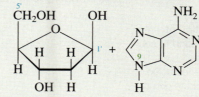
a.

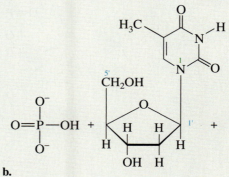
b.

11.8 Provide the products for each of the following condensation reactions:

a.

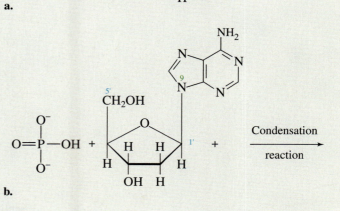
b.

11.2 Nucleic Acid Formation

11.2 Inquiry Question:
How are nucleic acids formed?

Many nucleotides linked together form **nucleic acids.** Much as amino acids in proteins are linked together by peptide bonds, nucleotides are linked together through **phosphodiester bonds** where the oxygens in the phosphate are connected between the 3′ and 5′ carbons of adjacent sugar molecules (see **Figure 11.4**).

Primary Structure: Nucleic Acid Sequence

The backbone of a nucleic acid consists of a sugar and a phosphate that alternate. The bases dangle from the sugar. Recall that a protein's primary structure is the sequence of amino acids. A nucleic acid's primary structure is indicated by its nucleotide sequence. The backbone of a nucleic acid runs directionally with the phosphates connected between the 3′ carbon of one sugar and the 5′ carbon of the neighboring sugar. This is analogous to a protein sequence running from N-terminus to C-terminus.

FIGURE 11.4 Condensation of nucleotides. Nucleotides bond to each other through a condensation reaction between the 3′ —OH of one nucleotide's sugar and the phosphate group (bonded to the 5′ —OH of the sugar) of a second nucleotide forming a dinucleotide.

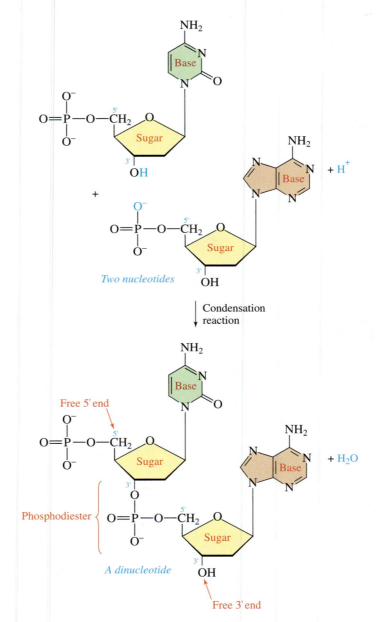

Nucleic acid sequence can be designated by one-letter base abbreviations. The convention is that if a single nucleic acid strand is shown as being drawn horizontally, the 5′ end of a nucleic acid is at the left end and the 3′ end is at the right end.

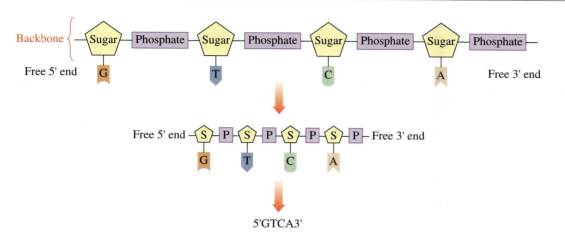

5'GTCA3'

A nucleic acid's sequence can be more simply represented by its one-letter base codes.

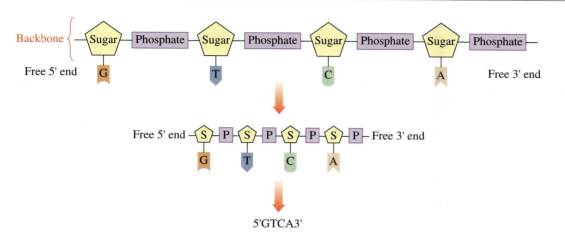

sample problem 11.4   Nucleic Acid Formation

Draw the RNA dinucleotide (two nucleotides joined by a phosphodiester) formed if two adenosine monophosphates undergo a condensation reaction. Label the 5' and 3' ends of your structure. Identify the phosphodiester bond.

Solution

The adenosines are joined between the 3' —OH of ribose from one adenosine and the phosphate —OH of the second adenosine forming a phosphodiester.

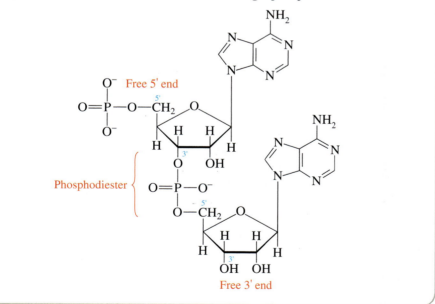

Nucleic Acid Abbreviations

Provide an abbreviation for the following nucleic acid sequence:

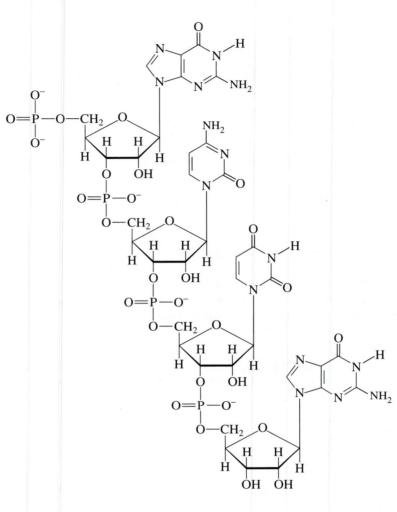

Solution

To solve this problem, you must locate the 5′ end of the nucleic acid and identify each nitrogenous base. The 5′ end of this molecule is at the top left. The sequence is abbreviated beginning with base letters at the 5′ end. This nucleic acid is abbreviated 5′GCUG3′, or simply GCUG.

11.9 How are nucleotides bonded together in a nucleic acid chain?

11.10 Describe the differences in the two ends of a nucleic acid.

11.11 Draw the dinucleotide GC that would be found in RNA. Label the 5′ and 3′ ends of your structure. Identify the phosphodiester bond.

11.12 Draw the dinucleotide AT that would be found in DNA. Label the 5′ and 3′ ends of your structure. Identify the phosphodiester bond.

Discovering the Concepts

? Inquiry Activity—The Unique Structure of DNA

Characteristics of DNA

- Double stranded—Two strands of nucleic acid line up in opposite directions (antiparallel) with the bases in the center.
- Hydrogen bonded—The bases hydrogen-bond between the strands like rungs in a ladder: G≡C, A=T. These pairings are referred to as *complementary base pairs*.
- Double helix—Because of the chirality of the sugars, the two antiparallel strands have a natural twist to them, forming a helical shape called a double helix.

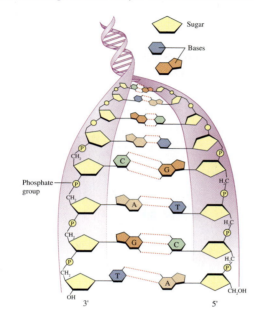

A section of double stranded DNA can be abbreviated as
5'ACTGTCAG3'
3'TGACAGTC5'

Questions

1. a. How would you designate the following nucleic acid in a one-letter abbreviation?

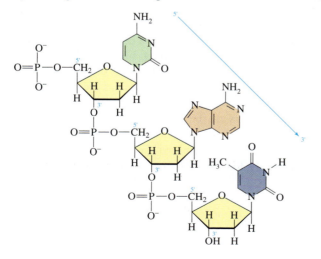

b. Is this a strand of DNA or RNA? How do you know?

c. Provide an abbreviation for the strand of DNA with complementary bases to the one shown in part a. Indicate the 5′ and 3′ ends.

2. Provide the complementary DNA strand to the following strand:
5′AATTCCGCTAACG3′

3. From your own experiences and the information presented, have your group list information on both the structure and function of DNA.

11.3 DNA

11.3 Inquiry Question:
What are the unique structural features of DNA?

The base sequences in nucleic acids stored as DNA in the cell's nucleus hold the code for cellular protein production. A few key discoveries beginning in the late 1940s led to what we now understand as the structure of DNA. Erwin Chargaff noted that the amount of adenine (A) is always equal to the amount of thymine (T) (A = T), and the amount of guanine (G) is always equal to the amount of cytosine (C) (G = C). This also implies that the number of purines equals the number of pyrimidines in DNA.

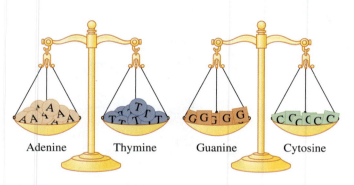

Adenine   Thymine   Guanine   Cytosine

In DNA the number of purines equals the number of pyrimidines.
A = T and G = C.

Secondary Structure: Complementary Base Pairing

DNA's secondary structure is described by the interaction of two nucleic acids to form a double helix. This was first proposed in 1953 by James Watson and Francis Crick. A double helix can be envisioned as a twisted ladder (see Figure 11.5). Imagine that the sugar–phosphate backbones make up the rails of the ladder and the bases dangling off the backbone interact as the rungs in the center of the ladder.

The two strands both have the bases in the center. Their backbones run in opposite directions, or as biochemists describe them, the backbones are *antiparallel* to each other. One strand goes in the 5′ to 3′ direction and the other strand goes in the 3′ to 5′ direction.

Each of the rungs in DNA's helical ladder contains one base from each of the strands. The two bases in each ladder-rung associate through hydrogen bonding. Watson and Crick realized that all the rungs are the same length, so they must contain one purine and one pyrimidine. The pairs A–T and G–C are called **complementary base pairs** (see Figure 11.6). Adenine and thymine form two hydrogen bonds, while guanine and cytosine form three hydrogen bonds. The DNA in one human cell contains about 3 billion of these base pairs!

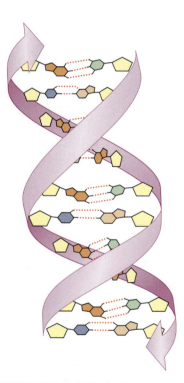

FIGURE 11.5 A DNA double helix.
The two backbones run antiparallel to each other (noted by arrows on ends) and the bases line up in the interior space.

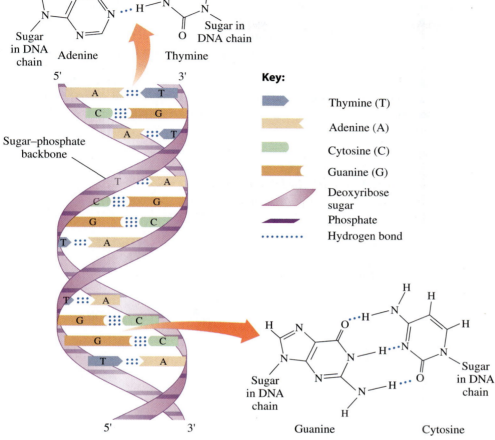

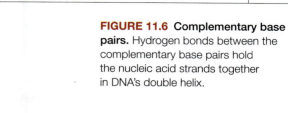

FIGURE 11.6 Complementary base pairs. Hydrogen bonds between the complementary base pairs hold the nucleic acid strands together in DNA's double helix.

Key:

Thymine (T)	
Adenine (A)	
Cytosine (C)	
Guanine (G)	
Deoxyribose sugar	
Phosphate	
Hydrogen bond	

_{sample}
problem
11.6 **Complementary Base Pairs**

Write the base sequence and label the 3′ and 5′ ends of the complementary strand for a segment of DNA with the base sequence 5′ACGATCT3′.

Solution

The complementary strand contains the complementary bases to those of the given strand. The two strands run opposite each other so

Given strand: 5′ACGATCT3′

Complementary strand: 3′TGCTAGA5′

Tertiary Structure: Chromosomes

Try this. Hold a rubber band vertically at the top and bottom and twist it. You can only twist it so many times before it starts to double up on itself. As in proteins, the tertiary structure refers to a large biomolecule's overall shape when it folds onto itself. Because DNA has a helical twist (the double helix), any further twisting (doubling it up on itself) that makes the DNA more compact constitutes its tertiary structure.

This further twisting is called **supercoiling.** The 3 billion base pairs of DNA in one human cell would stretch out as a double helix to about 6 feet in length. The DNA in a human cell is separated into 46 pieces (23 from mother, 23 from father) that are super-coiled around proteins called *histones*. These pieces of DNA wound about histones pack into **chromosomes** (see Figure 11.7). Chromosomes are an efficient package for lots of DNA information.

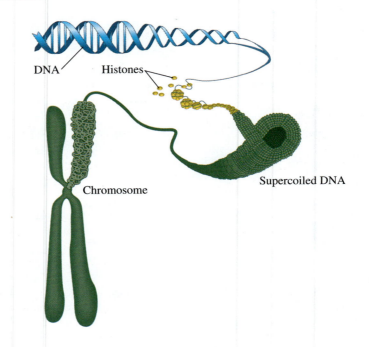

DNA   Histones   Supercoiled DNA

Chromosome

FIGURE 11.7 Chromosome structure.
The DNA is wound around proteins called
histones, which then wind into super-
coiled structures. These fibers are organized into the chromosome structure.

practice
problems

11.13 How are the two strands of nucleic acid in DNA held together?

11.14 Define complementary base pairing.

11.15 Write the base sequence and label the 3′ and 5′ ends of the complementary strand for a segment of DNA with the following base sequences:
  a. 5′AAAA3′
  b. 5′CCCCTTTT3′
  c. 5′ACATTGG3′
  d. 5′TGTGAACC3′

11.16 Write the base sequence and label the 3′ and 5′ ends of the complementary strand for a segment of DNA with the following base sequences:
  a. 5′AAAAAACC3′
  b. 5′GGGGGAT3′
  c. 5′AAAATTTT3′
  d. 5′CGCGATATTA3′

11.17 Fill in the following table with the analogous nucleic acid structures:

	Protein	Nucleic Acid
Primary structure	Sequence of amino acids	
Secondary structure	Local hydrogen bonding between backbone atoms	
Tertiary structure	Folding of backbone onto itself associates amino acids far away in sequence	

11.18 List the similarities and differences in the secondary structure of a protein and the secondary structure of DNA.

11.4 RNA and Protein Synthesis

RNA can be considered the "middleman" in the process of creating a protein from a gene in DNA. Like DNA, RNA is a string of nucleotides. Yet some important differences exist. These differences are listed in Table 11.2.

  One of the main differences is that RNA does not contain the base thymine. In RNA, the base uracil is substituted and it is complementary to adenine, forming two hydrogen bonds (A═U).

11.4 Inquiry Question:
What are the roles of RNA in protein synthesis?

TABLE 11.2 Differences Between RNA and DNA

Differences	RNA	DNA
Sugar	Ribose	Deoxyribose
Base	Uracil	Thymine
Stranding	Single strand of nucleic acid	Double strand of nucleic acid
Size	Much smaller	Larger

RNA Types and Where They Fit In

The three main types of RNA found in the cell are involved in transforming a DNA sequence into a protein sequence. The names of the three RNAs describe their role in this transformation: *messenger RNA*, *ribosomal RNA*, and *transfer RNA*.

Messenger RNA and Transcription

The process of making a protein from DNA has two key steps, the first of which is to make a gene copy from the DNA found in the cell nucleus. This step is called transcription. In **transcription,** DNA's double helix temporarily unwinds so that a complementary copy can be made from one of the strands. This complementary copy is the messenger RNA (**mRNA**). Messenger RNA is a single-stranded piece of RNA containing the bases complementary to those on the original DNA strand (see Figure 11.8). Gene copying is catalyzed by **RNA polymerase,** a complex enzyme containing several subunits that binds to the DNA. Once completed, the mRNA travels from the nucleus to an organelle called the ribosome where the mRNA sequence is processed into protein.

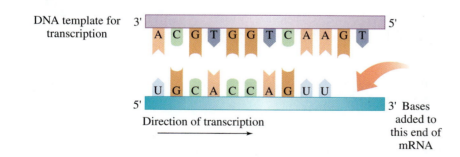

DNA template for transcription 3' A C G T G G T C A A G T 5'

U G C A C C A G U U

5' Direction of transcription → 3' Bases added to this end of mRNA

FIGURE 11.8 Transcription. In transcription, a messenger RNA is transcribed from one strand of genetic DNA (the template) as a complementary copy.

Ribosomal RNA and the Ribosome

The ribosome is one of many small structures found in cells called *organelles*. The ribosome can be thought of as a protein factory. It is composed of ribosomal RNA (**rRNA**) and protein. It is the place where the nucleotide sequence of mRNA is interpreted into an amino acid sequence. The ribosome has two rRNA/protein subunits called the small subunit and the large subunit. The general shape of each is shown in Figure 11.9. The mRNA strand fits into a groove on the small subunit with the bases pointing toward the large subunit.

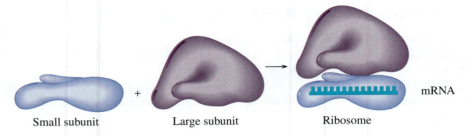

FIGURE 11.9 Ribosome structure. A typical ribosome consists of a small subunit and a large subunit. Both subunits contain rRNA and protein. The mRNA fits in a groove between them.

Transfer RNA and Translation

The second step in the process of making a protein from DNA occurs in the ribosome and is called **translation.** The mRNA sequence must be translated into a protein sequence. The facilitator for this process is the transfer RNA (**tRNA**). There are several areas on the tRNA sequence where complementary bases can hydrogen-bond with other bases on the tRNA. Doing so gives the tRNA a tightly compacted T-shaped structure (see **Figure 11.10**).

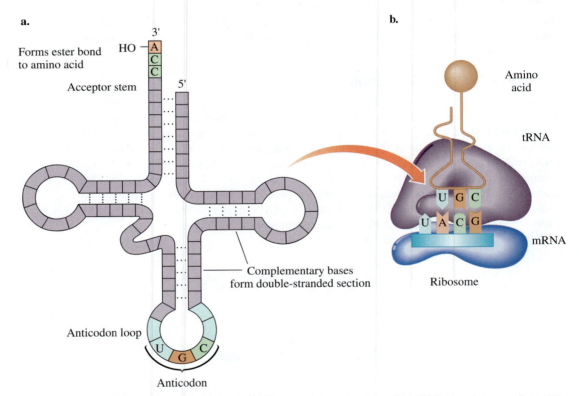

FIGURE 11.10 tRNA structure. (a) The secondary structure of the tRNA showing complementary base pairing. (b) The tRNA with amino acid and mRNA in the ribosome. A typical tRNA molecule has an anticodon loop that complements three bases on mRNA and has an acceptor stem at the 3' end of the nucleic acid where an amino acid attaches.

The tRNA has a three-base sequence (triplet) called an **anticodon** at its *anticodon loop*. When in the ribosome, the anticodon of a tRNA can hydrogen-bond to three complementary bases on the mRNA. The tRNA also has a place at the opposite end called the *acceptor stem* where it can bind an amino acid. The amino acid is joined to the tRNA through ester bond formation. (We saw this esterification reaction in triglyceride formation in Section 7.3.) The only way to get an amino acid incorporated into a growing protein chain is by bringing it to the ribosome bonded to the tRNA. Each of the 20 amino acids has one or more different tRNAs available to bring amino acids to the ribosome.

---

sample problem 11.7   Transcription

The sequence of bases in a DNA template strand is 5′CGATCA3′. What is the corresponding mRNA that is produced from this DNA?

Solution

To form the mRNA, the bases in the DNA template are paired with their complementary bases: G with C, C with G, T with A, and A with U.

DNA template:	5′CGATCA3′
Complementary mRNA:	3′GCUAGU5′

11.19 Name the three types of RNA and their functions.

11.20 List the mRNA bases that complement the bases A, T, G, and C in DNA.

11.21 In your own words, define the term *transcription*.

11.22 In your own words, define the term *translation*.

11.23 The sequence of bases in a DNA template strand is 5″TACGGCAAGCTA3′. What is the corresponding mRNA produced?

11.24 The sequence of bases in a DNA template strand is 5′CCGAAGGTTCAC3′. What is the corresponding mRNA produced?

Discovering the Concepts

? Inquiry Activity—The Genetic Code

Exploring Translation

The genetic code for amino acids found in Table 11.3 (facing page) can be used to translate a triplet of three nucleotides from a nucleic acid into one amino acid.

Codons in mRNA	5′GGU | GAC | CUA | AUC 3′				
Translation	↓	↓	↓	↓	
Amino acid sequence	Gly –	Asp –	Leu –	Ile	(three-letter abbreviation)
	G	D	L	I	(one-letter abbreviation)

Sample 1. Three codons of mRNA are translated to a tripeptide sequence.

Questions

1. How many nucleotide bases (A, G, C, U) are found in one codon?
2. Based on Table 11.3 and Sample 1, the mRNA sequence 5′UUU3′ codes for the amino acid phenylalanine. For what amino acid would the codon 5′GCA3′ code?
3. Name the amino acid corresponding to each of the following codons (use Table 11.3):
   a. 5′AUU3′
   b. 5′CAC3′
   c. 5′GGA3′
4. What is the sequence of amino acids coded by the following codons in mRNA?

   Codons in mRNA   5′CUA \| GUC \| UGC3′

   Amino acid sequence

5. The process of forming an amino acid sequence from mRNA as we did in the previous question is called *translation*. Is this term appropriate considering the process that is taking place? Describe how your group will remember this term.

Protein Synthesis

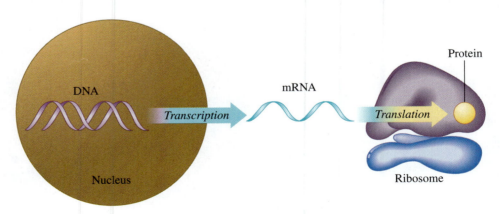

Scheme 1. The central dogma of molecular biology shown previously describes the transfer of sequence-specific information, in this case protein synthesis. Protein synthesis begins with transcription of gene DNA to a complementary copy of mRNA, which moves to the ribosome for translation. Amino acids are linked together at the ribosome until a stop codon on the mRNA is reached.

Questions

6. In what part of the cell does transcription take place?
7. In what part of the cell does translation take place?
8. Of the three molecules, DNA, mRNA, or protein, which moves from the nucleus to the ribosome during the process of protein synthesis?
9. During transcription, a messenger RNA (mRNA) molecule is made from a genetic piece of DNA. The mRNA is complementary to the DNA molecule from which it was copied. If a piece of the DNA has the sequence 5′ACGTAGTCACGT3′, what would the complementary RNA sequence be? Indicate the 5′ and 3′ ends on your sequence.
10. Using the genetic code in Table 11.3, what would the corresponding amino acid sequence be for the mRNA sequence in question 9?

11.5 Putting It Together: The Genetic Code and Protein Synthesis

If you were on a beach at night and saw a light in the ocean giving off three short flashes of light followed by three long flashes of light and three more short flashes of light, what would you do? Many of us would do nothing because we may not recognize that this is the Morse code distress signal for SOS. Someone who could translate Morse code would immediately get help if they saw this signal. Similarly, the genetic code from our gene sequences of DNA is a code transcribed into mRNA and decoded as a protein sequence. This *translation* involves tRNA and takes place at the ribosome.

11.5 Inquiry Question: How is the genetic code used to synthesize proteins?

The Genetic Code

The mRNA transcribed from the DNA contains a sequence of bases specifying the protein to be made. A given triplet called a **codon** in the mRNA translates to a specific amino acid. For example, the sequence UUU in the mRNA specifies the amino acid phenylalanine.

The **genetic code** (see Table 11.3) shows the codons of mRNA for the 20 amino acids. Sixty-four codon combinations are possible from the four bases A, G, C, and U.

The three codons UGA, UAA, and UAG are stop signals. When they appear in the mRNA sequence, this is a signal to stop adding amino acids to the growing protein chain.

TABLE 11.3 mRNA CODONS: The Genetic Code for Amino Acids

First Letter from 5′ End	Second Letter				Third Letter
	U	**C**	**A**	**G**	
U	UUU Phe / UUC Phe / UUA Leu / UUG Leu	UCU / UCC / UCA / UCG Ser	UAU Tyr / UAC Tyr / UAA STOP / UAG STOP	UGU Cys / UGC Cys / UGA STOP / UGG Trp	U / C / A / G
C	CUU / CUC / CUA / CUG Leu	CCU / CCC / CCA / CCG Pro	CAU His / CAC His / CAA Gln / CAG Gln	CGU / CGC / CGA / CGG Arg	U / C / A / G
A	AUU / AUC / AUA Ile / ᵃAUG Met/start	ACU / ACC / ACA / ACG Thr	AAU Asn / AAC Asn / AAA Lys / AAG Lys	AGU Ser / AGC Ser / AGA Arg / AGG Arg	U / C / A / G
G	GUU / GUC / GUA / GUG Val	GCU / GCC / GCA / GCG Ala	GAU Asp / GAC Asp / GAA Glu / GAG Glu	GGU / GGC / GGA / GGG Gly	U / C / A / G

ᵃCodon that signals the start of a peptide chain.
STOP codons signal the end of a peptide chain.

The triplet AUG has two roles in protein synthesis. If it is found within the mRNA, it codes for the amino acid methionine. If found at the 5′ end of a mRNA, the codon AUG represents the start codon initiating protein synthesis.

Codons in mRNA	5′UUU|GGG|CGC 3′			
Translation	↓	↓	↓	
Amino acid sequence	Phe –	Gly –	Arg	(three-letter abbreviation)
	F	G	R	(one-letter abbreviation)

sample problem 11.8 — Codons

What is the sequence of amino acids coded by the following codons in mRNA?

$$5′GUC|AGC|CCA3′$$

Solution

According to the genetic code in Table 11.3, GUC codes for valine, AGC for serine, and CCA for proline. The amino acid sequence is Val–Ser–Pro or VSP.

Protein Synthesis

Let's pause to review the entire process of protein synthesis from DNA through protein beginning with transcription (see **Figure 11.11** on the next page).

Transcription

DNA found in the nucleus unwinds at the site of a gene. A complementary copy of this DNA template, called mRNA, is created. See Figure 11.11 ①. *RNA polymerase* is the name of the enzyme complex that binds to the DNA and links nucleotides to create mRNA. The mRNA then travels out of the nucleus to the ribosome. See Figure 11.11 ②.

tRNA Activation

Before the tRNA can be used in the ribosome, an amino acid must be attached to its acceptor stem. An enzyme called tRNA synthetase attaches the correct amino acid to the acceptor stem of the tRNA. The amino acid is then ready for use in protein synthesis. See Figure 11.11 ③.

Translation

Protein synthesis begins when mRNA positions itself at the ribosome. The first codon in mRNA is the start codon, AUG. An activated tRNA with an anticodon of UAC and methionine attached enters the ribosome and hydrogen bonds to the mRNA. A second activated tRNA matching the next codon on the mRNA enters an adjacent position on the ribosome. The two amino acids then join, forming a peptide bond, and the methionine detaches from the first tRNA. The deactivated tRNA in the first position leaves the ribosome, and the second tRNA shifts into the first position with the dipeptide attached.

This shifting is called *translocation*. As the ribosome moves along the mRNA, a peptide bond joins each subsequent amino acid that enters. In this way, a growing polypeptide chain emerges. See Figure 11.11 ④. The tRNAs dislodge from the ribosome and return to the free tRNA pool to become recharged with another amino acid. See Figure 11.11 ⑤.

Termination

Eventually, the ribosome encounters a stop codon, and protein synthesis ends. The polypeptide chain is released from the ribosome. The initial amino acid methionine is often removed from the beginning of the polypeptide chain. The growing polypeptide chain folds into its tertiary structure, forming any disulfide links, salt bridges, or other interactions that make the polypeptide a biologically active protein.

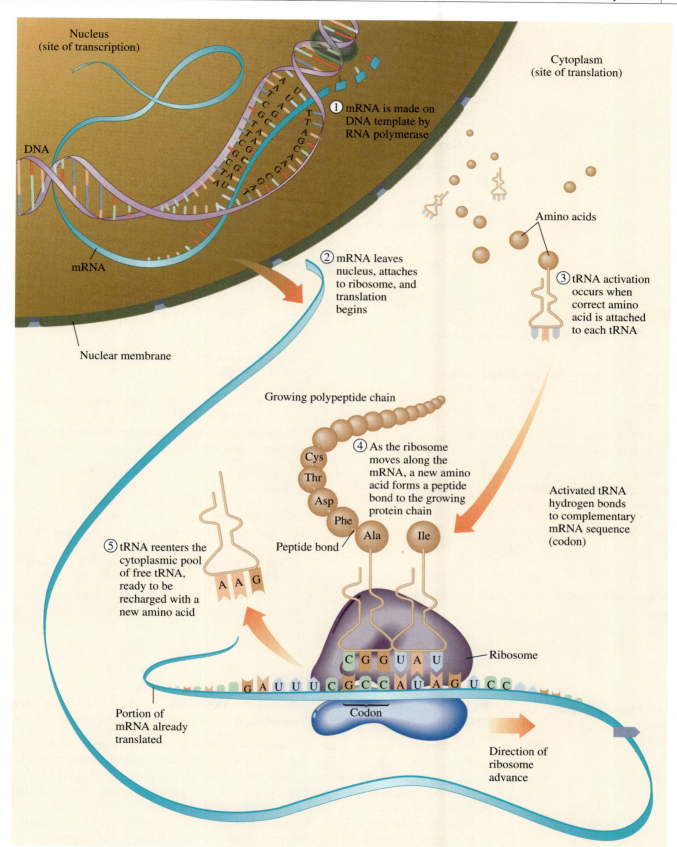

FIGURE 11.11 Protein synthesis. Protein synthesis begins with transcription of gene DNA to the mRNA (upper left), which moves to the ribosome for translation via activated tRNA molecules. Amino acids are linked until a stop codon on the mRNA is reached.

sample problem 11.9   **Translation**

What order of amino acids would you expect in a peptide made from the following mRNA sequence?

$$5'UCA|AAA|GCC|CUU3'$$

Solution

Each of the codons specifies a particular amino acid. Using Table 11.3, we write a peptide with the following sequence:

mRNA codons	$5'UCA|AAA|GCC|CUU3'$
Three-letter amino acid sequence	Ser–Lys–Ala–Leu
One-letter sequence	S K A L

practice problems

11.25 Why are there at least 20 tRNAs?

11.26 List the difference between a codon and an anticodon.

11.27 Provide the three-letter amino acid sequence expected from each of the following mRNA segments:
a. 5'AAA|CCC|UUG|GCC3'
b. 5'CCU|CGA|AGC|CCA|UGA3'
c. 5'AUG|CAC|AAA|GAA|GUA|CUU3'

11.28 Provide the three-letter amino acid sequence expected from each of the following mRNA segments:
a. 5'AAA|AAA|AAA3'
b. 5'UUU|CCC|UUU|CCC3'
c. 5'UAC|GGG|AGA|UGU3'

11.29 In your own words, define translocation.

11.30 In your own words, describe how a peptide chain is extended.

11.31 The following portion of DNA is in the template DNA strand:
$$3'TGT|GGG|GTT|ATT5'$$
a. Write the corresponding mRNA section. Show the nucleic acid sequence as triplets and label the 5' and the 3' ends.
b. Write the anticodons corresponding to the codons on the mRNA.
c. Write the three-letter and one-letter amino acid sequence that will be placed in a peptide chain.

11.32 The following portion of DNA is in the template DNA strand:
$$3'GCT|TTT|CAA|AAA5'$$
a. Write the corresponding mRNA section. Show the nucleic acid sequence as triplets and label the 5' and the 3' ends.
b. Write the anticodons corresponding to the codons on the mRNA.
c. Write the three-letter and one-letter amino acid sequence that will be placed in a peptide chain.

11.6 Genetic Mutations

11.6 Inquiry Question: How do genetic mutations occur?

What will happen to protein synthesis if the DNA sequence is changed? Any change in a DNA nucleotide sequence is called a **mutation.** Let's examine the possibilities.

- No change in protein sequence. Sometimes a change in a DNA base will have no effect. Only about 2.5% of the DNA in your chromosomes encodes for proteins. The rest of your DNA is nongene or commonly referred to as "junk" DNA. This DNA is likely not junk, in that it contains recognition sites for protein binding. Also, recall that there is more than one codon that codes for each amino acid. For example, if the codon UUU were changed to UUC, the amino acid phenylalanine would still be placed in the growing polypeptide chain. These types of mutations are called **silent mutations.**
- A change in protein sequence occurs, but it has no effect on protein function. If, for example, the codon AUU is mutated to GUU, then the amino acid isoleucine would be changed to valine. These two amino acids are similar in polarity and size, and it is likely that such a substitution would not have much of an effect on the protein function. This is another type of silent mutation.

• A change in protein sequence occurs and affects protein function. If, for example, the codon AUU were mutated to AAU, then isoleucine would be changed to asparagine. In this case, a nonpolar amino acid is replaced with a polar amino acid, which could affect the structure and the function of the protein. Other mutations that can have a negative effect on protein synthesis include mutating a codon into a stop codon, or inserting or deleting a base. The latter has the effect of shifting the triplets that are read in the mRNA and would change the identity of all subsequent amino acids after the insertion or deletion.

sample problem 11.10 Mutations

A mRNA has the sequence of codons 5'CCC|AGA|GCC3'. If a base substitution in the DNA changes the mRNA codon of AGA to GGA, how is the amino acid sequence affected in the resulting protein? Can you predict whether this might have an effect on the protein function?

Solution

The initial mRNA sequence of 5'CCC|AGA|GCC3' codes for the amino acids proline, arginine, and alanine. When the mutation occurs, the new sequence of mRNA codons, 5'CCC|GGA|GCC3', codes for proline, glycine, and alanine. Arginine is replaced by glycine, which changes a polar, charged amino acid into a nonpolar, uncharged amino acid. This change could have an effect on protein structure.

Sources of Mutations

How do mutations occur? Sometimes when DNA replicates itself, errors occur at random. This is considered a **spontaneous mutation.** Environmental agents like chemicals or radiation that produce mutations in DNA are referred to as **mutage**: Many mutagens can cause cancer and are designated **carcinogens.** Viruses can a cause mutations (Section 11.7).

integrating Chemistry

One common chemical mutagen is sodium nitrite ($NaNO_2$), which is used as a preservative in processed meats like hot dogs and bologna. In the presence of amines (found in proteins), sodium nitrite forms a compound called a nitrosamine, a known carcinogen in animals. Nitrosamines assist in the conversion of cytosine into uracil, which effectively converts a C–G base pair into a U–A base pair.

If it is a known mutagen, why is sodium nitrite still used to preserve meat? The amounts used in the preservation process are small, and the benefits to meat preservation outweigh the risk of mutation.

$$NaNO_2 \ + \ \text{Amine from protein} \longrightarrow \ \overset{R_1}{\underset{R_2}{>}}N-N=O$$

Sodium nitrite, food   Nitrosamine
preservative used in meats

If a mutation occurs in a **somatic cell** (any cell type other than egg or sperm), it affects only the individual organism and can cause conditions like cancer. Mutations that occur in **germ cells** (sperm or egg cells), however, can be passed on to future generations. A germ cell mutation can lead to malformation (and subsequent malfunction) of a particular protein. Germ cell mutations cause **genetic diseases.** More than 4,000 genetic diseases have been identified. See **Table 11.4**.

$$\text{DNA} \xrightarrow{\text{Mutation}} \underset{\text{of DNA}}{\text{Alteration}} \longrightarrow \underset{\text{protein}}{\text{Defective}} \longrightarrow \underset{\text{or cancer (somatic cells)}}{\text{Genetic disease (germ cells)}}$$

TABLE 11.4 Some Genetic Diseases

Genetic Disease	Protein or Chromosomal Defect	Disease Symptoms
Galactosemia	The transferase enzyme required for the metabolism of galactose-1-phosphate is absent.	Cataracts, mental retardation.
Cystic fibrosis	Mutation in a gene producing a protein that regulates salt transport in and out of cells	Thick secretions of mucus, difficulty breathing, blocked pancreatic function
Down syndrome	Formation of three chromosomes, usually number 21, instead of a pair of chromosomes	Heart and eye defects, mental and physical problems
Familial hypercholesterolemia	Mutation in a gene on chromosome 19 regulating cholesterol levels	High cholesterol levels, early coronary heart disease (age 30–40)
Muscular dystrophy (Duchenne)	One of 10 forms of MD. A mutation in the X chromosome results in the low or abnormal production of *dystrophin*.	Muscle-atrophy disease appears around age 5, with death by age 20. Occurs in about 1 of 10,000 males.
Huntington's disease (HD)	Mutation in a gene on chromosome 4, which can now be mapped to test people in families with HD.	Nervous tremors leading to total physical impairment
Sickle-cell anemia	Defective hemoglobin from a mutation in a gene on chromosome 11	Anemia from decreased oxygen-carrying ability of red blood cells
Hemophilia	One or more defective blood-clotting factors	Poor blood coagulation, excessive bleeding, and internal hemorrhages
Tay-Sachs disease	Defective hexosaminidase A	Accumulation of lipids in the brain, resulting in mental retardation, loss of motor control, and early death

practice
problems

11.33 In your own words, define mutagen.

11.34 In your own words, define a silent mutation.

11.35 Consider the following portion of mRNA produced by the normal order of DNA nucleotides:

5′CUU|AAA|CGA|GUU3′

a. Write the amino acid sequence that would be produced from this mRNA.
b. Write the amino acid sequence if a mutation changes CUU to AUU. Is this likely to affect protein function?
c. Write the amino acid sequence if a mutation changes CGA to AGA. Is this likely to affect protein function?
d. What happens to protein synthesis if a mutation changes AAA to UAA?
e. What happens to the protein sequence if an A is added to the beginning of the chain and the sequence changes to 5′ACU|UAA|ACG|AGU3′?
f. What happens to the protein sequence if the C is removed from the beginning of the chain and the sequence changes to 5′UUA|AAC|GAG3′?

11.36 Consider the following portion of mRNA produced by the normal order of DNA nucleotides:

5′ACA|UCA|CGG|GUA3′

a. Write the amino acid sequence that would be produced from this mRNA.
b. Write the amino acid sequence if a mutation changes UCA to ACA. Is this likely to affect protein function?
c. Write the amino acid sequence if a mutation changes CGG to GGG. Is this likely to affect protein function?
d. What happens to protein synthesis if a mutation changes UCA to UAA?
e. What happens to the protein sequence if a G is added to the beginning of the chain and the sequence changes to 5′GAC|AUC|ACG|GGU3′?
f. What happens to the protein sequence if the A is removed from the beginning of the chain and the sequence changes to 5′CAU|CAC|GGG3′?

11.7 Viruses

Viruses are small particles containing from 3 to 200 genes that can infect any cell type. Viruses are not considered cells because they cannot make their own proteins or their own energy. They contain only parts needed to infect a cell. Viruses have their own nucleic acid but use the ribosomes and RNA of the infected cell—also called the **host cell**—to make their proteins. Some infections caused by viruses invading human cells are listed in Table 11.5.

11.7 Inquiry Question:
How does a virus use
nucleic acids?

TABLE 11.5 Some Diseases Caused by Viral Infection

Disease	Virus
Common cold	Coronavirus (more than 100 types)
Influenza	Orthomyxovirus
Warts	Papovavirus
Herpes	Herpesvirus
Leukemia, cancers, AIDS	Retroviruses
Hepatitis	Hepatitis A virus (HAV), hepatitis B virus (HBV), hepatitis C virus (HCV)
Mumps	Paramyxovirus
Epstein–Barr	Epstein–Barr virus (EBV)

Viruses are very simple. Even though they have a variety of shapes, they all contain a nucleic acid (either DNA or RNA) enclosed in a protein coat called a **capsid.** Many viruses also contain an additional protective coat called an **envelope** surrounding the capsid. (See Figure 11.12.) The function of all viruses is the same: to monopolize the functions of the host cell for the benefit of the virus.

A virus infects a cell when an enzyme in the protein coat makes a hole in the host cell, allowing the viral nucleic acid to enter and mix with host cell material. If the virus contains DNA, the host cell begins to replicate the viral DNA in the same way it would replicate normal DNA. Viral DNA produces viral RNA, which proceeds to make the proteins for the virus. The completed virus particles are assembled and then released from the cell to infect more cells. This release often occurs by a process called budding (see Figure 11.13).

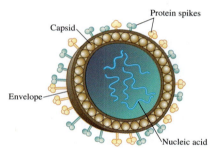

FIGURE 11.12 Virus structure. All viruses contain nucleic acid protected by a protein coat called a capsid. Many animal viruses also contain exterior spikes made of protein.

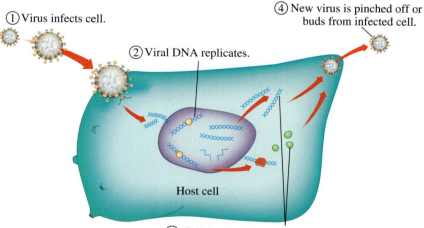

① Virus infects cell.
② Viral DNA replicates.
④ New virus is pinched off or buds from infected cell.
③ Cellular enzymes make a viral protein and viral DNA, which assemble into viruses.
Host cell

FIGURE 11.13 The life cycle of a virus. After a virus attaches to the host cell, it injects its viral DNA and uses the host cell's nucleic acids, enzymes, amino acids, and ribosomes to make viral mRNA, new viral DNA, and viral proteins. The newly assembled viruses are released to infect other cells.

Vaccines are often inactive forms of viruses that boost the immune response by causing the body to produce antibodies to fight the virus. Several childhood diseases such as polio, mumps, chicken pox, and measles can be prevented through the use of vaccines.

Retroviruses

A virus that contains RNA as the nucleic acid is called a **retrovirus**. Once retroviral RNA gets into the cell, it must first make viral DNA through a process known as *reverse transcription*. Retroviruses contain an enzyme called reverse transcriptase that uses the viral RNA to make complementary strands of viral DNA using the host nucleotides. The viral DNA joins the DNA of the host cell and uses the cell's enzymes and ribosomes to replicate virus particles (see **Figure 11.14**).

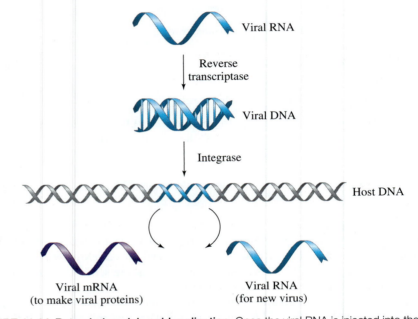

Viral RNA

Reverse transcriptase

Viral DNA

Integrase

Host DNA

Viral mRNA
(to make viral proteins)

Viral RNA
(for new virus)

FIGURE 11.14 Retroviral nucleic acid replication. Once the viral RNA is injected into the cell, it is first transcribed into DNA using the viral enzyme reverse transcriptase. The viral DNA gets incorporated into the host DNA by a viral enzyme integrase and is then transcribed to make both mRNA for viral protein synthesis and new viral RNA.

HIV-1 and AIDS

*H*uman *I*mmunodeficiency *V*irus type 1, or HIV-1, is a retrovirus responsible for the disease AIDS (*A*cquired *I*mmune *D*eficiency *S*yndrome). HIV-1 infects a type of blood cell known as a T4 lymphocyte that is part of the human immune system. T depletion of these immune cells caused by the HIV-1 infection reduces a person's ability to fight off other infections. AIDS patients are very susceptible to skin canc sarcomas, and bacterial infections like pneumonias.

To minimize damage to the host cell, viral therapies must be able to inactivate unique parts of the virus life cycle. Current AIDS therapies involve several drugs that attack HIV-1 at the points of reverse transcription and viral protein synthesis (see Figure 11.15).

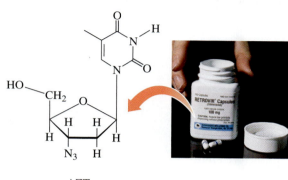

AZT
Azidothymidine

The first approved AIDS drug, called AZT (azidothymidine) and sold under the name Retrovir®, is similar to the nucleoside deoxythymidine, but it lacks a 3′ —OH group. If reverse transcriptase incorporates this modified nucleoside into the viral DNA, transcription halts because of this missing —OH. Since the introduction of AZT, several other drugs have been developed that also behave like a nucleoside. These chemically modified nucleosides are called **nucleoside analogs.** The array of treatments now available allows doctors to prescribe a "cocktail" of medicines that can extend the remission periods for AIDS patients.

A second set of drugs used to treat HIV-1 is called protease inhibitors and includes ritonavir, indinavir, and saquinavir. The target protein is HIV protease, and it is responsible for clipping the viral proteins down to size for viral assembly. Many of the HIV protease inhibitors are competitive inhibitors that block the active site, lowering enzyme activity.

Other classes of drugs being developed and approved to combat HIV-1 infection include cell entry drugs such as maraviroc and enfuvirtide, which block insertion of RNA into the host cell, and integrase inhibitors like raltegravir. Integrase is the enzyme involved in incorporating viral DNA into host DNA. Researchers are hopeful that combination therapies can keep the HIV infection in check.

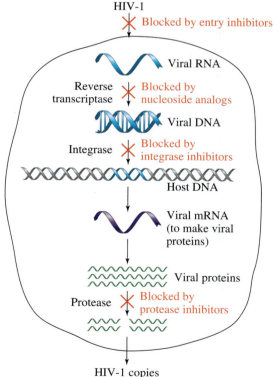

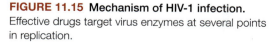

FIGURE 11.15 Mechanism of HIV-1 infection.
Effective drugs target virus enzymes at several points in replication.

Viruses

Why are viruses unable to replicate on their own?

Solution

Viruses contain only a protective protein coat and DNA or RNA and not the necessary replication machinery including enzymes, nucleosides, and ribosomes.

practice
problems

11.37 Why do viruses need to enter a host cell?

11.38 Name two components common to all viruses.

11.39 What is the purpose of a vaccine?

11.40 If a virus contains viral RNA,
    a. name the first step that must occur before DNA replication.
    b. name the class of virus.

11.41 How do protease inhibitors disrupt the life cycle of the HIV-1 virus?

11.42 How do nucleoside analogs disrupt the life cycle of the HIV-1 virus?

11.8 Recombinant DNA Technology

11.8 Inquiry Question: How can DNA be manipulated outside of the cell?

Since the discovery of DNA as the genetic material in all cells, scientists have been finding ways to manipulate DNA to produce proteins needed as medicines, proteins helpful in crop resistance to pests, and proteins capable of positive identification of individuals.

Just as the term sounds, **recombinant DNA** involves recombining DNA from two different sources. Often called *genetic engineering* or *gene cloning*, the genome of one organism is altered by splicing in a section of DNA containing a gene from a second organism. Why would anyone want to do that? Inserting a higher organism's gene into smaller organisms (like bacteria) with shorter life cycles produces larger amounts of the desired protein more quickly.

The idea of combining genes from two different organisms is not new. Humans have been crossbreeding plants and animals for centuries, exchanging DNA to produce desired traits. The recombinant DNA techniques developed in the mid-1970s just work more quickly, expand its usefulness, and are more predictable.

There are a few basic steps to recombining DNA and producing a protein in another organism (see **Figure 11.16**).

- *Identify and isolate a gene of interest.* A gene must be located on a chromosome and removed. This gene of interest is referred to as the *donor DNA.* The removal is done with enzymes called **restriction enzymes** that recognize specific DNA sequences four to eight bases in length. One of these enzymes is called *Eco*RI (pronounced "echo–R–one"), which recognizes the following DNA sequence and cuts the DNA as shown.

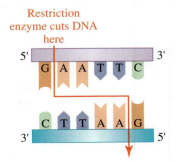

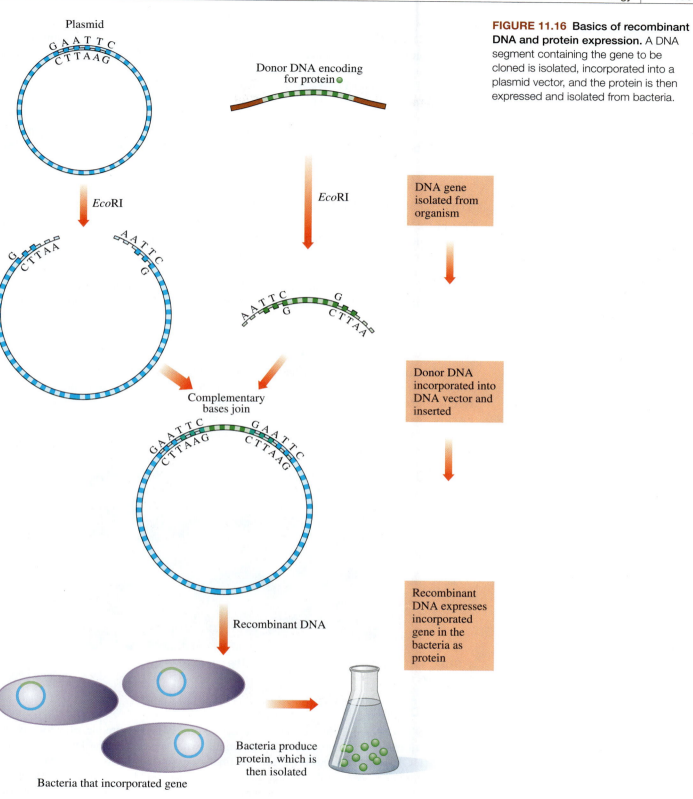

Plasmid

*Eco*RI

Donor DNA encoding
for protein

*Eco*RI

Complementary
bases join

Recombinant DNA

Bacteria produce
protein, which is
then isolated

Bacteria that incorporated gene

FIGURE 11.16 Basics of recombinant DNA and protein expression. A DNA segment containing the gene to be cloned is isolated, incorporated into a plasmid vector, and the protein is then expressed and isolated from bacteria.

DNA gene
isolated from
organism

Donor DNA
incorporated into
DNA vector and
inserted

Recombinant
DNA expresses
incorporated
gene in the
bacteria as
protein

- *Insert donor DNA into the organism DNA using a vector.* A **vector** is a transporter for the donor DNA. It can incorporate donor DNA into the genome of the organism. One common vector found in bacteria is a piece of circular DNA called a **plasmid.** Using the same restriction enzymes, the plasmid is opened and the isolated gene DNA can be inserted. The vector DNA can be incorporated into the bacterial DNA as the bacteria grow and divide.
- *Express the incorporated gene in the new organism.* Bacteria that have the donor DNA incorporated are induced to produce the protein from the gene of interest. This protein production of a nonnative gene is called **expression.** Then the protein can be subsequently isolated and purified for its intended use.

Therapeutic Proteins

One of the first applications of recombinant DNA technology incorporated the human insulin gene into bacteria called *Escherichia coli (E. coli)*, thereby allowing these bacteria to produce human insulin. Before 1982, pig or cow insulin was harvested for use by diabetics. The amino acid sequences of the pig and cow insulin differ in two or three amino acids, respectively, from human insulin. Although pig and cow insulin sequences are similar, impurities in the extraction process produced rejection in some people. Almost all insulin used today is expressed human insulin, thus causing fewer side effects.

Genetically Modified Crops

Recombinant DNA technology has also allowed the insertion of genes into food plants, affording crops advantages during growth. In the late 1990s, Monsanto introduced Roundup Ready® cotton to farmers. This variety of cotton has a gene inserted into it that gives resistance to Roundup®, a weed killer. When farmers apply Roundup to a Roundup Ready cottonfield, weeds are killed and the cotton plant survives. Many food crops—like the papaya—have been genetically modified to resist viral infection (see **Figure 11.17**).

FIGURE 11.17 Genetically modified crops. The genetically modified papaya (right) is resistant to the papaya ringspot virus that affects normal papaya (left), providing higher yields and larger fruits.

Genetic Testing

In the 1990s, U.S. government agencies in conjunction with private industry took on the task of sequencing the entire human genome in what was called the Human Genome Project (HGP). The last chromosome was finally completed in May 2006. The human genome contains over 3 billion base pairs and an estimated 20,000 to 25,000 genes. Because of the HGP we can now identify genes responsible for many genetic diseases. Through genetic testing, a person's DNA can be screened for mutated genes. Such tests can assist couples planning to have children, ascertain paternity, and predict certain cancers, Alzheimer's disease, and Huntington's disease.

Nuclear Transplantation—Cloning an Organism

Is cloning a gene the same thing as cloning an organism? No. The term **clone** means to make an exact copy. In one case, you are making an exact copy of a single gene to express a single protein. In the other, you are making an exact copy of an entire organism.

Cloning an organism creates a genetic copy of the original organism. This is done by taking the nuclear DNA from an adult cell (somatic cell) and transplanting it into an egg cell from which DNA has been removed. In some cases, such an egg cell with a full set of chromosomes behaves like a fertilized egg and will begin to divide, forming an embryo. The embryo can then be transplanted into a surrogate until it fully develops.

The first cloned mammal, Dolly the sheep, was born in 1996 and was the sole survivor of 276 attempts to create and implant an embryo. Dolly survived until 2003, about half the normal life expectancy. Since that time, a number of other animals have been successfully cloned, including the cow, horse, monkey, deer, and cat (see **Figure 11.18**).

FIGURE 11.18 Cloned animals. (a) Dolly the sheep was the first cloned animal. (b) In 2001, the first kitten—named Cc—was cloned.

---

sample problem 11.12 Recombinant DNA Technology

Describe how a gene from one organism can be incorporated into a second organism using recombinant DNA technology.

Solution

The steps in recombinant DNA include cutting the gene of interest (donor DNA) from an organism's genome using a restriction enzyme, inserting that gene into a vector, and incorporating the donor DNA into the genome of the host organism.

---

practice **problems**

11.43 In your own words, define gene expression.

11.44 In your own words, define recombinant DNA.

11.45 Describe the function of a vector.

11.46 Describe the structure of a plasmid.

SUMMARY

11.1 Components of Nucleic Acids

(Nucleotide) **11.1 Inquiry Question:** What are the components of nucleotides and nucleic acids?

Deoxyribonucleic acid (DNA) and ribonucleic acid (RNA) are nucleic acids. They consist of strings of nucleotides. A nucleotide has three components: a nitrogenous base, a five-carbon sugar, and a phosphate. A nucleoside consists of the nitrogenous base and the five-carbon sugar. The sugar deoxyribose is found in DNA, and the sugar ribose is found in RNA. The bases adenine (A), guanine (G), and cytosine (C) are found in both DNA and RNA. Thymine (T) is a fourth base found in DNA, and uracil (U) is a fourth base found in RNA. Nucleotides are named as the nucleoside + the number of phosphates (up to three) bonded to it. The components of nucleosides, as well as those of nucleotides, link together through condensation reactions.

Free 5' end —P—P—P—P— Free 3' end **11.2 Nucleic Acid Formation**

11.2 Inquiry Question: How are nucleic acids formed?

Each nucleic acid has a unique sequence that is its primary structure. Nucleic acids form when nucleotides undergo condensation linking sugar to phosphate. The 3′ —OH on the sugar of one nucleotide bonds with the phosphate on the 5′ end of a neighboring nucleotide. The backbone of a nucleic acid is formed from alternating sugar-phosphate-to-sugar-phosphate groups. The nitrogenous bases dangle from the backbone. Each nucleic acid has single free 5′ and 3′ ends.

11.3 DNA

11.3 Inquiry Question: What are the unique structural features of DNA?

A DNA molecule resembles a twisted ladder. It consists of two antiparallel strands of nucleic acid with bases facing inward. The two strands are held together through the bases (rungs of the ladder), which hydrogen-bond to complementary bases on the other strand, giving DNA its secondary structure. The base A forms two hydrogen bonds to T, and G forms three hydrogen bonds to C. In most plants and animals, DNA is found in the cell nucleus and is compacted into a tertiary structure called a chromosome. Chromosomes also contain a protein component called a histone, around which the DNA is supercoiled. Humans have 23 pairs of chromosomes in their cells.

11.4 RNA and Protein Synthesis

11.4 Inquiry Question: What are the roles of RNA in protein synthesis?

Direction of transcription

RNA differs from DNA in that it contains a ribose sugar instead of a deoxyribose. Instead of thymine, RNA contains uracil that can hydrogen-bond to adenine. RNA is smaller than DNA and is single stranded. The three types of RNA that are involved in transforming the DNA sequence into a protein sequence in the cell are messenger RNA (mRNA), ribosomal RNA (rRNA), and transfer RNA (tRNA). The messenger RNA is involved in transcribing a complementary copy from gene DNA in the cell nucleus and taking that copy to the ribosome. Ribosomes are cell organelles where protein synthesis takes place. They consist of rRNA and protein. The tRNA is a compact RNA structure that acts as a conduit between the code on the messenger RNA and an amino acid sequence.

tRNA has two features: the anticodon at one end that is complementary to the codon sequence on the mRNA and the acceptor stem where an amino acid can attach.

11.5 Putting It Together: The Genetic Code and Protein Synthesis

11.5 Inquiry Question: How is the genetic code used to synthesize proteins?

The genetic code is a series of base triplet sequences on mRNA specifying the order of amino acids in a protein. The codon AUG signals the start of transcription and the codons UAG, UGA, and UAA signal it to stop. Protein synthesis begins with transcription where a mRNA creates a complementary copy of a DNA gene. tRNA activation involves binding an amino acid to the tRNA, catalyzed by the enzyme tRNA synthetase. During translation, tRNAs bring the appropriate amino acids to the ribosome and peptide bonds form until termination when a stop codon is reached. The polypeptide becomes a functional protein upon release.

11.6 Genetic Mutations

Nitrosamine **11.6 Inquiry Question:** How do genetic mutations occur?

Base alterations from normal cell DNA sequences are called mutations. Some mutations are silent and have no effect on protein synthesis, and some may change the sequence but not alter protein function. Others can affect protein sequence, structure, and function. Mutations can be random during DNA replication or they can be caused by mutagens like chemicals or radiation. A mutation in a germ cell can be inherited. If such a mutation results in a defective protein, a genetic disease results.

11.7 Viruses

11.7 Inquiry Question: How does a virus use nucleic acids?

Viruses are particles containing DNA or RNA and a protein coat called a capsid. They invade a host cell and use the host cell's machinery to replicate more virus particles. Viruses containing RNA are called retroviruses. These viruses must go through an initial step of reverse transcription to make viral DNA from viral RNA. HIV-1 is a retrovirus. Several areas of retroviral replication have been studied and promising drugs have been developed in recent years to slow down HIV-1 infection in AIDS patients.

11.8 Recombinant DNA Technology

Restriction enzyme cuts DNA here **11.8 Inquiry Question:** How can DNA be manipulated outside of the cell?

Recombinant DNA technology involves the expression of a protein from one organism into a second organism. This can be accomplished after the gene of interest is isolated and clipped from a genome using a restriction enzyme. The gene of interest is incorporated into a vector that transports and incorporates the gene into the second organism. The host organism can then produce the protein of interest during transcription and translation. Gene cloning or making an exact copy of a gene is different from organism cloning where an exact copy of an organism is produced.

The study guide will help you check your understanding of the main concepts in Chapter 11. You should be able to

11.1 Components of Nucleic Acids

- Identify the five nitrogenous bases found in nucleic acids.
- Distinguish the sugars ribose and deoxyribose.
- Write nucleosides and nucleotides given their component parts.

11.2 Nucleic Acid Formation

- Write the product of a condensation of nucleotides.
- Abbreviate a nucleic acid using one-letter base coding.

11.3 DNA

- Characterize the structural features of DNA.
- Write the complementary base pairs for a single strand of DNA.

11.4 RNA and Protein Synthesis

- List three types of RNA and their role in protein synthesis.
- Translate a DNA strand into its complementary mRNA.

11.5 Putting It Together: The Genetic Code and Protein Synthesis

- Distinguish transcription from translation.
- Translate a mRNA sequence into a protein sequence using the genetic code.

11.6 Genetic Mutations

- Define genetic mutation.
- Determine changes in protein sequence if a mRNA sequence is mutated.

11.7 Viruses

- List the differences between a virus and a cell.
- List the structural components of a virus.
- Describe how a virus infects a cell.

11.8 Recombinant DNA Technology

- Apply knowledge of nucleic acid structure to DNA technology.

Key Terms

anticodon—The triplet of bases in the center loop of tRNA that is complementary to a codon on mRNA.

capsid—The protein coat on a virus enclosing nucleic acid.

carcinogen—A mutagen that causes cancer.

chromosome—A single piece of DNA wound around histones found in the nucleus of the cell.

clone—An exact copy.

codon—A sequence of three bases in mRNA that specifies a certain amino acid to be placed in a protein. A few codons signal the start or stop of transcription.

complementary base pair—Base pairing first described by Watson and Crick. In DNA, adenine and thymine are always paired (A–T or T–A) and guanine and cytosine are always paired (G–C or C–G). In RNA, adenine pairs with uracil (A–U or U–A).

DNA—Deoxyribonucleic acid; the genetic material of all cells containing nucleotides with deoxyribose sugar, phosphate, and the four nitrogenous bases adenine, thymine, guanine, and cytosine.

double helix—A shape that describes the double-stranded structure of DNA like a twisted ladder with rails as the sugar–phosphate backbone and base pairs as rungs.

envelope—An additional protein coat found on some viruses surrounding the capsid.

expression—In recombinant DNA technology, synthesis of a protein from one organism cloned into a second organism.

gene—The part of a piece of DNA that encodes for a particular protein.

genetic code—The information in DNA transferred to mRNA as a sequence of codons for the synthesis of protein.

genetic disease—Disease caused by an inherited mutation from a germ cell.

genome—The set of DNA in one cell.

germ cell—A sperm or egg cell.

host cell—The cell that a virus particle uses to reproduce itself.

mRNA—Messenger RNA; produced in the nucleus by DNA to carry the genetic information to the ribosomes for the construction of a protein.

mutagen—An environmental agent that causes a mutation.

mutation—A change in DNA sequence.

nucleic acid—Large molecules composed of nucleotides, found in cells as double stranded helical DNA and single stranded RNA.

nucleoside—A pentose sugar condensed with one of the five nitrogenous bases (adenine, thymine, guanine, cytosine, or uracil) at C1′.

nucleoside analog—A chemically modified nucleoside that can be incorporated into viral DNA halting transcription.

nucleotide—A pentose sugar condensed with one of the five nitrogenous bases (adenine, thymine, guanine, cytosine, or uracil) at C1′ and up to three phosphates at C5′.

phosphodiester bond—Bond linking adjacent nucleotides in a nucleic acid. This bond connects the 3′ —OH of one nucleotide to the phosphate linked to C5′ of a second nucleotide.

plasmid—A piece of circular DNA found in bacteria commonly used as a vector.

recombinant DNA—DNA recombined from two organisms.

restriction enzyme—Enzymes that cut DNA at specific DNA sequences.

retrovirus—A virus that contains RNA and uses the enzyme reverse transcriptase to make viral DNA from its RNA.

RNA—Ribonucleic acid; a type of nucleic acid that is a single strand of nucleotides containing adenine, cytosine, guanine, and uracil.

RNA polymerase—The enzyme complex that catalyzes the construction of mRNA from a DNA template.

rRNA—Ribosomal RNA; a major component of the cell structure called the ribosome.

silent mutation—A change in DNA sequence that has no effect on protein sequence.

spontaneous mutation—Mutations arising from random errors during DNA replication.

somatic cell—All cells except the reproductive cells, sperm or egg. In humans these cells have 23 pairs (46) of chromosomes.

supercoiling—The doubling up of a helical molecule to form a compact coiled structure.

transcription—The transfer of a gene copy from DNA via the formation of an mRNA.

translation—The interpretation of the codons in the mRNA as amino acids in a peptide.

tRNA—Transfer RNA; an RNA that places a specific amino acid into a peptide chain at the ribosome. There is one or more tRNA for each of the 20 different amino acids.

vector—A delivery agent for a gene from one organism for incorporation into another organism.

Additional Problems

11.47 Identify each of the following bases as a pyrimidine or a purine:
- **a.** cytosine
- **b.** adenine
- **c.** uracil
- **d.** thymine
- **e.** guanine

11.48 Indicate if each of the bases in Problem 11.47 is found in DNA only, RNA only, or both DNA and RNA.

11.49 Identify the base and sugar in each of the following nucleotides:
   a. CMP   **b.** dAMP   **c.** dGMP   **d.** UMP

11.50 Identify the base and sugar in each of the following nucleotides:
   a. dTMP   **b.** AMP   **c.** dCMP   **d.** GMP

11.51 How do the bases thymine and uracil differ?

11.52 How do the sugars ribose and deoxyribose differ?

11.53 Fill in the following table comparing structural similarities between proteins and nucleic acids:

	Protein	Nucleic Acid
Repeating unit	Amino acid	
Backbone repeat	$N-C_\alpha-C-N-C_\alpha-C$, etc.	
One-letter abbreviation	Name of amino acid	
Free left end	Amino or N-terminus	
Free right end	Carboxy or C-terminus	

11.54 Write the complementary base sequence for each of the following DNA segments. Indicate the 5′ and the 3′ ends.
   a. 5′GACTTAGGC3′
   b. 5′TGCAAACTAGCT3′
   c. 5′ATCGATCGATCG3′

11.55 Write the complementary base sequence for each of the following DNA segments. Indicate the 5′ and the 3′ end.
   a. 5′TTACGGACCGC3′
   b. 5′ATAGCCCTTACTGG3′
   c. 5′GGCCTACCTTAACGACG3′

11.56 Match the following statements with mRNA, rRNA, or tRNA:
   a. combines with proteins to form ribosomes
   b. carries the genetic information from the nucleus to the ribosome
   c. carries amino acids to ribosome for protein synthesis

11.57 Match the following statements with mRNA, rRNA, or tRNA:
   a. contains codons for protein synthesis
   b. contains anticodons
   c. found in the small subunit and the large subunit of the ribosome

11.58 List the possible codons for each of the following amino acids:
   a. threonine   **b.** serine   **c.** cysteine

11.59 List the possible codons for each of the following amino acids:
   a. valine   **b.** proline   **c.** histidine

11.60 Provide the amino acid corresponding to each of the following codons:
   a. ACG   **b.** CCA   **c.** GCA

11.61 Provide the amino acid corresponding to each of the following codons:
   a. UUG   **b.** CGG   **c.** AUC

11.62 Endorphins are polypeptides that reduce pain. What is the one-letter amino acid sequence formed from the following mRNA that codes for a pentapeptide that is an endorphin called Leu-enkephalin?
   5′AUG|UAC|GGU|GGA|UUU|CUA|UAA3′

11.63 What is the one-letter amino acid sequence formed from the following mRNA that codes for a pentapeptide that is an endorphin called Met-enkephalin?
   5′AUG|UAC|GGU|GGA|UUU|AUG|UAA3′

11.64 What is the anticodon on tRNA for each of the following codons in a mRNA?
   a. AGC   **b.** UAU   **c.** CCA

11.65 What is the anticodon on tRNA for each of the following codons in a mRNA?
   a. GUG   **b.** CCC   **c.** GAA

11.66 a. A base substitution changes a codon for an enzyme from GCC to GCA. Why is there no change in the amino acid order in the protein?
   b. In sickle-cell anemia, a base substitution in the hemoglobin gene replaces glutamate (a polar amino acid) with valine. Why does the replacement of one amino acid cause such a drastic change in biological function?

11.67 a. A base substitution for an enzyme replaces leucine (a nonpolar amino acid) with alanine. Why does this change in amino acids have little effect on the biological activity of the enzyme?
   b. A base substitution replaces cytosine in the codon UCA with adenine. How might this substitution affect the amino acids in the protein?

11.68 Discuss whether or not each of the following is a viable area for drug development to inactivate a retrovirus:
   a. reverse transcriptase inhibitors
   b. tRNA synthetase inhibitors
   c. cell entry blockers

11.69 Discuss whether each of the following is a viable area for drug development to inactivate a retrovirus:
   a. nucleoside analogs
   b. integrase inhibitors
   c. RNA polymerase inhibitors

11.70 How do integrase inhibitors disrupt the life cycle of the HIV-1 virus?

11.71 How do entry point inhibitors disrupt the life cycle of the HIV-1 virus?

11.72 List two societal benefits to recombinant DNA technology.

11.73 Distinguish between gene cloning and the cloning of an organism.

Challenge Problems

11.74 Oxytocin is a small peptide containing nine amino acids. How many nucleotides would be found in the mRNA for this protein?

11.75 A protein contains 35 amino acids. How many nucleotides would be found in the mRNA code for this protein?

11.76 a. If the DNA chromosomes in salmon contain 28% adenine, what is the percent of thymine, guanine, and cytosine?
b. If the DNA chromosomes in humans contain 20% cytosine, what is the percent of guanine, adenine, and thymine?

11.77 The DNA double helix can unwind or denature at temperatures between 90 °C–99 °C. Denaturing occurs when H bonds are broken. Which of the following strands of DNA would be expected to denature at a higher temperature? Provide an explanation.

DNA strand 1	5′ATTTTCCAAAGTATA3′
	3′TAAAAGGTTTCATAT5′
DNA strand 2	5′CGCGAGGGTCCACGC3′
	3′GCGCTCCCAGGTGCG5′

Answers to Odd-Numbered Problems

Practice Problems

11.1 a. pyrimidine **b.** pyrimidine

11.3 a. DNA **b.** both DNA and RNA

11.5 deoxyadenosine 5′-monophosphate (dAMP), deoxythymidine 5′-monophosphate (dTMP), deoxy-cytidine 5′-mono phosphate (dCMP), deoxyguanosine 5′-monophosphate (dGMP)

11.7 **a.**

b.

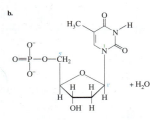

11.9 The nucleotides in nucleic acids are held together by phosphodiester bonds between the 3′-OH of a sugar (ribose or deoxyribose) and a phosphate group on the 5′-carbon of another sugar.

11.11

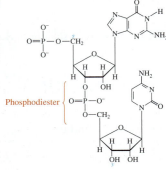

11.13 The two DNA strands are held together by hydrogen bonds between the bases in each strand.

11.15 a. 3′TTTT5′
b. 3′GGGGAAAA5′
c. 3′TGTAACC5′
d. 3′ACACTTGG5′

11.17

	Protein	Nucleic Acid
Primary structure	Sequence of amino acids	Sequence of nucleotides
Secondary structure	Local hydrogen bonding between backbone atoms	Hydrogen bonding between bases in nucleic acid strands
Tertiary structure	Folding of backbone onto itself associates amino acids far away in sequence	Supercoiling of DNA into tight, compact structures like chromosomes

11.19 messenger RNA—copies DNA and then moves from nucleus to ribosome
ribosomal RNA—structural component of the ribosome
transfer RNA—places a specific amino acid in a growing protein chain at the ribosome

11.21 In transcription, the sequence of nucleotides on one strand of a DNA template (the double helix is temporarily unwound) is used to produce a complementary mRNA copy using the enzyme RNA polymerase.

11.23 3′AUGCCGUUCGAU5′

11.25 Twenty amino acids are brought into the ribosome, each bonded to a different tRNA.

11.27 a. Lys–Pro–Leu–Ala
b. Pro–Arg–Ser–Pro–Stop
c. Met–His–Lys–Glu–Val–Leu

11.29 Translocation is the movement of the second tRNA to the spot vacated by the first tRNA, allowing for the next tRNA to bind to the ribosome. The amino acid is added to the growing peptide.

11.31 a. 5′ACA|CCC|CAA|UAA3′

    b. 3′UGU|GGG|GUU|AUU5′

    c. Three-letter code: Thr–Pro–Gln–Stop, one-letter code: TPQ-Stop

11.33 A mutagen is an environmental agent that produces a mutation (change in base sequence) in DNA.

11.35 a. LKRV. Three-letter code: Leu–Lys–Arg–Val, one-letter code: LKRV

    b. IKRV. Changing leucine to isoleucine, which are both nonpolar, will not have an effect on the protein function.

    c. LKRV. Nothing, because both codons code for arginine.

    d. Leucine will be the last amino acid incorporated into the growing protein chain because UAA is a stop codon.

    e. If A is added to the beginning of the chain, the codons would be shifted, and the new sequence would be: Thr–STOP–Thr–Ser. The growing chain would stop after incorporating the first threonine.

    f. If C is removed from the beginning of the chain, the codons would be shifted, and the new sequence would be Leu–Asn–Glu (LNE)

11.37 Viruses cannot replicate themselves without a host cell. They do not have the cell machinery.

11.39 A vaccine allows the body to mount an immune response by producing antibodies to a less active form of a virus. If an active form is encountered later, the body can fight against it more effectively.

11.41 Protease inhibitors inactivate an HIV-1 specific protease that is needed to process viral proteins.

11.43 Gene expression is producing a protein from a DNA gene sequence (usually of a non-native gene).

11.45 A vector is a transporter for donor DNA. It enables incorporation of the donor DNA into the organism DNA.

Additional Problems

11.47 a. pyrimidine   **b.** purine   **c.** pyrimidine

    d. pyrimidine   **e.** purine

11.49 a. cytosine, ribose   **b.** adenine, deoxyribose

    c. guanine, deoxyribose   **d.** uracil, ribose

11.51 Thymine contains a methyl group ($-CH_3$) that uracil lacks.

11.53

	Protein	Nucleic Acid
Repeating unit	Amino acid	Nucleotide
Backbone repeat	$N-C_\alpha-C-N-$ $C_\alpha-C$, etc.	sugar–phosphate– sugar–phosphate, etc.

	Protein	Nucleic Acid
One-letter abbreviation	Name of amino acid	Name of base
Free left end	Amino or N-terminus	5′ end
Free right end	Carboxy or C-terminus	3′ end

11.55 a. 3′AATGCCTGGCG5′

    b. 3′TATCGGGAATGACC5′

    c. 3′CCGGATGGAATTGCTGC5′

11.57 a. mRNA

    b. tRNA

    c. rRNA

11.59 a. GUU, GUC, GUA, GUG

    b. CCU, CCC, CCA, CCG

    c. CAU, CAC

11.61 a. leucine

    b. arginine

    c. isoleucine

11.63 YGGFM

11.65 a. CAC

    b. GGG

    c. CUU

11.67 a. Both alanine and leucine have small, nonpolar R groups. Both amino acids are found in similar environments in proteins so this substitution will unlikely affect protein structure or function.

    b. If a serine codon (UCA) is replaced with a stop codon (UAA), any remaining amino acids will not be linked to the growing polypeptide chain. This mutation can affect both the structure and function of the protein.

11.69 a. The development of nucleoside analogs that can be incorporated into viral DNA to stop reverse transcription is a viable means to inactivate viruses.

    b. Integrase is an enzyme found only in the virus so this would be a viable method for inactivating the virus.

    c. Because RNA polymerase is used to make all proteins in the cell, this method would not be viable in inactivating the virus.

11.71 Entry point inhibitors block sites on the outside of a host cell where a virus would attach and eventually enter. If the virus cannot enter the cell, the virus cannot replicate.

11.73 In gene cloning, a copy of a single gene is created. In organism cloning, many genes are copied and an exact copy of an organism is created.

11.75 3 nucleotides/amino acid + 1 start codon + 1 stop codon = (35 amino acids × 3) + 3 + 3 = 111 nucleotides

11.77 The more hydrogen bonds present, the more energy (as heat) must be applied to denature the DNA. Because strand 2 has more G–C base pairs with three hydrogen bonds each, it is predicted to have the higher denaturation temperature.

Our bodies use carbohydrates, lipids, and proteins in food to produce energy. Now that we have explored their chemical structures and basic reactivity, we are ready to apply these concepts to chemical reactions in the body to understand how the body produces and stores energy. Read Chapter 12 to learn more.

12

Food as Fuel—A Metabolic Overview

DO YOU KNOW people who claim to have a high metabolism? When they say this, they usually mean they can eat more than the average person and still maintain their body weight. Likewise, many nutrition products and weight-loss plans promise to "boost your metabolism." In this context, if your metabolism is "boosted," you will lose weight.

What exactly is metabolism? **Metabolism** refers to the chemical reactions occurring in the body that break down or build up molecules. Most of the time in biological systems, the chemical reactions of metabolism occur in a series of steps called a **metabolic pathway.**

In Chapter 5, we saw that chemical reactions can absorb or give off energy (endergonic or exergonic). In the body, this chemical energy is captured in the nucleotide **a**denosine **tri**phosphate or ATP, which then serves as the fuel for the processes of life. Among other things, ATP is used to contract muscles, transport molecules across membranes, and transmit nerve impulses. One of the ways our body produces ATP is by metabolizing the carbohydrate glucose.

In this chapter, we explore the absorption of biomolecules from food, metabolic pathways involved in glucose metabolism, controlling glucose levels in the blood, and the way other biomolecules, for example, proteins and lipids, can feed into the metabolic pathways.

Discovering the Concepts

Inquiry Activity—Reaction Pathways

Information

In the body, reactions occur in a series called a pathway. Each reaction in the pathway is catalyzed by a different enzyme. Most pathways have at least one irreversible reaction. These tend to be control points for pathways.

$$\mathrm{A} \xrightarrow[I]{E_1} \mathrm{B} \underset{II}{\overset{E_2}{\rightleftharpoons}} \mathrm{C} \underset{III}{\overset{E_3}{\rightleftharpoons}} \mathrm{D} \underset{IV}{\overset{E_4}{\rightleftharpoons}} \text{products}$$

SCHEME 1 A reaction pathway. This pathway contains four reactions labeled *I–IV*.

Questions

1. Label the items in Scheme 1 with the following terms if the reactions are occurring left to right.
   reactant **product** **enzyme** **equilibrium arrow** **nonequilibrium arrow**

12.1 How Metabolism Works
Describe features of metabolic reactions.

12.2 Metabolically Relevant Nucleotides
Distinguish the high- and low-energy forms of metabolically relevant nucleotides.

12.3 Digestion—From Food Molecules to Hydrolysis Products
Describe the products of food molecule hydrolysis.

12.4 Glycolysis—From Hydrolysis Products to Common Metabolites
Demonstrate the reactions of glycolysis.

12.5 The Citric Acid Cycle—Central Processing
Classify the reactions of the citric acid cycle.

12.6 Electron Transport and Oxidative Phosphorylation
Discuss the oxidation of reduced nucleotides and production of ATP.

12.7 ATP Production
Calculate the number of ATP produced during oxidative catabolism.

12.8 Other Fuel Choices
Examine how fatty acids and amino acids contribute to ATP production.

What's an Inquiry Question?
Inquiry Questions are designed to focus your reading on the main concepts by section. An Inquiry Question appears at the beginning of each section.

2. Are any of the reactions (*I–IV*) reversible? Irreversible? List them.
3. Can you tell the direction of an equilibrium reaction from Scheme 1? If yes, in which direction is that reaction occurring?
4. What would you have to know to decide Question #3?
5. Think back to Le Châtelier's principle (Chapter 9).
   a. If more A was added to the pathway, would more products be produced?
   b. If final products were added to the pathway, would more A be produced?
   c. If final products were added to the pathway, would more B be produced?
   d. If final products were removed from the pathway, what would happen to the amounts of A, B, C, and D?
6. Considering your answers to the previous questions, do you think that reaction pathways are reversible?

12.1 How Metabolism Works

12.1 Inquiry Question:
What is metabolism?

A bean burrito contains carbohydrates, proteins, and triglycerides that can be digested and metabolized for energy.

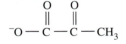

a. *Pyruvate*

b. *Acetyl group*

FIGURE 12.1 Common metabolites. (a) Pyruvate (a three-carbon molecule) and (b) acetyl groups (containing two carbons) found in acetyl coenzyme A are common intermediates in many biochemical pathways where carbon is metabolized.

All animals, including humans, must ingest food for fuel to carry out daily functions like circulation, respiration, and movement. How does eating food produce energy? How do our bodies convert food into useful energy? Animals can get energy (measured as Calories) from the covalent bonds contained in carbohydrates, fats, and proteins. As an example, let's look at what happens when someone eats a bean and cheese burrito.

In the first stage of metabolism, the large biomolecules in food are digested or broken down into their smaller units through hydrolysis reactions. Polysaccharides, like starch in the tortilla, are hydrolyzed into monosaccharide units, the triglycerides in the cheese are broken down to glycerol and fatty acids, and the proteins in the beans and cheese are hydrolyzed into their amino acid units. These hydrolysis products, the smaller molecules produced by the breakdown of the large biomolecules, are absorbed through the intestinal wall into the bloodstream and are eventually transported to different tissues for use by the cells.

Once in the cells, the hydrolysis products are broken down through oxidative processes into a few common metabolites containing two or three carbons (see **Figure 12.1**). **Metabolites** are chemical intermediates formed by enzyme-catalyzed reactions in the body. At this point, as long as the cells have oxygen and are producing energy, two-carbon acetyl groups can be broken down further to carbon dioxide (one carbon) through a series of chemical reactions called the citric acid cycle. This cycle works in conjunction with the electron transport and oxidative phosphorylation pathways to produce a lot of energy that is transferred through the cells by the molecules ATP, nicotinamide adenine dinucleotide (NADH), and flavin adenine dinucleotide (FADH$_2$). One burrito contains a lot of carbon atoms, but we have a lot of cells that require fuel (see **Figure 12.2**).

In our cells, when molecules from food oxidize, the energy is repackaged into molecules with high chemical potential energy instead of dissipating as heat. The chemical potential energy in ATP is transformed into kinetic energy or used to make other molecules in the body.

Chemical reactions that occur in living systems are called *biochemical reactions.* Most of the time in biological systems, chemical reactions occur in a series called a **metabolic pathway.** For example, the sugar molecule glucose (containing six carbons) is broken down to two molecules of pyruvate (three carbons each) through a series of chemical reactions collectively referred to as *glycolysis.*

Metabolism can be considered in two parts, catabolism and anabolism. **Catabolism** refers to chemical reactions in which larger molecules are broken down into a few common metabolites. Because catabolism breaks down molecules, we see many hydrolysis and oxidation reactions in catabolic pathways. These reactions tend to be exergonic $(-\Delta G)$.

Anabolism refers to chemical reactions in which metabolites combine to form larger molecules. In anabolic pathways, many of the reactions are condensation or reduction reactions. These reactions tend to be endergonic $(+\Delta G)$. These relationships are summarized in **Table 12.1**.

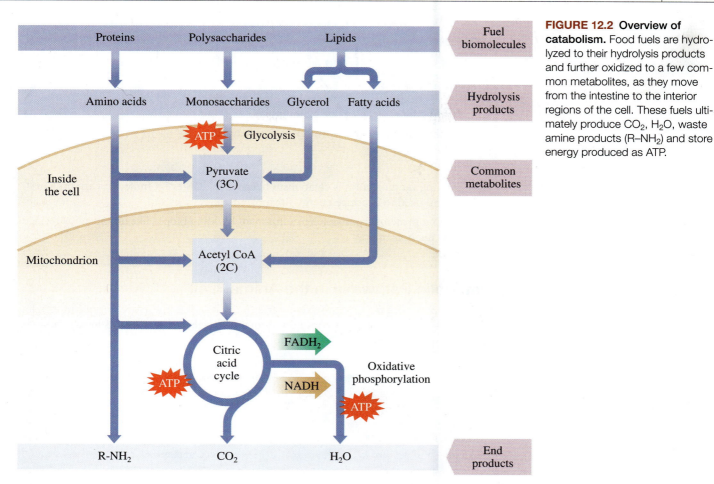

FIGURE 12.2 Overview of catabolism. Food fuels are hydrolyzed to their hydrolysis products and further oxidized to a few common metabolites, as they move from the intestine to the interior regions of the cell. These fuels ultimately produce CO_2, H_2O, waste amine products (R–NH_2) and store energy produced as ATP.

TABLE 12.1 Properties of Biochemical Reactions in Metabolism

Property	Catabolism	Anabolism
Chemical Reactions	Hydrolysis	Condensation
	Oxidation	Reduction
Energy	Exergonic (produce energy)	Endergonic (require energy)
Reaction Rates	Enzymes catalyze reactions	Enzymes catalyze reactions

The energy released during catabolic reactions is captured in molecules such as ATP and used to drive the anabolic reactions. ATP and other energy-rich molecules couple the reactions of catabolism and anabolism together. In this way, the energy given off in catabolism is transferred to the energy-requiring reactions in anabolism (see **Figure 12.3**).

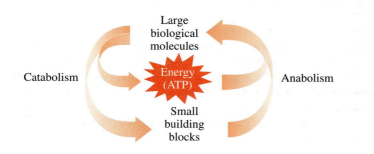

FIGURE 12.3 Catabolism and anabolism. Catabolic and anabolic reactions together define metabolism. Catabolism produces the energy-rich molecules and hydrolysis products that can be used as building blocks for anabolism (production of large biological molecules).

sample problem 12.1 **Distinguishing Anabolism and Catabolism**

Indicate if the following processes represent anabolism or catabolism.

a.   Burning fat molecules during an intense workout.

b.   Making proteins from component amino acids.

Solution

a.   Burning fat molecules is catabolism because the larger fat molecules are broken up and used to produce energy for movement.

b.   Making proteins from smaller component building blocks like amino acids represents anabolism.

Metabolic Pathways in the Animal Cell

Metabolic pathways occur in different parts of the cell. It is therefore important to know some of the major parts of a common animal cell (see **Figure 12.4**) to locate the main pathways in metabolism.

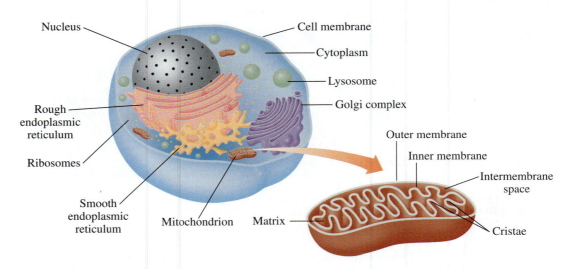

FIGURE 12.4 The cell. The major components of an animal cell showing the major organelles. The mitochondria shown in detail on the right is where most of the ATP is produced in the cell.

In animals, a cell membrane separates the materials inside the cell from the exterior aqueous environment. The structure of a cell membrane was discussed in Section 7.6. The *nucleus* contains DNA that controls cell replication and protein synthesis for the cell. The **cytoplasm** consists of all the material between the nucleus and the cell membrane. The **cytosol** is the fluid part of the cytoplasm. It is the aqueous solution of electrolytes and enzymes that catalyzes many of the cell's chemical reactions.

Within the cytoplasm are specialized structures called *organelles* that carry out specific functions in the cell. We have already seen (Chapter 11) that the ribosomes are the sites of protein synthesis. The **mitochondria** are the energy-producing factories of the cells. A mitochondrion consists of an outer membrane and an inner membrane, with an intermembrane space between them. The fluid section encased by the inner membrane is called the *matrix*. Enzymes located in the matrix and along the inner membrane catalyze the oxidation of carbohydrates, fats, and amino acids.

All of the oxidation pathways produce CO_2, H_2O, and energy, which are used to form energy-rich compounds. **Table 12.2** provides a summary of the functions of the cellular components in animal cells. Keep in mind that a cell can have more than one of the same organelle.

TABLE 12.2 Locations and Functions of Components in Animal Cells

Component	Description and Function
Cell membrane	Separates the contents of a cell from the external environment and contains structures that communicate with other cells
Cytoplasm	Consists of all the cellular contents between the cell membrane and nucleus
Cytosol	The fluid part of the cytoplasm that contains enzymes for many of the cell's chemical reactions
Endoplasmic reticulum (ER)	Processes proteins for secretion and synthesizes phospholipids (Rough ER); synthesizes fats and steroids (Smooth ER)
Golgi complex	Modifies and secretes proteins from the endoplasmic reticulum and synthesizes glycoproteins and cell membranes
Lysosomes	Contain hydrolytic enzymes that digest and recycle old cell structures
Mitochondria	Contain the structures for the synthesis of ATP from energy-producing reactions
Nucleus	Contains genetic information for the replication of DNA and the synthesis of protein
Ribosomes	Sites of protein synthesis using mRNA templates

sample
problem
12.2   **Metabolism and Cell Structure**

Identify the following as catabolic or anabolic reactions:

a.   digestion of polysaccharides

b.   synthesis of proteins

c.   oxidation of glucose to CO_2 and H_2O

Solution

a.   The breakdown of large molecules is a catabolic reaction.

b.   The synthesis of large molecules involves anabolic reactions.

c.   Oxidation of molecules such as glucose involves catabolic reactions.

practice
problems

12.1   How can you identify a catabolic reaction?

12.2   How can you identify an anabolic reaction?

12.3   Indicate if the following processes represent anabolism or catabolism.
   a. glycolysis
   b. synthesis of fats

12.4   Indicate if the following processes represent anabolism or catabolism
   a. conversion of glycolysis products to fats.
   b. generation of ATP from breakdown of fructose.

12.5   Name a chemical reaction that breaks down biomolecules into their component parts.

12.6   Pyruvate and acetyl groups are common chemical intermediates called _____.

12.7   Identify the cell organelle considered the energy-producing factory for the cell.

12.8   Name a molecule that transfers energy within the cell.

12.2 Metabolically Relevant Nucleotides

12.2 Inquiry Question:
What do the nucleotides that transfer energy look like?

In Chapter 11, we saw that nucleotides are the building blocks of the nucleic acids DNA and RNA. Nucleotides also have important metabolic functions in cells. They act as energy exchangers and can also be coenzymes.

All of these nucleotides have two forms: a high-energy form and a low-energy form. They consist of some basic components introduced in Chapter 11: the nucleoside adenosine, a phosphate, and a five-carbon sugar. Many of these molecules also have a vitamin within their structure.

Because the structures of the basic components were shown in Chapter 11, they are illustrated here using their component names. **Table 12.3** shows the structures of the metabolic nucleotides described in the following sections in both their high- and low-energy forms. The nucleotide portion (base, sugar, and at least one phosphate) and the vitamin portion are also indicated.

TABLE 12.3 Metabolically Relevant Nucleotides

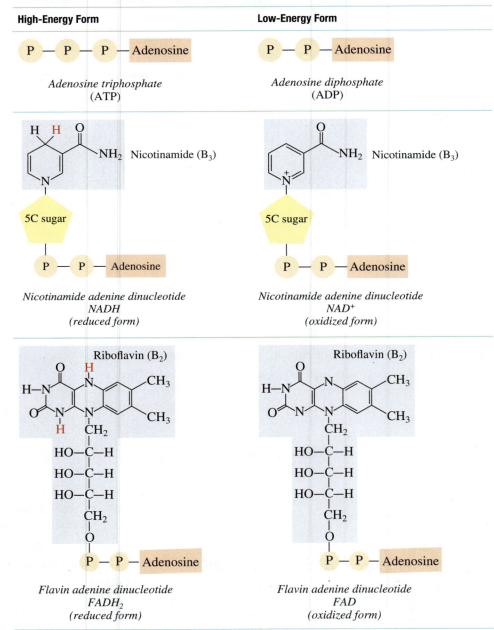

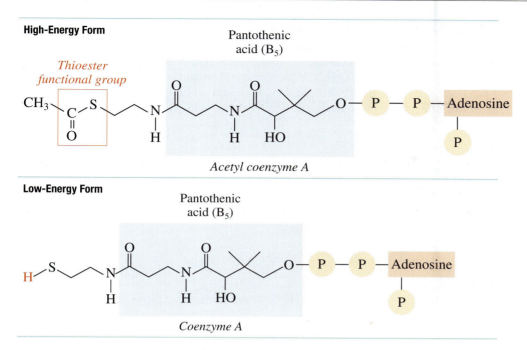

High-Energy Form

Pantothenic acid (B₅)

Thioester functional group

Acetyl coenzyme A

Low-Energy Form

Pantothenic acid (B₅)

Coenzyme A

ATP/ADP

The nucleotide ATP is often referred to as the energy currency of the cell. ATP can undergo hydrolysis to adenosine diphosphate (ADP) as shown in the following equation. The nucleotide components are shown beneath the chemical structure for simplicity. During hydrolysis, energy is released as a product, so in this case, ATP is the high-energy form and ADP is the low-energy form. The energy given off during the hydrolysis of ATP can be coupled to drive a chemical reaction that requires energy during anabolism.

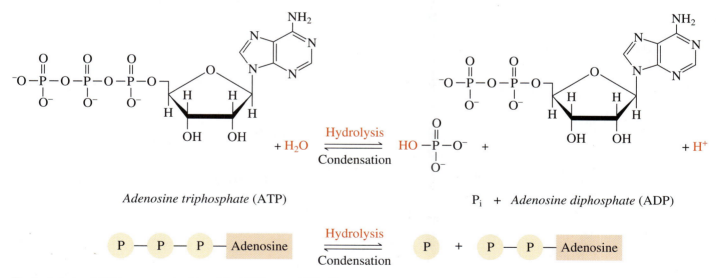

Adenosine triphosphate (ATP)

Hydrolysis ⇌ *Condensation*

P_i + *Adenosine diphosphate* (ADP)

The hydrolysis of ATP is commonly abbreviated ATP ⟶ ADP + P_i.

NADH/NAD⁺ and FADH₂/FAD

Nicotinamide adenine dinucleotide (NAD⁺) and flavin adenine dinucleotide (FAD) are two energy-transferring compounds with a high-energy form that is reduced (hydrogen added) and a low-energy form that is oxidized (hydrogen removed). The abbreviations for these forms are NADH (reduced form) and NAD⁺ (oxidized form) and FADH₂ (reduced form) and FAD (oxidized form). The active end of each molecule, the part that is being oxidized or reduced, contains a vitamin component. Nicotinamide is derived from the vitamin niacin (B₃), and riboflavin (B₂) is found in FAD.

Acetyl Coenzyme A and Coenzyme A

Another important energy exchanger containing a nucleotide is coenzyme A (CoA). The two forms of this compound are acetyl coenzyme A (high energy) and coenzyme A (low energy). Energy is released from acetyl coenzyme A when the C—S bond in the thioester functional group is hydrolyzed, producing an acetyl group and coenzyme A. CoA contains adenosine, three phosphates, and a pantothenic acid (vitamin B$_5$)–derived portion.

---

sample problem 12.3   Metabolic Nucleotides

Provide the abbreviation for the high-energy form of the following:

a.   nicotinamide adenine dinucleotide

b.   flavin adenine dinucleotide

Solution

The high-energy forms of these nucleotides are the reduced forms (contain H).

a.   NADH

b.   FADH$_2$

---

practice problems

12.9 Identify the metabolic nucleotide described by the following:
a. contains the vitamin riboflavin
b. contains a thioester functional group

12.10 Identify the metabolic nucleotide described by the following:
a. contains a form of the vitamin niacin
b. the main energy currency in the body

12.11 Match one of the following metabolic nucleotides to its description:

NADH	ATP
NAD$^+$	ADP
FADH$_2$	Acetyl CoA
FAD	CoA

a. the reduced form of nicotinamide adenine dinucleotide

b. exchanges energy when a C—S bond is hydrolyzed
c. the oxidized form of flavin adenine dinucleotide

12.12 Match one of the following metabolic nucleotides to its description:

NADH	ATP
NAD$^+$	ADP
FADH$_2$	Acetyl CoA
FAD	CoA

a. exchanges energy when a phosphate bond is hydrolyzed
b. the high-energy form of coenzyme A
c. the oxidized form of nicotinamide adenine dinucleotide

12.3 Digestion—From Food Molecules to Hydrolysis Products

12.3 Inquiry Question: How are food molecules digested?

When food enters the body, it begins to break down in a process called **digestion.** In digestion, large fuel biomolecules are hydrolyzed into smaller hydrolysis products. These hydrolysis products are then able to be absorbed by the body and delivered to the cells where they can be further catabolized.

Carbohydrates

Have you ever eaten a saltine cracker and, after swallowing most of it, had a sweet taste in your mouth? This happens because the starch (amylose and amylopectin) in the cracker begins to be digested in your mouth through the action of the enzyme alpha-amylase secreted in saliva. This salivary amylase hydrolyzes some of the α-glycosidic bonds in the starch molecules, producing glucose, the disaccharide maltose, and oligosaccharides. The slightly sweet taste is due to the presence of glucose in the hydrolysis products.

Only monosaccharides are small enough to be transported through the intestinal wall and into the bloodstream. To complete the digestion of starch, enzymes in the small intestine hydrolyze starch into monosaccharides. The disaccharides lactose, sucrose, and maltose are also hydrolyzed in the small intestine. These hydrolysis reactions were first discussed in Section 5.5 and are shown in Figure 12.5.

The first step to breaking down carbohydrates, proteins, and dietary fats is hydrolysis.

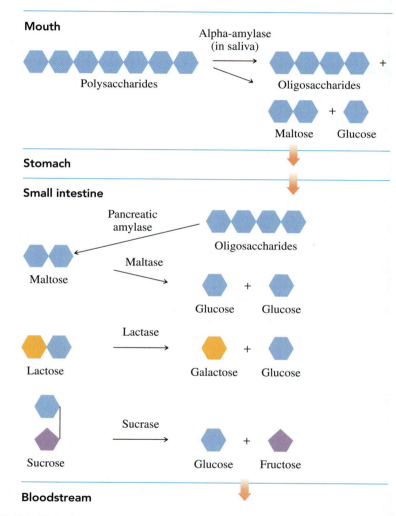

FIGURE 12.5 Digestion of carbohydrates. The digestion of carbohydrates begins in the mouth and is completed in the small intestine.

We eat other carbohydrates like cellulose that cannot be digested because we lack the enzyme cellulase that hydrolyzes its β-glycosidic bonds. These indigestible fibers are referred to as insoluble fibers. Although not useful as fuel for the body, they are important for a healthy digestive tract. Insoluble fiber stimulates the large intestine to help the body excrete waste.

Fats

Dietary fats such as the triglycerides and cholesterol are nonpolar molecules, so their digestion in aqueous digestive juices is a little tricky. To assist in the digestion of dietary fats, a substance called *bile* is excreted from the gall bladder into the duodenum, the upper portion of the small intestine, during digestion. Bile contains soap-like molecules called *bile salts*, which are amphipathic. They contain a nonpolar part and a polar part. The amphipathic bile salts place their nonpolar face toward the dietary fats and their polar face toward the water, forming micelles much like the soap micelles that break up a greasy dirt stain on clothing.

This process of breaking up larger nonpolar globules into smaller droplets (micelles) is called **emulsification.** The smaller micelles are able to move the dietary fats closer to the intestinal cell wall so cholesterol can be absorbed and triglycerides can be hydrolyzed via pancreatic lipase into free fatty acids and monoglycerides, which are then absorbed in the small intestine.

Once across the intestinal wall, free fatty acids and monoglycerides are reassembled as triglycerides while the cholesterol is linked to another free fatty acid forming a cholesterol ester. These are both repackaged with protein as a lipoprotein called a **chylomicron.** Chylomicrons ultimately transport triglycerides through the bloodstream to the tissues where they are used for energy production or are stored in the cells (see **Figure 12.6**).

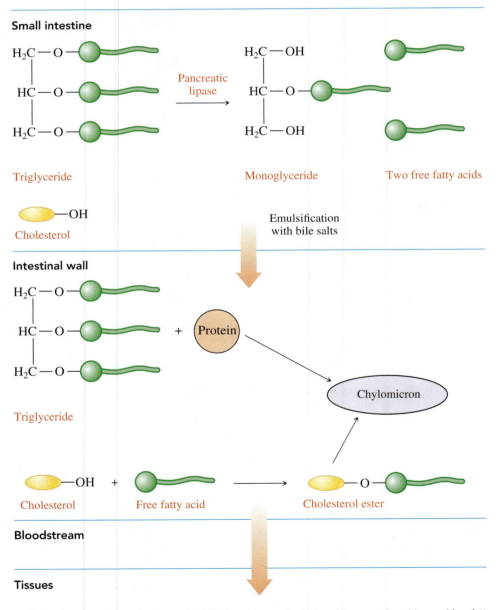

FIGURE 12.6 Digestion of dietary fats. Triglycerides are hydrolyzed to monoglycerides and free fatty acids for transport across the intestinal wall. They are reassembled into triglycerides, which combine with cholesterol ester and protein into droplets called chylomicrons that are delivered to the tissues.

Proteins

Protein digestion begins in the stomach where proteins are denatured (unfolded) by the acidic digestive juices. Digestive enzymes like pepsin, trypsin, and chymotrypsin hydrolyze peptide bonds as the denatured proteins begin their journey through the stomach and into the intestine. The resulting amino acids are absorbed through the intestinal wall into the bloodstream for delivery to the tissues. The digestion of proteins is depicted in **Figure 12.7**.

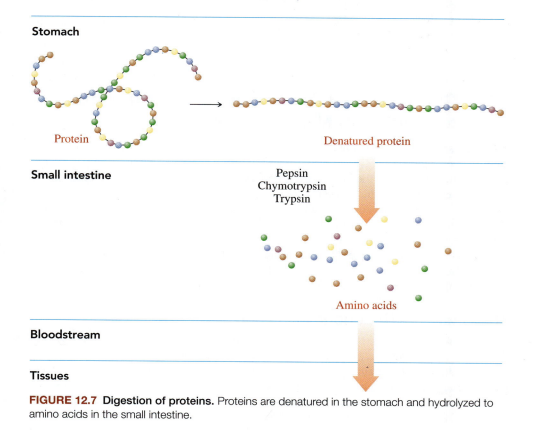

FIGURE 12.7 Digestion of proteins. Proteins are denatured in the stomach and hydrolyzed to amino acids in the small intestine.

sample problem 12.4 Digestion

Indicate the enzyme(s) involved in the digestion of each of the following:

a. lactose

b. triglyceride

c. protein

Solution

a. The disaccharide lactose is broken into the monosaccharides galactose and glucose by the enzyme lactase (see Figure 12.5).

b. A triglyceride is broken down into two free fatty acids and a monoglyceride by an enzyme called pancreatic lipase (see Figure 12.6).

c. A protein is broken up into component amino acids by digestive enzymes, some of which are pepsin, trypsin, and chymotrypsin (see Figure 12.7).

12.13 Name a carbohydrate (if any) that undergoes digestion in each of the following sites:
  a. mouth
  b. stomach
  c. small intestine

12.14 α-Amylase is produced in the _____ and it catalyzes _____ .

12.15 Describe how cholesterol is packaged after absorption in the intestine.

12.16 Explain the role of bile salts in the digestion of fats.

12.17 Name the end products for digestion of proteins.

12.18 Name the end products for digestion of starch.

12.4 Glycolysis—From Hydrolysis Products to Common Metabolites

? 12.4 Inquiry Question:
How is glucose catabolized through glycolysis?

The main source of fuel for the body is glucose, which is catabolized through a chemical pathway called glycolysis. Glucose is such an important fuel source that when there is not enough glucose entering the body during periods of sleeping or fasting, the body makes its own glucose through the anabolic process called **gluconeogenesis.**

As mentioned in Section 10.5, glucose enters the cell from the bloodstream through a glucose transporter protein in the cell membrane. In the cell, glycolysis occurs in the cytosol when the six-carbon monosaccharide glucose is broken down into two three-carbon molecules of pyruvate. Other hydrolysis products like fructose, amino acids, and free fatty acids can be used as fuel by the cells by entering glycolysis later in the pathway or by chemical conversion to one of the common metabolites.

The Chemical Reactions in Glycolysis

In the body, energy must be transferred in small amounts to minimize the heat released in the process. Reactions that produce energy are coupled with reactions that require energy, thereby helping to maintain a constant body temperature. In glycolysis, energy is transferred through phosphate groups undergoing condensation and hydrolysis reactions.

There are 10 chemical reactions in glycolysis that result in the formation of two molecules of pyruvate from one molecule of glucose. As shown in **Figure 12.8**, the first five reactions require an energy investment of two molecules of ATP, which are used to add two phosphate groups to the sugar molecule. This newly formed molecule is then split into two sugar phosphates. Reactions 6 through 10 generate two high-energy NADH molecules during the addition of two more phosphates and four ATP molecules when the four phosphates are removed from the sugar phosphates.

Through the 10 reactions in glycolysis, one molecule of glucose is converted into two molecules of pyruvate. Because two molecules of ATP were originally required in the energy-investment phase and four are produced in the energy-generation phase, the net energy output for one molecule of glucose is two NADH and two ATP. The net chemical reaction is shown.

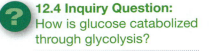

D-Glucose $+ 2NAD^+ + 2ADP + 2P_i \longrightarrow 2CH_3-\overset{\overset{\displaystyle O}{\|}}{C}-\overset{\overset{\displaystyle O}{\|}}{C}-O^- + 2NADH + 2H^+ + 2ATP$

Pyruvate

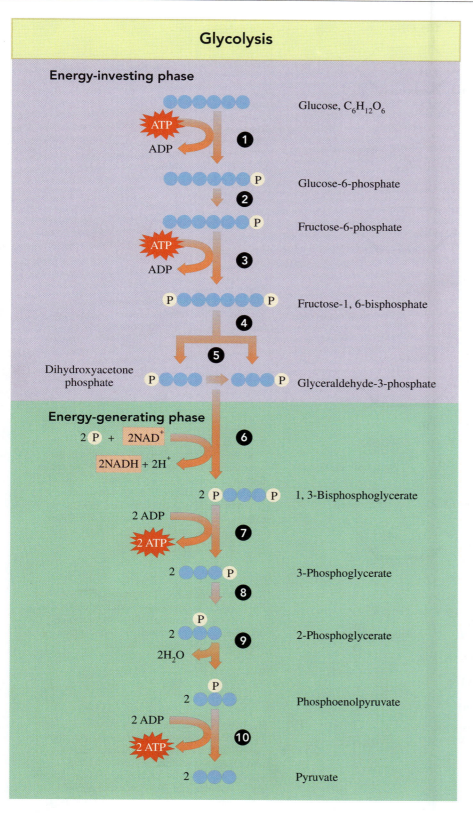

FIGURE 12.8 **Schematic of Glycolysis.** In glycolysis, a six-carbon (blue circles) glucose is catabolized into two three-carbon pyruvate molecules. One molecule of glucose produces two ATP (net) and two NADH molecules.

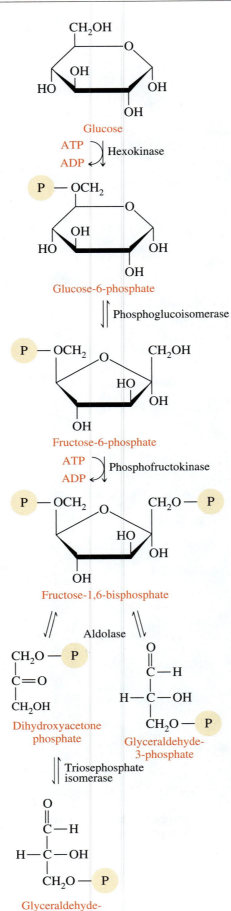

Reaction 1 Phosphorylation: First ATP invested
Glucose is converted to glucose-6-phosphate when ATP is hydrolyzed to ADP. The reaction is catalyzed by the enzyme *hexokinase*.

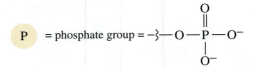

Reaction 2 Isomerization
The enzyme *phosphoglucoisomerase* converts glucose-6-phosphate, an aldose, to its isomer, fructose-6-phosphate, a ketose.

Reaction 3 Phosphorylation: Second ATP invested
A second ATP is hydrolyzed to ADP and the phosphate is transferred to fructose-6-phosphate forming fructose-1,6-bisphosphate. The word *bisphosphate* indicates that the phosphates are on different carbons in fructose. This reaction is catalyzed by the enzyme *phosphofructokinase*.

Reaction 4 Cleavage: Two trioses are formed
Fructose-1,6-bisphosphate is split or cleaved into two triose phosphates — dihydroxyacetone phosphate and glyceraldehyde-3-phosphate — catalyzed by the enzyme *aldolase*.

Reaction 5 Isomerization of a triose
In reaction 5, *triosephosphate isomerase* converts one of the triose products, dihydroxyacetone phosphate, to the other, glyceraldehyde-3-phosphate. Now all 6 carbon atoms from glucose are in two identical 3-carbon triose phosphates.

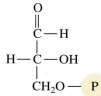

Glyceraldehyde-3-phosphate

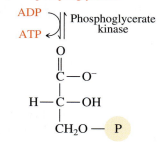

Reaction 6 First energy production yields NADH
The aldehyde group of glyceraldehyde-3-phosphate is oxidized and phosphorylated by *glyceraldehyde-3-phosphate dehydrogenase*. The coenzyme NAD^+ is reduced to the high energy compound NADH and H^+ in the process.

1,3-Bisphosphoglycerate

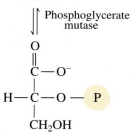

Reaction 7 Next energy production yields ATP
The energy-rich 1,3-bisphosphoglycerate now drives the formation of ATP when *phosphoglycerate kinase* transfers one phosphate from 1,3-bisphosphoglycerate to ADP.

3-Phosphoglycerate

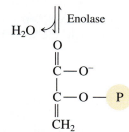

Reaction 8 Formation of 2-phosphoglycerate
A *phosphoglycerate mutase* transfers the phosphate group from carbon 3 to carbon 2 to yield 2-phosphoglycerate.

2-Phosphoglycerate

Reaction 9 Removal of water makes a high-energy enol
An *enolase* catalyzes the removal of water to yield phosphoenolpyruvate, a high-energy compound that can transfer its phosphate in the next step.

Phosphoenolpyruvate

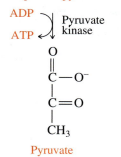

Reaction 10 Third energy production yields a second ATP
ATP is generated in this final reaction when the phosphate from phosphoenolpyruvate is transferred. The reaction is catalyzed by *pyruvate kinase*.

Pyruvate

Regulation of Glycolysis

As a dam regulates water output from a reservoir, most metabolic pathways have at least one reaction step that regulates the flow of reactants to final products through the pathway. The main step of regulation in glycolysis is step 3. The enzyme phosphofructokinase, which catalyzes the phosphorylation of fructose-6-phosphate to fructose-1,6-bisphosphate, is heavily regulated by the cells. ATP acts as an inhibitor of phosphofructokinase. If cells have plenty of ATP, glycolysis slows down.

Glycolysis

If one NADH is generated in step 6 of glycolysis, how are a total of two NADH generated from one molecule of glucose?

Solution

Two molecules of glyceraldhyde-3-phosphate (three-carbon molecules) are produced from one molecule of glucose, so this reaction occurs two times for every molecule of glucose going through the pathway.

The Fates of Pyruvate

The metabolite pyruvate generated from glycolysis can be catabolized further, producing more energy for the body. The fate of pyruvate depends on the availability of oxygen in the cell. When ample oxygen is available (referred to as **aerobic** conditions), pyruvate is oxidized further to acetyl coenzyme A. When oxygen is in short supply (referred to as **anaerobic** conditions), pyruvate is reduced to lactate.

Aerobic Conditions

When oxygen is readily available in the cell, pyruvate produces more energy for the cell. When pyruvate breaks down further, the carboxylate functional group of pyruvate is liberated as CO_2 during a process called **oxidative decarboxylation.** "Oxidative" tells us that this reaction is an oxidation–reduction reaction. In this case, the metabolite coenzyme A is oxidized (H removed) to its high-energy form acetyl CoA and the NAD^+ is reduced to NADH. "Decarboxylation" tells us that a carboxylate (COO^-) is removed as CO_2. Here the pyruvate is decarboxylated, producing the two-carbon acetyl group. The acetyl group binds to coenzyme A through a sulfur atom during the oxidation, creating a thioester functional group and acetyl CoA. This reaction occurs in the mitochondria in animals.

Anaerobic Conditions

During times of strenuous exercise, oxygen is in short supply in the muscles.

Under anaerobic conditions, the middle carbonyl in pyruvate is reduced (hydrogen added) to an alcohol functional group, and lactate is formed. The hydrogen (and energy) required for this reaction is supplied by NADH and H^+, producing NAD^+. Because no ATP is generated by pyruvate under anaerobic conditions, the NAD^+ produced funnels back into glycolysis to oxidize more glyceraldehyde-3-phosphate (step 6), providing a small but necessary amount of ATP. This reaction occurs in the cytosol.

Figure 12.9 provides a summary of the aerobic and anaerobic fates of pyruvate.

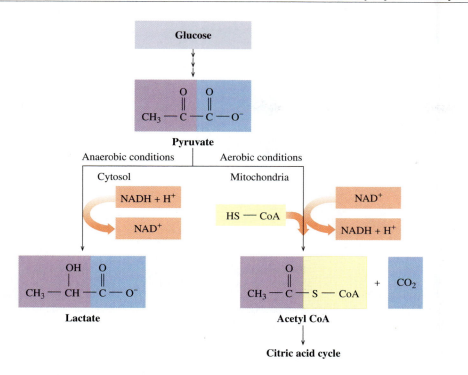

FIGURE 12.9 The fates of pyruvate. Pyruvate is converted to acetyl CoA under aerobic conditions and lactate under anaerobic conditions.

A familiar single-celled organism, yeast, converts pyruvate to ethanol under anaerobic conditions. This process is called **fermentation.** In the preparation of alcoholic beverages like wine, yeast produces pyruvate from glucose in grape juices and under low-oxygen conditions transforms pyruvate into ethanol.

$$CH_3-\overset{O}{\underset{\|}{C}}-\overset{O}{\underset{\|}{C}}-O^- \ + \ NADH \ + \ 2H^+ \longrightarrow CH_3-\overset{HO}{\underset{\underset{H}{|}}{C}}-H \ + \ CO_2 \ + \ NAD^+$$

Pyruvate   Ethanol

Fructose and Glycolysis

Can fructose enter glycolysis to produce energy? As we saw in Chapter 6, fructose tastes sweeter than an equivalent amount of glucose, so less can be used to produce the desired sweetness. In fact, refined sugars like high-fructose corn syrup use more fructose and less glucose to reach a given level of sweetness.

Fructose is readily taken up in the muscle and liver. In the muscles, it is converted to fructose-6-phosphate, entering glycolysis at step 3. In the liver, it is enyzmatically converted to the trioses dihydroxyacetone phosphate and glyceraldehyde-3-phosphate used in step 5 of glycolysis. Any fructose that enters a cell must flow from reaction 5 through 10. Because fructose uptake by the cells is not regulated by insulin, all fructose in the bloodstream enters the cells and is forced into catabolism.

Remember that glycolysis is regulated earlier in the pathway at step 3. The triose products created by fructose in the liver provide an excess of reactants that funnel into step 6, creating excess pyruvate and acetyl CoA that, if not required for energy by the cells, is converted to fat.

Food products containing excessive amounts of fructose like sugary drinks are harmful because the fructose enters the bloodstream (and then the cells) more quickly than fructose found in fruits does. Fructose in fruits is in combination with fiber that slows sugar absorption. Ingesting high levels of fructose can lead to excess end products in glycolysis and, eventually, fat synthesis. Maintaining a healthy balance of sugars in combination with fiber is necessary to preventing catabolic overload of sugar in the cells.

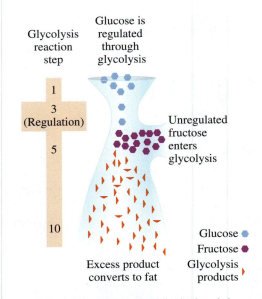

Glycolysis reaction step

Glucose is regulated through glycolysis

1

3 (Regulation)

5

10

Unregulated fructose enters glycolysis

Glucose ●
Fructose ●
Glycolysis products ▶

Excess product converts to fat

In the liver, the breakdown of glucose is regulated, while the breakdown of fructose is not. All fructose flows into glycolysis after the bottleneck (the regulation step), allowing all of it to be converted to end products. The result can be excess products, which, when unnecessary, are converted to fat.

---

sample problem 12.6 Fates of Pyruvate

Is each of the following products from pyruvate produced under anaerobic or aerobic conditions?

a. acetyl CoA   b. lactate   c. ethanol

Solution

a. aerobic conditions   b. anaerobic conditions   c. anaerobic conditions

---

practice problems

12.19 Name the starting reactant of glycolysis.

12.20 Name the end product of glycolysis.

12.21 How many ATP molecules are produced when one molecule of glucose undergoes glycolysis?

12.22 How many ATP molecules are invested in the energy investing stage of glycolysis for one molecule of glucose?

12.23 In terms of high-energy molecules, what is the net output for one molecule of glucose undergoing glycolysis?

12.24 How many NADH molecules are produced when one molecule of glucose undergoes glycolysis?

12.25 Name the coenzyme produced during the conversion of pyruvate to lactate.

12.26 Name the coenzyme produced during the conversion of pyruvate to acetyl CoA.

12.27 Name the products of fermentation.

12.28 The formation of lactate permits glycolysis to continue under anaerobic conditions. Explain.

12.29 What is the main regulation point of glycolysis? How does this explain why fructose is readily converted into fat?

12.30 Explain how the catabolism of fructose differs from that of glucose.

12.5 The Citric Acid Cycle—Central Processing

We have just seen how glucose begins its catabolism in the cell. What about amino acids and fatty acids? During aerobic catabolism when O_2 is present, all three of these biomolecules funnel into and out of the citric acid cycle. The **citric acid cycle** is a series of reactions that degrades the two-carbon acetyl groups from the metabolite acetyl CoA into CO_2 and generates the high-energy reduced molecules NADH and $FADH_2$. This cycle is also known as the Krebs cycle or the tricarboxylic acid cycle. We will use the term citric acid cycle in this book.

The name *citric acid cycle* originated because the first step in the cycle involves the formation of the six-carbon molecule citrate, the conjugate base of citric acid. This initial reaction is a condensation reaction between an entering acetyl CoA molecule and the four-carbon molecule oxaloacetate. The six-carbon molecule citrate loses first one and then a second carbon as CO_2, forming the four-carbon molecule succinyl CoA. These carbon–carbon bond-breaking reactions transfer energy and produce NADH from the coenzyme NAD^+. Succinyl CoA then runs through a set of reactions regenerating oxaloacetate, and the cycle begins again. The citric acid cycle is summarized in **Figure 12.10**.

Reactions of the Citric Acid Cycle

There are eight reactions in the citric acid cycle. Each is catalyzed by an enzyme. These reactions occur in the mitochondrial matrix, deep within the mitochondria (see Figure 12.4). The eight reactions are shown in **Figure 12.11** and described here beginning with the formation of citrate.

Reaction 1 Formation of Citrate
In reaction 1, the acetyl group from acetyl CoA (two carbons) combines with oxaloacetate (four carbons), forming citrate (six carbons) and CoA.

Reaction 2 Isomerization to Isocitrate
Examining the reactant and product of this reaction, you see that the two molecules are structural isomers. The —**OH** and one of the **H** atoms have been swapped in citrate to form isocitrate. Citrate contains a tertiary alcohol, whereas isocitrate contains a secondary alcohol (see Section 6.2). A secondary alcohol can be further oxidized to a carbonyl, whereas a tertiary alcohol cannot. This rearrangement is necessary because isocitrate is oxidized in the next reaction.

Reaction 3 First Oxidative Decarboxylation (Release of CO_2)
Similar to the decarboxylation of pyruvate to acetyl CoA discussed earlier, reaction 3 is an oxidative decarboxylation. In this reaction, an alcohol undergoes oxidation (two hydrogens removed) to a ketone called α-ketoglutarate. A corresponding reduction reaction also takes place. (Remember that oxidation and reduction always occur together.) The coenzyme NAD^+ is reduced to NADH, accepting the proton and electrons removed during the oxidation. The six-carbon isocitrate is decarboxylated to the five-carbon α-ketoglutarate.

Reaction 4 Second Oxidative Decarboxylation
A second decarboxylation occurs in this step, and a four-carbon molecule is produced. In this reaction, a second oxidation–reduction reaction also takes place. The thiol group of CoA is oxidized (loses a hydrogen), and another NAD^+ is reduced to NADH. Alpha-ketoglutarate (five carbons) is decarboxylated into a succinyl group (four carbons). The CoA is bonded to the succinyl group, thus producing succinyl CoA.

Reaction 5 Hydrolysis of Succinyl CoA
In this step, succinyl CoA undergoes hydrolysis to succinate and coenzyme A. The energy produced during the hydrolysis of the thioester is transferred to produce the high-energy nucleotide guanosine triphosphate or GTP from GDP and P_i. In the cell, GTP is readily converted to ATP through the transfer of a phosphate group.

$$GTP + ADP \longrightarrow GDP + ATP$$

12.5 Inquiry Question: What types of reactions occur during the citric acid cycle?

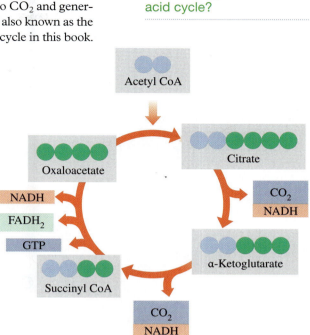

FIGURE 12.10 Fundamentals of the citric acid cycle. The two carbons of acetyl CoA (blue circles) condense with the four carbons of oxaloacetate (green circles) to form citrate. Two carbons are sequentially removed from citrate as CO_2, ultimately re-forming oxaloacetate. During the cycle the high-energy nucleotides NADH, $FADH_2$, and GTP are produced.

Reaction 6 Dehydrogenation of Succinate
In this oxidation, one hydrogen is eliminated from each of the two central carbons of succinate, forming a trans C=C bond, thus producing fumarate. These two hydrogens reduce the coenzyme FAD to FADH$_2$.

Reaction 7 Hydration of Fumarate
Water adds to the trans double bond of fumarate as —H and —OH forming malate.

Reaction 8 Oxidation of Malate
As in reaction 3, the secondary alcohol of malate is oxidized to a ketone, forming oxaloacetate, providing protons and electrons to reduce the coenzyme NAD$^+$ to NADH.

FIGURE 12.11 The reactions of the citric acid cycle. In one turn of the citric acid cycle, oxidative catabolism produces two CO$_2$ molecules, reduced coenzymes NADH and FADH$_2$, and a GTP molecule.

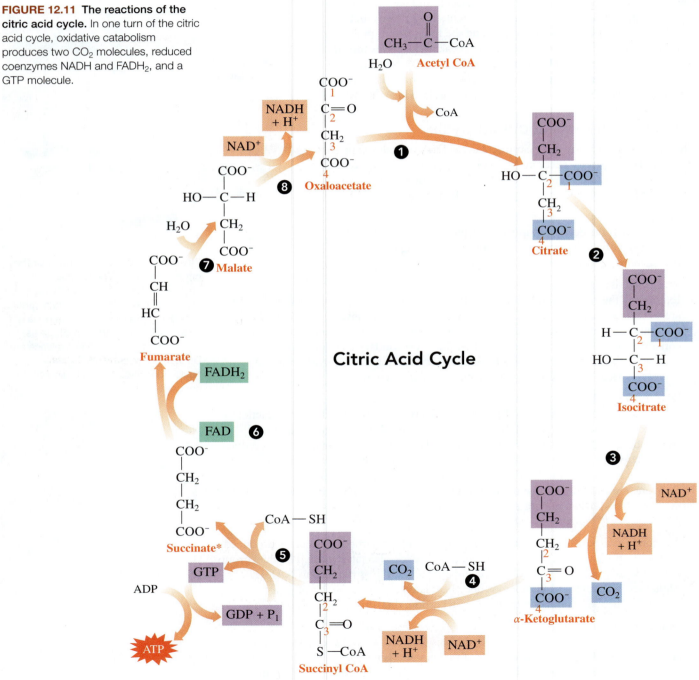

* Succinate is a symmetrical compound, so we can no longer track the acetyl group that entered the cycle as acetyl CoA.

Citric Acid Cycle Summary

One turn of the citric acid cycle produces a net energy output of three NADH, one $FADH_2$, and one GTP (which forms ATP). Two CO_2 and one CoA are also produced. Because the reactants in the cycle are regenerated, the net reaction for one turn of this eight-step cycle is

$$Acetyl\ CoA + 3NAD^+ + FAD + GDP + P_i + 2H_2O \longrightarrow$$
$$2CO_2 + 3NADH + 2H^+ + FADH_2 + CoA + GTP$$

sample problem 12.7   **Citric Acid Cycle**

When one acetyl CoA completes the citric acid cycle, how many of each of the following are produced?

a.   NADH

b. CO_2

c. $FADH_2$

Solution

a.   One turn of the citric acid cycle produces three molecules of NADH.

b.   Two molecules of CO_2 are produced by one turn of the citric acid cycle.

c.   One turn of the citric acid cycle produces one molecule of $FADH_2$.

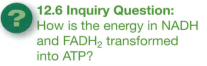
practice problems

12.31 Name the compounds required to start the citric acid cycle.

12.32 Name the products of one turn of the citric acid cycle.

12.33 Name the reactions in the citric acid cycle that involve oxidative decarboxylation.

12.34 Name the reaction in the citric acid cycle that is a condensation.

12.35 Name the reactions of the citric acid cycle that reduce NAD^+.

12.36 Name the reactions of the citric acid cycle that reduce FAD.

12.6 Electron Transport and Oxidative Phosphorylation

Let's review the number of ATP produced from the metabolism of a single molecule of glucose in glycolysis and the citric acid cycle. Two ATP are produced in glycolysis, and two ATP are produced in the citric acid cycle (as GTP) from the two pyruvate molecules that enter as acetyl CoA.

Organisms undergoing aerobic catabolism can produce more energy, but so far, we have not seen a substantial number of ATP produced. Just two ATP are produced in glycolysis and two ATP in the citric acid cycle. Where is all the energy? Remember that high-energy reduced forms of the nucleotides NADH and $FADH_2$ are also produced in glycolysis (two NADH per glucose), from pyruvate oxidation to acetyl CoA (two NADH per glucose), and in the citric acid cycle (six NADH and two $FADH_2$ per glucose).

? 12.6 Inquiry Question:
How is the energy in NADH and $FADH_2$ transformed into ATP?

Glycolysis	2 NADH and 2 ATP
Oxidation of 2 pyruvate	2 NADH
Citric acid cycle (2 acetyl CoA)	6 NADH, 2 $FADH_2$, and 2 ATP

As we will see, these high-energy reduced forms of the nucleotides transfer their electrons and hydrogens through the inner mitochondrial membrane and combine with oxygen to form H_2O. The energy generated as a result of this process is captured and used to drive the reaction of **ADP** + **P$_i$** to form ATP. The process of producing ATP using the energy from the oxidation of reduced nucleotides is called **oxidative phosphorylation.**

Electron Transport

Mitochondria are the ATP factories of the cell. Reduced nucleotides from the citric acid cycle are produced here, and their energy upon oxidation is used to generate ATP. The reactions of the citric acid cycle occur in the matrix of the mitochondria (refer to Figure 12.4), and the reduced nucleotides, NADH and FADH$_2$, begin their journey through the inner membrane here.

A set of enzyme complexes, commonly called complexes I through IV, are embedded in the inner membrane of the mitochondria and contain a set of electron carriers that transport the electrons and protons of NADH and FADH$_2$ through the inner mitochondrial membrane. Two of the electron carriers, coenzyme Q and cytochrome *c*, are not firmly attached to any one complex and serve to shuttle electrons between the complexes (see **Figure 12.12**). We examine each enzyme complex individually starting with complex I to see how electrons shuttle through the complexes.

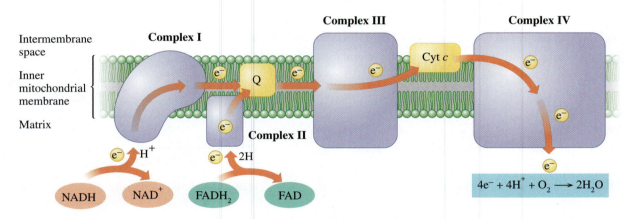

FIGURE 12.12 The electron transport chain. Electrons flow to molecular oxygen, ultimately producing water through a series of electron carriers embedded in proteins in the inner mitochondrial membrane.

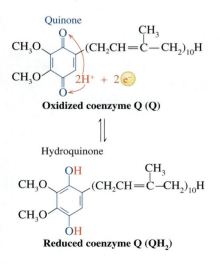

$$Q + 2H^+ + 2\ e^- \rightleftharpoons QH_2$$

Complex I NADH Dehydrogenase

At complex I, NADH enters electron transport. During its oxidation, two electrons and two protons are transferred to the electron transporter coenzyme Q, reducing its two ketone groups to alcohols (see figure at left). NAD$^+$ is regenerated and returns to a catabolic pathway as in the citric acid cycle. The overall reaction at complex I is

$$NADH + H^+ + Q \longrightarrow NAD^+ + QH_2$$

Complex II Succinate Dehydrogenase

FADH$_2$ enters electron transport at complex II after the reduced nucleotide is produced in the conversion of succinate to fumarate in the citric acid cycle. Two electrons and two protons from FADH$_2$ are also transferred to a coenzyme Q to yield QH$_2$. The overall reaction at complex II is

$$FADH_2 + Q \longrightarrow FAD + QH_2$$

Complex III Coenzyme Q Cytochrome *c* Reductase

At complex III, the reduced coenzyme Q (QH$_2$) molecules formed in complex I or II are reoxidized to coenzyme Q (Q), and the electrons pass through a series of electron acceptors until they arrive as single electrons in the mobile protein cytochrome *c*, which moves the electron from complex III to complex IV.

Complex IV Cytochrome *c* Oxidase

At complex IV, single electrons are transferred from cytochrome *c* through another set of electron acceptors to combine with hydrogen ions and oxygen (O_2) to form water. This is the final stop for the electrons.

$$4H^+ + 4e^- + O_2 \longrightarrow 2H_2O$$

---

sample problem

12.8 **Electron Transport**

Give the abbreviation for each of the following electron transporters:

a. the reduced form of flavin adenine dinucleotide

b. the reduced form of coenzyme Q

Solution

a. $FADH_2$ b. QH_2

Oxidative Phosphorylation

How does the movement of electrons through a set of enzyme complexes in the inner membrane generate ATP for the cell? Biochemist Peter Mitchell first proposed the **chemiosmotic model** (**Figure 12.13**), linking electron transport to the generation of a proton (H^+) difference or *gradient* on either side of the inner membrane and the resulting production of ATP. In this model, three of the complexes (I, III, and IV) span the inner membrane and pump (relocate) protons out of the matrix and into the intermembrane space as electrons are shuttled through the complexes. These protons are not generated directly from the oxidation of NADH and $FADH_2$ but are simply present in the matrix.

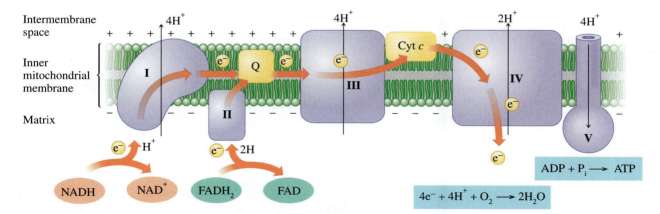

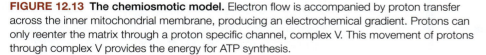

FIGURE 12.13 The chemiosmotic model. Electron flow is accompanied by proton transfer across the inner mitochondrial membrane, producing an electrochemical gradient. Protons can only reenter the matrix through a proton specific channel, complex V. This movement of protons through complex V provides the energy for ATP synthesis.

A difference in both charge (electrical) and concentration (chemical) of protons on either side of the membrane results in an **electrochemical gradient.** The formation of the proton gradient across the inner mitochondrial membrane provides the energy for ATP synthesis.

The inner mitochondrial membrane is a very tight membrane that is impermeable to protons. The only way for protons to move back into the matrix is through a protein complex, called complex V, or ATP synthase. The movement of protons from an area with many protons to an area with fewer protons releases energy. As protons flow back into the matrix through complex V, the resulting release of energy drives the synthesis of ATP.

$$ADP + P_i + energy \longrightarrow ATP$$

integrating chemistry

Thermogenesis—Uncoupling ATP Synthase

If the protons pumped during electron transport and their return to the matrix during ATP production is disrupted, ATP is not produced. If ATP cannot be produced, the energy that would have been harnessed as ATP is simply released as heat. This generation of heat in the body is called **thermogenesis.**

Some animals adapted to cold climates produce *thermogenin,* a protein that uncouples electron transport and oxidative phosphorylation allowing protons to leak back into the matrix, generating heat. Such animals can regulate their body temperature in part through thermogenesis. These animals have a higher amount of a tissue called *brown fat* that appears brown due to the high concentrations of mitochondria present. The cytochrome molecules present in mitochondria contain an iron ion that is responsible for the brown color. Newborn babies have higher levels of brown fat than do adults because newborns do not have much body mass to maintain body temperature. Brown fat deposits are located near major blood vessels that carry the warmed blood through the body, allowing a newborn to generate more heat to warm its body surface.

sample problem 12.9 Chemiosmotic Model

Provide another name for complex V and describe its function.

Solution

Complex V is also known as ATP synthase. It provides a channel for H^+ to flow back into the matrix (after being pumped out by complexes I, III, and IV) and is the site for ATP synthesis, which occurs in the matrix.

practice problems

12.37 Name the electron carrier that transports electrons from complex I to complex III.

12.38 Name the electron carrier that transports electrons from complex III to complex IV.

12.39 Name the complex where water is produced.

12.40 Name the complex where $FADH_2$ enters electron transport.

12.41 According to the chemiosmotic theory, how does the proton gradient provide energy to synthesize ATP?

12.42 How is the proton gradient established?

Discovering the Concepts

? Inquiry Activity—ATP Production

The reduced nucleotides NADH and $FADH_2$ are converted to ATP via the electron transport chain and oxidative phosphorylation. The most widely accepted values among biochemists are shown.

Nucleotide Input	ATP Output
NADH	2.5
$FADH_2$	1.5

Questions

1. The metabolites for one molecule of glucose through glycolysis is shown. Taking into consideration the conversion of reduced nucleotides to ATP, what is the total number of ATP produced by one glucose completing glycolysis?

$$\text{Glucose} \longrightarrow 2\text{Pyruvate} + 2\text{ATP} + 2\text{NADH}$$

2. The conversion of pyruvate to acetyl CoA is shown for one molecule of glucose. Taking into consideration the conversion of reduced nucleotides to ATP, what is the total number of ATP produced by one glucose (two pyruvate) completing this conversion?

$$2\text{Pyruvate} \longrightarrow 2\text{Acetyl CoA} + 2\text{NADH}$$

3. Each acetyl CoA that enters the citric acid cycle produces two CO_2, three NADH, one $FADH_2$, and one GTP (converted directly to one ATP in the cell). Taking into consideration the conversion of reduced nucleotides to ATP, what is the total number of ATP produced by one glucose (two acetyl CoA) going through the citric acid cycle?

4. Taking your answers to questions 1–3 into account, what is the total number of ATP produced during the complete oxidative catabolism of one molecule of glucose?

12.7 ATP Production

Oxidative phosphorylation couples the energy of electron transport to proton pumping and finally to ATP synthesis. How many ATP are synthesized for each reduced nucleotide entering electron transport? Because NADH enters the electron transport chain at complex I and $FADH_2$ enters at complex II, the number of protons pumped for these two molecules is different. Ten H^+ are pumped into the inner membrane space for every NADH entering electron transport and six H^+ are pumped into the inner membrane space for each $FADH_2$ transported (refer to Figure 12.13). The most widely accepted value for the number of H^+ flowing back into the matrix to synthesize one ATP is four. So, the number of ATP synthesized per NADH is 2.5 and the number of ATP synthesized per $FADH_2$ is 1.5.

> ### ? 12.7 Inquiry Question:
> How many ATP can be produced during oxidative catabolism?

Nucleotide Input	Protons (H^+) Pumped	ATP Output
NADH	10	2.5
$FADH_2$	6	1.5

sample problem 12.10 **ATP Production**

Why does the oxidation of NADH provide energy for the formation of 2.5 ATP molecules, whereas FADH$_2$ produces 1.5 ATP?

Solution

Electrons from the oxidation of NADH enter electron transport earlier through complex I than do those from FADH$_2$. For every four protons pumped through the inner membrane, one ATP can be synthesized. One NADH pumps 10 protons into the intermembrane, which can synthesize 2.5 ATP, whereas one FADH$_2$ pumps six protons, forming 1.5 ATP.

Counting ATP from One Glucose

How many ATP molecules can be generated from one molecule of glucose undergoing complete catabolism? As noted in Section 12.1, the end products for the oxidative catabolism for any fuel (foodstuff) are CO_2 and H_2O. For glucose ($C_6H_{12}O_6$), the overall balanced oxidation equation is

$$C_6H_{12}O_6 + 6H_2O \longrightarrow 6CO_2 + 6H_2O + Energy$$

This equation may look the same as one seen previously for the combustion of glucose. The net reaction is the same, but oxidative catabolism requires many steps and the energy is not given off as heat but is stored in nucleotides and ATP for further use by the body. Let's review the pathways of the biochemical oxidation of glucose and count the total number of ATP produced.

Glycolysis

In glycolysis, the oxidation of glucose produces two NADH molecules and two ATP molecules. Recall that glycolysis occurs in the *cytosol*, and electron transport draws NADH from the matrix inside the *mitochondria*. The two NADH from glycolysis must be shuttled into the matrix to enter electron transport. The direct shuttling of two NADH into the matrix results in the production of five ATP.

$$Glucose \longrightarrow 2Pyruvate + 2ATP + 2NADH \; (5ATP)$$

Oxidation of Pyruvate

After glycolysis, the two pyruvates enter the mitochondria, where they are further oxidized to produce a total of two acetyl CoA, two CO_2, and two NADH. The oxidation of two pyruvates thus leads to the production of five ATP.

$$2Pyruvate \longrightarrow 2Acetyl \; CoA + 2NADH \; (5 \; ATP)$$

Citric Acid Cycle

The two acetyl CoA produced from the two pyruvate next enter the citric acid cycle. Each acetyl CoA entering the cycle yields two CO_2, three NADH, one FADH$_2$, and one GTP (converted directly to ATP). Because two acetyl CoA enter the cycle from one glucose, a total of six NADH, two FADH$_2$, and two ATP are produced.

6 NADH	$\longrightarrow$	15 ATP
2 FADH$_2$	$\longrightarrow$	3 ATP
Directly in pathway	$\longrightarrow$	2 ATP

Total ATP for two acetyl CoA: 20 ATP

Total ATP from Glucose Oxidation

By summing the ATP produced from glycolysis, oxidation of pyruvate, and the citric acid cycle, a net number of ATP produced from one glucose can be estimated. A summary is shown in Table 12.4 and diagrammed in Figure 12.14.

TABLE 12.4 ATP Produced by the Complete Oxidation of Glucose

Pathway	Reduced Nucleotides Produced	ATP Yield
Glycolysis (Produced directly in pathway)	2 NADH$_{cytosol}$ $\longrightarrow$ 2 NADH$_{matrix}$	5 ATP
		2 ATP
2 Pyruvate $\longrightarrow$ 2 Acetyl CoA	2 NADH	5 ATP
Citric acid cycle	6 NADH	15 ATP
Two turns of the cycle accommodate two acetyl CoA (Produced as GTP in pathway)	2 FADH$_2$	3 ATP
		2 ATP
	TOTAL ATP	**32 ATP**

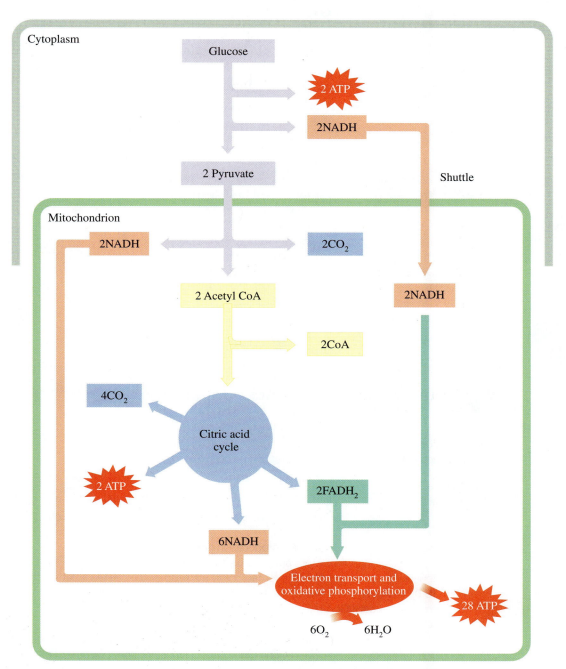

FIGURE 12.14 Complete glucose oxidation. The complete oxidation of glucose to CO_2 and H_2O yields a total of 32 ATP.

sample problem 12.11 **ATP Production**

Indicate the total number of ATP produced by the following oxidations:

a.   one pyruvate to one acetyl CoA

b.   one acetyl CoA turning through the citric acid cycle

SOLUTION

a.   The oxidation of pyruvate to acetyl CoA produces one NADH, which yields 2.5 ATP.

b.   One turn of the citric acid cycle produces

$$1 \text{ GTP} + 3 \text{ NADH} + 1 \text{ FADH}$$
$$\downarrow \qquad\qquad \downarrow \qquad\qquad \downarrow$$
$$1 \text{ ATP} + 7.5 \text{ ATP} + 1.5 \text{ ATP}$$
$$\downarrow$$
$$10 \text{ ATP}$$

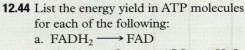
practice problems

12.43 List the energy yield in ATP molecules for each of the following:
   a. NADH $\longrightarrow$ NAD$^+$
   b. glucose $\longrightarrow$ 2 pyruvate
   c. 2 pyruvate $\longrightarrow$ 2 acetyl CoA + 2CO$_2$
   d. acetyl CoA $\longrightarrow$ 2CO$_2$

12.44 List the energy yield in ATP molecules for each of the following:
   a. FADH$_2$ $\longrightarrow$ FAD
   b. glucose + 6O$_2$ $\longrightarrow$ 6CO$_2$ + 6H$_2$O
   c. glucose $\longrightarrow$ 2 lactate
   d. pyruvate $\longrightarrow$ lactate

Discovering the Concepts

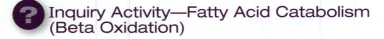

? Inquiry Activity—Fatty Acid Catabolism (Beta Oxidation)

Information
Examine Figure 12.15 to answer the questions below.

Questions
1.   How is the fatty acyl CoA at the top of the cycle different from the fatty acyl CoA at the bottom of the cycle? (*Acyl* is used instead of *acid* because the fatty acid is attached to another molecule, in this case a CoA.)
2.   How many reduced nucleotides are produced in one cycle? Name them.
3.   How many acetyl CoA molecules are produced in one cycle of β oxidation?
4.   a.   How many cycles would it take to catabolize a stearic acid molecule (a fatty acid, [18:0]) into acetyl CoA units?
     b.   How many acetyl CoA molecules would be produced?
     c.   How many reduced nucleotides would be produced?
5.   If a molecule of glucose produces a net 32 ATP when completely catabolized, which do you think will produce more energy, one molecule of glucose or one molecule of stearic acid? Justify your answer.

12.8 Other Fuel Choices

Glucose is our main source of fuel. When there is more than enough to take care of the energy requirements of the cell, glucose is stored in the liver and muscles as glycogen. Hydrolysis of glycogen (called *glycogenolysis*) produces glucose when glucose concentrations are low (during sleeping and fasting). Once glycogen stores are depleted, glucose can be synthesized from noncarbohydrate sources (originating from proteins and triglycerides) that feed into *gluconeogenesis*. Amino acids and fatty acids can also feed into oxidative catabolism at different points, allowing ATP production to continue in the absence of glucose.

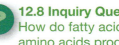

12.8 Inquiry Question:
How do fatty acids and amino acids produce ATP?

Energy from Fatty Acids

If glycogen and glucose are not available, cells that still require ATP can oxidize fatty acids to acetyl CoA through a degradation pathway known as **beta oxidation (β oxidation).** In β oxidation, carbons are removed two at a time from an activated fatty acid. An activated fatty acid is a fatty acid bonded to coenzyme A and is called **fatty acyl CoA.** The removal of two carbons during β oxidation produces an acetyl CoA and a fatty acyl CoA shortened by two carbons. The four-step cycle repeats until the original fatty acid is completely degraded to two-carbon acetyl CoA units. The acetyl CoA can then enter the citric acid cycle. This reaction cycle occurs in the mitochondrial matrix.

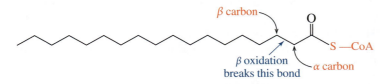

Fatty acyl CoA containing 18 carbons

The Beta Oxidation Cycle

β Oxidation (see **Figure 12.15**) includes a cycle of four reactions that convert the —CH$_2$— of the β carbon to a β ketone. Once this ketone is formed, the two-carbon acetyl group splits from the fatty acyl carbon chain, shortening the fatty acid. One cycle of β oxidation yields one FADH$_2$ and one NADH.

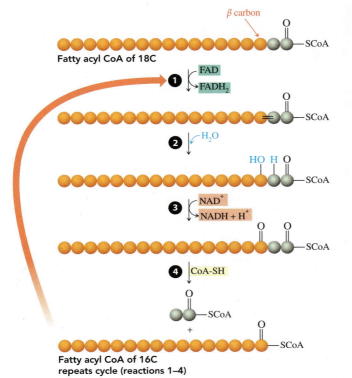

FIGURE 12.15 β oxidation. A fatty acyl CoA is broken up two carbons at a time into acetyl CoA units through a four-step cycle known as β oxidation. One FADH$_2$, one NADH, and one acetyl CoA are produced during each turn of the cycle.

Reaction 1 Oxidation (Dehydrogenation)

The first reaction removes one hydrogen from the alpha and beta carbons, and a double bond is formed. These hydrogens are transferred to FAD to form $FADH_2$.

Reaction 2 Hydration

In reaction 2, water is added to the α and β carbon double bond as $-H$ and $-OH$, respectively.

Reaction 3 Oxidation (Dehydrogenation)

The alcohol formed on the β carbon is oxidized to a ketone. As we have seen before in the citric acid cycle, the hydrogen from the alcohol reduces NAD^+ to NADH.

Reaction 4 Removal of Acetyl CoA

In the fourth reaction of the cycle, the bond between the α and β carbon is broken and a second CoA is added, forming an acetyl CoA and a fatty acyl CoA shortened by two carbons. The fatty acyl CoA can be run through the cycle again.

The net reaction for one cycle of β oxidation is summarized as

Fatty acyl $CoA_{n\ carbons} + NAD^+ + FAD + H_2O + CoA \longrightarrow$

Fatty acyl $CoA_{n-2\ carbons} + Acetyl\ CoA + NADH + H^+ + FADH_2$

Cycle Repeats and ATP Production

How many cycles of β oxidation does one fatty acid go through and how many acetyl CoA are produced? Let's consider the saturated fatty acid stearic acid (18 carbons) as it undergoes β oxidation. Every two carbons will produce an acetyl CoA, so nine acetyl CoA are produced. This will take eight turns of the β oxidation cycle because the last turn produces two acetyl CoA. How many ATP will one stearic acid molecule produce?

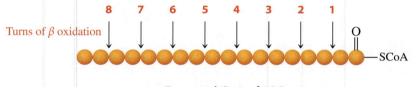

Fatty acyl CoA of 18C

Solving a Problem

Producing ATP from β Oxidation

Calculate the total number of ATP produced from the β oxidation of one stearic acid (saturated C18) entering the β oxidation cycle as an acyl CoA.

Solution

STEP 1: Determine the total number of acetyl CoA. Each acetyl CoA contains 2 carbons, so for a fatty acid with 18 carbons, a total of 9 acetyl CoA will be produced.

STEP 2: Determine the number of cycles of β oxidation. If 9 acetyl CoA are produced, then the stearic acid went through 8 cycles since the last cycle produces two acetyl CoA from the final 4 carbons.

STEP 3: Calculate the total number of ATP produced from NADH, $FADH_2$, and the acetyl CoA generated. In Section 12.7, we calculated that each turn of the citric acid cycle ultimately produces 10 ATP. Also, 1 NADH and 1 $FADH_2$ are produced per cycle of β oxidation, giving a total of 8 NADH and 8 $FADH_2$ per stearic acid molecule.

ATP Production from β Oxidation for a Stearic Acid (18C) Molecule

9 Acetyl CoA	
9 Acetyl CoA × 10 ATP/acetyl CoA	90 ATP
8 turns of β oxidation	
8 NADH × 2.5 ATP/NADH	20 ATP
8 FADH$_2$ × 1.5 ATP/FADH$_2$	12 ATP
Total	**122 ATP**

sample problem **12.12** **Producing ATP from β Oxidation**

Calculate the number of ATP produced from the β oxidation of one myristic acid (saturated C14) entering the β oxidation cycle as an acyl CoA.

Solution

STEP 1: Determine the total number of acetyl CoA. Each acetyl CoA contains 2 carbons, so for a fatty acid with 14 carbons, a total of 7 acetyl CoA will be produced.
STEP 2: Determine the number of cycles of β oxidation. If 7 acetyl CoA are produced, then the myristic acid went through 6 cycles. The last cycle produces 2 acetyl CoA from the final 4 carbons.
STEP 3: Calculate the total number of ATP produced from NADH, FADH$_2$, and the acetyl CoA generated. In Section 12.7, we calculated that each turn of the citric acid cycle ultimately produces 10 ATP. In addition, 1 NADH and 1 FADH$_2$ are produced per cycle of β oxidation, producing a total of 6 NADH and 6 FADH$_2$ per stearic acid molecule.

7 Acetyl CoA	
7 Acetyl CoA × 10 ATP/acetyl CoA	70 ATP
6 turns of β oxidation	
6 NADH × 2.5 ATP/NADH	15 ATP
6 FADH$_2$ × 1.5 ATP/FADH$_2$	9 ATP
Total	**94 ATP**

Too Much Acetyl CoA—Ketosis

In the absence of carbohydrates, the body breaks down its body fat through β oxidation to continue ATP production. This breaking down may seem efficient to the dieter considering a low-carb diet, yet some tissue types like the brain need glucose for energy. Even in the absence of carbohydrates, the liver will produce glucose from pyruvate through gluconeogenesis for tissues like the brain.

The oxidation of large amounts of fatty acids can cause acetyl CoA molecules to accumulate in the liver. When accumulation occurs, the two-carbon acetyl units condense in the liver, forming the four-carbon ketone molecules β-hydroxybutyrate and acetoacetate and the molecule acetone. These are collectively referred to as *ketone bodies*. Their formation, termed *ketogenesis*, is outlined in **Figure 12.16**. The condition known as **ketosis** occurs when an excessive amount of ketone bodies is present in the body (and cannot be efficiently metabolized). This condition is often seen in diabetics who have difficulty getting glucose into the cells for energy, individuals on low-carbohydrate or high-fat diets, or individuals undergoing starvation.

Because two of the ketones are carboxylic acids, the excessive formation of ketone bodies can lower the blood pH and cause *metabolic acidosis*. Acetone vaporizes easily, giving someone suffering from ketosis an odd, sweet-smelling breath upon exhalation similar to that of someone who has been drinking alcohol.

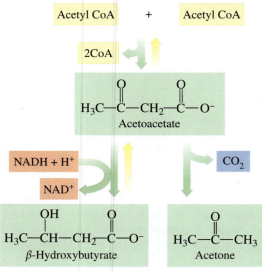

FIGURE 12.16 Ketogenesis. Excess acetyl CoA molecules not necessary for ATP production combine to produce the ketone bodies acetoacetate, β-hydroxybutyrate, and acetone.

Energy from Amino Acids

Amino acids from proteins produce nitrogen when metabolized in the body. When excess protein is ingested, amino acids must be degraded. The α-amino group of an amino acid is removed, yielding an α-keto acid through a process called **transamination.** The α-keto acid produced can then be converted into intermediates for other metabolic pathways.

The excess ammonium ions produced in this process must be excreted from the body because they are toxic if allowed to accumulate. A series of reactions called the **urea cycle** converts ammonium ions (NH_4^+) into urea, which can then be excreted in the urine.

Various amino acids offer a way to replenish the intermediates in the citric acid cycle and therefore have the ability to generate ATP. Which intermediate they replenish depends on the number of carbons in the amino acid structure. Amino acids like alanine, containing three carbons, can enter the pathways as pyruvate. Amino acids with four carbons are converted to oxaloacetate, and five-carbon amino acids are converted to α-ketoglutarate. Some amino acids can enter at more than one point depending on cellular requirements (see **Figure 12.17**).

Urea

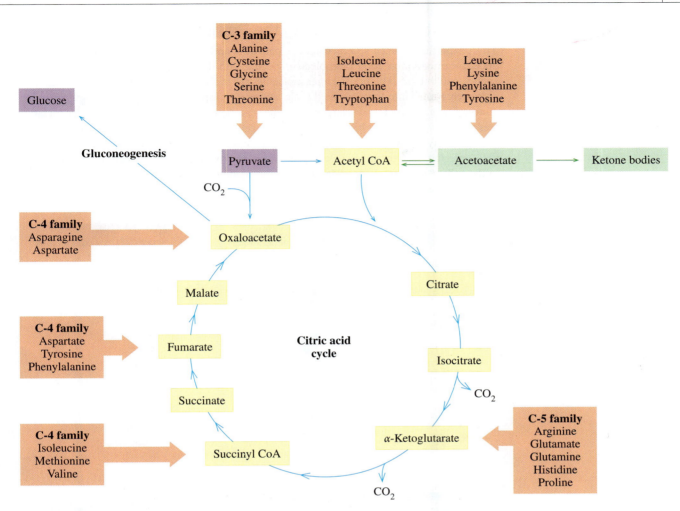

FIGURE 12.17 Amino acids feed into oxidative catabolism. Carbon atoms from degraded amino acids are converted to the intermediates of the citric acid cycle and other pathways.

In this way, we get some energy from amino acids, but it is only about 10% of the required energy under normal conditions. More energy is extracted from amino acids in conditions like fasting or starvation when carbohydrate and fat stores are depleted. Under these conditions, proteins in body tissues are degraded for fuel.

sample problem 12.13   Carbon Atoms from Amino Acids

Which amino acids provide carbon atoms that enter the citric acid cycle as α-ketoglutarate?

Solution

Those amino acids with five carbons feed into α-ketoglutarate. These are arginine, glutamate, glutamine, histidine, and proline.

Putting It Together: Linking the Pathways

This chapter introduced several catabolic pathways and the high-energy molecules they produce. Degradation of food biomolecules begins with digestion, and when the cell requires energy, and oxygen is plentiful, larger molecules are metabolized into smaller metabolites that ultimately funnel into the citric acid cycle, electron transport, and oxidative phosphorylation. Through anabolic pathways, larger molecules can be synthesized from the smaller metabolites when necessary. The biological hydrolysis products can be shifted into anabolic or catabolic pathways depending on the requirements of the cell. Glucose can be degraded to acetyl CoA entering the citric acid cycle to produce energy or be converted to glycogen for storage in the cells. Amino acids provide nitrogen for anabolism of nitrogen compounds, but their carbons can enter the citric acid cycle as α-keto acids if necessary. **Figure 12.18** gives a visual summary of the material discussed in this chapter.

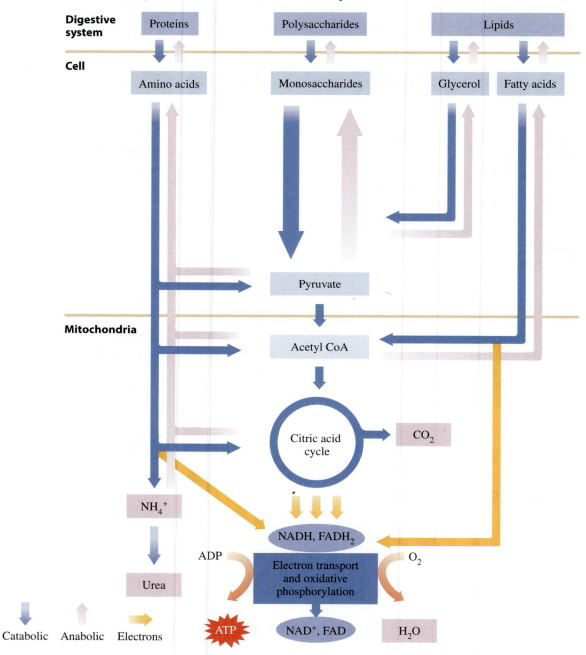

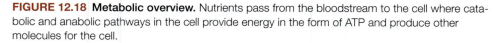

FIGURE 12.18 Metabolic overview. Nutrients pass from the bloodstream to the cell where catabolic and anabolic pathways in the cell provide energy in the form of ATP and produce other molecules for the cell.

practice
problems

12.45 Name the location in the cell where β oxidation takes place.

12.46 Name the coenzymes necessary for β oxidation.

12.47 Capric acid is a saturated C10 fatty acid.
   a. Draw fatty acyl capric acid activated for β oxidation.
   b. Identify the α and β carbons in fatty acyl capric acid.
   c. Write the overall equation for the complete β oxidation of capric acid.

12.48 Arachidic acid is a saturated C20 fatty acid.
   a. Draw fatty acyl arachidic acid activated for β oxidation.
   b. Identify the α and β carbons in fatty acyl arachidic acid.
   c. Write the overall equation for the complete β oxidation of arachidic acid.

12.49 Under what conditions are ketone bodies produced in the body?

12.50 Explain why diabetics produce high levels of ketone bodies.

12.51 Why does the body convert NH_4^+ into urea?

12.52 Draw the structure of urea.

12.53 Name the metabolic substrate(s) that can be produced from the carbon atoms of each of the following amino acids:
   a. alanine
   b. aspartate
   c. valine
   d. glutamine

12.54 Name the metabolic substrate(s) that can be produced from the carbon atoms of each of the following amino acids:
   a. leucine
   b. asparagine
   c. cysteine
   d. arginine

12.1 How Metabolism Works

12.1 Inquiry Question: What is metabolism?

Metabolism refers to biochemical reactions occurring in the body. Catabolism refers to reactions that break down larger molecules into smaller ones. Catabolic reactions are exergonic overall, and the processes are oxidative. Anabolism refers to reactions that synthesize larger biological molecules from smaller ones. Anabolic reactions are endergonic overall and are reductive. In the body, biochemical reactions are usually grouped into pathways. Biochemical reactions that produce energy tend to be coupled to reactions requiring energy. Metabolic pathways tend to be compartmentalized in different parts of the cell. The mitochondria are the energy-producing factories of the cells.

P — P — P — Adenosine

12.2 Metabolically Relevant Nucleotides

12.2 Inquiry Question: What do the nucleotides that transfer energy look like?

Nucleotides are used in metabolism to transfer energy throughout the cell. ATP is considered the main energy currency of the cell and produces energy when hydrolyzed to **ADP + P$_i$**. NADH and FADH$_2$ contain a nucleotide portion and a vitamin portion. They are important coenzymes that transport hydrogens and electrons in the cell. Coenzyme A (CoA) also contains a nucleotide and vitamin portion. Each of these nucleotides has a high-energy and low-energy form.

12.3 Digestion—From Fuel Molecules to Hydrolysis Products

12.3 Inquiry Question: How are food molecules digested?

Food molecules are broken down into their component parts through hydrolysis prior to absorption into the bloodstream. Monosaccharides and amino acids travel directly through the bloodstream to the cells for absorption whereas triglycerides are packaged into lipoproteins called chylomicrons for delivery.

12.4 Glycolysis—From Hydrolysis Products to Common Metabolites

12.4 Inquiry Question: How is glucose catabolized through glycolysis?

Glycolysis is a series of 10 reactions that catabolizes a six-carbon glucose molecule to two three-carbon pyruvate molecules. These 10 reactions yield two ATP and two NADH molecules per glucose. Under aerobic conditions, pyruvate is further oxidized in the mitochondria to acetyl CoA. In the absence of oxygen, pyruvate is reduced to lactate and regenerates NAD$^+$ so that glycolysis can continue. Glucose can be stored as glycogen in the liver and muscle for later use. Fructose can undergo glycolysis. In the liver, its catabolism is unregulated and can produce excess products that become stored in the body as fat.

12.5 The Citric Acid Cycle—Central Processing

12.5 Inquiry Question: What types of reactions occur during the citric acid cycle?

The citric acid cycle occurs in the mitochondrial matrix and combines acetyl CoA (two carbons) with oxaloacetate (four carbons), producing citric acid. Citric acid undergoes oxidation, decarboxylation, dehydrogenation and a hydration yielding two CO$_2$, GTP, three NADH, and one FADH$_2$. The cycle regenerates oxaloacetate to begin again. GTP readily converts to ATP in the cell.

12.6 Electron Transport and Oxidative Phosphorylation

12.6 Inquiry Question: How is the energy in NADH and FADH$_2$ transformed into ATP?

The reduced nucleotides NADH and FADH$_2$ become oxidized, transporting H$^+$ and electrons through a series of enzyme complexes in the inner mitochondrial membrane. The final electron acceptor in this process is O$_2$, which combines with H$^+$ to form H$_2$O. The complexes act as proton pumps, moving protons from the matrix to the inner membrane space during electron transport. This produces an electrochemical gradient. The protons can return to the matrix through complex V, ATP synthase, which generates ATP. This process is known as oxidative phosphorylation. When ATP and proton transport are uncoupled, the energy in the proton gradient is released as heat, which assists in maintaining body temperature in a process called thermogenesis.

Nucleotide Input	Protons (H$^+$) Pumped	ATP Output
NADH	10	2.5
FADH$_2$	6	1.5

12.7 ATP Production

12.7 Inquiry Question: How many ATP can be produced during oxidative catabolism?

For every four H$^+$ pumped into the inner membrane space and returned to the matrix, one ATP can be synthesized. The oxidation of one NADH provides enough energy to synthesize 2.5 ATP. One FADH$_2$ provides energy to synthesize 1.5 ATP. Under aerobic conditions, the complete oxidation of one molecule of glucose produces a total of 32 ATP.

Turns of β oxidation 8 7 6 5 4 3 2 1 — SCoA

Fatty acyl CoA of 18C

12.8 Other Fuel Choices

12.8 Inquiry Question: How do fatty acids and amino acids produce ATP?

Fatty acids produce ATP when glucose supplies are low. Fatty acids link to coenzyme A, forming activated fatty acyl CoA which is transported to the mitochondria for catabolism in a reaction called β oxidation. Fatty acyl CoA is oxidized, producing a new fatty acyl CoA that is two carbons shorter and one molecule of acetyl CoA. Each turn of β oxidation produces one NADH and one FADH$_2$. High levels of acetyl CoA in the cell activate the ketogenesis pathway, forming ketone bodies that can lead to ketosis and acidosis. Amino acids can produce ATP when other fuel supplies are low and the cell does not require other nitrogen-containing compounds. When amines are removed from amino acids as ammonium ions, they are converted to urea for excretion. The carbons from amino acids can feed into oxidative catabolism as different intermediates depending on the amino acid.

The study guide will help you check your understanding of the main concepts in Chapter 12. You should be able to

12.1 How Metabolism Works

- Distinguish catabolism from anabolism.
- Identify reactions as catabolic or anabolic.
- Name the parts of a cell associated with metabolism.

12.2 Metabolically Relevant Nucleotides

- Identify the metabolically relevant nucleotides.
- Distinguish the low-energy and high-energy forms of the relevant nucleotides.

12.3 Digestion—From Food Molecules to Hydrolysis Products

- Compare digestion of carbohydrates, lipids, and proteins.

12.4 Glycolysis—From Hydrolysis Products to Common Metabolites

- Follow a molecule of glucose through the ten reactions of glycolysis.
- Discuss anaerobic and aerobic fates of pyruvate.
- Contrast glycolysis for glucose and for fructose.

12.5 The Citric Acid Cycle—Central Processing

- Identify the reactions in the citric acid cycle.
- List the energy output of the citric acid cycle.

12.6 Electron Transport and Oxidative Phosphorylation

- Describe the function of each enzyme complex (I–IV) during electron transport.
- Discuss the function of coenzyme Q and cytochrome c.
- Describe the production of ATP at complex V using the chemiosmotic model.

12.7 ATP Production

- Convert the number of reduced nucleotides produced (NADH, $FADH_2$) to a corresponding number of ATP.
- Calculate the number of ATP produced during the oxidative catabolism of a molecule of glucose.

12.8 Other Fuel Choices

- Calculate the number of ATP produced from a saturated fatty acid undergoing β oxidation.
- Describe the metabolic pathways of β oxidation, transamination, and the urea cycle.
- Identify catabolic and anabolic pathways in the cell.

Key Terms

aerobic—In the presence of oxygen.

anabolism—Metabolic chemical reactions in which smaller molecules are combined to form larger ones. These reactions tend to be reductive and require energy.

anaerobic—In the absence of oxygen.

beta (β) oxidation—The degradation of fatty acids by removing two-carbon segments from a fatty acid at the oxidized β carbon.

catabolism—Metabolic chemical reactions in which larger molecules are broken down into smaller ones. These reactions tend to be oxidative and produce energy.

chemiosmotic model—The conservation of energy from the transfer of electrons in the electron transport chain by pumping protons into the intermembrane space to produce a proton gradient that provides the energy to synthesize ATP.

chylomicron—Droplet of dietary fat and protein that circulates from the intestine to the tissues for absorption.

citric acid cycle—A series of reactions that degrades two-carbon acetyl groups from acetyl CoA into CO_2, generating the high-energy molecules NADH and $FADH_2$ in the process.

cytoplasm—The material in a cell between the nucleus and the cell membrane.

cytosol—The aqueous solution of the cytoplasm.

digestion—The breakdown of large molecules into smaller components for absorption, usually through hydrolysis.

electrochemical gradient—A difference in both concentration and electrical charge of an ion across a membrane; the driving force for oxidative phosphorylation.

emulsification—Breaking up large insoluble globules into smaller droplets via an amphipathic molecule.

fatty acyl CoA—A fatty acid bonded to coenzyme A through a thioester bond.

fermentation—The anaerobic production of ethanol from pyruvate.

gluconeogenesis—The anabolic process of synthesizing glucose.

ketone bodies—Ketone compounds formed from the condensation of excess acetyl CoA in the liver.

ketosis—A condition in which not all ketone bodies are metabolized, leading to low blood pH.

metabolic pathway—A series of chemical reactions leading to a common product.

metabolism—The sum of all chemical reactions occurring in an organism.

metabolite—A chemical intermediate formed during enzyme-catalyzed metabolism. Some common ones are pyruvate and the acetyl group.

mitochondrion—Cell organelle where energy-producing reactions take place.

oxidative decarboxylation—An oxidation in which a carboxylate is released as CO_2. The corresponding reduction forms a reduced nucleotide like NADH.

oxidative phosphorylation—The production of ATP from ADP and P_i using energy generated by the oxidation reactions of electron transport.

thermogenesis—The generation of heat. In the mitochondria, it can be caused by the uncoupling of ATP synthase from the electron transport chain.

transamination—The transfer of an amino group from an amino acid, thereby producing an α-keto acid.

urea cycle—The process by which excess ammonium ions from the degradation of amino acids are converted to urea for excretion.

Summary of Reactions

Hydrolysis of ATP

$$ATP \longrightarrow ADP + P_i + energy$$

Formation of ATP

$$ADP + P_i + energy \longrightarrow ATP$$

Digestion of Proteins

$$Protein + H_2O \xrightarrow[Enzyme]{H^+} Amino\ acids$$

Glycolysis

Hydrolysis of Disaccharides

$$Lactose + H_2O \xrightarrow{Lactase} Galactose + glucose$$

$$Sucrose + H_2O \xrightarrow{Sucrase} Glucose + fructose$$

$$Maltose + H_2O \xrightarrow{Maltase} Glucose + glucose$$

Digestion of Triglycerides

$$Triglycerides + 3\ H_2O \xrightarrow{Lipases} Glycerol + 3\ Fatty\ acids$$

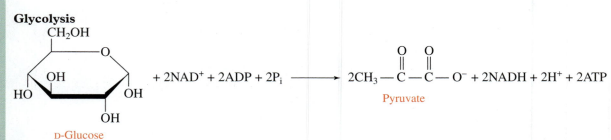

Oxidation of Pyruvate to Acetyl CoA

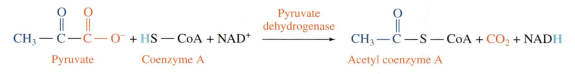

Reduction of Pyruvate to Lactate

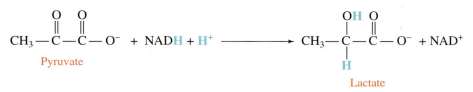

Citric Acid Cycle

$$\text{Acetyl CoA} + 3NAD^+ + FAD + GDP + P_i + 2H_2O \longrightarrow 2CO_2 + 3NADH + 2H^+ + FADH_2 + CoA + GTP$$

Complete Oxidation of Glucose

$$C_6H_{12}O_6 + 6H_2O \longrightarrow 6CO_2 + 6H_2O$$

β Oxidation of Fatty Acid

$$\text{Fatty acyl CoA}_{n \text{ carbons}} + NAD^+ + FAD + H_2O + CoA \longrightarrow \text{Fatty acyl CoA}_{n-2 \text{ carbons}} + \text{Acetyl CoA} + NADH + H^+ + FADH_2$$

Additional Problems

12.55 Name the location in the cell where the following catabolic processes take place:
   a. glycolysis   **b.** citric acid cycle
   c. ATP synthesis

12.56 Name the location in the cell where the following catabolic processes take place:
   a. pyruvate oxidation
   b. electron transport
   c. β-oxidation

12.57 Identify the type of food—carbohydrate, fat, or protein—that gives each of the following digestion products:
   a. glucose   **b.** fatty acid
   c. maltose   **d.** glycerol
   e. amino acids

12.58 How and where does sucrose undergo digestion in the body? Name the products.

12.59 How and where does lactose undergo digestion in the body? Name the products.

12.60 If glycolysis occurs in the cytosol and the citric acid cycle occurs in the mitochondrial matrix, how do the products of glycolysis get inside the mitochondrial matrix?

12.61 Which of the following reactions in glycolysis produce ATP or NADH?
   a. 1,3-bisphosphoglycerate to 3-phosphoglycerate
   b. glucose-6-phosphate to fructose-6-phosphate
   c. phosphoenolpyruvate to pyruvate

12.62 Which of the following reactions in glycolysis produce ATP or NADH?
   a. glucose to glucose-6-phosphate
   b. glyceraldehyde-3-phosphate to 1,3-bisphosphoglycerate
   c. dihydroxyacetone phosphate to glyceraldehyde-3-phosphate

12.63 Which of the reactions given in Problem 12.61 represent isomerizations where the reactants and products are structural isomers?

12.64 Which of the reactions given in Problem 12.62 represent isomerizations where the reactants and products are structural isomers?

12.65 After running a marathon, a runner has muscle pain and cramping. What might have occurred in the muscle cells to cause this?

12.66 After eating a large, starchy meal, what does the body do with all the glucose from starch?

12.67 Under what condition is pyruvate converted to lactate in the body?

12.68 When pyruvate is used to form acetyl CoA, the product has only two carbon atoms. What happened to the third carbon?

12.69 Refer to the diagram of the citric acid cycle to answer each of the following:
   a. Name the six-carbon compounds.
   b. Name the five-carbon compounds.
   c. Name the oxidation reactions.
   d. Name the reactions where secondary alcohols are oxidized to ketones.

12.70 Refer to the diagram of the citric acid cycle to answer each of the following:
   a. Name the four-carbon compounds.
   b. Name the reactant that undergoes a hydration reaction.
   c. Name the reaction that is coupled to GTP formation.
   d. Provide the number of CO_2 molecules produced per turn of the citric acid cycle.

12.71 If there are no reactions in the citric acid cycle that use oxygen, O_2, why does the cycle operate only in aerobic conditions?

12.72 What products of the citric acid cycle are used in electron transport?

12.73 Identify the following as the reduced or oxidized form:
   a. NAD^+　　**b.** $FADH_2$　　**c.** QH_2

12.74 Identify the following as the reduced or oxidized form:
   a. NADH　　**b.** FAD　　**c.** Q

12.75 During electron transport, H^+ are pumped out of the _____, across the inner membrane, and into the _____.

12.76 During ATP synthesis, H^+ move from the _____, across the inner membrane, and into the _____.

12.77 In the chemiosmotic model, how is energy provided to synthesize ATP?

12.78 What is the effect of proton accumulation in the intermembrane space?

12.79 How many ATP are produced when glucose is oxidized to pyruvate compared to when glucose is oxidized to CO_2 and H_2O?

12.80 What metabolic substrate(s) can be produced from the carbon atoms of each of the following amino acids?
   a. histidine　　　　**b.** isoleucine
   c. methionine　　　**d.** phenylalanine

12.81 Consider the complete oxidation of capric acid, a saturated C10 fatty acid.
   a. How many acetyl CoA units are produced?
   b. How many cycles of β oxidation occur?
   c. How many ATP are generated from the complete oxidation of capric acid?

12.82 Consider the complete oxidation of arachidic acid, a saturated C20 fatty acid.
   a. How many acetyl CoA units are produced?
   b. How many cycles of β oxidation occur?
   c. How many ATP are generated from the complete oxidation of arachidic acid?

Challenge Problems

12.83 Identify all steps in the oxidative catabolism of glucose that can be considered oxidative decarboxylations.

12.84 How many turns of the citric acid cycle does it take for the two carbons from an acetyl group entering the citric acid cycle to be liberated as CO_2?

12.85 Lauric acid, a saturated C12 fatty acid, is found in coconut oil.
   a. Draw fatty acyl lauric acid activated for β oxidation.
   b. Identify the α and β carbons in fatty acyl lauric acid.
   c. Write the overall equation for the complete β oxidation of lauric acid.
   d. How many acetyl CoA units are produced?
   e. How many cycles of β oxidation occur?
   f. Account for the total ATP yield from β oxidation of lauric acid (C12) by completing the following table:

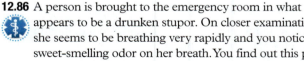

_____ Acetyl CoA × 10 ATP/acetyl CoA		_____ATP
_____ NADH × 2.5 ATP/NADH		_____ATP
_____ $FADH_2$ × 1.5 ATP/$FADH_2$		_____ATP
	Total	

12.86 A person is brought to the emergency room in what appears to be a drunken stupor. On closer examination, she seems to be breathing very rapidly and you notice a sweet-smelling odor on her breath. You find out this person has not been drinking alcohol. What is her likely condition?

12.87 Can acetyl CoA feed into gluconeogenesis producing glucose for the body? Can fatty acids be used to produce glucose?

Answers to Odd-Numbered Problems

Practice Problems

12.1 In metabolism, a catabolic reaction breaks apart molecules, releases energy, and is an oxidation.

12.3　**a.** catabolism
　　　　b. anabolism

12.5 hydrolysis

12.7 mitochondrion

12.9 a. FAD/FADH$_2$ **b.** acetyl CoA

12.11 a. NADH   **b.** acetyl CoA   **c.** FAD

12.13 a. starch   **b.** none   **c.** starch, oligosaccharides, disaccharides

12.15 Cholesterol is esterified and packaged into lipoproteins.

12.17 amino acids

12.19 D-glucose

12.21 4 ATP (net 2 ATP)

12.23 2 NADH and 2 ATP

12.25 NAD$^+$

12.27 ethanol, CO$_2$, and NAD$^+$

12.29 The main regulation point of glycolysis is phosphofruc-tokinase, the enzyme involved in step 3. Because the products of the metabolism of fructose enter glycolysis at step 5, they bypass the main regulation point, which leads to the production of excess pyruvate and acetyl CoA not required in the cells, which is ultimately converted to fat.

12.31 acetyl CoA and oxaloacetate

12.33 isocitrate $\longrightarrow$ α-ketoglutarate (reaction 3) and α-ketoglutarate $\longrightarrow$ succinyl CoA (reaction 4)

12.35 isocitrate $\longrightarrow$ α-ketoglutarate (reaction 3), α-keto glutarate $\longrightarrow$ succinyl CoA (reaction 4), and malate $\longrightarrow$ oxaloacetate (reaction 8)

12.37 Coenzyme Q

12.39 complex IV

12.41 As protons flow through ATP synthase, energy is released to produce ATP.

12.43 a. 2.5 ATP   **b.** 7 ATP
c. 5 ATP   **d.** 10 ATP

12.45 mitochondrial matrix

12.47 a. and **b.**

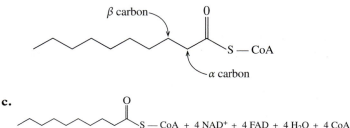

β carbon   α carbon

c.

S — CoA + 4 NAD$^+$ + 4 FAD + 4 H$_2$O + 4 CoA

5 Acetyl CoA + 4 NADH + 4 H$^+$ + 4 FADH$_2$

12.49 Ketone bodies form when excess acetyl CoA results from the breakdown of large amounts of fat.

12.51 NH$_4^+$ is toxic if it accumulates in the body.

12.53 a. pyruvate   **b.** oxaloacetate, fumarate
c. succinyl CoA   **d.** α-ketoglutarate

Additional Problems

12.55 a. cytosol
b. mitochondrial matrix
c. mitochondrial matrix

12.57 a. carbohydrate   **b.** fat
c. carbohydrate   **d.** fat
e. protein

12.59 Lactose undergoes digestion in the small intestine to yield galactose and glucose.

12.61 a. produces ATP   **b.** neither   **c.** produces ATP

12.63 b.

12.65 The runner's muscles may be switching over to anaerobic catabolism to keep ATP production going. The muscles produce lactate and H$^+$ during this process, causing soreness.

12.67 anaerobic (low oxygen) conditions

12.69 a. citrate and isocitrate
b. α-ketoglutarate
c. isocitrate $\longrightarrow$ α-ketoglutarate (reaction 3), α-ketoglutarate $\longrightarrow$ succinyl CoA (reaction 4), succinate $\longrightarrow$ fumarate (reaction 6), and malate $\longrightarrow$ oxaloacetate (reaction 8)
d. isocitrate $\longrightarrow$ α-ketoglutarate (reaction 3) and malate $\longrightarrow$ oxaloacetate (reaction 8)

12.71 O$_2$ is used during electron transport. The coenzymes NADH & FADH$_2$ produced in the citric acid cycle are oxidized to NAD$^+$ and FAD during electron transport.

12.73 a. oxidized   **b.** reduced   **c.** reduced

12.75 matrix, intermembrane space

12.77 Energy released as protons flow down a concentration gradient through ATP synthase back to the matrix is utilized for the synthesis of ATP.

12.79 The oxidation of glucose to pyruvate produces 7 ATP, 5 from NADH, whereas 32 ATP are produced from the complete oxidation of glucose to CO$_2$ and H$_2$O.

12.81 a. 5   **b.** 4   **c.** 66 ATP

12.83 Oxidation of pyruvate: pyruvate $\longrightarrow$ acetyl CoA

Reaction 3 of citric acid cycle: isocitrate $\longrightarrow$ α-ketoglutarate

Reaction 4 of the citric acid cycle: α-ketoglutarate $\longrightarrow$ succinyl CoA

12.85 a. and **b.**

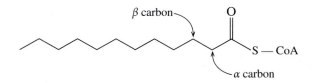

β carbon   S — CoA   α carbon

c.

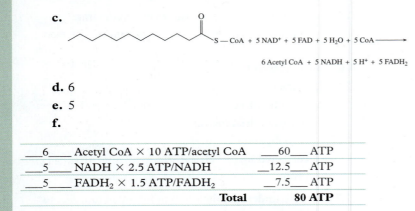

$$S—CoA + 5 NAD^+ + 5 FAD + 5 H_2O + 5 CoA \longrightarrow$$

$$6 \text{ Acetyl CoA} + 5 \text{ NADH} + 5 H^+ + 5 FADH_2$$

d. 6

e. 5

f.

__6___ Acetyl CoA × 10 ATP/acetyl CoA	___60___	ATP
__5___ NADH × 2.5 ATP/NADH	__12.5___	ATP
__5___ FADH$_2$ × 1.5 ATP/FADH$_2$	__7.5___	ATP
Total	**80 ATP**	

12.87 Only oxaloacetate or pyruvate are starting points for gluconeogenesis (see Figure 12.18). Acetyl CoA can not be converted to pyruvate, so fatty acids cannot generate glucose.

CREDITS

Front Matter

p. iii Laura D. Frost. p. iii Todd Deal.

Chapter 1

p. 2 Corbis / SuperStock. p. 5 *top, left*: Josef Bosak / Shutterstock. p. 5 *top, middle left*: Pearson Education. p. 5 *top, middle right*: Titelio / Shutterstock. p. 5 *top, right*: Pearson Education. p. 5 *right, top*: Li Wa / Shutterstock. p. 5 *right, middle*: Mark Evans / iStockphoto. p. 5 *right, bottom*: Rd / Shutterstock. p. 9 Richard Megna / Fundamental Photographs. p. 11 *left*: Big Pants Production / Shutterstock. p. 11 *middle*: Garsya / Shutterstock. p. 11 *right*: Edyta Pawlowska / Shutterstock. p. 13 Iofoto / Shutterstock. p. 16 STOCK4B GmbH / Alamy. p. 19 MarFot / Shutterstock. p. 20 IAN HOOTON / Science Photo Library / Alamy. p. 21 *top*: Ron Kloberdanz / Shutterstock. p. 21 *bottom*: Pearson Education / Eric Schrader. p. 23 Michael Wright. p. 26 jacomstephens / Getty Images. p. 27 shock / Shutterstock. p. 36 Michael C. Gray / Shutterstock. p. 37 Pearson Education. p. 40 *top*: Mark Evans / iStockphoto. p. 40 *bottom, left*: STOCK4B GmbH / Alamy. p. 40 *bottom, right*: Pearson Education.

Chapter 2

p. 48 Media Minds / Alamy. p. 50 *left, top*: Elen / Shutterstock. p. 50 *left, middle*: Pearson Education. p. 50 *left, bottom*: Lawrence Berkeley National Laboratory. p. 50 *middle*: Pearson Education. p. 51 *left-right*: Pearson Education. p. 57 Tomaz Levstek / iStockphoto. p. 58 Stanford Dosimetry, LLC. p. 63 Gusto Images / Science Photo Library, Ltd / Photo Researchers, Inc. p. 65 Simon Fraser / Photo Researchers, Inc. p. 66 *left, top*: Myo Han. p. 66 *left, bottom*: Michael Tobin. p. 66 *right*: Editorial Image, LLC / Alamy. p. 67 Dr. Robert Friedland / Science Photo Library, Ltd / Photo Researchers, Inc. p. 68 *left*: Tomaz Levstek / iStockphoto. p. 68 *right, top*: Gusto Images / Science Photo Library, Ltd / Photo Researchers, Inc. p. 68 *right, bottom*: Michael Tobin.

Chapter 3

p. 74 silver-john / Shutterstock. p. 82 Kurhan / Shutterstock. p. 90 *top*: Laitr Keiows / Shutterstock. p. 90 *bottom*: Mark Kostich / iStockphoto. p. 97 Pearson Education. p. 99 Pearson Education / Eric Schrader. p. 103 Pearson Education / Eric Schrader. p. 114 Pearson Education / Eric Schrader.

Chapter 4

p. 122 Shebeko / Shutterstock. p. 124 Big Cheese / Photolibrary. p. 131 Pearson Education / Eric Schrader. p. 139 Pearson Education / Eric Schrader. p. 140 *top*: Pearson Education / Eric Schrader. p. 140 *middle*: Glen Jones / iStockphoto. p. 140 *bottom*: Steffen Foester Photography / Shutterstock. p. 141 Leonid Nyshko / iStockphoto. p. 146 moodboard / Alamy. p. 152 *left*: monticello / Shutterstock. p. 152 *middle, top*: *middle, bottom*: Kurhan / Shutterstock, p. 152 *right, top*: Bochkarev Photography / Shutterstock. p. 152 *right, bottom*: Wiktory / Shutterstock. p. 156 *top, left*; *top, right*; *bottom, left*; *bottom, right*: Pearson Education / Eric Schrader. p. 160 *left*: Photos.com. p. 160 *right*: Vladimir Kolobov / iStockphoto. p. 161 *top-bottom*: Pearson Education / Eric Schrader. p. 168 Big Cheese / Photolibrary.

Chapter 5

p. 178 Bartosz Hadyniak / iStockphoto. p. 179 P.B. Mann. p. 180 Pearson Education / Eric Schrader. p. 181 Greg Stanfield / Shutterstock. p. 183 Pearson Education / Eric Schrader. p. 184 Pearson Education / Eric Schrader. p. 187 AFP / Getty Images / Newscom. p. 190 *left*: i love images / Alamy. p. 190 *right*: Rich Legg / Getty Images. p. 191 Sevenke / Shutterstock. p. 200 Belmonte / age fotostock. p. 202 Pearson Education / Eric Schrader.

Chapter 6

p. 212 Anton Prado Photo / Shutterstock. p. 214 Picsfive / Dreamstime. p. 215 Rick Wilking / Reuters. p. 216 *left*: Wojtek Kryczka / iStockphoto. p. 216 *right*: Baloncici / iStockphoto. p. 217 Marcelo Wain / iStockphoto. p. 224 *top*: Freeze Frame Studio / iStockphoto. p. 224 *middle*: Elena Elisseeva / iStockphoto. p. 224 *bottom*: Kelly Cline / iStockphoto. p. 233 *top*: Pearson Education. p. 233 *bottom*: Jürgen Fälchle / Fotolia. p. 234 Pearson Education / Eric Schrader. p. 239 Jon Larson / iStockphoto. p. 240 *top-bottom*: Pearson Education. p. 244 Pearson Education. p. 245 *top, left*: Danny E Hooks / Shutterstock. p. 245 *top, right*: David Toase / Getty Images. p. 245 *bottom*: Biophoto Associates / Photo Researchers, Inc. p. 246 nicholashan / Fotolia. p. 250 *top, left*: Picsfive / Dreamstime. p. 250 *top, right*: Pearson Education. p. 250 *bottom*: Freddie Vargas / iStockphoto.

Chapter 7

p. 260 Miqul / Fotolia. p. 264 Pearson Education / Eric Schrader. p. 275 *top*: Michaela Stejskalova / Shutterstock. p. 275 *bottom*: mygueart / iStockphoto. p. 280 Filipe B. Varela / Shutterstock. p. 283 Pearson Education. p. 285 Vanessa Davies / Dorling Kindersley. p. 289 *left-right*: Pearson Education / Eric Schrader. p. 293 Pearson Education / Eric Schrader. p. 298 Filipe B. Varela / Shutterstock.

Chapter 8

p. 306 Glenda Powers / Fotolia. p. 307 Pearson Education. p. 308 *left, top*: DouglasFreer / iStockphoto. p. 308 *left, middle top*: Monia / Fotolia. p. 308 *left, middle bottom*: Pearson Education / Eric Schrader. p. 308 *left, bottom*: Meliha Gojak / Fotolia. p. 309 Klaus Guldbrandsen / Photo Researchers, Inc. p. 310 *left-right*: Pearson Education. p. 311 *top*: Dr. P. Marazzi / Photo Researchers, Inc. p. 311 *bottom*: remik44992 / Shutterstock. p. 312 *left-right*: Pearson Education. p. 314 Pearson Education / Eric Schrader. p. 315 *top*; *middle*; *bottom*: Pearson Education. p. 320 Pearson Education / Eric Schrader. p. 323 *top-middle*: Pearson Education / Eric Schrader. p. 323 *bottom*: soupstock / Fotolia. p. 324 pkline / iStockphoto. p. 326 *top, left*; *top, middle*; *top, right*; *bottom, left*: Pearson Education. p. 329 Sandia National Laboratories. p. 330 *left*; *middle*; *right*: Sam Singer. p. 332 Picsfive / Shutterstock. p. 336 *left, top*: Pearson Education / Eric Schrader. p. 336 *left, middle*: Pearson Education. p. 336 *left, bottom*: Pearson Education / Eric Schrader. p. 336 *right, top*: soupstock / Fotolia. p. 336 *right, bottom*: Pearson Education.

Chapter 9

p. 344 Pete Saloutos / Fotolia. p. 348 *top*: Pearson Education. p. 348 *bottom*: Pearson Education / Eric Schrader. p. 350 Blue Lemon Photo / Fotolia. p. 364 *left*: Richard Megna / Fundamental Photographs. p. 364 *right*: Alina Hart / iStockphoto. p. 372 Nina Shannon / iStockphoto.

Chapter 10

p. 380 Pearson Education / Eric Schrader. p. 396 *left*: Lee Torrens / Shutterstock. p. 396 *mid-left*: rendering by L. Frost from The Protein Data Bank/ RCSB. p. 396 *right*: AJPhoto / Photo Researchers, Inc. p. 397 Medical-on-Line / Alamy. p. 400 Pearson Education.

p. 402 *left*: rendering by L. Frost from The Protein Data Bank/RCSB. p. 404 rendering by L. Frost from The Protein Data Bank/RCSB. p. 404 rendering by L. Frost from The Protein Data Bank/RCSB. p. 405 *left-right*: rendering by L. Frost from The Protein Data Bank/RCSB.

Chapter 11

p. 424 Ryan McVay / Getty Images. p. 425 SPL / Photo Researchers, Inc. p. 454 Will & Deni MC / Photo Researchers, Inc. p. 457 Stephen A. Ferreira, University of Hawaii at Manoa. p. 458 *top-bottom*: Associated Press.

Chapter 12

p. 466 Supri Suharjoto / Shutterstock. p. 468 Hannes Eichinger / Shutterstock. p. 475 Jaimie / Duplass / Shutterstock. p. 482 Diego Cervo / iStockphoto. p. 490 Laura D. Frost.

TEXT CREDITS

Chapter 1

p. 45 *top-bottom*: Timberlake, Karen, C., General, Organic, and Biological Chemistry: Structures of Life, 4th Ed., ©2013. Reprinted and Electronically reproduced by permission of Pearson Education, Inc., Upper Saddle River, New Jersey.

Chapter 5

p. 182 Timberlake, Karen, C., General, Organic, and Biological Chemistry: Structures of Life, 4th Ed., ©2013. Reprinted and Electronically reproduced by permission of Pearson Education, Inc., Upper Saddle River, New Jersey.

Chapter 9

p. 358 Data from General Chemistry Online! Antoine.frostburg.edu/chem/senese/101/index.shtml.

METRIC UNITS AND SOME USEFUL CONVERSION FACTORS

Length
meter (m)

1 meter (m) = 100 centimeters (cm)
1 meter (m) = 1000 millimeters (mm)
1 cm = 10 mm
1 kilometer (km) = 0.621 mile (mi)
1 inch (in.) = 2.54 cm (exact)

Volume
cubic meter (m³)

1 liter (L) = 1000 milliliters (mL)
1 mL = 1 cc = 1 cm³
1 tsp = 5 mL
1 L = 1.06 quart (qt)
1 qt = 946 mL

Mass
kilogram (kg)

1 kilogram (kg) = 1000 grams (g)
1 g = 1000 milligrams (mg)
1 kg = 2.20 lb
1 lb = 454 g
1 mole = 6.02×10^{23} particles

Water
density = 1.00 g/mL

Temperature
kelvin (K)

$$°F = \left(\frac{1.8\ °F}{1\ °C} \times °C \right) + 32\ °F$$

$$°C = (°F - 32\ °F) \times \frac{1\ °C}{1.8\ °F}$$

Kelvin (K) = Celsius (°C) + 273 °C

Pressure
pascal (Pa)

1 atmosphere = 760 mmHg
1 atm = 14.7 psi

Energy
calorie (cal)

1 kcal = 1000 calories (cal)
1 nutritional calorie (Cal) = 1 kcal

METRIC PREFIXES

Prefix	Abbreviation	Relationship to Standard Unit
kilo-	k	1000 ×
Base unit (has no prefix)		1 × (gram, liter, meter)
deci-	d	÷ 10
centi-	c	÷ 100
milli-	m	÷ 1000
micro-	mc or μ	÷ 1,000,000

SOLUTION EQUATIONS

Concentration Equations

$$\text{Concentration} = \frac{\text{amount of solute}}{\text{amount of solution}}$$

Molarity, M $M = \dfrac{\text{Mole solute}}{\text{L Solution}}$

Percent concentration

$$\%\ \text{concentration} = \frac{\text{parts of solute}}{100\ \text{parts of solution}}$$

$$\%\ (m/v) = \frac{\text{g solute}}{\text{mL solution}} \times 100\%$$

$$\%\ (m/m) = \frac{\text{g solute}}{\text{g solution}} \times 100\%$$

$$\%\ (v/v) = \frac{\text{mL solute}}{\text{mL solution}} \times 100\%$$

$$ppm = \frac{\text{g solute}}{\text{mL solution}} \times 1,000,000$$

$$ppb = \frac{\text{g solute}}{\text{mL solution}} \times 1,000,000,000$$

Dilution Equation

$$C_{\text{initial}} \times V_{\text{initial}} = C_{\text{final}} \times V_{\text{final}}$$

ATOMIC MASSES OF THE ELEMENTS

Name	Symbol	Atomic Number	Atomic Mass[a]	Name	Symbol	Atomic Number	Atomic Mass[a]
Actinium	Ac	89	(227)	Mendelevium	Md	101	(258)
Aluminum	Al	13	26.98	Mercury	Hg	80	200.6
Americium	Am	95	(243)	Molybdenum	Mo	42	95.94
Antimony	Sb	51	121.8	Neodymium	Nd	60	144.2
Argon	Ar	18	39.95	Neon	Ne	10	20.18
Arsenic	As	33	74.92	Neptunium	Np	93	(237)
Astatine	At	85	(210)	Nickel	Ni	28	58.69
Barium	Ba	56	137.3	Niobium	Nb	41	92.91
Berkelium	Bk	97	(247)	Nitrogen	N	7	14.01
Beryllium	Be	4	9.012	Nobelium	No	102	(259)
Bismuth	Bi	83	209.0	Osmium	Os	76	190.2
Bohrium	Bh	107	(264)	Oxygen	O	8	16.00
Boron	B	5	10.81	Palladium	Pd	46	106.4
Bromine	Br	35	79.90	Phosphorus	P	15	30.97
Cadmium	Cd	48	112.4	Platinum	Pt	78	195.1
Calcium	Ca	20	40.08	Plutonium	Pu	94	(244)
Californium	Cf	98	(251)	Polonium	Po	84	(209)
Carbon	C	6	12.01	Potassium	K	19	39.10
Cerium	Ce	58	140.1	Praseodymium	Pr	59	140.9
Cesium	Cs	55	132.9	Promethium	Pm	61	(145)
Chlorine	Cl	17	35.45	Protactinium	Pa	91	231.0
Chromium	Cr	24	52.00	Radium	Ra	88	(226)
Cobalt	Co	27	58.93	Radon	Rn	86	(222)
Copernicium	Cn	112	(285)	Rhenium	Re	75	186.2
Copper	Cu	29	63.55	Rhodium	Rh	45	102.9
Curium	Cm	96	(247)	Roentgenium	Rg	111	(272)
Darmstadtium	Ds	110	(271)	Rubidium	Rb	37	85.47
Dubnium	Db	105	(262)	Ruthenium	Ru	44	101.1
Dysprosium	Dy	66	162.5	Rutherfordium	Rf	104	(261)
Einsteinium	Es	99	(252)	Samarium	Sm	62	150.4
Erbium	Er	68	167.3	Scandium	Sc	21	44.96
Europium	Eu	63	152.0	Seaborgium	Sg	106	(266)
Fermium	Fm	100	(257)	Selenium	Se	34	78.96
Flerovium	Fl	114	(289)	Silicon	Si	14	28.09
Fluorine	F	9	19.00	Silver	Ag	47	107.9
Francium	Fr	87	(223)	Sodium	Na	11	22.99
Gadolinium	Gd	64	157.3	Strontium	Sr	38	87.62
Gallium	Ga	31	69.72	Sulfur	S	16	32.07
Germanium	Ge	32	72.64	Tantalum	Ta	73	180.9
Gold	Au	79	197.0	Technetium	Tc	43	(98)
Hafnium	Hf	72	178.5	Tellurium	Te	52	127.6
Hassium	Hs	108	(269)	Terbium	Tb	65	158.9
Helium	He	2	4.003	Thallium	Tl	81	204.4
Holmium	Ho	67	164.9	Thorium	Th	90	232.0
Hydrogen	H	1	1.008	Thulium	Tm	69	168.9
Indium	In	49	114.8	Tin	Sn	50	118.7
Iodine	I	53	126.9	Titanium	Ti	22	47.87
Iridium	Ir	77	192.2	Tungsten	W	74	183.8
Iron	Fe	26	55.85	Uranium	U	92	238.0
Krypton	Kr	36	83.80	Vanadium	V	23	50.94
Lanthanum	La	57	138.9	Xenon	Xe	54	131.3
Lawrencium	Lr	103	(260)	Ytterbium	Yb	70	173.0
Lead	Pb	82	207.2	Yttrium	Y	39	88.91
Lithium	Li	3	6.941	Zinc	Zn	30	65.41
Livermorium	Lv	116	(292)	Zirconium	Zr	40	91.22
Lutetium	Lu	71	175.0	—	—	113	(284)
Magnesium	Mg	12	24.31	—	—	115	(288)
Manganese	Mn	25	54.94	—	—	117	(294)
Meitnerium	Mt	109	(268)	—	—	118	(293)

[a]Values in parentheses are the mass number of the most stable isotope.